Synthesis Series in Biomedical Engineering

Volume 1

Sensory Organ Replacement and Repair
Gerald E. Miller
2006

Artificial Organs
Gerald E. Miller
2006

Signal Processing of Random Physiological Signals
Charles S. Lessard
2006

Image and Signal Processing for Networked E-Health Applications
Ilias G. Maglogiannis, Kostas Karpouzis, Manolis Wallace
2006

Volume 2

Fundamentals of Respiratory Sounds and Analysis
Zahra Moussavi
2006

Basic Probability Theory for Biomedical Engineers
John D. Enderle, David C. Farden, Daniel J. Krause
2006

Intermediate Probability Theory for Biomedical Engineers
John D. Enderle, David C. Farden, Daniel J. Krause
2006

Advanced Probability Theory for Biomedical Engineers
John D. Enderle, David C. Farden, Daniel J. Krause
2006

Volume 3

Capstone Design Courses: Producing Industry-Ready Biomedical Engineers
Jay R. Goldberg
2007

BioNanotechnology
Elisabeth S. Papazoglou, Aravind Parthasarathy
2007

Bioinstrumentation
John D. Enderle
2006

Volume 4

Multimodal Imaging in Neurology: Special Focus on MRI Applications and MEG
Hans-Peter Müller, Jan Kassubek
2007

Estimation of Cortical Connectivity in Humans: Advanced Signal Processing Techniques
Laura Astolfi, Fabio Babiloni
2007

Brain–Machine Interface Engineering
Justin C. Sanchez , José C. Principe
2007

Introduction to Statistics for Biomedical Engineers
Kristina M. Ropella
2007

Volume 5

Introduction to Continuum Biomechanics
Kyriacos A. Athanasiou, Roman M. Natoli
2008

The Effects of Hypergravity and Microgravity on Biomedical Experiments
Thais Russomano, Gustavo Dalmarco, Felipe Prehn Falcão
2008

A Biosystems Approach to Industrial Patient Monitoring and Diagnostic Devices
Gail Baura
2008

Volume 6

Neural Interfacing: Forging the Human-Machine Connection
Thomas D. Coates Jr.
2008

Quantitative Neurophysiology
Joseph V. Tranquillo
2008

Tremor: From Pathogenesis to Treatment
Giuliana Grimaldi, Mario Manto
2008

Volume 7

Introductory Medical Imaging
A. A. Bharath
2008

Lung Sounds: An Advanced Signal Processing Perspective
Leontios J. Hadjileontiadis
2008

An Outline of Informational Genetics
Gérard Battail
2008

Volume 8

Basic Feedback Controls in Biomedicine
Charles S. Lessard
2009

Understanding Atrial Fibrillation: The Signal Processing Contribution
Luca Mainardi , Leif Sörnmo , Sergio Cerutti
2008

Volume 9

Landmarking and Segmentation of 3D CT Images
Graham Boag
2009

Phonocardiography Signal Processing
Rasha Bassam
2009

Introduction to Biomedical Engineering: Biomechanics and Bioelectricity Vol 1
Christensen
2009

Introduction to Biomedical Engineering: Biomechanics and Bioelectricity Vol 2
Christensen
2009

Synthesis Series in Biomedical Engineering 7

Introductory Medical Imaging
A. A. Bharath

Lung Sounds
An Advanced Signal Processing Perspective
Leontios J. Hadjileontiadis

An Outline of Informational Genetics
Gerard Battail

ISBN: 9781608451906

www.morganclaypool.com

A Publication in the Morgan & Claypool Publishers

SYNTHESIS SERIES IN BIOMEDICAL ENGINEERING

Series Editors: John D. Enderle

Synthesis Series in Biomedical Engineering 7

This volume includes:

Introductory Medical Imaging

A.A.Bharath

Lung Sounds: An Advanced Signal Processing Perspective

Leontios J. Hadjileontiadis

An Outline of Informational Genetics

Gérard Battail

Introductory Medical Imaging

A. A. Bharath
Imperial College, London

MORGAN & CLAYPOOL PUBLISHERS

ABSTRACT

This book provides an introduction to the principles of several of the more widely used methods in medical imaging. Intended for engineering students, it provides a final-year undergraduate- or graduate-level introduction to several imaging modalities, including MRI, ultrasound and X-Ray CT. The emphasis of the text is on mathematical models for imaging and image reconstruction physics. Emphasis is also given to sources of imaging artefacts. Such topics are usually not addressed across the different imaging modalities in one book, and this is a notable strength of the treatment given here.

KEYWORDS

Medical Imaging, Ultrasonic Imaging, X-Ray Imaging, Magnetic Resonance Imaging, Computer Aided Tomography (CAT), Mathematical Models for Imaging, Image Formation Physics, Image Reconstruction.

Contents

CHAPTER 1

Introduction

This book began as a series of one-term lectures on an MSc in Physical Sciences and Engineering in Medicine which I taught on between 1993 and 2000. During that time, I had the immense pleasure of conveying the underlying physical principles of medical instrumentation to over 200 students. The purpose of the notes was to provide a comprehensive, all inclusive overview of the techniques of medical imaging. These notes include levels of detail that are beyond what would constitute an introductory course; they are also, however, somewhat below the level of what would be considered a specialist course in any *particular* modality.

One of the main challenges with a course of this sort, is that the breadth of physical principles and terminology to be covered is huge. Confusion in terminology often arises. The feedback from engineering students has always been appreciative, and this book is targeted really at the engineer who wishes to learn more about the physical principles underlying medical imaging systems.

It should therefore be admitted (and I am the first to admit it!) that no book on medical imaging can be complete: it is a fast moving discipline, in which the individual modalities contain a high degree of specialisation, and have necessarily evolved a thick and nearly impenetrable jargon to the newcomer. Notably, these notes have not been kept entirely up to date in a number of technical areas that are rapidly becoming more and more significant. Indeed, one of my aims in getting these notes out into a wider community is that it will then catalyse and motivate the refinement and updating of the basic material with more up-to-date coverage of technologies, particular in the areas of magnetic resonance imaging, ultrasound, spiral CT and digital radiography. Also, these notes do not yet cover the very substantial area of nuclear medicine, including PET and SPCET, which remain its biggest omission.

I have organised the chapters as follows: Chapter 2 covers diagnostic X-ray imaging, planar x-ray imaging to be precise. I have not covered instrumentation and equipment in great detail here; my primary objective has been to convey the physical principles and the nature of contrast mechanisms in such a way that it is easily appreciated by an "engineering" mind. These principles are used in the following chapter to explore tomographic imaging via X-ray CT. In this chapter, one that I think is closest to maturity in terms of layout and flow, is strongly inspired by a paper by Webb (1987) covering the generations (in terms of technology) of CT scanners. Chapter 4 covers the basic physics of ultrasonics in perhaps, far too much detail from a theoretical point of view. Nevertheless, I think that most of the principles of ultrasound can be derived from the treatment that has been given. I regret that details on transducer design have not been incorporated. I hope these will emerge at a later date, but Chapter 5 is largely to be refined along these lines, as the technology continues to develop very quickly. Chapter 6 contains details on Doppler instrumentation, and is a quite good treatment of the basic Doppler equations. I have not been able to augment this chapter yet to incorporate more information on colour-flow mapping systems, but this will come in time. Chapter 7 on MRI provides more quantum mechanics than most people are happy with, and there is much room to develop, but I think the foundations are as I wish them to be. More material on fMRI, fast pulse sequences and k-space sampling is needed, and this is in the process of being collated.

The appendices contain a few pieces of material that are always being sought by inquisitive engineering minds so I have included them here: Appendix A provides a very intuitive derivation of the homogenous wave equation, and Appendix B provides mathematical conventions used in the Chapters on CT and MRI. Note that there may yet be differences between the conventions used in different chapters; the master plan is eventually to unify these so that there is one consistent notation for the whole. For historical reasons, this can be time consuming to achieve.

My thanks, first and foremost, go to the legions of student that have passed through my course who have encouraged me to get these notes disseminated on a wider scale - to highlight any one person would be by definition exclusionary. Secondly, I would like to thank Dr. Janet de Wilde for her many useful discussions on magnetic resonance imaging; Dr. Keith Straughan also provided very useful pointers to my early collation of material for the chapters on X-ray imaging and CT. Professor Richard Kitney is responsible for introducing me to the area of ultrasound, and in particular Doppler ultrasonic instrumentation, my first area of entry into medical instrumentation. Huge thanks also go to Professor Caro for keeping a perceptive eye on me - the benefits of a good family physician. Finally, I should like to thank the Hayward Foundation for generously funding my medical imaging lectureship post during the years 1993-2003.

On a professional level, I should like to dedicate this book to the memories of Professor M. J. Lever and my father, Samuel Bharath; both of these men were supportive of my attempts to make cohesive a range of disparate topics to students of engineering…

On a personal level, I dedicate this book to André and Paula, for their grounding and support. Without them, I would be lost.

CHAPTER 2

Diagnostic X-Ray Imaging

The contents of this chapter are, essentially, the following:

- Basic principles of X-Ray Imaging: Introduction to terminology and mode of operation.
- X-Ray physics: Quantum nature; X- Ray interactions with matter; X-Ray spectra; energy range for diagnostic use.
- X-Ray image quality parameters: general requirements on physical parameters; image Signal/Noise (S/N); image contrast.
- X-Ray production: functional design requirements;tube construction; target selection.
- Image receptors: functional design requirements; film-based receptors; image intensifiers.
- Patient dosage: trade-offs with image quality; noise and dosage; effective dose equivalent.

2.1 BASIC PRINCIPLES OF X-RAY IMAGING

In the simplest case, an X-ray imaging system requires

- X-Ray Source
- A patient to image
- Film (image receptor)
- Radiologist/Diagnostician

2.1.1 IDEAL DESCRIPTION OF IMAGING PROCESS

X-Rays are generated within the tube, and they are directed towards the patient. As the x-ray photons pass through the patient, some are absorbed, others scattered, and some pass through the patient with no interaction. The transmitted photons i.e., those which do not interact with the patient) are detected (received) by the photon receptor, usually based around film. The formation of an image on the film is dependent on the number of photons which are captured (detected) by the receptor: areas of the film which are dark have received a large number of photons; brighter areas have received fewer. The distribution of the light and dark areas on film is approximately a projection onto a two-dimensional map of the three-dimensional distribution of attenuating structures within the patient. As we shall see, there are many aspects which complicate the simplistic, ideal situation:

- Statistical arrival of photons (Poisson process).
- Photon scatter.
- Lines of projection are not parallel i.e., one has beam divergence).
- Photon detection is inefficient.
- Beam hardening.
- X-rays represent a form of ionizing radiation - there are health risks associated with prolonged or repetitive exposures.

2.2 RELEVANT PHYSICS

2.2.1 ATOMIC STRUCTURE

All atoms have a similar structure, in that they consist of small, dense nuclei, which have a radius of $\approx 10^{-14} m$, with a positive charge given by $Z \times 1.6 \times 10^{-19}$C, where Z is the atomic number of the atom i.e., the number of protons in the nucleus). In unionised form, the atom is electrically neutral: the nucleus is orbited by Z electrons, each of which has a negative charge of 1.6×10^{-19}C, and a mass of 9×10^{-31}kg, around 0.05% of the mass of each proton. Nuclei also contain uncharged particles, known as neutrons, of mass almost equal to that of protons, which provide short-range attractive forces that bind the protons together. The number of protons plus the number of neutrons is the mass of the atom.

2.2.2 NATURE OF X-RAYS

X-rays represent electromagnetic radiation in the frequency range of about 10^{18}-10^{20} Hz. At this frequency range, the free space wavelength is of the order of

$$\begin{aligned} \lambda &= c/f \\ &= 3 \times 10^{8}/10^{18} \\ &= 3 \times 10^{10}\text{m} \end{aligned} \tag{2.1}$$

which is very small indeed. At such dimensions of wavelength, the quantum nature of electromagnetic phenomena becomes significant and, indeed, sets fundamental limits on imaging. The primary significance of this quantum nature is that the electromagnetic radiation is delivered in discrete lumps of energy, known as quanta, or photons. The quantity of energy in each photon is related to the wavelength of the radiation according to

$$E = hf = hc/\lambda \tag{2.2}$$

where h is Planck's constant, 6.626×10^{-34}Js. Thus, the energy of 1 quantum of x-ray radiation i.e., 1 X-ray photon) at 10^{20} Hz is about

$$6.626 \times 10^{-34} \times 10^{20}\text{J} = 6.626 \times 10^{-14}\text{J}\,. \tag{2.3}$$

Instead of quoting the photon energy in Joules, it is more common to use the relationship 1eV$= 1.6 \times 10^{-19}$J to express the energy in eV's. Using this relation yields an energy of approximately 414keV.

We shall approach the subject of photon detection later in the course, but two aspects are immediately important. First, note the very small amount of energy present in each photon. The detection of single photons is difficult, and some form of amplification is necessary if single photons are going to contribute to a visible image. Secondly, the quantum nature of photons implies some uncertainty in localising photons simultaneously in space, and in wavelength[1].

Indeed, any description of photon arrival has to be treated in statistical terms. The arrival of X-ray photons at a detector may be treated as obeying Poisson statistics. An example of the form that this takes is to consider the number of photons arriving within a given time T at a perfect photon receiver:

$$\text{Probability of } k \text{ photons arriving in time } T = \frac{a^k T^k}{k!} e^{-aT} \tag{2.4}$$

where the mean number arriving in time T is aT, with a being a constant reflecting the photon fluence and detector area.

We shall examine the significance of this when we consider the physics of the X-ray photon detection and image formation.

2.2.3 X-RAY GENERATION

There are two primary physical processes that underly X-ray production in diagnostic radiology. Recall that the manner in which we can generate an electromagnetic wave is to wriggle an electron

[1] In the early 20th century, the curiosities of quantum physics were very much debated in society. A fairly well known sceptic of time, Ronald Knox, poked fun at one aspect of the theory, which is now known through the "thought experiment" of Schrödinger's Cat:

There was a young man who said, 'God
must think it exceedingly odd
if he finds that this tree
continues to be
When there's no one about in the Quad.'

To which an equally gifted wit, Berkeley, replied:

Dear Sir,
Your astonishment's odd:
I am always about in the Quad.
And that's why the tree
Will continue to be,
Since observed by
Yours faithfully,
GOD

around. This is the basis of radio transmission (e.g., dipole antenna). The oscillation of an electron may be thought of as subjecting an electron to accelerative motion. So, in order to produce x-rays, one method would be to subject electrons to very rapid accelerative motion.

"Braking" Radiation

The "Bremsstrahlung," or "braking" mechanism, is the principal mechanism of x-ray production in diagnostic radiology. Braking radiation is the radiation released as an electron is rapidly decelerated by a nucleus in a target material. It arises through bombarding a material of high atomic number nucleus with fast-moving electrons (See Figures 2.1 and 2.2). Feynman's lectures on Physics has a description of this type of radiation.

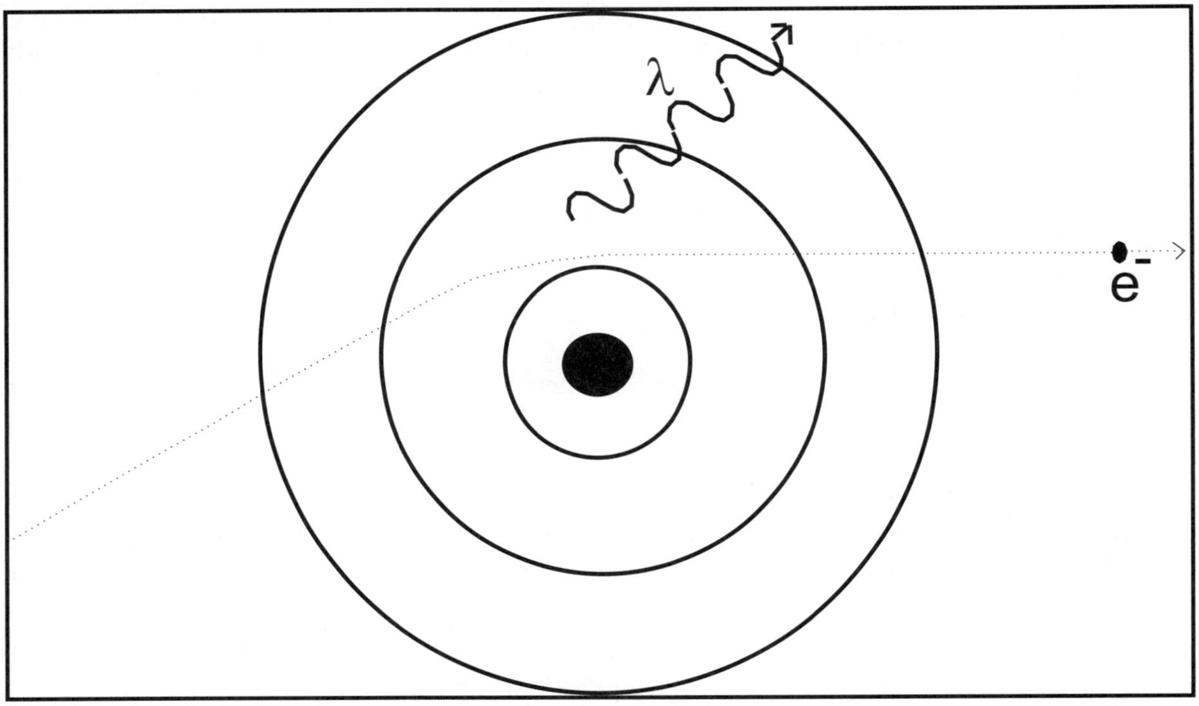

Figure 2.1: Mechanism of Bremsstrahlung Radiation.

In order to bombard a target of atoms with a stream of high energy electrons, the arrangement shown in Figure 2.2 is usually used. Note that this is a functional diagram, and does not show all of the details of an x-ray tube.

Characteristic X-Rays

If we bombard a suitable target material with a stream of high energy electrons, as we do in the generation of Bremsstrahlung radiation described above, we also experience another mechanism of x-ray production, arising from energy level transitions. The x-rays arising from this mechanism

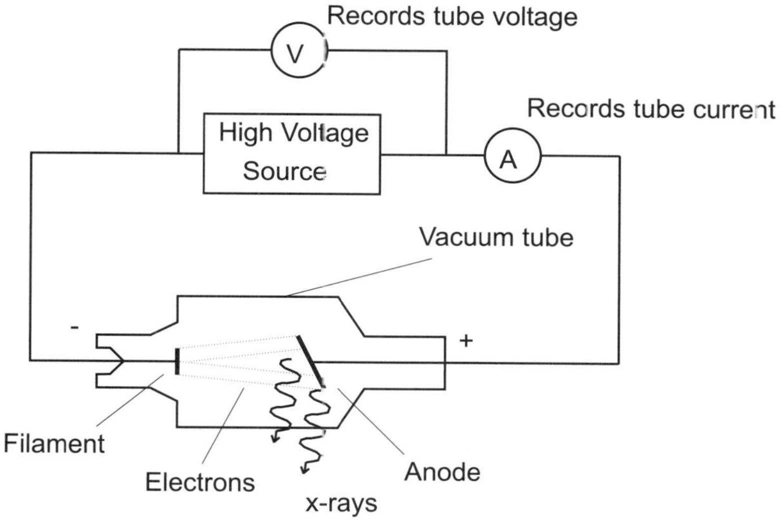

Figure 2.2: Simplified X-Ray generation.

are known as characteristic x-rays. Characteristic x-rays are so called because they are very much dependent on the target material, i.e., the electronic configuration of the atom. One finds, for example, that whilst Tungsten anodes have characteristic lines at 69.5 and 59.3 keV energies, Molybdenum anodes have lines at 20.0 and 17.3 keV.

Characteristic x-rays are formed by electrons from the L and higher bands reverting to the K-shell, after a K shell electron has been ejected by the arrival of an electron of high kinetic energy. The wavelength of the photon produced by the L⟶K or M⟶K transition is dependent on the precise energy levels between the bands.

2.2.4 X-RAY SPECTRA

An x-ray spectrum is simply a plot of intensity vs frequency of radiation in the emission spectra of an x-ray source. One must be careful in interpreting such spectra: rather than being the direct emission as a result of the target bombardment, they are generally sketched for the case of photons having left the tube, and therefore having been subjected to some *beam filtration* (through the envelope of the tube, at the very least). The shape of such spectra is fairly easy to understand. The lower cut off in energy is the value of energy below which no photons have sufficient energy to get past the glass tube of the x-ray source. The upper cut-off point, often labelled by kVp, is determined by the maximum voltage across the anode/cathode. The photon energy[2] at which the peak beam energy of the smooth curve, due to brehmsstrahlung, occurs is known as the effective energy (E_{eff}) of the beam.

[2]Think frequency! This is confusing, but remember that the horizontal axis is essentially frequency, the vertical axis is number of photons.

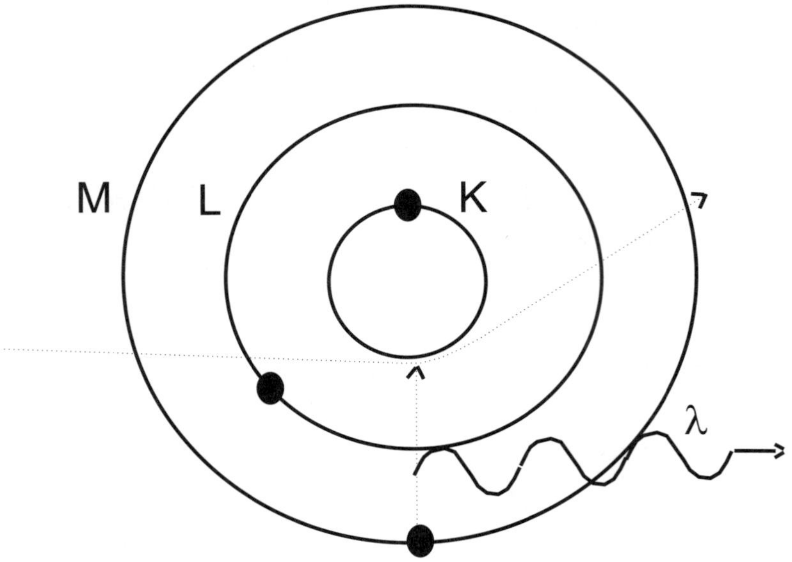

Figure 2.3: Mechanism Underlying Characteristic X-rays.

The overall intensity of the entire curve can be increased by increasing the number of electrons bombarding the target. This scales the height of the curve, thereby altering the number of emitted photons at each frequency, but keeps the shape of the beam spectrum otherwise unchanged. Physicists say that the beam quality remains the same. The quantity of photons increases linearly with tube current. If the tube current is only kept flowing for a limited time, then the quantity of photons generated is also linearly dependent on this time. Other factors that directly affect the yield of photons are the atomic number of the target material (linear increase) and the applied (tube) voltage (quadratic increase).

If the tube voltage is changed, the shape of the Brehmsstrahlung portion of the curve also changes, generally stretching to reach up to higher frequencies. The effective energy, E_{eff} also shifts towards higher photon energies. This change in shape of the brehmsstrahlung component of the spectrum is described as a change in *beam quality*. However, the position of any characteristic spectra remains the same (unless new transitions become energetically possible).

2.2.5 X-RAY INTERACTIONS WITH MATTER

When a thin beam of X-ray photons passes through matter, it becomes weaker and is said to be attenuated as photons are removed from the forward direction of propagation. Attenuation takes place through the action of two processes:

- Scattering

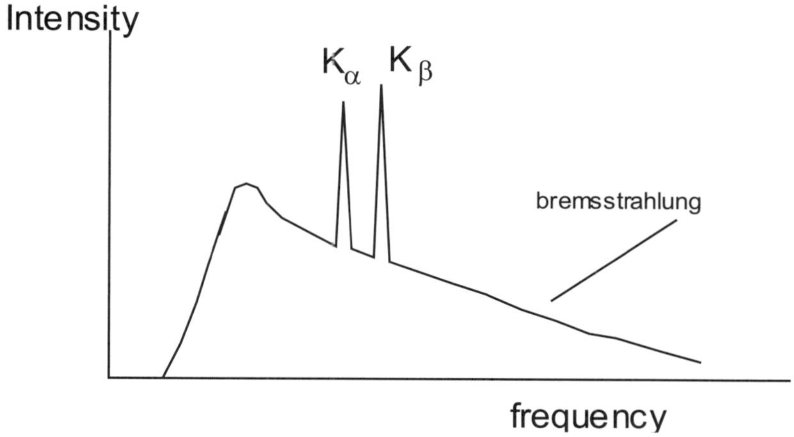

Figure 2.4: A typical x-ray spectrum.

- Absorption

Scattering losses are energy losses due to photons that are *redirected* from the direction of the primary beam by various possible scattering events. There are two main types of scattering events that we shall consider: elastic and inelastic scattering. Absorption losses refers to the energy removed from the primary beam and transferred to the local medium in the form of heat, typically via electron recoil (Compton scattering) and by the kinetic energy imparted to an ejected electron during the photoelectric effect, and in pair production. In the next sections, we consider the various interactions between x-ray photons and matter, which are most significant in diagnostic radiology. The physical principles are treated by considering the likelihood of various events occurring between a photon in an x-ray beam passing through some material, and the atoms of that (homogeneous) material.

Elastic Scattering
In elastic scattering, the electrons of the atom in the material remain in their orbits. The incoming photon interacts with the electrons through a quantum resonance phenomenon, but there is no energy loss to the incoming photon. According to some reports, the scattering is primarily in the forward direction; this, however, may depend on the type of material (e.g., polarised or not), and it is more likely in tissue that all directions are equally favoured. The probability of this event is higher at lower photon energies and increases with increasing atomic number of the atom in the material. Also known as Rayleigh scattering, this effect only accounts for about 10% of the interactions occurring at diagnostic energies in tissue.

The Photoelectric Effect
This is an absorption effect. A primary photon is annihilated on collision with a bound electron. Most of the photon energy is transferred to the electron, which is ejected, and the momentum of the photon is transferred to the molecule or atom to which the electron was bound. The kinetic energy

of the ejected electron is given by

$$E_k = hf - E_{ionisation} \tag{2.5}$$

where $E_{ionisation}$ is the ionisation energy of the electron. The kinetic energy of the electron is dissipated locally. If the event occurs in tissue, the electron is quickly neutralised. Ionization of K-shell electrons is about 4 times more probable than of L-shell interactions (given that the photon energy is sufficient for both!), so that K-shell interactions are of greater significance. The ejection of an electron from a lower shell will clearly leave the atom in an excited state, and so there is a release of a characteristic photon - this is clearly quite similar to the production of characteristic X-rays at the anode of the x-ray tube, but in this case, of course, the incoming particle is an x-ray photon. The component of the photoelectric effect which involves emission of a photon is known as fluorescence, and the yield of such photons turns out to be higher in atoms of high atomic number. It is a significant effect in the choice of materials for performing beam filtration (after the photons leave the tube, but before they reach the patient).

In atoms of lower atomic number, such as might be found in tissue, Auger electrons released from the outer shell are more probable during the photoelectric effect. These electrons have low kinetic energies, and are absorbed very quickly in a small region of tissue. This can lead to high local energy densities, with consequent radiobiological damage.

The likelihood of a photoelectric event generally decreases with photon energy *until one crosses a transition energy threshold.* This happens when the photon has sufficient energy to eject an electron from a shell. If one plots likelihood of photoelectric event vs photon energy, one then has a "sawtooth" appearance to the graph. The sharp edges in this plot are known as "absorption edges," and correspond to these transition energies. They are of importance in beam filtering.

Compton Scattering

A photon collides with a free or loosely bound electron; kinetic energy is given to the electron, causing it to recoil. A second photon, of lower energy is created as a result of this interaction.

The momentum of a photon of electromagnetic radiation at frequency f is given by de Broglie's relationship as hf/c, where h is Planck's constant. By conservation of momentum, we may derive the Compton shift, which is the difference in wavelength between the primary beam photon, and the secondary photon produced by the interaction. Let the momentum of the incident photon be directed along the $\vec{z}$ direction. The sum of the momenta in the $\vec{z}$ direction after collision is then (See Feynman Lectures on Physics):

$$\frac{hf'}{c}\cos(\theta) + mv\cos(\phi) \tag{2.6}$$

where f' denotes the frequency of the photon after collision (called the *secondary* photon). Now, conservation of the momentum vector along the $\vec{z}$ direction requires that the expression above be

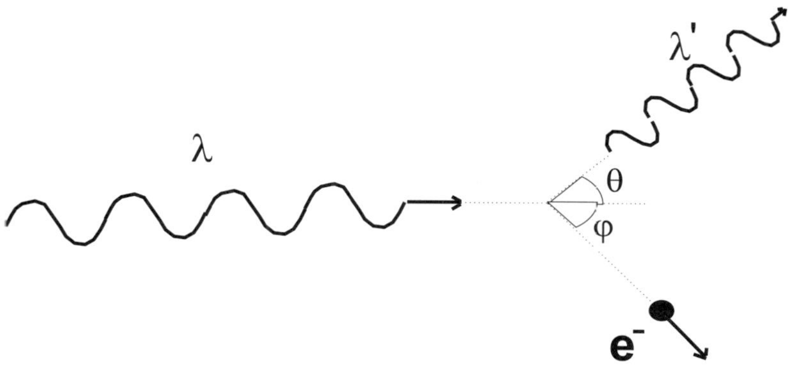

Figure 2.5: Compton scattering.

equal to the total momentum in the $\vec{z}$ direction *before* the collision, so

$$\frac{hf}{c} = \frac{hf'}{c}\cos(\theta) + mv\cos(\phi) \tag{2.7}$$

where the electron is at rest before collision, and is moving with velocity magnitude v after the collision, in a direction shown in Figure 2.5. Looking at momentum along the perpendicular direction leads to

$$0 = \frac{hf'}{c}\sin(\theta) - mv\sin(\phi)\ . \tag{2.8}$$

The energy of the entities in the system should be conserved through the collision. Using the obscure[3] relationship, $E = mc^2$, leads to,

$$hf + m_0c^2 = hf' + mc^2 \tag{2.9}$$

where the left-hand side of Equation (2.9) represents the energy of the system before collision, and the right-hand side is the energy after collision. m_0 is the mass of the electron before it has collided, m is the mass of this electron after collision. Squaring this equation, and using

$$m = m_0(1 - v^2/c^2)^{-1/2} \tag{2.10}$$

leads to

$$(h\delta f)^2 + 2m_0c^2h\delta f = m_0^2c^4\frac{v^2}{c^2 - v^2} \tag{2.11}$$

where $\delta f = f - f'$. By squaring, Equation (2.7) and Equation (2.8), and using the identity

$$\cos^2(\phi) + \sin^2(\phi) = 1 \tag{2.12}$$

[3]Not!

and eliminating ϕ leads to

$$h^2\{(\delta f)^2 + 2f(f - \delta f)(1 - \cos(\theta)\} = m_0^2 c^4 \frac{v^2}{c^2 - v^2} . \tag{2.13}$$

But the right-hand sides of Equations (2.11) and (2.13) are equivalent, so that

$$h^2 f(f - \delta f)(1 - \cos(\theta)) = m_0 c^2 h \delta f . \tag{2.14}$$

Leading to a Compton shift given by

$$\delta\lambda = \lambda - \lambda' = \frac{h}{m_0 c}(1 - \cos(\theta)) . \tag{2.15}$$

This says that the wavelength of the secondary photon is shifted with respect to the first. In fact, the wavelength of the secondary photon is greater than that of the incident photon. As might be expected, because some kinetic energy is imparted to the recoil electron, the energy of the secondary photon will be lower than that of the primary. The energy of the secondary photon depends on the angle at which it travels away from the scattering event. This fact is used, not in X-ray imaging, but in Nuclear Medicine, where the energy of photons is analysed to reduce effects of scatter.

In X-ray imaging, the secondary photons are simply a nuisance: they contain tissue information out of the straight-line direction of photon propagation, and contribute very much to imaging noise. Their effect can be somewhat reduced by employing anti-scatter grids near to the film, which remove photons that are not travelling in straight-line paths.

Compton scatter is a "mixed" event, as it includes some absorption (energy given to the recoil electron) and scatter (the secondary photon). The proportion of these coefficients varies with the angle of the secondary photon. The likelihood of a Compton event does not vary significantly (at diagnostic frequencies) with the energy of the primary photon, nor does it significantly vary with the atomic species of the material through which the beam is passing.

Pair Production

This occurs at high energies, when the photon has in excess of 1.02 MeV. As the photon passes close to a heavy nucleus, its energy creates, simultaneously, a positron and an electron. Since these have equal mass, in the absence of relativistic effects, one has

$$hf = 2m_0 c^2 + \text{kinetic energy} . \tag{2.16}$$

The electron has a short range in tissue, contributing to dose, whilst the positron heads off to combine with a free electron and undergoes annihilation to produce two antiparallel gamma rays, each of 0.51MeV. Pair production is favoured by atoms of high atomic number, and clearly at quite high photon energies. It is of limited importance in the diagnostic use of X-rays.

Conclusion

Of the four types of interaction we have met, by far the most significant are the Compton effect and the photoelectric effect. Rayleigh scattering and pair production have peripheral significance, at extreme energies, or outside the radiological field.

Finally, note that whilst the Compton effect is pretty much constant with changes in both photon energy and atomic number, the photoelectric effect is quite dependent on both atomic number (Z) and the incident photon energy. At low energies, the photoelectric effect dominates, providing good contrast due to different atomic species in tissue; at higher diagnostic energies, the Compton effect dominates, leading to greater dependence on the mass density of tissue, rather than the atomic number. We now need to investigate measures of these effects in different types of tissue. This brings us on to the idea of *collision cross section*.

2.3 ATTENUATION

2.3.1 THE BASICS

Since there are so many effects that can contribute to the absorption and scattering of X-ray photons travelling through a material, it is difficult to "get a handle on them" simultaneously. Any description of interaction also needs to be easily related to physically verifiable results. This is where the ideas of collision cross section and linear attenuation coefficient come in. Linear attenuation coefficient is a good empirical measure, well correlated with tissue type, whilst collision cross section is a solid construct from statistical physics. It is a measure of the fraction of photons that get removed from the beam. A superb discussion on collision cross sections, and scattering cross sections, is to be found in the Feynman Lectures on Physics [17]. For now, let us imagine a beam of photons incident on a material. On passing through that material, some of the photons will be removed from the direction of propagation by absorption and scatter. In fact, the loss of photons can be regarded as a certain fraction of the cross sectional area of the beam that has been removed. One may describe different materials as having different collision cross sections.

Let us assume that a particular sample of material, of length dz, is placed in the path of an x-ray beam. We measure the incident beam intensity, and we measure the exit beam intensity (see Figure 2.6).

We can then express the exit intensity as

$$I(z+dz) = I(z) + \frac{\partial I}{\partial z}dz = I_0 + \frac{\partial I}{\partial z}dz \,. \tag{2.17}$$

The fractional difference between the exit and incident beam intensity is then

$$\frac{I_0 - \left(I_0 + \frac{\partial I}{\partial z}dz\right)}{I_0} = \frac{-\frac{\partial I}{\partial z}}{I_0}dz \,. \tag{2.18}$$

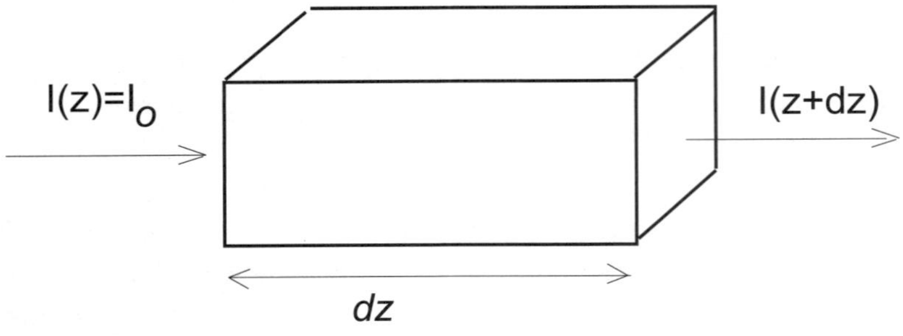

Figure 2.6: Beam Attenuation.

Let us denote the fractional loss per unit length by the symbol μ. Then, the loss over the length dz is μdz. We therefore have

$$\frac{-\partial I(z)}{\partial z} = \mu I_0 \, . \tag{2.19}$$

This represents a differential equation, having a solution

$$I(z) = I_0 \exp(-\mu \, (z) \tag{2.20}$$

where μ is known as the linear attenuation coefficient of the material. This attenuation coefficient has dimensions of $length^{-1}$, and in radiology is usually stated as cm^{-1}. Equation (2.20) is one form of *Beer's Law*.

Let us now see how μ is related to the collision cross section of a material. If one squashes the volume shown in Figure 2.6, so that its length is halved, we may wonder what Beer's law tells us about the exit intensity. If we assumed that μ was constant, one might think that the exit intensity would be higher (since the path length is smaller). In fact, the measured intensity would be pretty much the same. The reason for this is that the number of scattering elements is constant, and is dependent on the total mass of material present in the irradiated volume (since mass is due to the number of atoms, and atoms cause the absorption and scattering). To distinguish between different types of material, we can think about the quantity referred to as *mass attenuation coefficient*, which is the attenuation coefficient per unit mass of material. This is represented as μ/ρ and has units of m^2kg^{-1}. This, therefore, represents an *area* per unit kilogram of mass which is removed from the beam: the collision cross section![4].

Some Examples of Linear Attenuation Coefficient

- μ_{muscle} = $0.180 \, cm^{-1}$

- μ_{blood} = $0.178 \, cm^{-1}$

[4]There are different subtle variations of collision cross section; this measure is per unit mass, but there are also measures which are per unit volume, per unit length, etc.

- $\mu_{bone} = 0.48\ cm^{-1}$

If we have three equal path lengths of an x-ray beam through three different types of tissue, muscle, bone and blood, that are at the same depths, the emergent beam (given that the entry beam to all tissues is of equal intensity) from bone will be attenuated to a greater extent than the beam through either muscle or through blood. Moreover, the difference in exit beam intensities between muscle and blood will be relatively small (less than 1%!). This great difference between the linear attenuation coefficients for soft tissue and bone is the reason that x-ray imaging is so superlative at imaging broken bones; this directly explains its importance in orthopaedics and dentistry.

2.3.2 VARIATION OF LINEAR ATTENUATION COEFFICIENT

For equal path lengths, the variation of linear attenuation coefficient is what is responsible for our contrast in x-ray imaging. It is therefore important to understand how this quantity varies. In some cases, it is useful to relate this variation back to the underlying physical principles, so we sometimes split the linear attenuation coefficient into different components. Luckily, these components can be added together to find the net attenuation coefficient.

Variation with Atomic Species

One finds that the Compton component of linear attenuation coefficient, μ_C, increases approximately linearly with the mass density of the sample, and is rather independent of atomic species, i.e., packing density of the atoms is more important. The photoelectric component of absorption μ_{photo} is linearly dependent on the 4th power of the atomic number, and linearly dependent on the mass density of the sample. The pair-production component of the absorption increases with the square of the atomic number of the species.

Variation with Photon Energy

At diagnostic frequencies, the linear attenuation coefficient tends to decrease with increasing photon energy, primarily due to the dependence of the photoelectric effect with photon energy. Thus, a typical curve of variation of attenuation coefficient in tissue might look something like Figure 2.7.

2.3.3 BEAM HARDENING

Remember that a true x-ray beam will typically contain a range of photon energies (range of wavelengths). What happens to this spectrum as the photons propagate through tissue? Let us model an x-ray spectrum using the solid trace of Figure 2.8. If one passes a beam with such an intensity profile through a slab of attenuating material (such as tissue), whose *variation of attenuation coefficient with photon energy* is as in Figure 2.7, then the photons of lower energy will experience a higher attenuation than the photons of higher energy. This means that, *relative to the incident x-ray beam*, there will be a greater proportion of photons with higher photon energies (higher frequencies, lower wavelengths) in the exit beam. This *does not* mean that the overall beam energy has increased: it just means that the distribution of photon energies is shifted towards higher energy photons: E_{eff} in particular is increased (refer back to Section 2.2.4). Note, from the dotted trace of Figure 2.8,

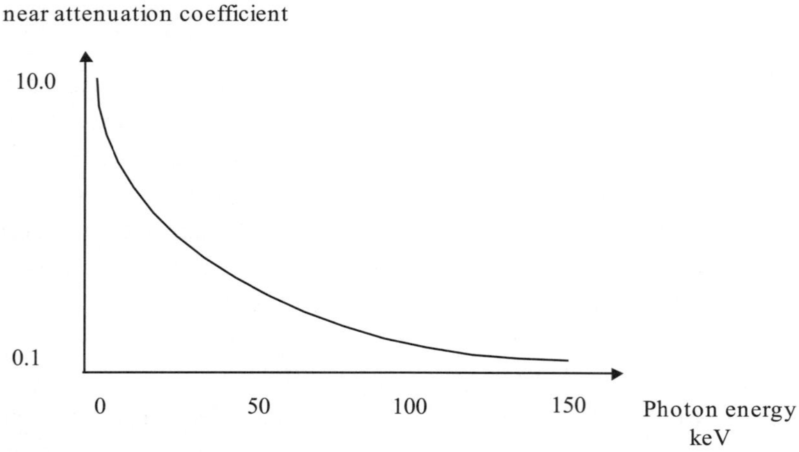

Figure 2.7: Variation of generic μ with photon energy.

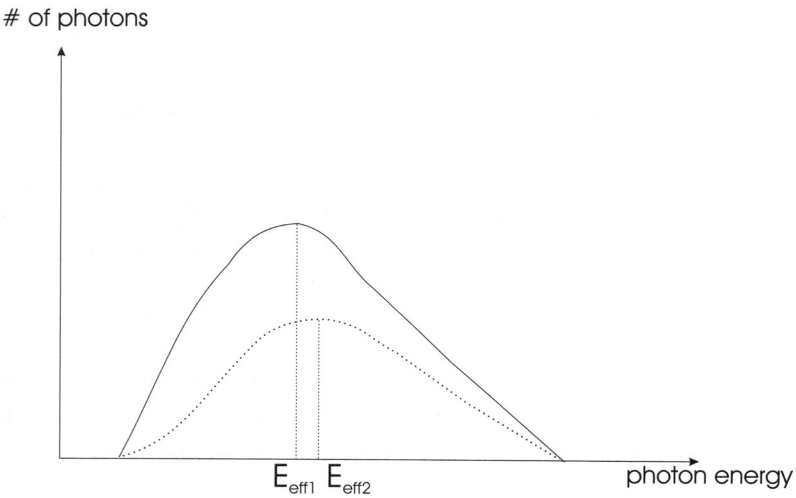

Figure 2.8: Exaggerated sketch depicting the difference between incident (solid curve) and exit x-ray (dotted curve) beam spectra. Note the difference between the effective energies.

there is an apparent shift in the peak energy (and also the average energy) of the beam.

Suppose, now, that one has a particular of homogeneous tissue. For the effective energy of the beam, there will be an associated "effective" attenuation coefficient for that piece of tissue. If, however, the effective energy of the beam is shifted towards higher photon energies (by passing through, say bone, prior to entering the tissue in question), the effective attenuation coefficient of the tissue will be lower (since μ decreases with photon energy). Beam hardening can cause a loss of image contrast if there is a highly attenuating material in the path of the beam, near to the entrance point into the body i.e., "early" in the photon path).

2.4 IMAGE FORMATION PHYSICS

2.4.1 FILM

X-Ray film is required to perform the task of converting an electromagnetic form of radiation into a visible representation of the variation in of the intensity of this radiation over a two-dimensional plane. In practice, the main mechanism of this conversion is the photoelectric effect that takes place within the "active" part of the film. The active part of the film is a photographic emulsion, consisting of grains of silver halide crystals embedded in a gelatine matrix. The emulsion itself coats an acetate or polymer film base.

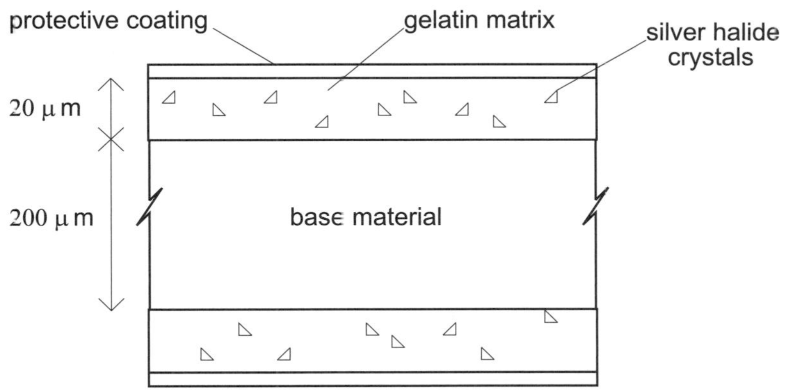

Figure 2.9: X-ray film structure.

On the capture of x-ray photons, electrons from the halide are produced by the photoelectric effect. These electrons are then captured by sensitivity specks (electron traps) within the lattice of the halide crystals. Such negative traps act as cluster centres for mobile silver ions, which are positively charged. The effect is to deposit individual atoms of silver within the grains. The neutral halides produced by this effect drift into the emulsion of the film and remain there.

During the process of film development, those halide grains which have a minimum number of silver atoms become converted into metallic silver; those with fewer than the required number of neutralised silver atoms are not converted, and are removed during the fixation part of the development process. The deposits of metallic silver render the film opaque to visible light, and it is the distribution of these opaque regions which represents the x-ray image. Note that one may view the action of the initial photoelectric effect and subsequent ion neutralisation (before development) as providing a sort of catalyst for the subsequent reduction of the grains to "large" groups of silver atoms. This is thus an amplification process, and it means that a particular grain may be converted into a little deposit of silver (containing $\sqrt{A}$ atoms, typically, where A is Avogadro's number) by the reception of as few as two x-ray photons. Remember this in terms of the points raised at the beginning of the course about the detection of the *extremely* small quantities of energy represented by single photons. Finally, note that x-ray films are designed to be viewed under transillumination i.e., they are placed over a radiographer's light box), and so the opaque regions are those which appear dark. Hence, relative to the intensity of the received x-rays, the final radiological image can be considered as photonegative.

The precise relationship between the number of photons captured by film (and hence the intensity distribution of the emergent x-ray beam), and the final visual intensity produced is of a very complex nature, requiring considerations into quantum efficiency, multiple photon interactions with matter, and the uniformity of silver halide crystals. Luckily, the tools and terminology necessary to handle this are to be readily found in the field of photography!

2.4.2 MODELLING FILM CHARACTERISTICS

We shall consider a measure of the transmittance of the developed film to *visible* light. Define the film *optical density* after exposure as

$$D = -\log(T) \tag{2.21}$$

where T represents the fraction of transmitted light when the developed film is placed in front of a light box and viewed by an observer:

$$T = \frac{I_{obs}^{visible}}{I_0^{visible}} \tag{2.22}$$

$I_{obs}^{visible}$ is the observed transmitted light intensity after passing through film, whilst $I_0^{visible}$ is the intensity of visible light produced by the light box.

Where the optical density is high, there have been many grains of silver halide crystals which have been reduced to metallic silver, and in these regions, there is little visible light transmitted in viewing. The optical density has a well-known relationship to the logarithm of the incident energy (See Figure 2.10). The range of exposures for which the density characteristic is approximately linear is the regime that one usually wishes to work in. In this region, the slope of the curve is refereed to as the γ characteristic[5] of the film, and it is given by

$$\gamma = \frac{D_2 - D_1}{\log(E_2) - \log(E_1)} \tag{2.23}$$

where E_1 and E_2 are two exposures. You can think of *exposure*, a term inherited from photography, as being proportional to the intensity of the x-ray beam as it impinges on the film, multiplied by the time over which it impinges on the film. In practical terms, this time is the time for

[5] Pronounced "gamma characteristic"!

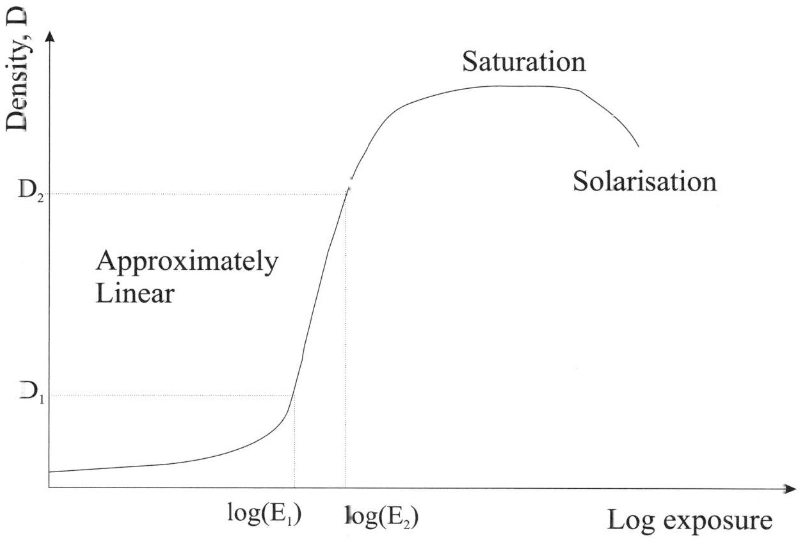

Figure 2.10: Variation of Optical Density with exposure (number of photons received).

which the tube current is kept on. This is not to be confused with *radiation exposure*, to be defined later.

Within this linear region, one can therefore express the relationship between the density of the film and the incident x-ray exposure, E_{exp} by

$$D = \gamma \log(E_{\text{exp}}) + D_0 \ . \tag{2.24}$$

We will use this simple film model to understand the basis of image contrast.

2.5 X-RAY IMAGE QUALITY

If one's intention is to diagnose illness by looking for abnormalities in a patient's anatomy, it is easy to formulate (in a purely academic way) some requirements for the criteria under which an image can be described as "high quality." A primary requirement (which does not necessarily follow from the physics of the imaging process) is to obtain a map of attenuating tissue structure which bears as close a resemblance to the actual distribution of tissues as possible. While this is an obvious requirement, it is difficult to express this in a manner which is useful in the design of diagnostic equipment, as there is often no precedent for what the image should look like.

2.5.1 BROAD IMAGE QUALITY GOALS

- Improve detectability for small objects.
- Minimize geometric distortion and beam unfocussing.

- Work within the constraints of patient dosage.

Detectability of small objects depends on four main factors: the size of the object, its density relative to surrounding tissue, the degree of blurring introduced by the system, and the amount of noise in the image. In the case of x-ray imaging, we shall approach the issue of defining image quality by considering the physics of the imaging process, and conducting a "thought experiment" involving the use of a tissue phantom, some simple physics, and a bit of common sense. This will also allow us formulate some concrete ideas about aspects of contrast and detectability.

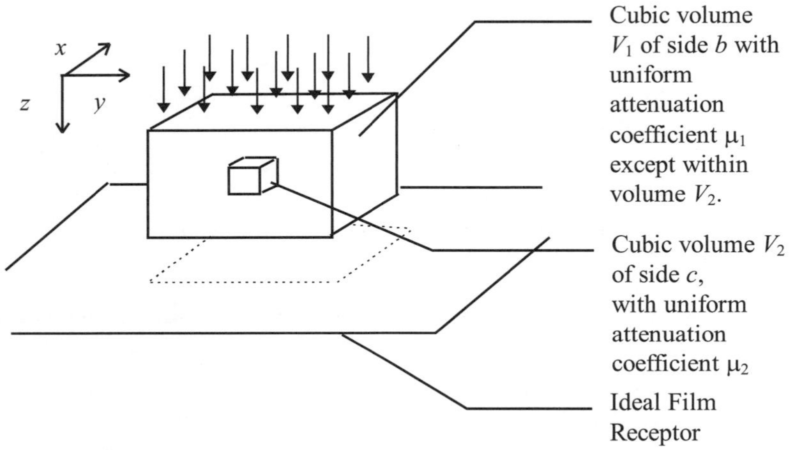

Figure 2.11: X-ray cubic phantom.

Consider the x-ray "phantom" shown in Figure 2.11. This phantom consists of two nested cubes of tissue-like material. First, assume that the x-ray system we are going to use to image this phantom is approximately monochromatic. The inner cube has attenuation coefficient μ_2 at this particular beam energy, and the outer cube has attenuation coefficient μ_1, where $\mu_1 < \mu_2$. We will also assume that the x-ray system is set to neither magnify nor shrink the dimensions of the phantom on film. An ideal x-ray image[6] of such a phantom will look as shown in Figure 2.12. Using Beer's law, and our simple model for film, we will be able to get an idea of these intensities.

First, let us work out what the relative intensities of the x-ray beam leaving the phantom are. We can do this using Beer's law. There are two different intensities that will be found across the x-ray beam: areas where the photons have passed through *only* V_1, and areas where the photons have passed through both V_2 and V_1. Let us represent these exit *x-ray* intensities I_1^x and I_2^x respectively.

[6]Assume that we are viewing this in transillumination mode, and can measure the light intensity using a light meter.

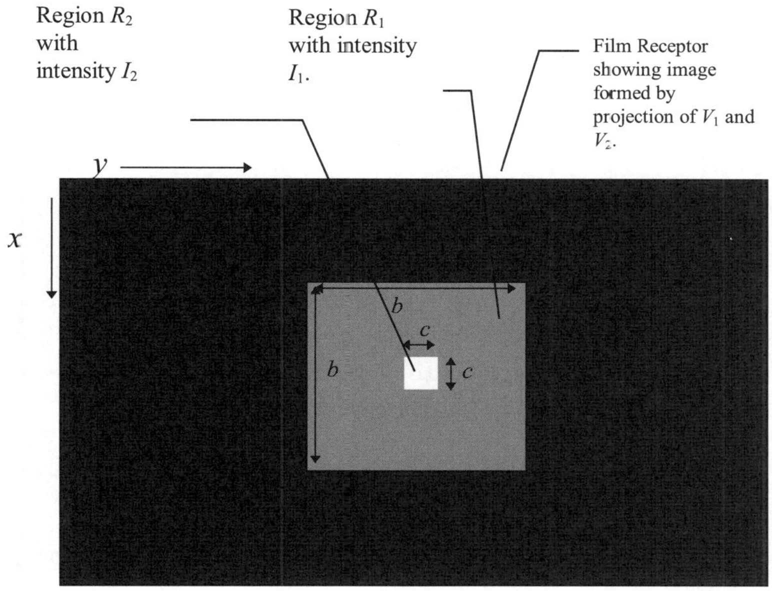

Figure 2.12: Image of cubic phantom. I_1 and I_2 are visible light intensities when viewed by transillumination!

These intensities can be determined by applying Beer's law[7] as follows:

$$I_1^x = I_0^x e^{-\mu_1 b} \tag{2.25}$$

and

$$I_2^x = I_0^x e^{-\mu_1(b-c)-\mu_2 c} \tag{2.26}$$

where I_0^x is the incident x-ray intensity.

Let us now look at the film response. The time for which the x-ray beam is on will be denoted by t. Combining Equations (2.24) and (2.25), we can write

$$\begin{aligned} D_1 &= \gamma \log(I_1^x t) + D_0 \\ &= \gamma \log(I_0^x e^{-\mu_1 b} t) + D_0 \\ &= \gamma \log(I_0^x) - \gamma \mu_1 b + \gamma \log(t) + D_0 \\ &= \gamma(\log(I_0^x) + \log(t) - \mu_1 b) + D_0 \,. \end{aligned} \tag{2.27}$$

[7]Beer's law does not include the photons that are due to Compton scattering - we are therefore assuming that these are negligible, or can be distinguished from primary photons.

Similarly,

$$\begin{aligned} D_2 &= \gamma \log(I_2^x t) + D_0 \\ &= \gamma \log(I_0^x e^{-\mu_1(b-c)-\mu_2 c} t) + D_0 \\ &= \gamma \log(I_0^x) - \gamma(\mu_1(b-c) + \mu_2 c) + \gamma \log(t) + D_0 \\ &= \gamma(\log(I_0^x) + \log(t) - \mu_1(b-c) - \mu_2 c) + D_0 \, . \end{aligned} \tag{2.28}$$

The absolute difference between the film densities is given by

$$\begin{aligned} C_D &= D_1 - D_2 \\ &= \gamma(\mu_2 - \mu_1)c \, . \end{aligned} \tag{2.29}$$

This can be thought of as a measure of the visual contrast, although the actual intensity of visible light contains an exponentiation (from the definition of T with respect to D):

$$I_1^{visible} = I_0^{visible}(I_0^x t)^{-\gamma} e^{-D_0} e^{\gamma \mu_1 b} \tag{2.30}$$

and

$$I_2^{visible} = I_0^{visible}(I_0^x t)^{-\gamma} e^{-D_0} e^{\gamma(\mu_1(b-c)+\mu_2 c)} \, . \tag{2.31}$$

A more sensible definition of contrast where visual intensities is concerned is *relative* contrast, which in this case would yield

$$\begin{aligned} C_R &= \frac{I_2^{visible} - I_1^{visible}}{I_1^{visible}} \\ &= e^{\gamma(\mu_2-\mu_1)c} - 1 \, . \end{aligned} \tag{2.32}$$

One finds that the difference between both visible light intensity and film density is dependent on γ, c, and the *difference* between the attenuation coefficients. Films with higher γ values therefore produce greater contrast; larger path lengths through the inner cube also lead to increased contrast, as do bigger differences between μ_2 and μ_1.

The Ideal Image

Let us consider the characteristics of the image developed by our "thought experiment:"

- We have *exact* predictions of the intensities of the areas of the film, which are derived from Beer's law, from geometrical considerations, and from the film characteristics.
- The images of the cubic phantoms are precisely of the same dimension as the cross-sections of the respective cubes.
- The projections of the cuboids are "solid" in the sense that the intensity is uniform within each of the regions R_1 and R_2.

In what ways will a real x-ray image be different from that of the ideal image?

2.6 THE REAL IMAGING PROCESS

2.6.1 GEOMETRICAL CONSIDERATIONS

There are significant geometrical distortions that can arise in a real x-ray imaging system. Some of these are deliberately introduced to magnify areas of tissue, thereby making it easier to see small structures. Others are products of the acquisition system.

Magnification or Minification

If beam magnification is desired, then nonparallel x-ray projections result. Relative to the ideal image, this yields two main effects. First, the dimensions of the areas on the film, which are shown in the "ideal" image of Figure 2.12 are multiplied (scaled up) by the magnification factor. The way that this factor is determined by geometry is shown in Figure 2.12.

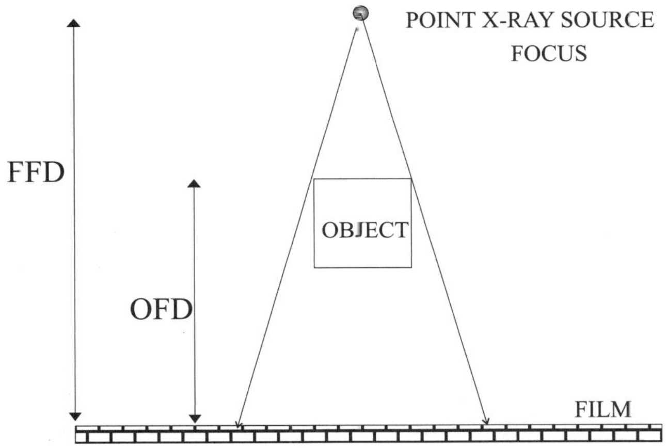

Figure 2.13: Magnifying geometry.

Objects in a plane at the same distance from the x-ray source will experience a magnification that is given by

$$m = \frac{FFD}{FFD - OFD} . \tag{2.33}$$

"Patient" Unsharpness

The eagle-eyed among you will notice something else: parts of the object below this plane will have a different magnification which depends on the distance from the x-ray source! The effect of this is best illustrated by looking at a "zoomed" in sketch of the photon paths through this cube (see Figure 2.14).

The result of this geometric magnification (diverging ray paths) is that the edge of the cube is blurred: neighbouring rays 1,2,3, and 4 gradually pass through greater path lengths of the cube. Many purists lament the fact that patients are not infinitely thin.

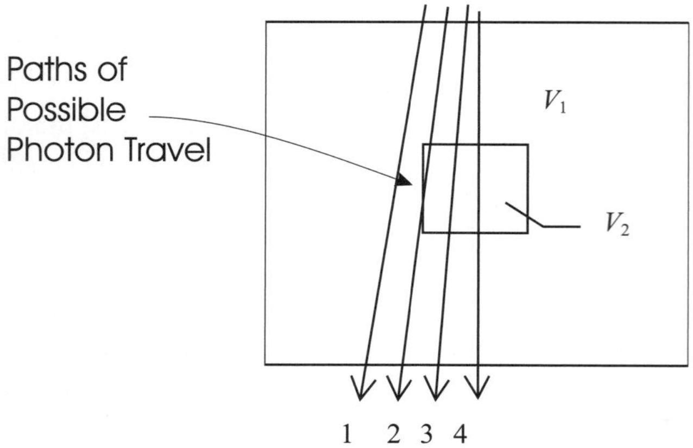

Figure 2.14: Illustrating "patient" unsharpness.

Beam Focal Size

Sometimes referred to as simply *focal spot size*, this simply reflects the fact that the impact area of the electron beam onto the x-ray anode has a finite (and sometimes, quite considerable) width. This means that the small point sketched in Figure 2.13, representing the source of x-rays, really needs to be broadened; the (exaggerated) effect of this is depicted in Figure 2.15.

If one considers a small perfect attenuator, there are now several ray paths that can be sketched from the source to the film which pass through the same edge point of the object; this provides a blurring effect which is in addition to any "patient" unsharpness effects. An expression for the unsharpness measure is now given as follows (using similar triangles):

$$B = \frac{OFD.FSS}{FFD - OFD} \tag{2.34}$$

where B is the width of the transition region for the edge on the image plane. Since, however, we have defined the magnification factor according to Equation (2.33), we can write this as

$$B = FSS(m - 1)\ . \tag{2.35}$$

The value of B is sometimes used as a measure of *geometric unsharpness*, U_g. However, this measure of blurring varies with magnification: as the object-film distance grows, so does the value of B, irrespective of the local spot size (the real source of this problem). Sometimes, therefore, the measure U_g is defined instead as

$$U_g = B/m = FSS(1 - 1/m)\ . \tag{2.36}$$

Be aware of this possible difference in definitions. We shall use the former definition for calculations on a given x-ray/patient configuration.

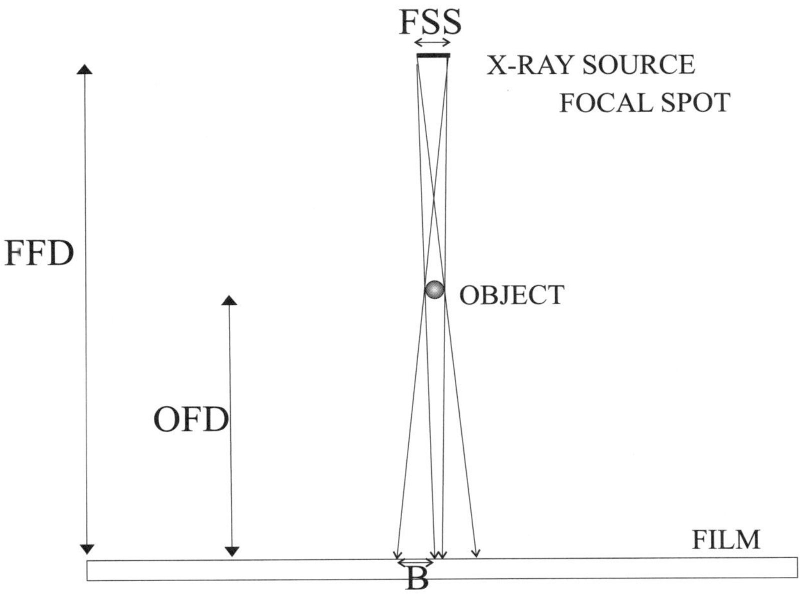

Figure 2.15: Illustrating geometric unsharpness.

2.6.2 QUANTUM (PHOTON) CONSIDERATIONS

Statistical Arrival of Photons

The arrival of photons is a statistical process. Beer's laws only tells us about the *expected* number of photons over a given small area within a fixed time. As the length of time over which we count photons increases, we get closer to the predictions of Beer's law[8]. In general, we can say that Beer's law only tells us about the *average* number of photons that we could expect, if we were able to take several x-rays of the same phantom, and take the average intensities at each point in space over all the x-rays. For a given x-ray, there will be fluctuations away from the ideal, which are partly responsible for the "mottled" appearance of the real image.

Photon Detection

More bad news, I am afraid: the proportion of photons that is actually detected (captured) by the film is very small. At best, for pure film, it is about 1-2%. The proportion that are captured give a measure of the quantum detective efficiency of the film. *Film-screen* combinations can improve this measure dramatically, but at a price (see later). In principle, this inefficiency affects the image appearance indirectly: it means that if the exposure time is too short (i.e., t used first in Equation (2.27)) the amount of quantum mottle in the image will increase (more image noise). Allowing t to increase reduces the mottle, but with a subsequent increased radiation dosage to the patient (phantom).

[3]Ignoring (temporarily) effects such as contributions by secondary photons, and possible beam hardening.

Photon Scatter

Secondary photons, produced mainly by Compton interactions will lead to some contribution to film blackening which is not of interest to us. This has two effects: the background level of film darkening is increased, thus reducing contrast, and the quantum fluctuation on the film is increased. The effects of Compton scatter can be minimised by the use of either moving or stationary anti-scatter grids.

2.6.3 BEAM HARDENING

Beam hardening will alter the effective attenuation coefficients of the material in the phantom. The effects of beam hardening are quite complex to describe, and its visible effect depends on the effective beam energy on entry to the cubic phantom, and the value of photon energy at which μ_1 and μ_2 are quoted. However, it can be said that the more pronounced the effect of beam hardening, the *lower* will be the final intensities on the film, and in general. This happens because the attenuation coefficients are always dependent on the photon energy. As a beam hardens, therefore, and its effective energy, E_{eff} increases, the effective attenuation coefficient will decrease. Consider the typical linear attenuation coefficient curves shown in Figure 2.16, which are for cortical bone and soft-tissue

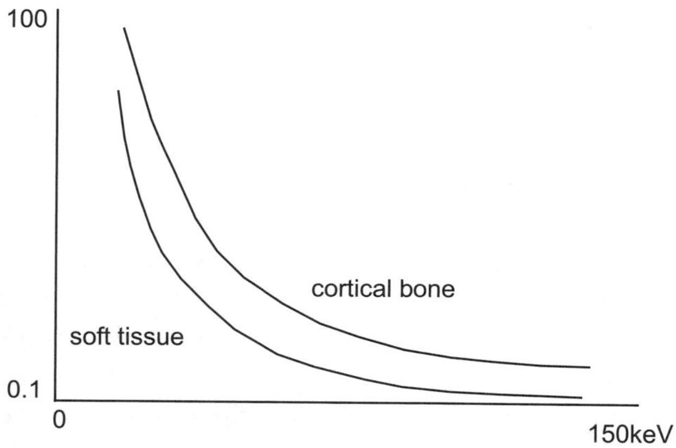

Figure 2.16: Consider the difference in attenuation coefficients as photon energy increases. If beam hardening occurs, the visual result is decrease in image contrast.

Because of the shape of the curves, higher photon energies are associated with *lower differences* in attenuation coefficients. Therefore, contrast will decrease (check with Equations (2.29) and (2.32)).

2.6.4 FILM EFFECTS

In addition to the poor quantum detective efficiency of film described above, there will also be some effects on film that contribute both to image mottle and unsharpness. First, the distribution of silver halide crystals on film will be nonuniform. This means that given two patches of film which are exposed to the equal beam intensities, the patch of film having the higher silver-halide matrix density will most likely be darker. Sometimes, the distribution of silver halide crystals is unacceptably poor, leading to very blotchy results when viewed at high magnifications. The other effect is that there is, of course, a definite size below which structures cannot be resolved. This would be the size of an individual deposit of atomic silver on the film. If the grain size before development is large, the size of the clumps of metallic silver set limits on the attainable spatial resolution, although this is of limited practical significance. High-quality X-ray film can allow the resolution of 5-8 line pairs per millimeter (more on this unit of measurement later). What is of somewhat greater practical importance is the loss of spatial resolution when a *film-screen* combination is used.

2.6.5 GROUPING THE EFFECTS OF UNSHARPNESS

The effects of unsharpness due to any magnification, finite focal spot size and film considerations all lead to blurring of the edges of the structures in the final image. By far, the most significant degrees of blurring in a system employing pure film (as opposed to a film screen combination) are the geometrical effects. Of these, there is not much that can be done about the nature of the object geometry in a planar x-ray system: the patient looks as he/she does!

For a given film, focal spot size and magnification will determine the final amount of blurring in the image. For a given desired magnification, we are stuck with focal spot size (FSS) as the one parameter that sets limits on the degree of unsharpness in the image. What stops us from making the focal spot size arbitrarily small? The answer lies in the design of the x-ray tube.

2.7 QUANTITATIVE MEASURES OF IMAGE QUALITY

In order to quantitatively measure the quality of an imaging system, which incorporates all of the effects that contribute to noise, blurring or loss in contrast, it is useful to adopt a "systems" level approach. Thus, rather than dealing with individual mechanisms of blurring, we consider one measure of the degree of blurring in the system, incorporating all the effects. This also has the advantage of being easily linked with empirical measures.

2.7.1 MEASURES OF SPATIAL RESOLUTION

Details of these measurements are defined in other parts of the course material.

- The Point Spread Function
- Full Width Half Maximum value
- The Modulation Transfer Function

2.7.2 MEASURES OF CONTRAST

If we define the small cube (V_2) to be the 'target' tissue that we are interested in observing, detection of this target area is related to the relative intensities of $I_1^{visible}$ and $I_2^{visible}$. We can define a measure of visual contrast, C_R, by the ratio:

$$C_R = \frac{I_2^{visible} - I_1^{visible}}{I_1^{visible}} \tag{2.37}$$

where it is assumed that $I_2^{visible} > I_1^{visible}$. This is dimensionless, and is known to be well correlated with human visual perception. Remember that all measures of contrast are dependent on an exact definition of what is considered to be the 'image' and what is considered the 'background,' so do not be worried if you find a different definition elsewhere.

Measures of Quantum Noise (Mottle)

The Signal-to-Noise Ratio (SNR) is a measure of the fluctuation of the image intensity relative to its mean (expected) level. It is very easy to suggest a method of determining the signal-to-noise ratio for an x-ray image, using the statistical idea of *expectation*. Let $< I(x, y) >$ be the expected intensity of the final x-ray image. A reasonable root-mean-square (RMS) measure of SNR would be

$$SNR = \sqrt{\int_{(x,y)\in Image} \left(\frac{< I(x, y) >}{I(x, y) - < I(x, y) >} \right)^2 dxdy} \, . \tag{2.38}$$

Here, $< I(x, y) >$ could be determined in one of two ways: for a phantom, it could be calculated using Beer's law, corrected for the effects of scatter and beam hardening. Alternatively, it could be obtained by taking several x-rays of the phantom, and averaging the intensities on the film (you would need to digitise all the films to perform this operation). This then represents the "signal." The noise is the fluctuation away from this, and it varies from image to image. In a single image, one could determine the noise by subtracting expected pixel intensities from the actual intensities. This "difference" image is then a measure of image noise. Squaring and integrating this noise image gives a measure of the noise energy. In the equation above, this is expressed *relative* to the signal energy (hence, SN *ratio*).

A contrast to noise ratio could also be defined for *specific* phantom images, and is sometimes used as an indicator of the so-called density resolution limit due to quantum noise. For the case of our phantom defined earlier, consisting of a white square on a dark background, then a simple form could be

$$CNR = \frac{I_2^{visible} - I_1^{visible}}{\sqrt{\int_{(x,y)\in R_1} (I(x, y) - I_1^{visible})^2 dxdy + \int_{(x,y)\in R_2} (I(x, y) - I_2^{visible})^2 dxdy}} \, . \tag{2.39}$$

2.8 DOSAGE

Biological damage can result when x-rays are absorbed by living tissue. This damage results either from electron recoil (during the Compton effect), or from the photoelectric effect. Both represent an injection of energy into the patient.

2.8.1 EXPOSURE

It is useful to define a property of the x-ray beam, known as the *exposure*, which describes the ability of the beam to produce biological damage. Exposure, then, can be thought of as a quantitative measure of the ability of the beam to produce ionisation. Exposure is defined by

$$X = \frac{\Delta Q}{\Delta m} \tag{2.40}$$

where ΔQ is the absolute value of the total charge of ions of one sign produced in air when all the electrons liberated by the x-ray photons in a volume element of air, of mass Δm, are completely stopped that volume. It has units of Coulombs kg^{-1}. The old unit of exposure is known as the Roentgen and it is given by 1 Roentgen = 2.58x10-4 Coulombs kg^{-1}. Why is such an odd definition used? Several requirements were laid down on the desirable characteristics of a standard system for radiation measure

- It must be accurate and unequivocal.
- It must be very sensitive.
- It must be reproducible.
- The measurement should be easily normalised: if intensity I is applied for time t, the result should be the same as for intensity $2I$ over time $t/2$.
- It must be usable to describe both large and small dosage quantities.
- It must be usable at all radiation energies (in a practical sense).
- It must be easily related to tissue absorbed dose (see later).

Ionisation in air is an extremely sensitive measure; if a 100keV photon is completely absorbed in a region of air, the ionisation results in very many ion pairs. Furthermore, in terms of biological damage, ionisation is the most relevant effect of radiation. This measure focuses primarily on the ionisation capability of the beam, and is therefore the measure though to be most appropriate for patient safety considerations.

The measurement of this property of an x-ray beam can be performed by using ionising chambers or the GM (Geiger-Muller) tube. Of these, the GM tube is more sensitive, compact and lighter, and can detect lower energies of radiation; however, it does not have a very direct relationship with the measure of exposure defined above, and this can limit its relevance (and use) in diagnostic radiation monitoring. Other types of radiation detectors and monitors are covered elsewhere in the MSc course.

2.8.2 ABSORBED DOSE

Dose is a measure of actual energy absorbed by an object from the x-ray beam. It is measured in units of energy per mass, such as J kg^{-1}. Dose is related to the amount of ionisation in biological tissue. Actual biological damage is, of course, very hard to define, since any assessment of damage must consider the biological function of the tissue, and how that function is affected by radiation. It is known that some cells, due to their biomolecular structure, are quite radiation resistant, while others are radiation sensitive. The absorbed dose, D, is defined by

$$D = \frac{\Delta E}{\Delta m} \tag{2.41}$$

where ΔE is the mean energy imparted by ionizing radiation to a sample of matter of mass Δm. The current unit of absorbed dose is the Gray, which is equal to 1 J kg^{-1}, and is also equivalent to 100 rad, the old unit of dose.

2.8.3 KERMA

One problem with absorbed dose is that it is biased towards the photoelectric effect. An increasingly used measure is the KERMA. KERMA stands for (k)inetic (e)nergy (r)eleased per unit (ma)ss, and is dependent on the type of material that the beam is incident on. It is intended to include the absorption due to any photon scattering processes. However, at diagnostic photon energies, it is quite similar to the absorbed dose. We shall, therefore, not consider it further.

2.8.4 CONVERTING EXPOSURE TO ABSORBED DOSE IN AIR

In order to convert exposure in air to an absorbed dose in a particular tissue, we need to understand how dosage relates to the ion pairs produced by ionisation. Let exposure be X. The photon energy required to produce 1 ion pair is about 34 eV. Thus, the dose in air, D_A is given by

$$D_A = 34X \quad \text{Grays .} \tag{2.42}$$

This therefore relates the absorbed energy in air to the degree of ionisation in air, a measureable quantity.

2.8.5 DOSE IN AIR VS DOSE IN TISSUE

For a specific tissue, we need a conversion factor to go from dose in air to dose in that tissue. For the purposes of example, we shall consider muscle as the tissue we are concerned about. The conversion factor is related to the absorption coefficients of tissue: recall the linear attenuation coefficient we met earlier; the linear *absorption* coefficient, μ^{abs}_{muscle} is similar, but excludes the effects of scatter. The *mass* absorption coefficient is given by

$$\frac{\mu^{abs}_{muscle}}{\rho_{muscle}} \tag{2.43}$$

and is used implicitly in the aforementioned normalisation. The dosage in muscle is given by the following equation:

$$D_{muscle} = D_A \frac{\mu_{muscle}^{abs} \rho_{air}}{\mu_{air}^{abs} \rho_{muscle}} \tag{2.44}$$

and has units of Grays. It is used as a crude estimate of the dosage given to particular tissues in a patient during a certain x-ray procedure.

2.8.6 GENETIC & EFFECTIVE DOSE EQUIVALENTS

One of the big problems with radiation dosimetry is that risk to the patient is a statistical one. That is, the information that one has (from empirical evidence) is that if, say, N persons receive a certain exposure, a certain fraction of them, x will develop cancer, or pass on a genetic mutation to their offspring. Moreover, the risk of cancer is organ-dependent. In order to try to convert these empirical risks to a number that is associated with a given patient in a given imaging procedure, two measures can be used: the *genetic dose equivalent* and the *effective dose equivalent.*

The Genetic Dose Equivalent is a means of trying to quantify the statistical nature of the genetic risk on receiving radiation doses prior to conception. It is defined as the dose which, if given to the whole population, would produce the same total genetic injury as the actual dosage received by the individual(s) concerned.

Effective Dose Equivalent This is a quantity which represents a dosage which, if given uniformly to the whole body, would produce the same total somatic detriment (for all purposes, risk of carcinogenesis) as the actual dosage received by the patient. This 'normalizes' measures of dosage relative to the particular target organ(s).

2.8.7 DOSE AND IMAGE CONTRAST

We already have the following facts about X-ray interactions with matter and with image formation:

- as photon energy decreases, the attenuation coefficient increases.
- as photon energy decreases we get improved tissue contrast.

We are always seeking to enhance the contrast between tissue types in an image, so an incident x-ray spectrum weighted towards lower photon energies (corresponding to a low E_{eff}) should be a good thing! But if the attenuation of the beam is increasing, where is the beam energy going to? Remember that at low photon energies, we have strong contributions from the photoelectric process. Attenuation is therefore very large at low beam photon energies. Moreover, this attenuation effect is due very much to ionisation of atoms in the tissue. So although we would like to use beam photon energies which are quite low to obtain good contrast, decreasing the beam energy increases the patient dosage. There is therefore a trade-off in the choice of photon energies in diagnostic imaging; in most cases, photon energies in the region of 50keV-150keV are used.

2.8.8 DOSE AND SIGNAL/NOISE RATIO

In considering the relationship between dose and signal/noise ratio, we need to look at the statistics of the photon fluctuations. We have already introduced the idea of signal/noise ratio as a measure of the mean (expected) image intensity level relative to the mean square fluctuation of the image intensity from the expected levels.

For a given tube current (which determines the intensity of the beam), as exposure time is increased (that is, as the net *fluence* of photons is increased), the *relative* fluctuation away from the expected levels decreases. This results in an *increased* signal-to-noise ratio with increasing exposure time[9].

Roughly speaking, given equal numbers of photons arriving at the film, lower photon energies are generally associated with increased contrast/noise ratios. However, for a given tube current, if we lower the tube voltage, then because of the higher attenuation that would result from the consequent lowering of the effective beam energy, a longer exposure time is required to maintain the signal-to-noise ratio.

The result is that for a given contrast/noise ratio, we need a certain minimum exposure time, and hence dosage, to the patient. The minimum exposure time varies, depending on beam characteristics (photon energies), the anatomical site being examined, the film characteristics and the degree of blurring (which equate to spatial averaging). However, in general, one can say that as exposure time increases, contrast to noise ratio also increases.

2.9 PRACTICAL ISSUES

2.9.1 THE X-RAY SOURCE

The X-ray source is carefully designed to produce radiation which is appropriate to the clinical application of the imaging system. For example, in mammography, where one is often attempting to find very small microcalcifications, signal-to-noise ratio is critical, and depends both on the thickness of the tissue and the beam spectrum. The design of the X-ray source can alter the spectrum. The anode target material affects the radiation. As we know, characteristic radiation will depend on the atomic configuration of the anode material. However, Brehmsstrahlung yield is also somewhat affected by the target material: higher atomic number materials provide greater Brehmsstrahlung yields, higher conversion efficiency and, for a given tube potential and current, higher beam intensity.

The photon energy distribution is also affected by the kinetic energy of the electrons hitting the target material. Thus, controlling the potential difference across the tube is another simple, practical way of altering the spectrum. In clinical equipment, one also finds that changeable x-ray filters are included, which remove photons from the spectrum that possess energies unlikely to contribute to image contrast. Typical filter materials would be aluminum or copper. In some

[9]We will consider an easier to understand example of this in X-ray CT.

applications, molybdenum is used. This filtering reduces the net dosage to the patient by removing very low-energy photons.

The vacuum tube in which the anode and cathode sit is a thick-walled glass tube, containing a slightly thinner region which acts as a window for the emitted x-rays. Because convection cooling is clearly not possible within the tube, heat dissipation at the anode of the tube is a major engineering challenge. As electrons collide into the target material, 99% of their energy ends up as heat! Target materials of high atomic weight are therefore preferred, as they tend to have higher melting points, and better thermal conductivity and heat capacity. Typical target materials include tungsten and molybdenum. Still, it is often necessary to rotate the target material, so that the electrons impinge on different areas on the surface over time. Typical rates of revolution are between 3000 and 10000 rpm. The surface of the anode is often bevelled to increase the surface impact area of electrons, whilst keeping the cross-sectional area of the beam small. Further cooling is provided by a surrounding oil bath; this encases all or part of the vacuum tube. Clearly, such mechanical complexity adds to the design problem.

The electrical supplies of such tubes are also highly customised.

2.9.2 SPATIAL DISTRIBUTION OF X-RAY PHOTONS

The focal spot size (the width of the x-ray photon beam leaving the tube)is determined by several factors. First, the width of the electron beam which hits the anode is clearly important. However, the *angle* of the anode contact surface is also important. Figure 2.17 shows the relationship between the incident electron beam width and the effective exit x-ray photon width.

We can say that

$$D_x = D_e \tan(\alpha) \tag{2.45}$$

where D_x is the effective width of the x-ray beam, and D_e is the effective width of the incident electron beam. The angle α is known as the anode angle.

This is only an approximate relationship; in fact, the x-ray beam tends to be quite nonuniform across the cross-section of the exit x-ray beam. In order to understand why this is, have a look at the point labelled A in Figure 2.17. Many electrons do not give up their energy until relatively deep into the target material (when they approach close enough to a nucleus). If we consider the two possible directions of emission of an x-ray photon from point A in the anode, it is clear that the path lengths through the anode material will be different for each possibility. This difference in exit paths for a large proportion of photons results in the so-called *heel effect*: the beam is of lower intensity as one moves across x-ray the beam, along the direction of electron travel. Thus, in general, it is difficult to characterise the precise width of the beam: it depends on what the "cut-off point" is in terms of intensity. Thus, the relationship between D_e and D_x is an approximate one.

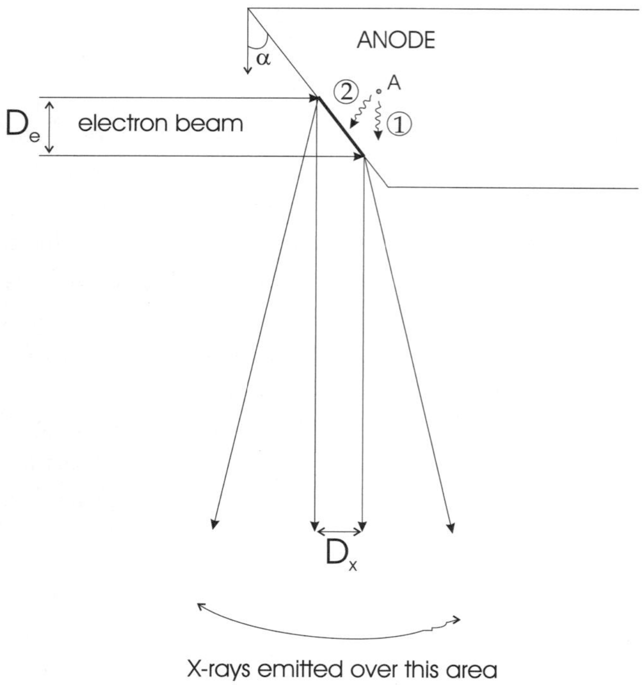

Figure 2.17: X-ray beam width, electron beam width, and anode angle.

2.9.3 RECEPTORS

As mentioned earlier, the Detective Quantum Efficiency (DQE) of film is very low. Typical estimates are of the order of 1-2%. Where it is important to keep dosage down as much as possible, one favours the use of more efficient X-ray photon detection. Examples include the screen-film combinations, which are a simple form of image intensifier. Phosphor coated screens are placed in front of a film emulsion. The phosphor emits fluorescent light photons under X-ray photon absorption, and these light photons are detected by the film. Although there is considerable amplification of the single X-ray photon into a cascade of up to 1500 light photons (energy of about 1-5eV), this increased sensitivity suffers from poorer spatial resolution than straight X-ray film. For x-ray film alone, the spatial resolution is of the order of 5-8 line pairs/mm for standard radiographic film. In mammography, a much higher quality film is used, sometimes permitting the resolution of 40 line pairs/mm. For a film-screen combination, the resolution limit will be set by the screen type[10] and will (typically) be between 3 and 8 line pairs/mm.

The purpose of a screen is primarily to improve the contrast to noise ratio of the image: in order to do this, the screen is performing a type of spatial averaging, thereby reducing the amount of fluctuation of the image, whilst simultaneously sacrificing spatial resolution.

[10]Although film type can also play a role.

Practically, the way that these screens work is that they consist of a cassette containing two phosphor screens. The film is placed in between these, and the cassette is closed tightly. The poorer the contact between the screens and the film surface, the poorer the spatial resolution of the resulting image. This has to do with the fact that the light photons are largely undirected, and travel outwards in all directions from the point of primary x-ray photon capture. Therefore, if they have to travel a longer distance before meeting film (due to poor contact), they will traverse a greater lateral distance.

2.9.4 DOSAGE & CONTRAST ISSUES

In order to obtain the best contrast while minimizing the risk to the patient, several factors need to be taken into account. These include the thickness of the body part to be imaged, the effective dose equivalent values of the target organs and the range of tissues to be imaged. Other specific parameters that can affect contrast include the portion of the film characteristic curve that is being utilised. In fact, by operating on different parts of the film characteristic curve, different contrasts will be obtained, because the effective γ changes. In order to understand how dosage is related to contrast, remember that in a given imaging time, there will always be a certain amount of intensity fluctuation. In the extreme case, these fluctuations can obscure the visibility of small, low contrast 'objects' embedded in a 'background.' One can attempt to increase the contrast, usually by reducing average beam energies. Alternatively, one can try to reduce the quantum fluctuations of the image. This can be done by some form of averaging: typically by acquiring the image over a longer period of time. Both of these approaches usually increase patient dosage. In addition, one can use a more efficient detector, such as a film-screen combination, with the accompanying loss of spatial resolution.

For any particular examination, and in using particular x-ray machines, it is essential that specific radiological training be undertaken to learn about the particular accepted dosage and operational guidelines.

2.9.5 CONTRAST AGENTS

In certain applications, conventional x-ray imaging can not provide enough contrast at acceptable dosage levels. In some applications, we can, however, increase intrinsic contrast by making use of a suitable contrast agent which is administered to the patient before examination.

A typical clinical application involving the use of a contrast agent is in angiography. This is used for visualising blood vessels, and for looking for regions of vessel occlusion. Without the use of a contrast agent, the attenuation coefficient of blood is very close to that of tissue. Blood/tissue contrast is therefore too low to be able to see blood vessels on a standard x-ray. The solution is to introduce a radio-opaque contrast agent (typically into the arterial circulation via a catheter) into the patient's blood vessels. Iodine compounds are typically used.

For surgical visualisation, it is also commonplace to use the technique as a real-time visualisation system, known as cine-angiography, in which fast image intensifiers, coupled to camera systems, are used to display the angiograms. This is one specific application of what is known as *fluoroscopy*.

A further contrast enhancement technique, again common in angiography, is known as Digital Subtraction Angiography (DSA), where contrast is further enhanced by subtracting the image before contrast injection, from images acquired after the contrast agent has reached the field of view (or the other way round!). This further improves the visibility of small blood vessels.

2.9.6 SAFETY

Perhaps the single most important safeguard available in an x-ray clinic is the presence of specially trained radiological technicians (known as radiographers), specifically for the purpose of running the equipment and performing the scans. For a given type of clinical work, the radiographers will have undergone training particular to the field of imaging (for example dentistry, orthopaedics etc.). This training should also be supplemented by training by the manufacturer of the specific device(s) to be used at the site of working.

Including calibration checks into a clinical routine is important. Generally, one finds that a daily scheme of checking the machine (pre-scan checks before seeing any patients the first thing each morning) is sufficient. Checks will include image quality, so that tube outputs, film density quality and beam focal size will be examined. Image quality checks are necessary, because it reduces the need to re-scan the patient though poor performance or tuning of any part of the system.

Once a month a qualified radiographer should fully check all the calibration on the X-ray scanner, by using *independent equipment* to verify beam output, tube voltage and exposure timings. You will meet more of this under the radiation dosimetry and advanced x-ray imaging course.

Summary

It is essential to underscore that radiation dosimetry cannot be left to chance, and these notes in no way prepare you to work in this area. Specialised courses, often run (appropriately) by radiological protection bodies or active hospital safety units exist, and the student who is considering a career in this area must seek appropriate formal training and certification.

CHAPTER 3

X-Ray CT

3.1 PLANAR X-RAYS:REVIEW

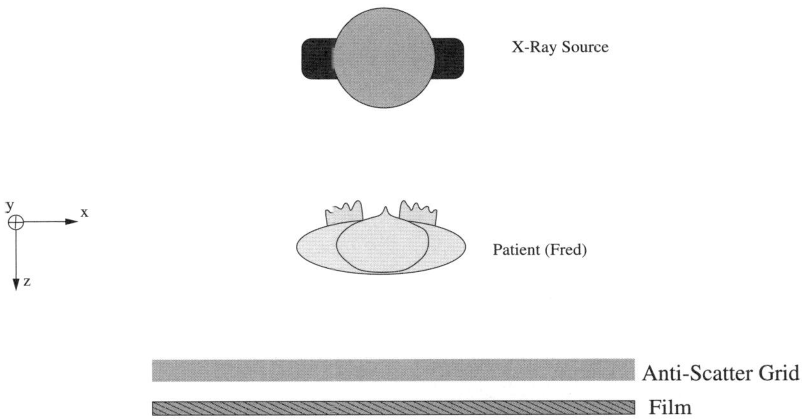

Figure 3.1: Acquiring an X-ray image of Fred.

Remember that in obtaining a 2D x-ray of Fred (the patient), we are obtaining a projection of what is really a 3D distribution of x-ray attenuating structures:

$$< \tilde{I}^{xray}(x, y) > \quad \propto \quad e^{-\int \mu(x,y,z)dz} \tag{3.1}$$

where $\mu(x, y, z)$ is the three-dimensional distribution of linear attenuation coefficients in the patient. We have used $< \cdot >$ to emphasise that Beer's law tells us about the *expected* value of the projected intensities[1] $\tilde{I}^{xray}$ is stochastic in nature.

Remember, that in order to convert $\tilde{I}^{xray}(x, y)$ to an image, we still need to convert the x-ray radiation to some visible form.

3.1.1 LIMITATIONS

In order to review the problems of the projection, consider a simplified example of a 1cm^3 volume of muscle, next to a 1cm^3 volume of bone.

Let us look at the exit intensity of x-ray radiation through a the (x, y) plane:

[1] $< \cdot >$ is equivalent to $E\{\cdot\}$.

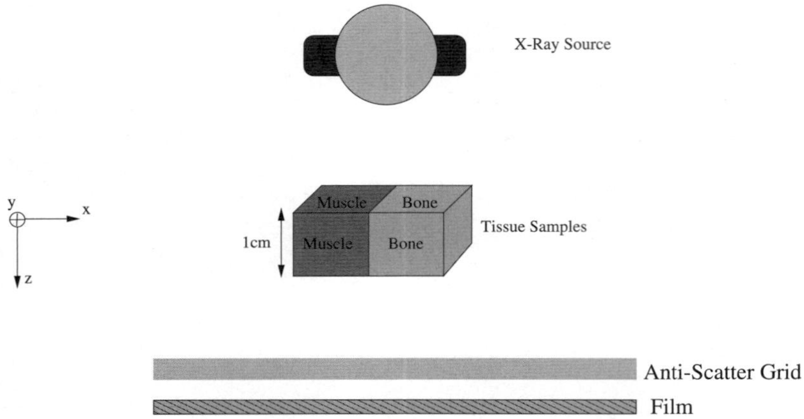

Figure 3.2: Simple contrast experiment.

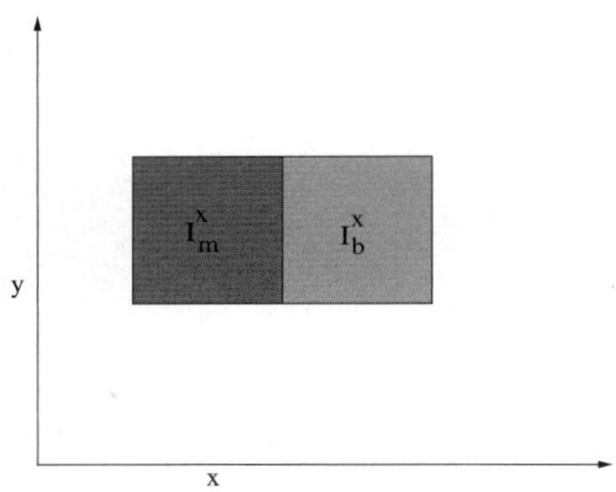

Figure 3.3: X-ray intensities; degree of shading is *not* representative of actual intensity.

Now, let us apply Beer's law to determine the relative expected intensities I_m^x and I_b^x. Using Equation (3.1), and typical figures for μ_m (the attenuation coefficient of muscle) of 0.1801cm^{-1}:

$$\begin{aligned} I_m^x &= I_0 e^{-\int \mu(x,y,z)dz} \\ \frac{I_m^x}{I_0} &= e^{-\int \mu_m dz} \\ &= e^{-\mu_m} \\ &= 0.8352\,. \end{aligned} \tag{3.2}$$

For bone, we use a μ_b of 0.4801cm^{-1}:

$$\begin{aligned} I_b^x &= I_0 e^{-\int \mu(x,y,z)dz} \\ \frac{I_b^x}{I_0} &= e^{-\int \mu_b dz} \\ &= e^{-\mu_b} \\ &= 0.6187\,. \end{aligned} \tag{3.3}$$

From this, we can define a relative x-ray contrast (ignoring the detector characteristics) of

$$C = \frac{0.8352 - 0.6187}{0.8352} \times 100\% \tag{3.4}$$

or approximately 25%.

Consider a second physical experiment, where we now have the little cube of bone replaced by a cube of blood:

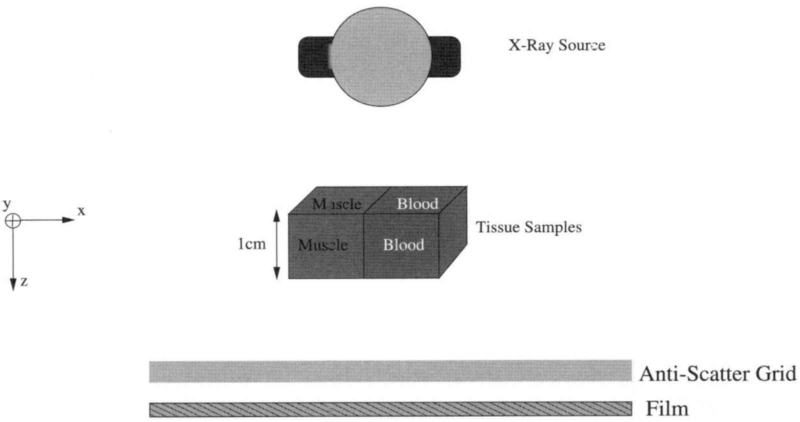

Figure 3.4: Another contrast experiment.

Apply Beer's law to determine the relative x-ray intensities of I_m^x and I_{bl}^x:

$$\frac{I_m^x}{I_0} = 0.8352 \tag{3.5}$$

and, using μ_{bl}=0.1781cm^{-1}

$$\begin{aligned} \frac{I_{bl}^x}{I_0} &= e^{-\int \mu_{bl} dz} \\ &= e^{-\mu_{bl}} \\ &= 0.8369 \end{aligned} \tag{3.6}$$

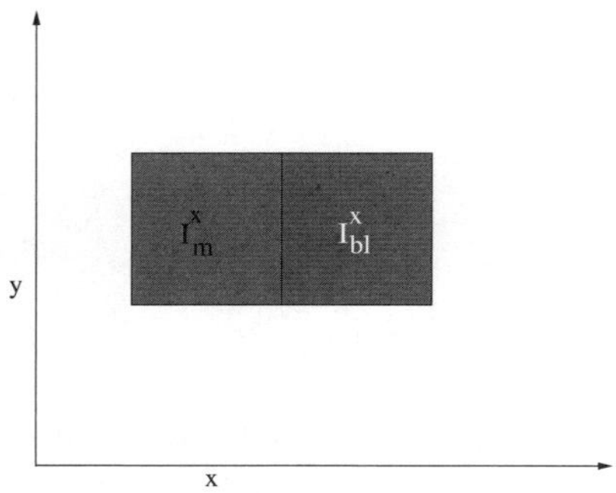

Figure 3.5: X-ray intensities.

the relative contrast is now only 0.2%.

So, we find that while the intrinsic x-ray contrast is quite large between bone and soft tissue, it is weak between tissue and blood. To visualise vessels, we could use a contrast agent (such as an iodine compound), but this is mildly invasive. Potentially, we can also improve the *image* contrast by careful selection of a film characteristic.

Let us look at another related problem of planar x-ray technology. Consider the following distribution of attenuating volumes:

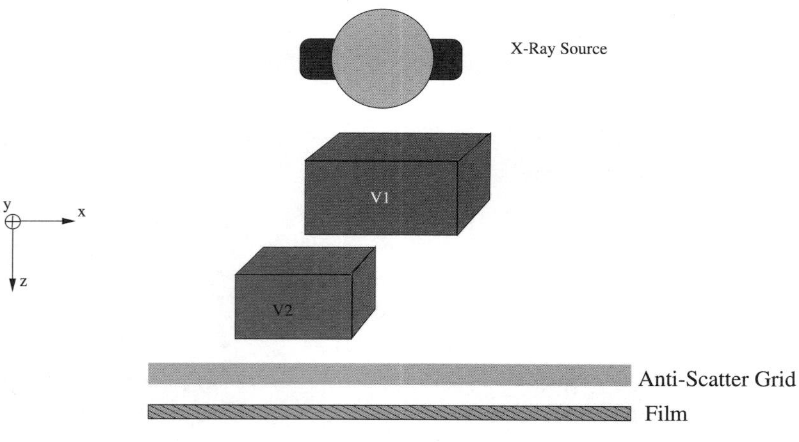

Figure 3.6: Depth Collapse problem.

First, let us assume that volume V_1 has attenuation coefficient μ_1 and that volume V_2 has attenuation coefficient μ_2. Note that due to the loss of z-direction information in the projected x-ray image, it is not possible to tell from the final developed film that

- Only 2 attenuating volumes are present
- V_1 and V_2 are at different z-positions

In fact, the result is *exactly* the same as if three distinct attenuating volumes of differing attenuation coefficients were present along the source-detector path.

3.1.2 SOLUTIONS TO CONTRAST AND DEPTH COLLAPSE

Contrast Agents

One solution to the contrast issue is to use a contrast agent in order to improve the differential attenuation between different types of soft tissue. Clinical examples where contrast agents are of use include the visualisation of the gastro-intestinal tract, where a barium meal is administered to the patient before the x-ray is taken, allowing high contrast images to be obtained. From these enhanced images, obstructions can be easily seen.

Another typical example is in the visualisation of blood vessels in the body. Iodine compounds are typically used to increase the intrinsic attenuation coefficient of blood. These are administered usually by catheter, and so there is a degree of morbidity with their use. This application is known as *angiography*.

Digital Techniques and Dynamic Range

One of the problems with image contrast has to do not only with the small differences in linear attenuation that exists between different tissue types, but also with the acquisition precision of the x-ray receptor. In principle, it is possible to enhance small differences in attenuation by using a more sensitive film. However, the *gray-scale* dynamic range of the human visual system in a typical viewing environment is not much more than 10 bits, so that even if *precision* is increased, we usually have some problem in being able to make it visually perceptible.

The use of grey-scale *windowing* can be a real advantage here, as it allows us to select regions of subtle intensity variation, and to remap these onto the perceptible visible range. In fact, a visit to your local MR clinic will show you that there is a windowing control on the display console; MR scanners can acquire and reconstruct to 12 bit resolution, and so if the full data *precision* is to be perceptible, only a portion of the image's *full* dynamic range can be shown.

In order to achieve this sort of thing with planar x-rays, the dynamic range of the receptor will have to be very high. Film could achieve this, but the subsequent windowing operations require specialist A/D conversion, and full digital representation and display.

Conventional Tomography

Conventional tomography goes some way toward solving the problem of loss of depth resolution. The approach is to use a moving source and film in order to hold only one *imaging plane* in focus over time. Regions of the patient which are out of the focal plane become blurred, and do not yield a significant coherent image. Conventional tomography has been largely totally replaced by *computed tomography*.

Summary
So, we see that there are a couple of theoretical limitations which make x-ray imaging rather less than ideal. Using computed tomography allows us to get around the loss of depth localisation. However, it turns out that solving the problem of depth localisation also goes some way towards improving the contrast between different types of soft tissue, and the relationship between contrast and patient dosage.

3.2 SLICING FRED

What would be ideal would be to take a slice out of Fred, and to be able to examine the tissue density distribution across that slice. Unfortunately, this solution is not appealing to Fred, so we try to find some way of *virtually* slicing Fred into sections in which we can determine attenuation coefficients, and hence get better tissue characterisation and localisation capability. The idea is illustrated in Figure 3.7, with an example of an image slice shown in Figure 3.8.

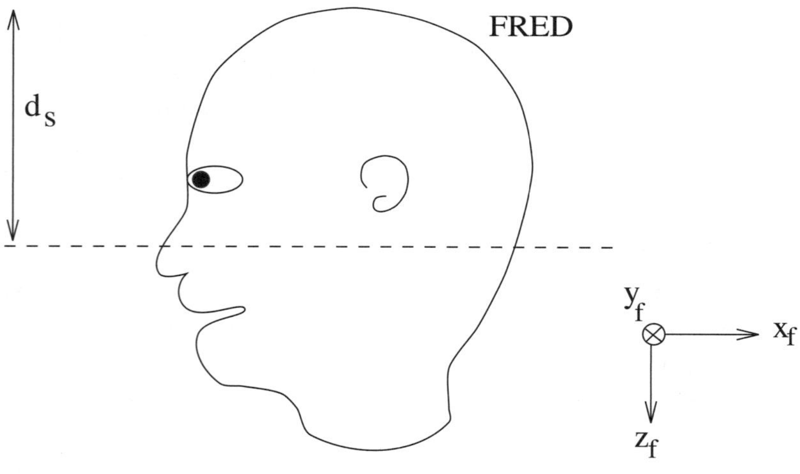

Figure 3.7: Imaging a Slice of Fred.

In order to acquire this slice of Fred, we shall show that we can use computed tomography. This involves collecting linear scans of Fred from several different directions in the (x_f, y_f)-plane.

3.2.1 LINEAR PROJECTIONS

The process of taking linear projections is illustrated in Figure 3.9.

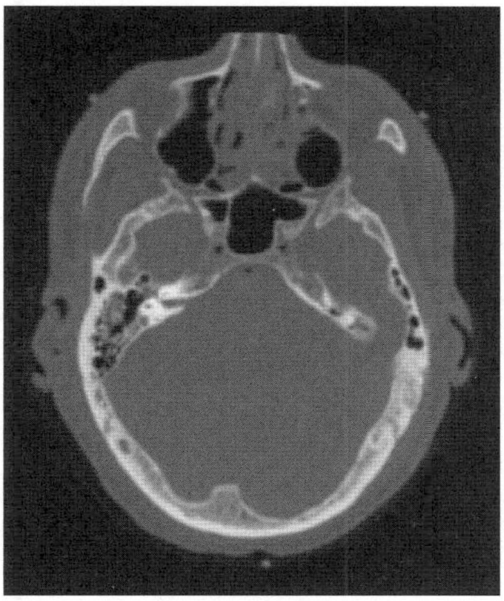

Figure 3.8: X-Ray CT slice through a human head.

The source and detector are assumed to be very highly collimated, so that the beam of x-rays is infinitesimally thin. (x_ϕ, y_ϕ) define a rotating co-ordinate system. We can collect a linear projection scan, $I_\phi^{xray}(x_\phi)$ at a particular angle ϕ. It is defined by sliding the source and detector in tandem along parallel paths at an angle ϕ to the x_f axis. Because the source and detector now act as *point* source and *point* detector, we have the following form for Beer's law:

$$I_\phi(x_\phi) = I_0 e^{-\int_S^D \mu(x_f, y_f) dy_\phi} \tag{3.7}$$

where we have dropped the "x-ray" superscripts for convenience. The operation represented by this equation is known as a simple *linear projection* of Fred at a particular angle ϕ. The result is a function of the rotating x_ϕ axis. Because the x_ϕ axis may be discretised, we sometimes refer to the set of data at one angle as a *parallel projection set at angle* ϕ.

Having defined this operation, we can now give a simple description of the principle of projection tomography.

3.2.2 BASIC PRINCIPLE OF CT

By collecting linear projection scans, $I_\phi(x_\phi)$ for a sufficiently large number of angles, ϕ, we can recover the information lost in a single linear scan. The operation is known as *slice reconstruction from projections*, or simply *reconstruction from projections*.

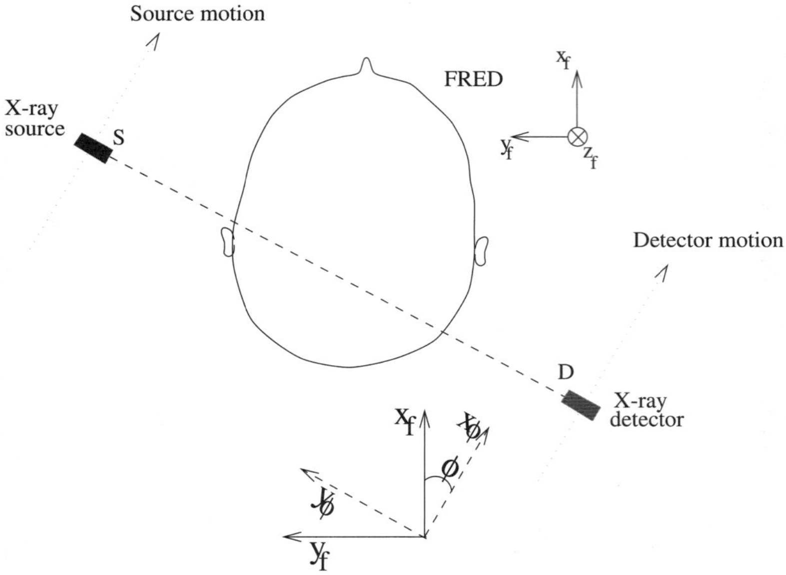

Figure 3.9: Linear Parallel Projections.

3.3 ALGEBRAIC INTERPRETATION

As a prelude to the details of image reconstruction from projections, we will introduce the algebraic interpretation of the problem. We will find this useful in order to explain certain restrictions of the method, although as a theory, it is not as powerful (or as practically useful) as the Fourier interpretation we shall introduce later.

Recall that a digital image may be represented by a matrix of values. We require that each element of the matrix bear some significance to the x-ray linear attenuation coefficient values in the patient, Fred.

Let us, for simplicity, assume that Fred's head is really made up of arbitrarily small volumes of distinct attenuating structures. We shall label each element by its position on a 2D grid, and assign an attenuation coefficient unknown to each grid element. The action of x-ray projection can now be defined along specific straight-line paths at various angles. Consider, first, the parallel projection data, along linear projection axes as shown in Figure 3.10. Using Beer's law, we can write expressions for the exit beam intensity as a function of location at given angles.

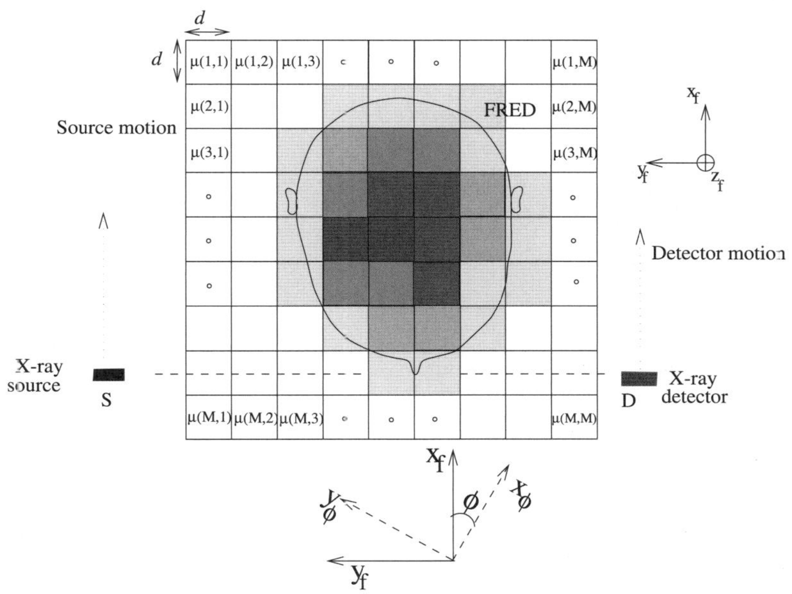

Figure 3.10: Algebraic Interpretation of Reconstruction from Projections: Acquiring $I_{0^\circ}(M-1)$.

For example, at $\phi = 0^\circ$ we can write

$$\begin{aligned} I_{0^\circ}(M) &= I_0 \prod_{j=1}^{M} e^{-d\mu(M,j)} \\ I_{0^\circ}(M-1) &= I_0 \prod_{j=1}^{M} e^{-d\mu(M-1,j)} \\ \vdots \quad &\vdots \quad \vdots \\ I_{0^\circ}(2) &= I_0 \prod_{j=1}^{M} e^{-d\mu(2,j)} \\ I_{0^\circ}(1) &= I_0 \prod_{j=1}^{M} e^{-d\mu(1,j)} . \end{aligned} \tag{3.8}$$

Rearranging, and taking $-\log_e$, we get

$$\begin{aligned}
-\ln\left(\frac{I_{0^\circ}(M)}{I_0}\right) &= d\sum_{j=1}^{M}\mu(M, j) \\
-\ln\left(\frac{I_{0^\circ}(M-1)}{I_0}\right) &= d\sum_{j=1}^{M}\mu(M-1, j) \\
\vdots \quad & \vdots \quad \vdots \\
-\ln\left(\frac{I_{0^\circ}(2)}{I_0}\right) &= d\sum_{j=1}^{M}\mu(2, j) \\
-\ln\left(\frac{I_{0^\circ}(1)}{I_0}\right) &= d\sum_{j=1}^{M}\mu(1, j)\,. \qquad (3.9)
\end{aligned}$$

All of the M terms on the left-hand side are measurable quantities, so this represents a set of M equations in the M^2 unknowns, $\mu(i, j), i = 1, 2..M, j = 1, 2..M$. Clearly not soluble! However, this represents only *one* view angle of the data. If we acquire parallel projections along different directions, then we can obtain a sufficient number of independent equations to be able to solve for the unknown attenuation coefficients. For example, at 90°, we obtain the following set of equations[2]:

$$\begin{aligned}
-\ln\left(\frac{I_{90^\circ}(1)}{I_0}\right) &= d\sum_{i=1}^{M}\mu(i, 1) \\
-\ln\left(\frac{I_{90^\circ}(2)}{I_0}\right) &= d\sum_{i=1}^{M}\mu(i, 2) \\
\vdots \quad & \vdots \quad \vdots \\
-\ln\left(\frac{I_{90^\circ}(M-1)}{I_0}\right) &= d\sum_{i=1}^{M}\mu(i, M-1) \\
-\ln\left(\frac{I_{90^\circ}(M)}{I_0}\right) &= d\sum_{i=1}^{M}\mu(i, M)\,. \qquad (3.10)
\end{aligned}$$

So, if we acquire a large enough number of views, we could obtain M^2 equations in M^2 unknowns.

Unfortunately, even if we had just the right number of independent equations to solve for the unknowns, we can't use our usual matrix inversion methods to solve for the $\mu(i, j)$, because the measurements are imperfect. Remember that Beer's law tells us only about *expected* x-ray intensities,

[2]For angles which are not parallel to the rows or columns, we need to include a few geometrical factors to weight the coefficients; theoretically, these are easily computable.

and individual measurements (experiments) will fluctuate from these.

There are a number of ways to solve for the $\mu(i, j)$, and we shall look at some of these in the Advanced Image Reconstruction Course. It happens that a very practical solution for the $\mu(i, j)$ is obtained via Fourier theory, and we will introduce this method in this module. The underlying theory for Fourier-based reconstruction is known as the *Central Slice Theorem*, and its derivation is quite different from the algebraic approach given above. At times, however, we will use the algebraic interpretation of the problem to describe some of the requirements and limitations of reconstruction from projections.

3.4 THE CENTRAL SLICE THEOREM

The *Central Slice Theorem* states that the Fourier Transform, $\Lambda_\phi(\omega)$, of the log of the normalised projection data, $\lambda_\phi(x_\phi)$ is one diametric slice of the two-dimensional Fourier Transform of the density distribution, $\mu(x_f, y_f)$.

3.4.1 DEMONSTRATION

Begin with Beer's law for a finely collimated source and detector pair, projecting along the y direction:

$$I(x) = I_0 e^{-\int \mu(x,y)dy} \tag{3.11}$$

or, for a rotating co-ordinate system, with respect to a fixed system, (x_f, y_f), we have

$$I_\phi(x_\phi) = I_0 e^{-\int_S^D \mu(x_f,y_f)dy_\phi} . \tag{3.12}$$

As we have seen in our simple algebraic treatment, the $\log_e$ operation can reduce the problem substantially:

$$-\ln\left(\frac{I_\phi(x_\phi)}{I_0}\right) = \lambda_\phi(x_\phi) = \int_S^D \mu(x_f, y_f)dy_\phi . \tag{3.13}$$

We shall refer to the quantity, $\lambda_\phi(x_\phi)$ as the natural logarithm of the normalised projection data. In fact, this quantity, in any real scan, must be a random variable, since the projected intensity is a random variable. So, $\lambda_\phi(x_\phi)$, really represents the *expected value* of the log-normalised projection data.

In order to proceed, it is useful to be able to express the linear projection of Equation (3.13) in terms of an integral with respect to the *fixed* coordinate system, (x_f, y_f). We can use the *sifting* property of the Dirac-δ function to achieve this:

$$\lambda_\phi(x_\phi) = \int\int \mu(x_f, y_f)\delta(x_f \cos\phi + y_f \sin\phi - x_\phi)dx_f dy_f . \tag{3.14}$$

You should confirm that the equation

$$x_f \cos\phi + y_f \sin\phi = k \tag{3.15}$$

describes a line in a rotated coordinate system parallel to the y_ϕ axis, running through k on the x_ϕ axis.

Let us now see what happens when we take the cartesian Fourier Transform of $\lambda_\phi(x_\phi)$:

$$\begin{aligned}
\Lambda_\phi(\omega) &= \int\int\int \mu(x_f, y_f)\delta(x_f \cos\phi + y_f \sin\phi - x_\phi)dx_f dy_f e^{-j\omega x_\phi} dx_\phi \\
&= \int\int\int \mu(x_f, y_f)\delta(x_f \cos\phi + y_f \sin\phi - x_\phi)e^{-j\omega x_\phi} dx_\phi dx_f dy_f \\
&= \int\int \mu(x_f, y_f)e^{-j\omega(x_f \cos\phi + y_f \sin\phi)} dx_f dy_f
\end{aligned} \tag{3.16}$$

where the last line follows by the sifting property.

Introducing the auxiliary variables, $u(\phi)$ and $v(\phi)$ as

$$u(\phi) = \omega \cos\phi \tag{3.17}$$
$$v(\phi) = \omega \sin\phi \tag{3.18}$$

then Equation (3.16) reduces to

$$\Lambda_\phi(\omega) = \int\int \mu(x_f, y_f)e^{-j(u(\phi)x_f + v(\phi)y_f)} dx_f dy_f \tag{3.19}$$

which is the 2D Fourier Transform of $\mu(x_f, y_f)$, evaluated along a line passing through the origin of Fourier space at an angle ϕ. Thus, for each and every view angle, ϕ, we have defined a function, $\Lambda_\phi(\omega)$, which represents diametric slices through the two-dimensional Fourier space of the attenuation coefficient distribution. We can also associate the variables ϕ and ω with a polar co-ordinate representation of 2D Fourier space.

We must now ask the following question: given that we can estimate slices of the Fourier space of the attenuation coefficient spatial distribution, how would we recover an estimate, $\hat{\mu}(x_f, y_f)$ of $\mu(x_f, y_f)$ itself? From the Central Slice Theorem, and Figure 3.11, one might deduce that we can use the 2D Inverse Fourier Transform to recover $\mu(x_f, y_f)$. However, our data has been acquired in a polar fashion. Thus, to use the well-known tool of the Inverse Discrete Fourier Transform (IDFT), we must map the polar representation onto a cartesian grid. This involves a considerable amount of very careful interpolation *between* the radial slices in the frequency domain.

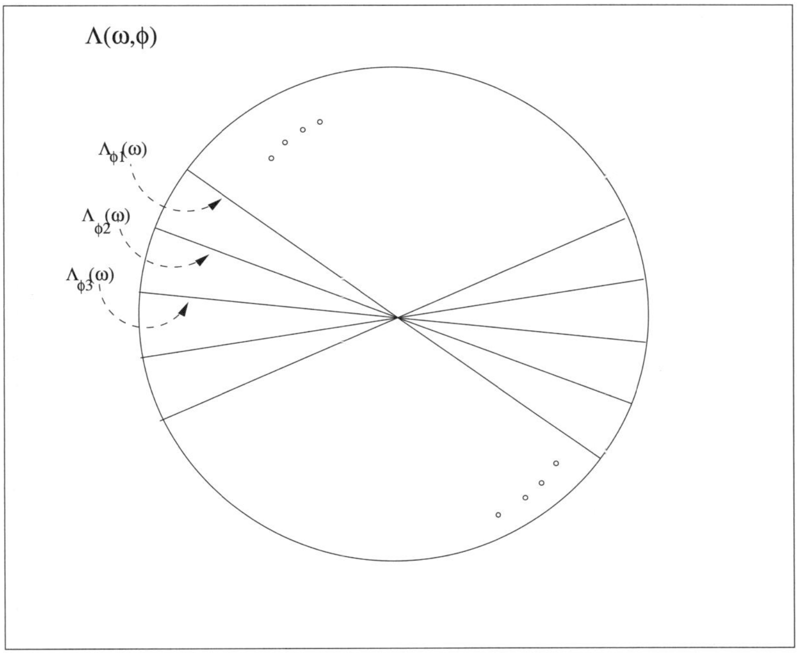

Figure 3.11: Central Slice Theorem.

3.5 CONVOLUTION BACKPROJECTION ALGORITHM

A very efficient reconstruction method an be derived if we proceed a little further with the analysis used above in deducing the central slice theorem. First, we make a couple of assumptions:

- our projections are true line projections
- we have an infinite number of view angles

Based on our second assumption, we shall alter our notation so that the Fourier transform of our projection data is viewed as a continuous function of ϕ as well as ω. Write Equation (3.16) as

$$\Lambda(\omega, \phi) = \int \int \mu(x_f, y_f) e^{-j\omega(x_f \cos\phi + y_f \sin\phi)} dx_f dy_f \,. \tag{3.20}$$

Taking the *inverse polar Fourier Transform* of both sides leads to

$$\begin{aligned} \mu(x_f, y_f) &= \frac{1}{4\pi^2} \int_0^{\pi} \int_{-\infty}^{\infty} \Lambda(\omega, \phi) e^{j\omega(x_f \cos\phi + y_f \sin\phi)} |\omega| d\omega d\phi \\ &= \frac{1}{4\pi^2} \int_0^{\pi} \int_{-\infty}^{\infty} \Lambda(\omega, \phi) e^{j\omega x_\phi} |\omega| d\omega d\phi \,. \end{aligned} \tag{3.21}$$

Equation (3.21) may be rewritten in terms of the following two simple expressions:

$$\mu(x_f, y_f) = \int_0^{\pi} \lambda'(x_\phi, \phi)\Big|_{x_\phi = x_f \cos\phi + y_f \sin\phi} d\phi \tag{3.22}$$

where

$$\lambda'(x_\phi, \phi) = \frac{1}{4\pi^2} \int_{-\infty}^{\infty} \Lambda(\omega, \phi)|\omega| e^{j\omega x_\phi} d\omega \,. \tag{3.23}$$

Equation (3.22) is known as the *backprojection* operation, and has a very easy physical interpretation. The backprojection operation is performed upon the function $\lambda'(x_\phi, \phi)$, which is known as the *filtered* projection data. The reason for this name becomes quite clear on considering Equation (3.23), which defines $\lambda'(x_\phi, \phi)$ in terms of the inverse *one-dimensional* Fourier transform of the product of two frequency (ω) domain functions, i.e.,

$$\lambda'(x_\phi, \phi) = \mathfrak{F}^{-1}\left\{\Lambda(\omega, \phi) \cdot \frac{|\omega|}{2\pi}\right\} \,. \tag{3.24}$$

Clearly, we can apply the convolution equivalence theorem to note that $\lambda'(x_\phi, \phi)$ is obtained merely by convolving (filtering) $\lambda(x_\phi, \phi)$ with an appropriate FIR filter. For the case of a finite number of view angles, we can see that the reconstruction operation may be performed without performing calculations in the Fourier domain at all, simply by

- Filtering the observed $\lambda_\phi(x_\phi)$ with a filter, $h(x)$
- Backproject $\lambda'_\phi(x_\phi)$ to form $\hat{\mu}(x_f, y_f)$

Table 3.1: Summary of Relationships

Fourier Domain	Spatial Domain
$\Lambda_\phi(\omega)\frac{\lvert\omega\rvert}{2\pi}$	$\lambda'_\phi(x_\phi)$
$\Lambda_\phi(\omega)$	$\lambda_\phi(x_\phi)$
$\frac{\lvert\omega\rvert}{2\pi}$	$? = h(x)$

3.5.1 BACKPROJECTION

The backprojection operation is not the precise inverse of the forward projection operation, because of the loss of the spatial information along the projection path. Rather, as discussed in [7], a backprojection operation is a "smearing out" of the original projection data back along the path of projection. A graphical illustration of the backprojection operation[3] is shown in Figure 3.12. Remember, in addition, in order to obtain a reasonable reconstruction, we must backproject the *filtered* projection data, and all of the filtered back-projected data must be summed.

[3] For the sake of simplicity, this is illustrated for the case of a discrete number of view angles.

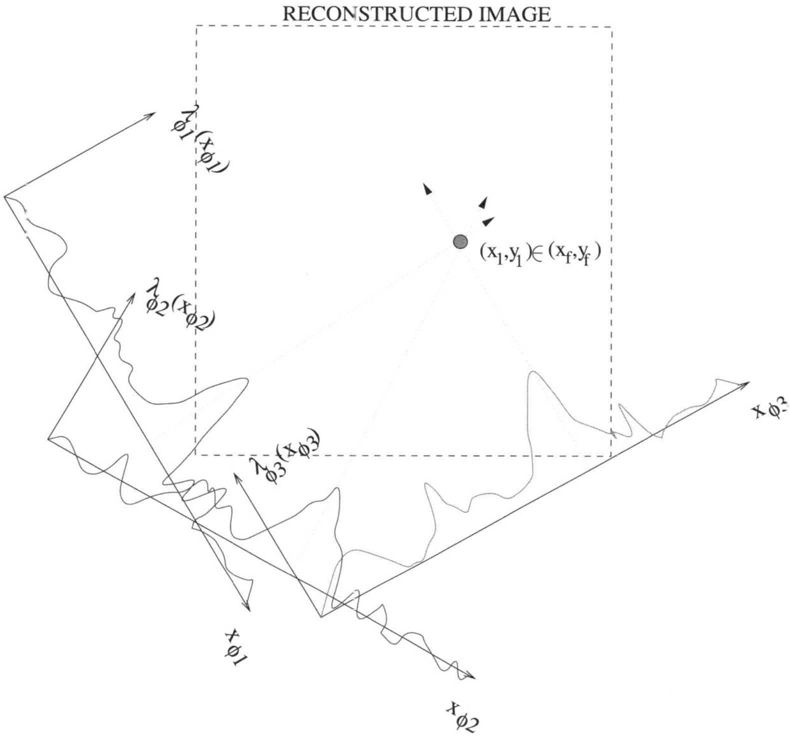

Figure 3.12: The Backprojection Operation.

3.5.2 DETERMINING $h(x)$

From Table 3.1, we can deduce that the filter which we need to perform the reconstruction should have a frequency domain magnitude response of $|\omega|/(2\pi)$ as shown in Figure 3.13.

However, it is clear that such a filter specification is not reasonable, since it is of infinite extent in the Fourier domain, and also implies that infinitely high spatial frequencies are given infinite amplification. In fact, the reason for this impractical filter specification comes from the assumptions made earlier in deriving the central slice theorem. Recall that we assumed that the projection data consisted of *line* integrals through a two three-dimensional space, and this projection operation corresponds to a measurement of spatial frequencies. In fact, this is impossible to achieve. Practical limits on collimation are set by quantum fluctuation, and so our integrals should be along paths with finite thickness. Rather than go into the details here, it suffices to say that the result of finite width integration paths leads to a simple alteration on the required frequency domain specification of Figure 3.13 to that shown in Figure 3.14, and denoted by $H_c(\omega)$. The choice of the filter cutoff

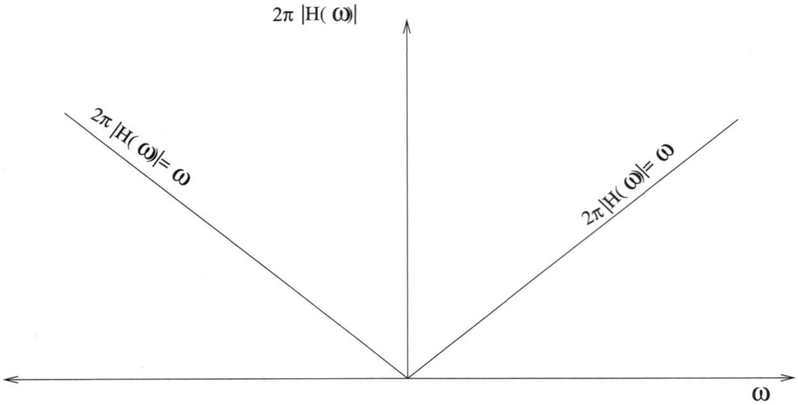

Figure 3.13: $2\pi H(\omega)$.

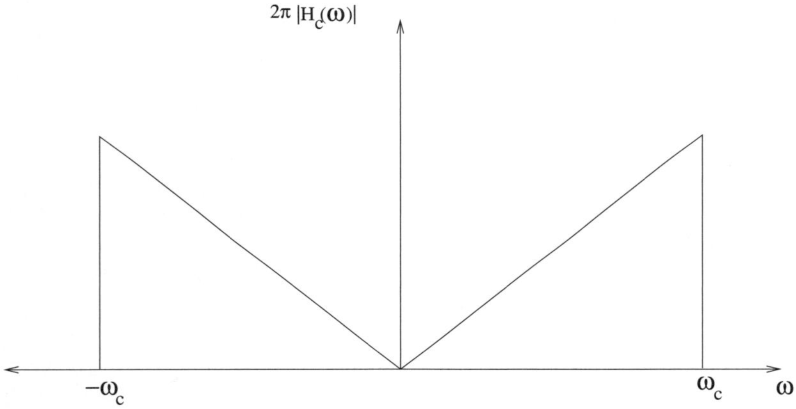

Figure 3.14: $2\pi H_c(\omega)$.

point ω_c will be addressed later. The inverse Fourier Transform of $H_c(\omega)$ yields the function

$$h_c(x) = \frac{\omega_c^2}{4\pi^2}\left(2\text{sinc}(\omega_c x) - \text{sinc}^2\left(\frac{\omega_c x}{2}\right)\right) \tag{3.25}$$

which looks pretty much like a single sinc() function.

3.6 SCANNING CONFIGURATIONS AND IMPLEMENTATION

3.6.1 INTRODUCTION

So far, we have looked in some detail at the concept of reconstructing an image, whose grey level intensities represent the linear tissue attenuation coefficient of matter in the centre of the scanning plane. We have considered the reconstruction from projections in a theoretical sense; we now turn to look at the acquisition of the linear projections in practice.

It is useful to segment the various types of existing and new CT scanners into a number of generations which are loosely ordered chronologically in the development of this tomographic technique.

3.6.2 FIRST GENERATION SCANNERS

As we have seen, the central slice theorem prescribes that we acquire a series of linear projections, $I_\phi(x_\phi)$, of the x-ray beam from a large number of different view angles, ϕ. The most literal interpretation of this requirement is to be found in the so-called first-generation scanners, in which a mechanically mounted gantry supported a very highly collimated source-detector pair. A diagram is shown in Figure 3.15. To collect each parallel projection set, the source and detector would be moved in a discrete number of steps across the patient. This operation would be repeated for a large number of view angles.

Because of the high degree of collimation employed, these first generation scanners had superb scatter rejection characteristics, and thus intrinsically better contrast/noise ratios than the later generation scanners. (This was, however, eventually offset by improved detector sensitivity in the later scanners). In addition, because a single source and detector are used, I_0, the transmitted x-ray intensity from the source is not a function of projection angle. Thus, calibration was unnecessary to the degree required by later multidetector systems.

The principal drawback of these early generation CT scanners was that the time taken to acquire a single slice comprising 300 views or so, is in the order of 4-5 minutes. This is acceptable for imaging the head of a not-too-unhealthy patient, but is less than adequate for abdominal imaging, where the patient would have to breath- hold in order to reduce motion artefact.

A further advantage of the first-generation scanners which is worth mentioning for historical reasons is that, because of its very literal implementation of the translate-rotate acquisition, image reconstruction is simplified, and is virtually as presented earlier i.e., convolution back-projection).

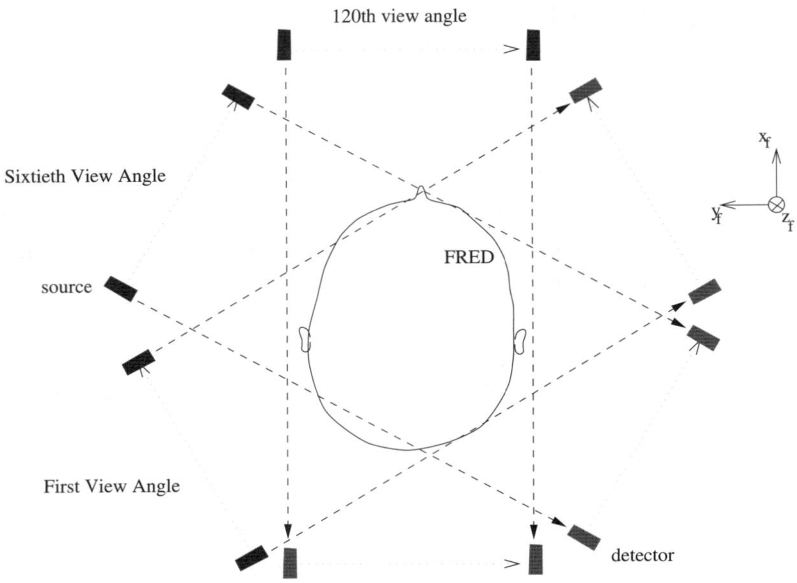

Figure 3.15: First Generation Scan Geometry.

3.6.3 SECOND GENERATION SYSTEMS

These are similar to the first generation system, in that a translate-rotate approach to acquiring the parallel projection data is used. However, in this case, a fan- shaped source of x-rays is used, and an array of detectors is mounted on the gantry.

Because a diverging beam was used, with a limited divergence angle, one is effectively, at each step, collecting a number of linear projections from slightly different angles. The improvement of the second generation scanner over the first generation scanner comes about through a reduced number of mechanical rotations: in the both scanners, it is typically necessary to acquire 300 or more views. In the second generation scanner, the number of gantry rotations required is of the order of 50. See Figure 3.16 for an illustration.

The first drawback is that now there is effectively a number of different source/ detector pairs, so that the *effective* I_0 can vary across different viewing angles. The detectors need to be calibrated with respect to each other.

The reconstruction approach is much the same as for that of the first-generation scanner, with the minor requirement that the projections from different viewing angles need to be correctly unscrambled.

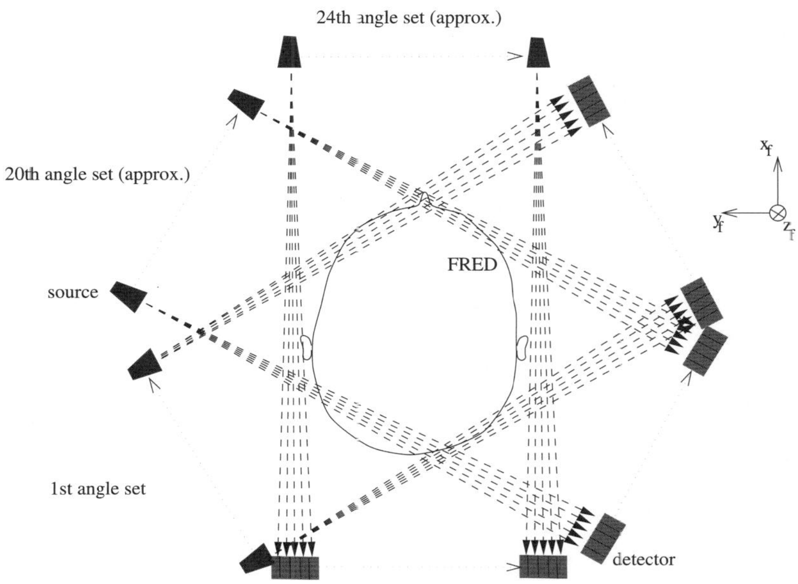

Figure 3.16: Second Generation Geometry.

Acquisition times for these systems, due to reduced mechanical rotation, is about 20 seconds, so that abdominal imaging on most patients is possible.

3.6.4 THIRD GENERATION SCANNERS

This is the first of the rotate-only scanners, which dramatically reduced mechanical complexity, while leaving plenty of opportunity for artefact!

The removal of the linear parallel projection acquisition mode required modifications to the reconstruction algorithm initially devised for the translate-rotate geometry. The scanner, depicted in Figure 3.17 employed a very wide fan beam, with a few hundred detectors. Note that for requirements of stable, high packing density of the detectors, it was necessary to operate the detectors in a pulsed mode, i.e., the x-ray source transmits bursts of x-rays.

The necessity of having balanced, stable detectors was critical in these systems; unbalanced or malfunctioning detectors lead to circular (ring) artefacts.

Because of the removal of gantry translation, the acquisition time for these systems was down to the order of 5 seconds per slice, which is quite manageable for abdominal and chest imaging of even critically ill patients.

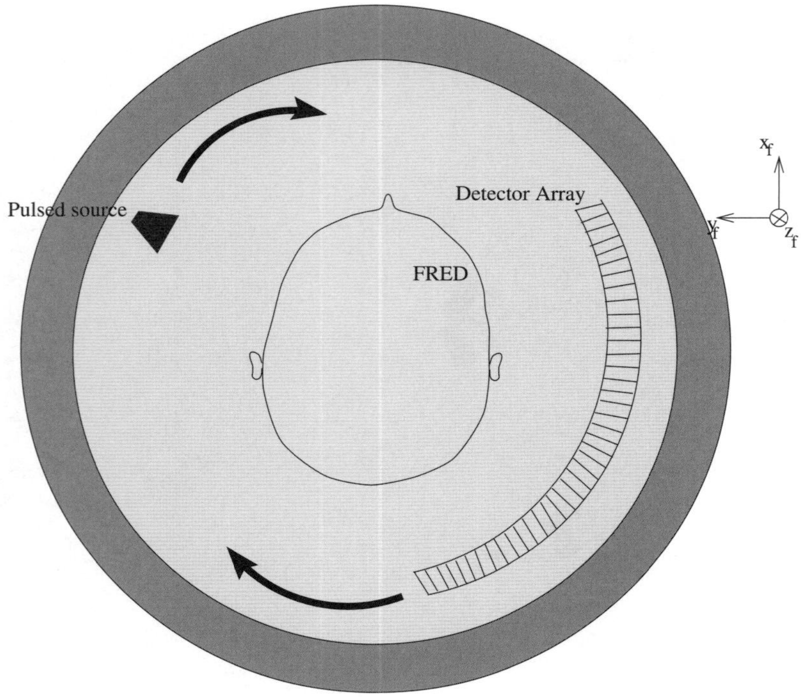

Figure 3.17: Third Generation Geometry.

3.6.5 FOURTH GENERATION SCANNERS

These were the first systems to employ stationary detector banks. These scanners had the largest number of detectors at the time, typically of the order of 1000. However, the stability requirements on the detectors, which presented such a problem in the third-generation scanners, could be slightly relaxed, because the stationary detector ring coupled with a very wide-angle rotating source allowed real-time monitoring of the detectors. Thus, one could compensate for variations in the effective reference intensity, and could determine if a detector was malfunctioning.

In addition, the less stringent demands on the detectors meant that continuous mode fan beams, which are easier to construct, could be used as the source.

Acquisition times for these systems were of the order of 4-5 seconds, and thus comparable to that of third generation scanners. Overall image quality was improved. See Figure 3.18.

3.6.6 FIFTH GENERATION SCANNERS

(Almost) no moving parts! Instead, a ring of target material is used, and a ring of 1000 detectors or so is employed. An electron beam is swept around to impinge upon the target material at different

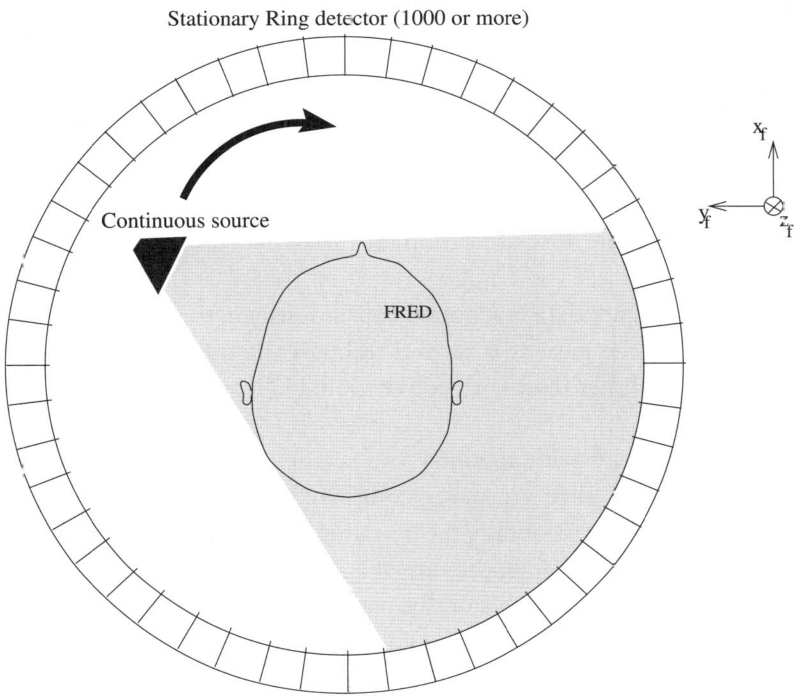

Figure 3.18: Fourth Generation Geometry.

points around the ring, so that the x-ray source is effectively moving (See [4], Figure 5).

Scan times are reduced to the order of milliseconds, because of the lack of mechanically moving parts. These systems allow the acquisition of slice data at a sufficiently high rate to enable cine mode imaging of cardiac motion. In addition, by using multiple target and detector rings, one can obtain multiplanar image slices.

This scanner was greatly "hyped" around the end of the 1980's. There was silence for about a decade, where the company developing the scanner floundered. Recently, there has been a resurgence of the technology through some serious clinical applications of the very fast imaging that such a scanner can permit. Also, the technology has been re-labelled as "Electron Beam Computed Tomography", a very misleading title, but clearly a marketing ploy that has worked. Expect to see some more on this very fast X-ray CT scanner in targeted applications, including cardiac imaging.

3.6.7 6TH GENERATION

In this section we include what is effectively a new generation of CT scanner. The method is known as *Spiral CT* and the name derives from the relative patient/gantry geometry of scanning.

Many consider this scanning method to be the state-of-the-art in x-ray computed tomography[4]. Using 4th generation CT X-ray systems, it is possible to acquire a series of slices at different axial locations, and stacking these up to provide volumetric data. There are a number of reasons whilst this is undesireable. Three main reasons are that (i) if the slices are not exactly spaced, and of the right thickness, then it is possible for small lesions to be missed in scanning, and (ii) the time taken in between scanning each slice can be large (since the table on which the patient is positioned needs to be advanced). (iii) Another problem is that the dose utilisation for such 3D volumes is fairly poor.

Helical or spiral CT addresses these issues: data is acquired continuously in the axial direction, and scanning is performed continuously as the patient is moved through the bore of the scanner. Dose utilisation for multi-slice data is also higher than with conventional CT.

Spiral, or helical, CT slowly moves the patient along the scanner axis as the slices are being acquired, so that the projections of the slice are never completely obtained in one plane. An interpolation scheme is then used in order to obtain the required plane of projections. The advantages of helical scanning are many fold. First of all, the rate of acquisition of 3D voxel space is much faster than with a conventional scanning system. Secondly, patient dosage is minimized, for a given volumetric field of view. Thus, in a manner similar to the advantage of CT over conventional X-ray radiology, dose utilization is improved.

There are several machine requirements for successful spiral CT. First, because of the axial relative motion of the patient, the imaging time that any slice plane experiences is very small. Thus, the beam power needs to be substantially higher than for planar slice CT systems. Because the scanning might extend (radially) over the length of the body, continuous rotation of the X-ray source and/or detectors is usually required through angles greater than 360 degrees, i.e., the rotation is continuous. The problem here is that there must be no cabling to source or detectors! Slip ring and brush systems for sources and optical transmission systems for detectors can solve these problems.

The final requirement is the image reconstruction. The method of reconstruction for spiral CT may be formulated so that simple modifications of existing algorithms may be used to effect reconstruction. Specifically, helically scanned data is resampled or interpolated onto a cylindrical coordinate system.

3.6.8 SPIRAL RECONSTRUCTION

The easiest understanding of the operation of the helical reconstruction system comes from considering the axial variation of attenuation coefficients in the patient as a smoothly changing function. We then assume that we are, by nature of the axial motion of the patient, sampling the *projection* data.

[4]Although some reserve this position for EBCT. Take your pick!

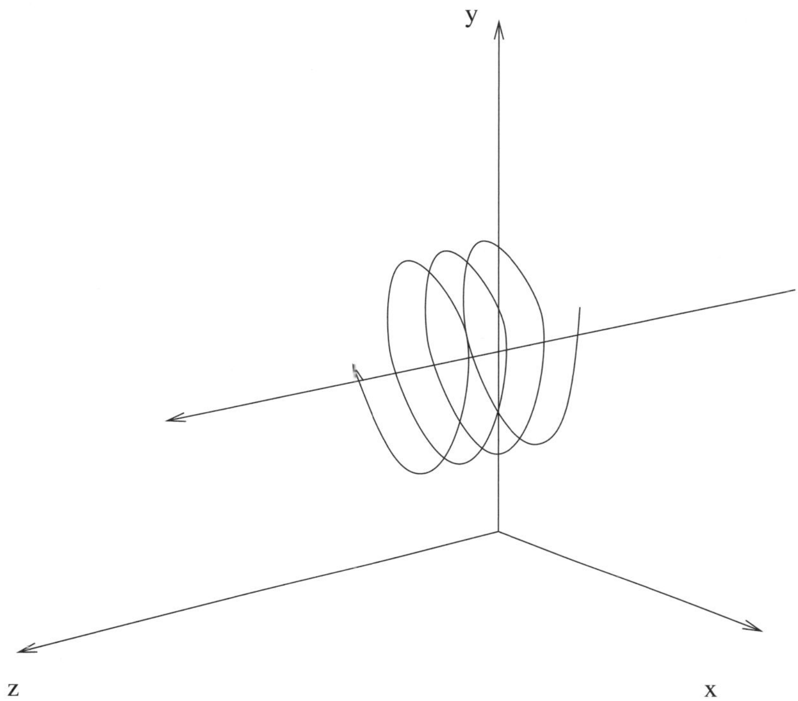

Figure 3.19: Spiral/Helical CT geometry.

In order for the interpolation of the system to make sense, we must assume that the density distribution in the volume of the system, $\mu(x, y, z)$ is band limited. This simply means that if we take the 3D Fourier Transform of the 3D density distribution, we would find that there is a definite concentration of the nonzero frequency components to a finite length cylinder.

In practice, one can set a reasonably hard limit to the spatial frequencies in the projection set because of the thickness of the X-ray beam along the axial direction. If we assume that this bandlimiting of spatial frequencies is a valid assumption, then one can simply apply the Nyquist criteria, based on some assumed upper limit of spatial frequency support (cycles/mm for the z axis, cycles/ radian around the diameter of the scan volume), and one can select an appropriate helical pitch for the scanning system. This is dealt with in detail in the thesis of Skrabacz (1988, Univ of Illinois; see also [10]), with the result that most of the detail of the *projections* of $\mu(x, y, z)$ may be recovered by interpolation. Thus, one can interpolate the data acquired in between the pitches of the helical scan to construct the projection data for a desired slice. Multiple slices can also be recovered in this way.

3.7 IMAGE QUALITY

We can attempt to describe image quality with reference to two aspects of the image:

- Resolution
- Freedom from Artefact

Under the heading of image resolution, we can identify two main sub-headings

- Spatial resolution
- Density resolution

For 5th generation scanners, yielding time dependent images, we can also consider *temporal* resolution, but shall not expand on this in this course. Rather, we shall simply define time resolution as the time taken to acquire and reconstruct an image slice.

We shall treat the topics of Image resolution and Image Artefact in detail in the next two sections of the notes. Now, we shall consider some general concepts used in quantifying image quality.

3.7.1 SPATIAL RESOLUTION

Spatial resolution is a measure of the ability to resolve two closely spaced structures within the field of view. The definition is normally based on resolving structures which have the same attenuation coefficient.

Point Spread Function

As with any approximately linear imaging system, we can describe the performance of an x-ray CT scanner in terms of its Point Spread Function (PSF). In the case of x-ray CT, we generally describe the PSF in the plane of the scan by use of the Full-Width, Half Maximum (FWHM) parameter.

An x-ray CT phantom consisting of a long, cylindrical solid bar, of high attenuation coefficient μ_m is defined as shown in Figure 3.20. This is placed into the bore of an X-ray CT scanner, and an image slice is acquired.

The *ideal image* is shown in Figure 3.21 (top). Consider a one-dimensional intensity profile through the ideal CT image as shown in Figure 3.21(bottom).

Because of the finite resolution nature of the image scan projections, we have an image profile which will look more like that shown in Figure 3.22.

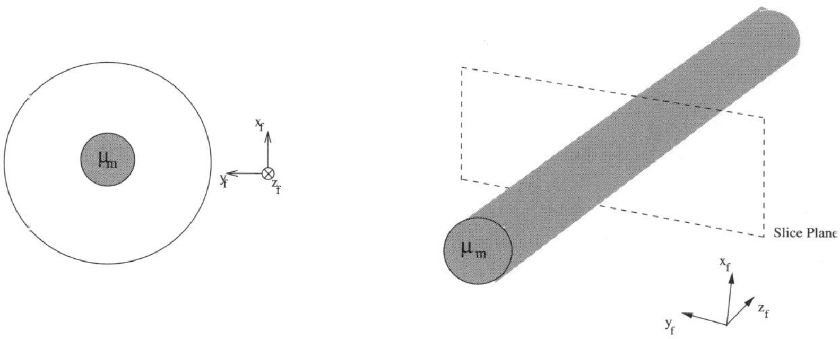

Figure 3.20: Cylindrical phantom.

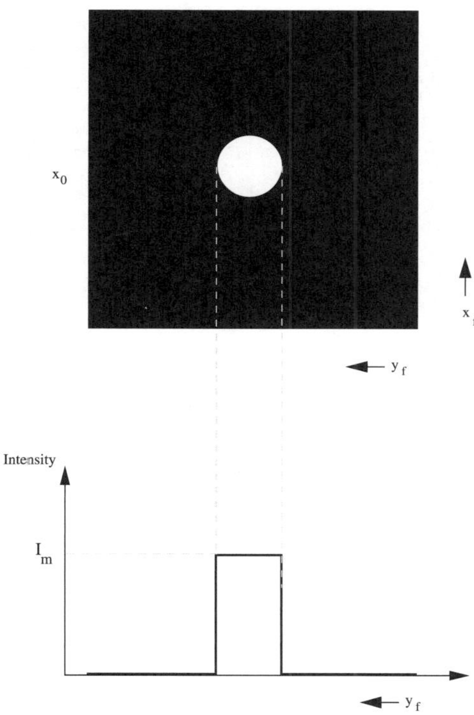

Figure 3.21: Ideal Image of phantom (top) and profile (bottom).

Provided that the point spread function is isotropic, we can represent it uniquely by $PSF(r)$, and the FWHM is defined as shown in Figure 3.23.

Modulation Transfer Function
Another popular measure of image quality in x-ray CT imaging is known as the *modulation transfer function* (MTF). The MTF has its roots in optics. In particular, it can be used for the evaluation of

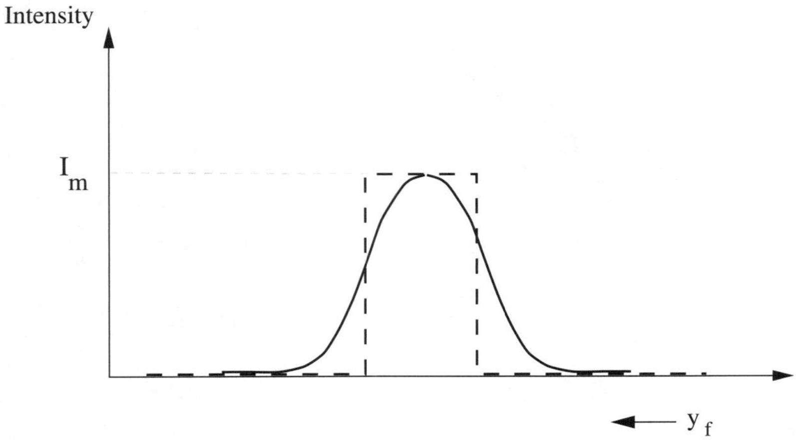

Figure 3.22: Ideal and Actual Intensity Profiles.

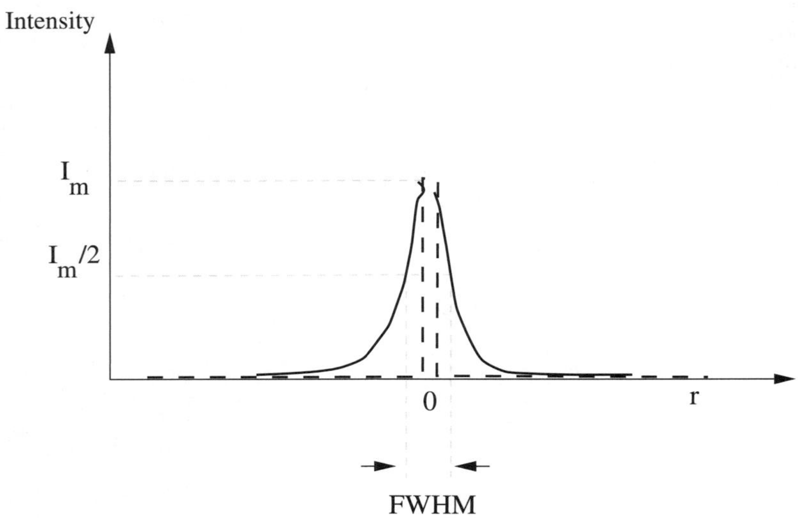

Figure 3.23: Definition of FWHM.

overall response of multistage imaging systems.

For a simple optical system, the MTF would be measured by "imaging" an appropriate phantom. In the case of an X-ray CT scanner, such a phantom would consist of closely spaced bars of equal, and relatively high-attenuation material. The spacing between the bars gradually changes across the phantom, from wider to smaller.

The difference in image intensity between the light (=bars) and dark (=spaces) in the reconstructed image is known as the *response* of the imaging system[5], and a plot of the response, R, versus bar spacing yields a curve looking something like that shown in Figure 3.24.

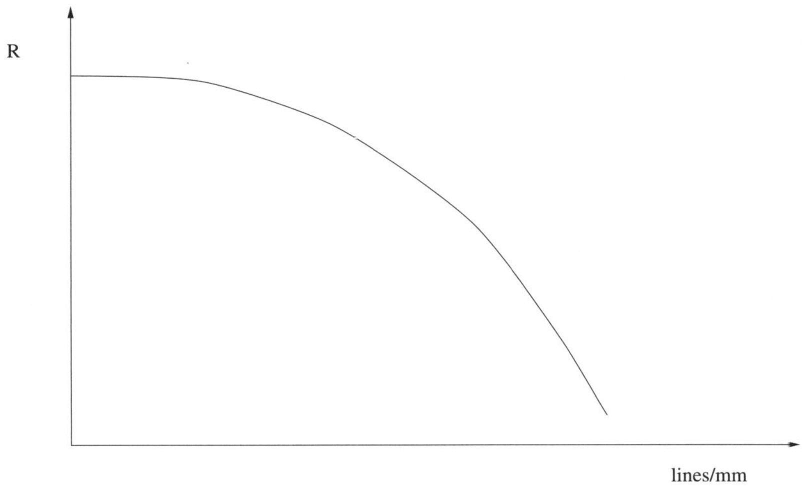

Figure 3.24: Typical Modulation Transfer Function.

This curve of R vs phantom bar (or line) density became known as the MTF. Note that it explicitly shows a drop in *contrast* with increasing spatial frequency. This has some significance for the detectability of very small objects in an image.

It can also be shown that the modulation transfer function is simply the normalised Fourier transform of the point spread function:

$$MTF(\omega) = \frac{\int PSF(r)e^{-i\omega r}dr}{\int PSF(r)dr} . \tag{3.26}$$

Because the MTF measures are essentially scaled frequency domain responses, under the assumption of linearity, multistage imaging systems may be analysed by cascading the MTF's of each system.

Experimentally, it would be easier to estimate the MTF by using the scaled Fourier transform of the point spread function, as in Equation (3.26).

[5]Not to be confused with the *impulse* response, which is equivalent to the PSF. Rather confusing, but these are accepted terms.

3.8 SPATIAL RESOLUTION

3.8.1 PHYSICAL FACTORS IN SPATIAL RESOLUTION

Is spatial resolution in X-ray CT isotropic or not? Well, in collecting single-slice image data, it is clear that we have two axes for the image plane, x_f, and y_f. With ultrasound, the method of acquiring B-mode images contains fundamental differences in the mechanisms which limit (determine) resolution along the two-axes of the image. In an X-ray CT image slice, the method of acquisition is via a multiview approach, which leads to an image reconstruction from projections. Thus, the spatial resolution is isotropic (or as isotropic as it can be on a rectangular sampled grid).

The resolution in the *axial* direction i.e., the z_f direction) is another matter. Clearly, some axial spatial averaging must occur, since the X-ray source, whether fan beam or collimated, cannot be infinitely thin. In practice, each image corresponds to a slice of 1-10mm in thickness.

We first deal with the *axial* scanner resolution.

Aperture Size & Axial Resolution

In order to understand this fundamental limit on axial resolution, we must return to the nature of x-ray transmission and detection.

X-rays are subject to the laws of quantum physics, and x-ray energy is delivered in packets or quanta, known as photons. Photon emission and detection are subject to statistical fluctuations, occasionally referred to as quantum noise.
In a conventional, planar x-ray system, quantum fluctuations lend a "mottled" appearance to the image. The same is true of CT, though the statistics of the noise are different, since every pixel receives a contribution from a number of view angles.

As the slice thickness is decreased, we perform a smaller amount of spatial averaging (in the z_f direction) in each of our projection measurements, and so we get a net increase in the effect of quantum fluctuation in the image.

If one wishes to decrease the slice thickness, thus increasing the axial spatial resolution, then it is necessary to use longer exposure times in order to maintain the signal to noise ratio of the system. This increases the scan time, and also increases the radiation dosage to both the skin and tissue.

So, the achievable axial resolution is constrained by any minimum on the signal/noise ratio, and by the patient dosage.

Image Spatial Resolution

We now return to look at the spatial resolution of the scanner in the plane of the slice (x_f, y_f) plane. We can identify many more factors which affect the spatial resolution in the plane of the slice. First of all, we have the effect of source and detector collimation.

As previously mentioned, the line integral described by the $3D \to 1D$ version of Beer's law is really an integral performed along a solid tube, as the X-ray beam has a finite (nonzero) thickness, both in the z_f direction, and along the x_ϕ direction.

For first generation scanners, the effective "thickness" of the tube is controlled by both the source and aperture collimation. For scanners employing fan beams, the detector collimation is the factor which determines the effective overall aperture[6].

When reducing aperture size, however, one again runs into the problem of increased mottle due to reduced spatial averaging of the quantum fluctuations. Thus, once more, we are limited by our desired signal/noise ratio and patient dosage.

Reconstruction Filter

As we have remarked, the choice of filter for the convolution- backprojection algorithm greatly affects the spatial resolution of the resulting image.

Recall that the back-projected image is derived from *filtered* projection data:

$$\lambda_\phi(x_\phi) \overset{\text{filter}}{\Longrightarrow} \lambda'_\phi(x_\phi) \tag{3.27}$$

$$\lambda'_\phi(x_\phi) \overset{\text{backproject}}{\Longrightarrow} \hat{\mu}(x_f, y_f) \tag{3.28}$$

where the filter used in the convolution has impulse response $h_c(x)$. Remember that in the frequency domain, we can think of increasing ω as corresponding to increasing spatial frequency. The effect of having a larger value of cutoff, ω_c, is that we have more high frequency components in the reconstructed image, and therefore higher spatial resolution. The effect of a smaller value of ω_c results in a lower spatial resolution. One can also deduce this from the convolution operation: as we increase the value of ω_c, the filter, $h_c(x)$, shrinks correspondingly and thus averages over fewer of the linear parallel projections during the convolution operation. As the value of ω_c *decreases*, the filter $h_c(\omega)$ *increases* in width, and so the effect of convolution is to "average" together

[6]There is a fundamental difference between the aperture dependent resolution effects in CT and those in ultrasound. In CT, finer aperture is generally associated with higher spatial resolution. At the frequencies used for medical applications, ultrasound is a diffraction limited process, requiring a larger aperture for better lateral resolution. The difference is contained in the aperture/wavelength ratio.

more of the linear parallel projections at a given angle, resulting in a more blurred reconstructed image.

As with all engineering issues, there is a tradeoff. If we increase the value for the filter cutoff design, ω_c, $h_c(x)$ shrinks. But the convolution action, in addition to being vital for the backprojection to succeed, also performs the useful operation of smoothing over neighbouring parallel projection samples (in a single view), thus reducing the quantum fluctuations, in almost the same way that a larger aperture would.

Thus, increasing ω_c to increase the high frequency content of the image also results in an increase in high frequency quantum fluctuation, or noise, in the reconstruction. In fact, where it is important to have high *density* resolution, it is useful to reduce ω_c. We shall return to this point later.

Finally, note that although in theory a larger value of ω_c implies a smaller convolution kernel, in practice the range of useful values of ω_c do not significantly affect reconstruction times on today's hardware.

Number of Views and Projection Samples

The algebraic interpretation of the reconstruction process suggests that if we require a reconstructed image matrix of size $M \times M$, then at least M^2 independent projections must be obtained. An approximate value for the number of view angles would be M. Because the number of degrees of freedom in a data set can be removed by imposing constraints to eliminate unlikely solutions, we *can* obtain reasonable reconstructions with a smaller number of views (say, $M/2$). However, if one uses a Fourier based reconstruction algorithm, it is more likely that a larger number of views is needed. An approximate figure for the number of view angles is provided by Webb [3] as $D\pi/(2d)$ where D is the diameter of the space to be reconstructed i.e., Field Of View (FOV), and d is the spatial resolution required. For *each* view angle, the sampling resolution i.e., along the x_ϕ direction) would need to be roughly equivalent to that of the required spatial resolution. Thus, for each view angle, we would require D/d samples.

As a rough example, a 50cm FOV diameter, with a 1mm resolution would require about 800 views, and each view angle, about 500 projection readings. This yields about a 512×512 reconstructed image[7], requiring the "solution" of about $500 \times 800 = 400000$ simultaneous equations!

As we have previously mentioned, for a *fixed* field of view, an increase in spatial resolution calls for an increase in the number of "unknowns" in the image reconstruction. The relationship is defined in Figure 3.25.

Clearly, a larger number of projection profiles (views) will also increase the patient dosage. Further considerations on the number of view angles are given in the section on image artefact.

[7]For a simpler digital hardware architecture, we seek maximum address spaces of integer powers of two; 512 is the nearest to 500.

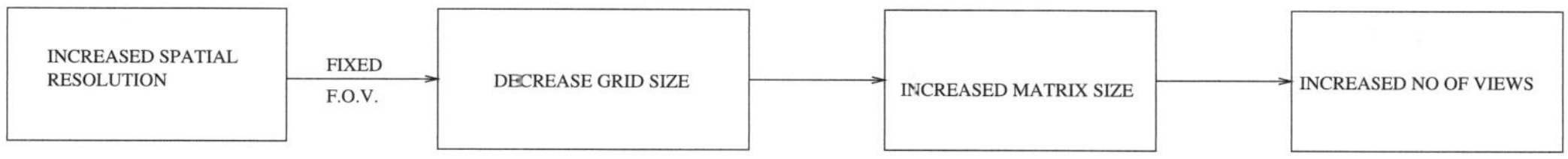

Figure 3.25: Increasing Spatial Resolution within a given Field of View.

Matrix Size
On older scanners, this was a limiting factor for the spatial resolution obtainable over a fixed field of view. The corresponding relationship is depicted in Figure 3.26.

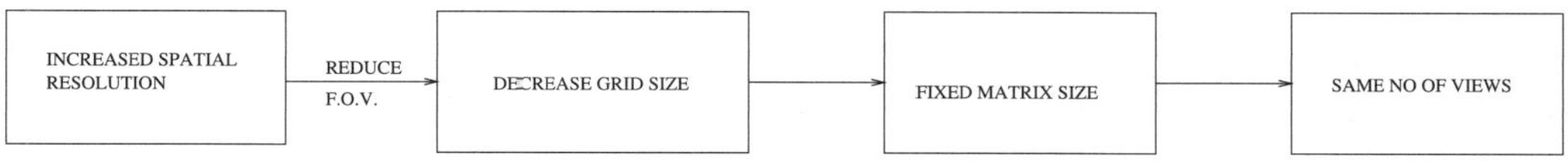

Figure 3.26: Increasing Spatial Resolution with Fixed Matrix (image) Size.

Relative Object/Background Contrast
Although the spatial resolution may be thought of as being quite distinct from density resolution, there are some practical considerations which link the two. As will be clear from the form of the response function of the modulation transfer function (MTF), objects which have a high contrast (relative to the background, or some other object) are more easily resolved. There is thus some dependence of resolving capability on available density contrast.

3.8.2 DENSITY RESOLUTION

Density resolution refers to the ability to distinguish between two regions of different attenuation coefficients which lie close to each other. In X-ray CT, the ultimate limitation to density resolution is the amount of quantum noise (mottle) in the reconstructed image.

Factors which determine the amount of mottle in a reconstructed image are

- Photon flux
- Quantity of scatter
- Voxel size
- Reconstruction Filter

Photon Flux
Smaller aperture sizes and shorter exposure times yield a larger proportional fluctuation in the observed projection measurements. As a rule of thumb, one can say that the relative amount of

mottle in the image is inversely proportional to the square root of the number of detected photons. So, allowable patient dosage and desired spatial resolution become controlling parameters.

X-Ray Scatter
Secondary photons travel in all directions, and lead to noisy projection data sets, particularly in multi-detector systems. For this reason, the scatter rejection of high-collimation, single detector systems of the first generation scanners was quite desirable.

Voxel Size
This is related to the absolute difference in average beam intensity after travelling through tissues of differing attenuation coefficients. Look at the situation depicted in Figure 3.27, using the algebraic model. Consider the relative differences in expected transmitted beam intensities, ignoring beam hardening and assuming a pencil-thin beam.

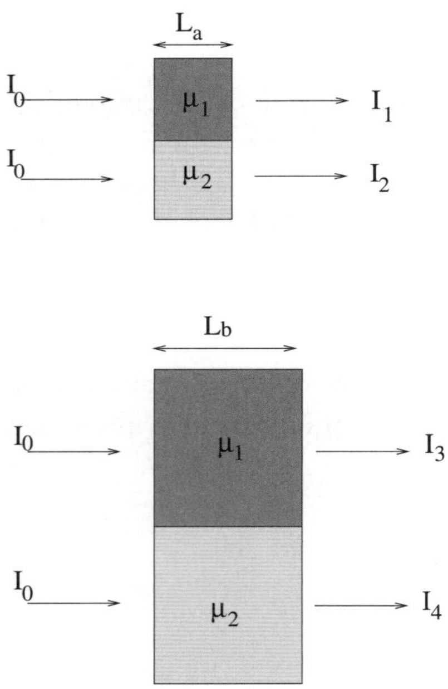

Figure 3.27: Effect of Voxel Size on Contrast.

It is clear that since $L_b > L_a$, then $I_4 - I_3 > I_2 - I_1$. In other words, the intrinsic contrast between the projections of the voxel pairs is greater when the voxels are larger in size. Greater contrast in the projected data yields greater contrast in the reconstructed image.

Dosage and Signal/Noise Ratio
We shall give a simplified treatment of dosage considerations, and the relationship with signal/noise ratio. If we define the signal/noise ratio of a CT scanner by

$$SNR = \sqrt{\frac{Signal Power}{Noise Power}} \tag{3.29}$$

then the dosage, u, is *proportional to*

$$\frac{1}{\eta}\frac{(SNR)^2}{\epsilon^3 b} \tag{3.30}$$

where b is the thickness of the CT slice, ϵ is the spatial resolution in the plane of the slice and η is the quantum efficiency of the X-ray detector.

3.9 CT IMAGE ARTEFACTS

We move on to consider the second factor which affects the quality of X-Ray CT images, namely the existence of a wide range of CT artefacts[8]. Many of these artefacts are unlike what one might find in other modalities, such as planar x-ray imaging or ultrasound, and are as a direct result of the reconstruction process.

For a radiologist, it is important to be able to recognise artefact, so that its occurrence is not mistaken for some pathological condition.

3.9.1 STREAK & RING ARTEFACT

On multidetector systems, streak and ring artefacts are usually associated with an uncalibrated x-ray detector, or with a detector which has simply failed. In a rotate-only system, a malfunction in any one detector would back-project onto an arc in the image space. This might produce a region of higher than average apparent attenuation coefficient, in the shape of a ring.

On a translate-rotate system, a malfunction of one detector in a multidetector system would lead to an incorrect backprojection for each translation of that angle. The artefact is one of streaks in the image. In addition, on a stationary, multidetector system, detector malfunction will result in streaking, but in this case, the streaks will be fan-shaped.

Misalignment of x-ray tube and detector can also result in error; streak or ring artefacts result if the misalignment is sufficiently large.

[8] Worried about the spelling of artefact? Don't be. The Oxford English Dictionary declares both spellings (artifact, artefact) as equivalent!

In general, it is safe to say that most of the machine related artefacts are easily recognised as being such, since they are usually not very subtle.

3.9.2 PATIENT-RELATED ARTEFACTS

These artefacts can indeed be subtle, and the radiologist must be aware of their cause and nature.

Among the most common, particularly in the early generation scanners, are the motion-related artefacts. The backprojection algorithm assumes that the patient does not move between the acquisition of the different patient "views."

Unlike a simple camera scene acquisition, where motion deblurring can be quite easily done by deconvolution techniques, motion compensation in CT should really be done on the original projection data. Remember that the effect of motion in CT is that it renders inconsistent the set of simultaneous equations that we acquire about the attenuation distribution.

The presence within the patient of objects which have abnormally high or low attenuation coefficients can lead to problems if the dynamic range of the detectors is exceeded during the acquisition of the projection data.

Patients who have metallic surgical pins within the field of view (FOV) of the slice acquisition can often yield streaks in the image due to this dynamic range issue. Modern x-ray detectors, of dynamic range exceeding 10^6 can alleviate this problem.

3.9.3 X-RAY CT INHERENT

Partial Volume Artefact

When tissues of differing attenuation coefficient occupy the same voxel, the reconstructed attenuation coefficient may then be quite misleading. One site where this occurs commonly is the lung diaphragm interface; increased apparent density may be due to the lung base.

Another partial volume artefact stems from the divergence of fan-beam sources, in which the projection measured in one direction may not be the same as the projection measured in the reverse direction. This observation, in fact, points to the solution to this problem, which is really an issue of acquisition geometry. Acquiring over 360 degrees rather than 180 degrees helps to reduce this source of geometric error.

Beam Hardening Artefacts

These result from the preferential absorption of low-energy photons from the beam. The predominant mechanism for photon absorption is through the photo-electric process.

As an x-ray beam progresses through the body, the preferential absorption of low energy photons leads to an increase in the *average* beam energy. But remember, attenuation coefficient is a function of energy, and in fact tends to decrease with increasing photon energy. In addition, the form of this energy dependency is such that there is also lower *differential* attenuation between any two tissues with increasing photon energy.

After passage through tissue, the beam becomes relatively *less* affected by the attenuating structures, and is said to have been *hardened.*

Reconstruction algorithms assume that any change in the beam intensity is due to tissue attenuation effects along the line of projection, and so substantial reconstruction error may result.

The effect is most noticeable near to highly attenuating structures, such as bone, although it occurs to some extent in all x-ray CT images. Particularly common anatomic sites for observing the effect include just below the ribs, and near the *posterior fossa* of the skull. A common manifestation is of a loss in image detail near to highly attenuating structures.

See [18] for more information, including jargon, such as "dishing"!

Spatial Aliasing

The process of projection acquisition may be likened to one of a sampling process. If we try to reconstruct an image with a high spatial frequency content, we must have a sufficient number of view angles. Failure to acquire a sufficient number of views for the attempted reconstruction results in *spatial aliasing,* analogous to the aliasing phenomenon which one meets in the sampling of time-varying signals. The appearance is dependent on the geometry of projection acquisition. In translate-rotate geometries, the artefact is star-shaped.

As an example, we illustrate aliasing effects where a phantom is reconstructed from differing numbers of view angles. The log-normalised projection data can be analytically defined by using Beer's law

$$\lambda_\phi(x'_\phi) = 2\mu_m\sqrt{r^2 - (x'_\phi)^2} \tag{3.31}$$

where x'_ϕ is distance measured from the projection of the axis of the cylinder onto the x_ϕ axis (see Figure 3.28), and r is the radius of the phantom.

The radius is taken to be approximately 5 pixels, and the attenuation coefficient is set to $\frac{1}{2}$. We can easily sketch the projection data (Figure 3.29) which is the same at all view angles, if the

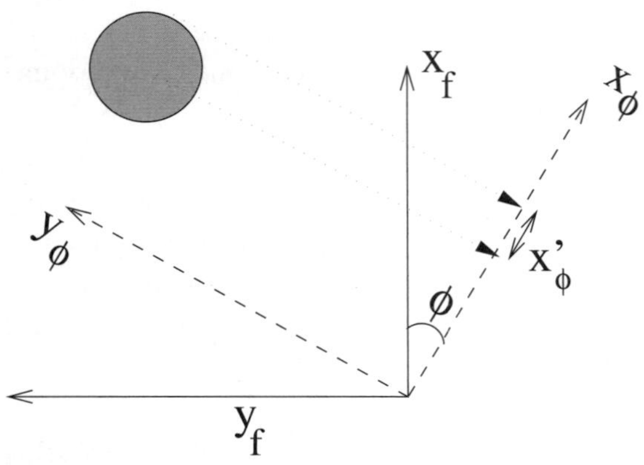

Figure 3.28: Linear Parallel Projection of Cylinder, and x'_ϕ.

central axis of the cylinder coincides with the central axis of scanning.

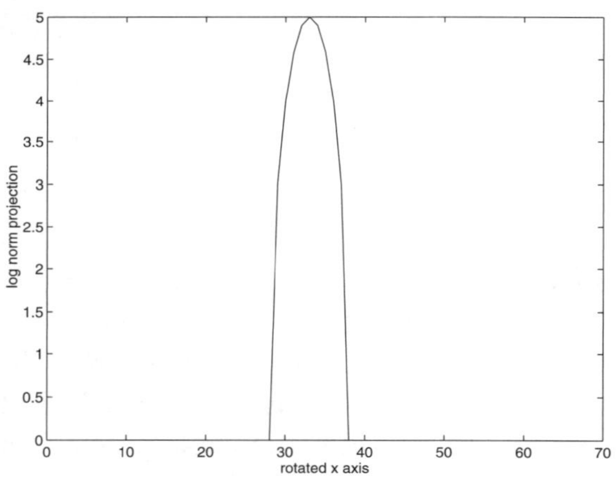

Figure 3.29: Sketch of analytic log-normed projection data.

A filtered backprojection using only 5 view angles is illustrated in Figure 3.30, and the effect of increasing the number of views to 20 is illustrated in Figure 3.31. The reconstructed image from a set of 50 views is illustrated in Figure 3.32. Note the reduction in artefact with increasing number of views. By way of comparison, we illustrate the effect of backprojecting *without* filtering in Figure 3.33. Although this may appear cleaner than the filtered backprojected image, this data is unrepresentative, as the projections are analytically defined, and are true line projections. Nevertheless, the loss of the edge definition is apparent in Figure 3.33, compared to Figure 3.32. In operating on real projection data, the difference is more marked. Furthermore, increasing the number of views reduces the amplitude of the reconstruction artefact in Figure 3.32, whilst the poor edge definition of Figure 3.33 remains.

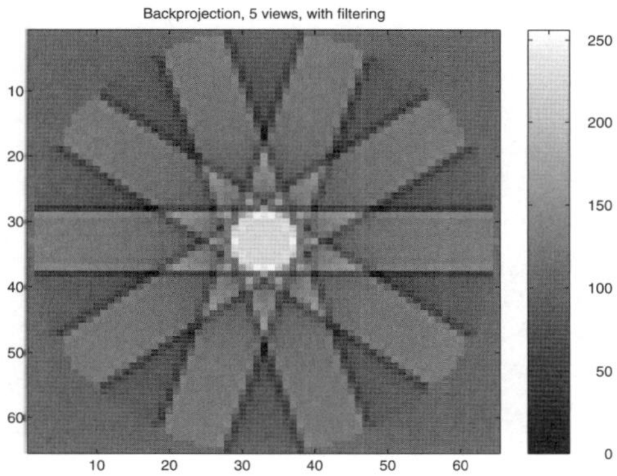

Figure 3.30: Convolution Backprojection using 5 views.

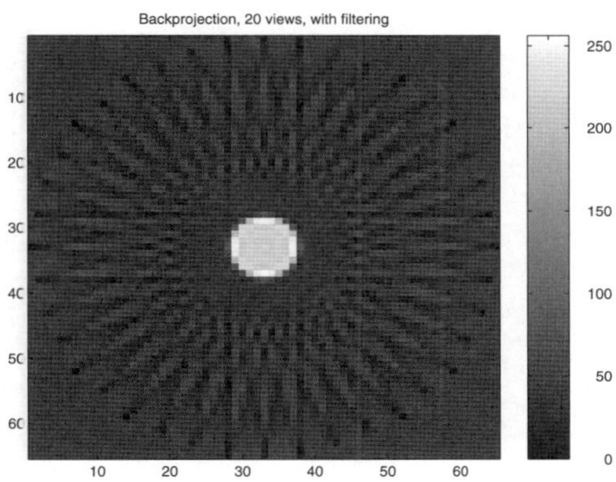

Figure 3.31: Convolution Backprojection using 20 views.

3.10 DIGITAL IMAGE MANIPULATION

We briefly mention a few of the image processing tools which are useful in the clinical context of CT. To the radiologist, a scanning console which provides a huge selection of image processing functions is not necessarily desirable. Radiologists tend to (currently) use only the following tools regularly.

3.10.1 GREY-SCALE WINDOWING

Although, within the dynamic range of possible pixel intensities detectable by a CT scanner, 1% differences in contrast can be measured, the human eye can only appreciate contrast differences

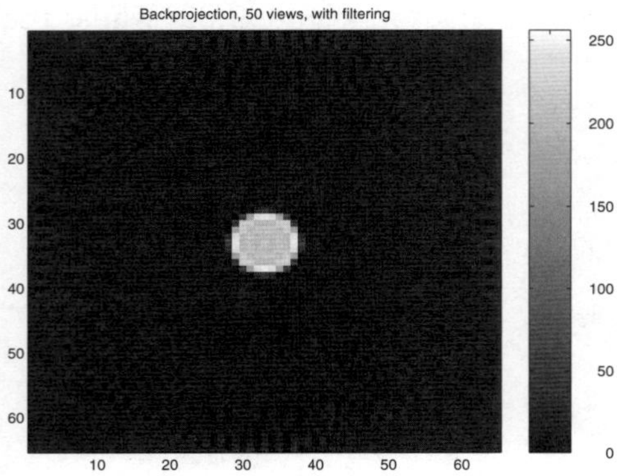

Figure 3.32: Convolution Backprojection using 50 views.

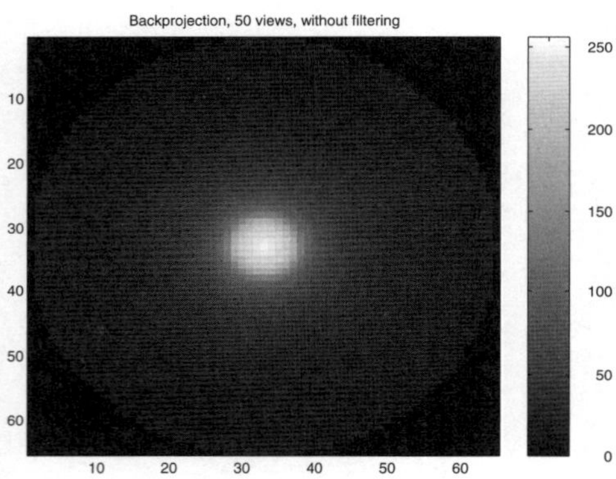

Figure 3.33: Backprojection using 50 views, no filtering.

of greater than 10% if the full range of pixel intensities is displayed in typical radiological viewing conditions.

As mentioned previously, if a small range of pixel values (CT numbers) is mapped to the intensity range of the console, smaller contrast differences may be appreciated.

The range of CT numbers which is mapped to the visible display intensities is generally referred to as the *window width* in radiological parlance, with the central pixel intensity value of this window being referred to as the *window level*.

3.10.2 ROI SELECTION

Region of Interest (ROI) definition and selection is a particularly powerful facility if it is linked with zoom reconstruction techniques.

A common method of implementing an ROI selector is via a trackball and cross-hairs, which allow a square ROI to be defined. The original projection data is then refiltered, using a filter with a different ω_c and some interpolation may be applied after backprojection.

Related functionality includes the ability to label and to annotate ROI's with diagnostic notes.

CHAPTER 4

Ultrasonics

Ultrasonic scanners are among the cheapest, fastest and most widely used of diagnostic imaging equipment. Normally given a bad reputation because of poor image quality, more recent scanners are capable of providing imaging spatial resolution which is comparable to MRI. Ultrasonic scanners using electronic steering can acquire and display information at a sufficient rate to enable dynamic two-dimensional information to be obtained. Furthermore, the acquisition of information is performed using a localised hand-held probe. Whilst this may seem less appealing than a whole body scanner (such as MRI, CT or PET), the sonographer has the ability to subtly alter the probe position and orientation. This direct interaction is one of the most difficult aspects to quantify, but is an essential part of the successful application of this technology to medical imaging.

This course will consist of 6 lectures, roughly organised into the following structure:

- Introductory Physics
- Pulse-Echo Imaging
- Doppler Ultrasonic Velocimetry

4.1 BASIC PHYSICS

In order to understand the principles of ultrasonic imaging, it is necessary to have some familiarity with the physical principles underlying the image formation process. In ultrasonics, we are dealing with the physics of wave propagation. A key equation is, therefore, the homogenous wave equation, which we provide a derivation for in Appendix A. This equation can be expressed as

$$\nabla^2 p(x, y, z, t) - \rho\kappa \frac{\partial^2 p(x, y, z, t)}{\partial t^2} = 0 \tag{4.1}$$

where $p(x, y, z, t)$ is the spatio-temporal pressure distribution away from the source, rho is known as the mass density of the physical medium, and κ is the compressibility. ∇^2 is the Laplacian operator (see Appendix A for more details).

In much of what follows, we will rely on a simple solution to this equation which is valid a long distance from a point source being excited with simple harmonic motion at angular frequency ω_0. This is known as the *infinite plane wave* solution to the HWE:

$$p_r(r, t) = A_1 e^{j\omega_0(t-r/c)} \tag{4.2}$$

where the wave a principle direction of propagation along r.

Equation (4.2) is sometimes written in the form

$$p_r(r, t) = A_1 e^{j(\omega_0 t - kr)} \tag{4.3}$$

where $k = \omega_0/c = 2\pi/\lambda$ is known as the propagation constant. Recall that λ, the wavelength of the harmonic wave in the medium is given by $\lambda = c/f$.

4.1.1 THE INTENSITY OF A PLANE WAVE

We now introduce a useful measure of acoustic wave propagation, the *wave intensity*. The intensity of a propagating wave is a measure of how much energy is being propagated (or "transported") by the wave. Its units are Wm^{-2}.

It is defined as the *average* rate of energy flow through an area normal to the direction of wave propagation. It is determined by considering the kinetic and potential energies of the wave. An elemental fluid volume has a particular mass and velocity, and thus possesses kinetic energy according to:

$$E_k = \frac{1}{2}\rho u^2 A dx \ . \tag{4.4}$$

Since the velocity of a fluid element changes over time, this quantity will also change with time, in a manner dictated by the source excitation.

The potential energy is an internal energy, and arises from the work done in compressing the elemental volume under a pressure gradient:

$$E_p = -Adx\kappa \int p dp \tag{4.5}$$

which reduces to

$$E_p = \frac{1}{2}\frac{p^2}{\rho c^2} \cdot Adx \tag{4.6}$$

under suitable approximations [2], [1].

The total energy, E_T, is then given by $E_T = E_k + E_p$.

We are interested in the *intensity* which is the average rate of energy flow along the direction of wave propagation over unit area cross section. This is most conveniently defined in terms of D_E, the energy density:

$$I = \left\langle \frac{\partial D_E}{\partial t} \right\rangle . \tag{4.7}$$

where D_E is defined as

$$D_E = \frac{E_T}{Adx} \ . \tag{4.8}$$

This might seem confusing[1], but what we are really doing is separating the energy into two components - a group component (describing, for example the motion of the envelope of excitation), and a phase component (describing the wriggling of the waveform within the envelope). The group rate of energy flow is always equal to the velocity of propagation, c, so that we get

$$I = c \langle D_E \rangle \ . \tag{4.9}$$

The reason we still have a temporal average here is that the exact oscillatory nature of the excitation will affect the way that the energy density varies in the volume over time.

If we consider the particular form for excitation corresponding to the infinite plane wave with source at $r = 0$, (propagating in the $+r$ direction),

$$p(r, t) = P_0 e^{j\omega(t-r/c)} \tag{4.10}$$

and I is then simply given by

$$I = \frac{P_0^2}{2\rho c} \ . \tag{4.11}$$

Note the dependence on the amplitude of wave oscillation (expected) and on the properties of the medium (perhaps not expected!). An alternative derivation can be found in [1].

4.1.2 THE ACOUSTIC IMPEDANCE

The quantity

$$Z = \rho c \tag{4.12}$$

appears very frequently in acoustic computations, particularly those dealing with energy flows or local momentum. For this, and historical reasons, the quantity Z is given its own name: the acoustic impedance. Its value varies in human soft tissue, and indeed, the existence of tissue structures in the body implies (acoustically) a layering of media with slightly different acoustic impedances. Most of the difference in acoustic impedance is actually due to different values of c, rather than variations in ρ. Since $c = (\rho\kappa)^{-\frac{1}{2}}$ one can deduce that the variations in tissue impedance are largely due to variations in the compressibility.

4.1.3 PROPAGATION OF HPW ACROSS ACOUSTIC INTERFACE

Consider the physical aspects of the following problem: a harmonic plane wave (HPW) is incident normally on an interface between media of differing acoustic properties, as shown in Figure 4.1.

In the diagram, we specify (as a general solution) a forward and reverse propagating wave on the L.H.S of the interface, and a forward propagating wave on the R.H.S of the interface. We

[1] An analogy is to compute the average number of people arriving at the top of an up escalator per second. To do this, we need two quantities, the speed of the escalator, $c_{escalator}$, in ms^{-1}, and the *average* density of people, D_{people}, in $N_{people} \cdot m^{-1}$. The average rate of people-arrival will then be $c_{escalator} D_{people}$. The analogy improves if you allow people to shuffle around on the escalator.

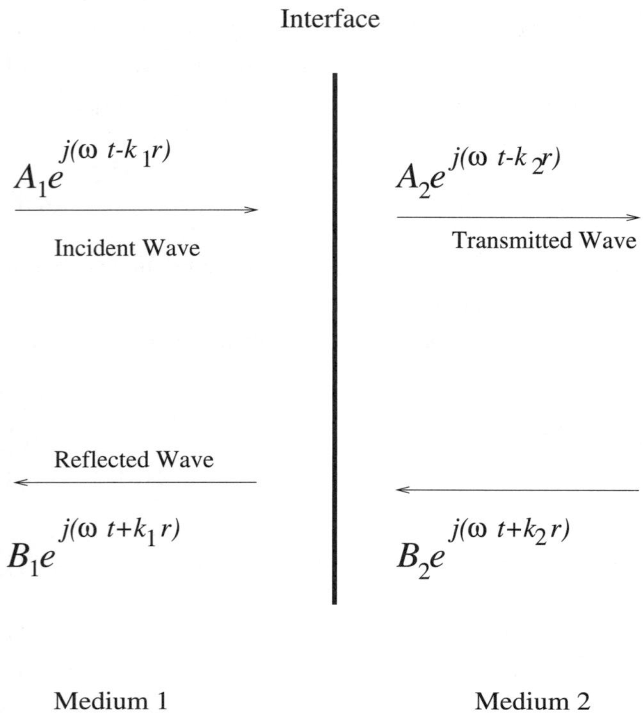

Figure 4.1: Physical description and general solution of interface problem.

assume the solution to be in the form of infinite plane waves of different amplitudes. Again, we must construct our solution to this problem by considering appropriate boundary conditions. The important boundary to consider is at the interface itself, which we define to be $r = 0$.

At the interface, we would expect that the pressure would be the same, whether one was infinitesimally left or right of the boundary. Thus, we equate the expressions for pressure on either side of the boundary:

$$A_1e^{j\omega t} + B_1e^{j\omega t} = A_2e^{j\omega t} \tag{4.13}$$

which should hold for all time instants, so that

$$A_1 + B_1 = A_2 \ . \tag{4.14}$$

In addition, we expect that the normal components of velocity across the interface will always be equal, and in phase; violation of this would cause the interface to fall apart. For this, we need to relate pressure to velocity.

Starting from

$$\rho\frac{\partial u_n}{\partial t} = -\frac{\partial p}{\partial r} \tag{4.15}$$

where u_n is the normal component of fluid velocity, we obtain

$$u_n = -\frac{1}{\rho}\int \frac{\partial p}{\partial r} dt \ . \tag{4.16}$$

Since the pressure variation is of complex exponential form (for a monochromatic infinite plane wave)

$$u_n = -\frac{1}{jp\omega}\frac{\partial p}{\partial r} \ . \tag{4.17}$$

Equating the velocities on both sides of the boundary leads to

$$\frac{A_1}{\rho_1 c_1} - \frac{B_1}{\rho_1 c_1} = \frac{A_2}{\rho_2 c_2} \ . \tag{4.18}$$

Note the appearance of the "magical" Z! Solving for the amplitude of the reflected wave, we find

$$B_1 = A_1 \frac{Z_2 - Z_1}{Z_2 + Z_1} \tag{4.19}$$

this shows that the amplitude of the transmitted wave is proportional to that of the incident wave *and* the ratio $(Z_2 - Z_1)/(Z_2 + Z_1)$. In tissue, as acoustic impedances tend to be around the same order of magnitude, then the principle effect of this ratio tends to be a proportionality between the reflected wave amplitude and the *difference* between the acoustic impedances of Media 1 and 2.

Similarly, one may show that

$$A_2 = A_1 \frac{2Z_2}{Z_2 + Z_1} \ . \tag{4.20}$$

The eagle-eyed among you may realise that the transmitted wave amplitude, A_2, *can* be larger than the incident amplitude, A_1. We must be careful in jumping to conclusions about the relationship between reflected and transmitted waves. In particular, we should really evaluate intensities of the transmitted and reflected waves, with respect to the incident wave.

Indeed, if we compare the energies of incident and transmitted waves, we see that

$$\begin{aligned} \frac{I_2}{I_1} &= \frac{A_2^2/Z_2}{A_1^2/Z_1} \\ &= \frac{4Z_1Z_2}{(Z_1 + Z_2)^2} \end{aligned} \tag{4.21}$$

which will always be less than one.

A final comment is in order here. Whilst we have considered the reflection of an acoustic plane wave at an interface, the analysis can also be applied to planar pulse waveforms, i.e., plane waves which have finite temporal extent. This follows through superposition, and the use of Fourier theory, whereby any temporal excitation may be represented in terms of a summation of harmonic terms.

4.1.4 SUMMARY

So far, we have

1. derived the acoustic wave equation. We remark that the specification of the acoustic wave equation is general enough to consider virtually any combination of fluid/fluid and fluid/scatterer interface, although finding an analytic description of the resulting pressure wave is quite difficult in all but the simplest acoustic systems.

2. demonstrated that the spherically symmetric solution to a sinusoidal excitation at some position r = 0, leads to a diverging wave, which, as the distance from the source increases, looks increasingly like an infinite plane wave.

3. considered the energy transfer of an infinite plane wave and its reflection and transmission when incident normal to an interface between media of different acoustic impedances.

4. showed that time-of-flight of a pulse is related to distance of travel.

5. mentioned that the amplitude of a reflected pulse will also information on the "hardness" of an interface, just as the amplitude of a HPW does.

4.2 FINITE APERTURE EXCITATION

We have previously considered a simple complex exponential pressure excitation at $r = 0$ as a source excitation. A source which physically produces such an excitation (very idealised, very small) is known as a monopole source. (Audiophiles: some Hi-Fi enthusiasts have actually attempted to make loudspeakers which approximate acoustic monopoles!). The monopole source is a useful model, as it can be extended very easily for the purposes of modelling and analysis.

For imaging applications, monopoles are not very useful practical sources, since they do not transmit much acoustic energy, and, furthermore, they radiate energy in all directions. A more useful ultrasonic source for imaging must have some directionality to it. It turns out that one can model a *finite aperture* source by placing an infinite number of monopoles very close together within a defined area, and by employing the principle of superposition to determine the resultant pressure field. This sort of approach is a very general one, which can be extended to nonplanar (curved) and distributed sources.

Consider a flat sheet of many monopoles, all placed very closely together, and oscillating in phase with each other at angular frequency ω_0. It can be shown that such a sheet of monopoles, called an aperture, produces a distinctly *directional* transfer of energy, normal to the aperture plane.

In order to show this, we can assume that the medium that the sheet of monopoles is placed in is linear, and simply sum the responses from each monopole. We first need to generalise the expression that we have for the pressure field from a monopole placed at $r = 0$ to the more general case of a monopole at $r = r_s$. Luckily, since monopole acoustic radiation is isotropic, this is easy!

$$p_r(\vec{r}_o|\vec{r}_s, t) = A_s \frac{e^{j(\omega_0 t - k|\vec{r}_o - \vec{r}_s|)}}{|\vec{r}_o - \vec{r}_s|} \tag{4.22}$$

where $\vec{r}_s$ and $\vec{r}_o$ denote the vector locations of source and observer respectively.
For source points distributed in the plane of an aperture, $\mathcal{A}$, the net response from all monopoles becomes

$$p_r(\vec{r}_o|\vec{r}_s \in \mathcal{A}, t) = \int_{\vec{r}_s \in \mathcal{A}} p_r(\vec{r}_o|\vec{r}_s, t) d\vec{r}_s \ . \tag{4.23}$$

The result of this integral is generally quite complicated, but a number of well-known approximate forms can be identified. The two most common are the Fraunhofer and Fresnel approximations, and they are valid in different regions away from the face of the transducer.

4.2.1 THE FRAUNHOFER APPROXIMATION

The Fraunhofer approximation relies on considering an observation plane parallel to the transducer face, at a sufficient distance away. The geometry is shown in Figure 4.2.

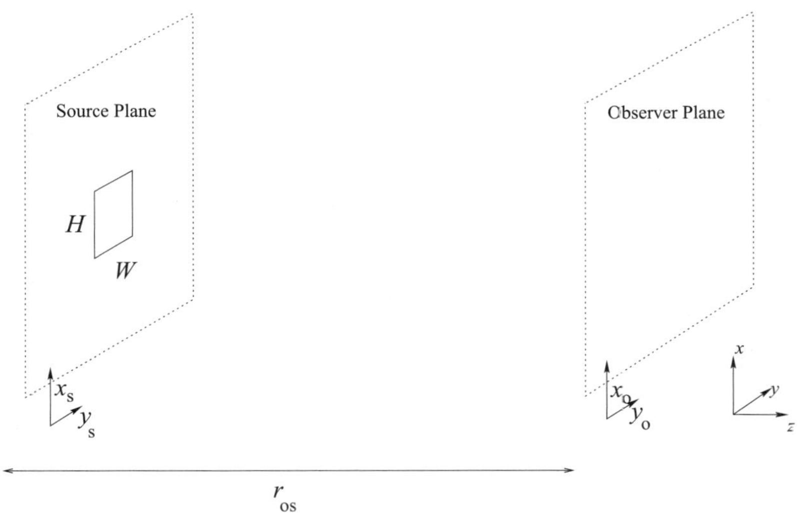

Figure 4.2: Planar Source/Observer Geometry.

For uniform excitation over a sheet of monopoles of width W and height H, Equation (4.23) becomes

$$p_r(\vec{r}_o|\vec{r}_s \in \mathcal{A}, t) = \int\int_{|x_s|<H/2, |y_s|<W/2} \frac{A_s}{\sqrt{z_{os}^2 + (x_o - x_s)^2 + (y_o - y_s)^2}} e^{j\omega_0 t} e^{-jk\sqrt{z_{os}^2+(x_o-x_s)^2+(y_o-y_s)^2}} dx_s dy_s \; . \tag{4.24}$$

Factoring by z_{os}^2 and using the binomial expansion [20], we have

$$\begin{aligned} p_r(\vec{r}_o|\vec{r}_s \in \mathcal{A}, t) &\approx \frac{e^{j(\omega_0 t - kz_{os})}}{z_{os}} \int\int_{|x_s|<H/2, |y_s|<W/2} A_s e^{-jk\frac{1}{2}\left\{\frac{(x_o-x_s)^2}{z_{os}} + \frac{(y_o-y_s)^2}{z_{os}}\right\}} dx_s dy_s \\ &\approx \frac{e^{j(\omega_0 t - kz_{os})}}{z_{os}} \int\int_{|x_s|<H/2, |y_s|<W/2} A_s e^{-j\frac{k}{2z_{os}}\{x_o^2 - 2x_o x_s + x_s^2\}} \\ &\quad \cdot e^{-j\frac{k}{2z_{os}}\{y_o^2 - 2y_o y_s + y_s^2\}} dx_s dy_s \\ &\approx \frac{e^{j(\omega_0 t - kz_{os})}}{z_{os}} e^{-j\frac{kr_o^2}{2z_{os}}} \int\int_{|x_s|<H/2, |y_s|<W/2} A_s e^{jk\left\{\frac{x_o x_s}{z_{os}} + \frac{y_o y_s}{z_{os}}\right\}} \\ &\quad \cdot e^{-j\frac{kr_s^2}{2z_{os}}} dx_s dy_s \; . \end{aligned} \tag{4.25}$$

Where

$$r_s^2 = x_s^2 + y_s^2 \tag{4.26}$$

and

$$r_o^2 = x_o^2 + y_o^2 \; . \tag{4.27}$$

This expression contains a number of tricky terms to evaluate. Let us pull the expression apart, and see what the important pieces are. The leftmost exponential term is very similar to the expression for the plane wave that we met in Section 4.1, and it is reassuring to see the familiar $e^{j\omega_0 t}$ term here. The next exponential is known as the *quadratic phase term*. This term is maintained in certain forms of the solution (corresponding to the Fresnel zone) in which the observation point is relatively close to the plane of the aperture.

Under the Fraunhofer conditions, the rightmost exponential term under the integral sign actually contributes very little to the integral kernel. This is because the magnitude of this term is quite small for small apertures. The cross-term is the interesting one. For values of z_{os} of the order of D/λ, and greater, this term tends to dominate in the kernel of the transform.
In the Fraunhofer region, and where r_o is small compared to r_{os}, Equation (4.25) can then be written

$$p_r(\vec{r}_o|\vec{r}_s \in \mathcal{A}, t) = \frac{e^{j(\omega_0 t - kz_{os})}}{z_{os}} \int\int_{|x_s|<H/2, |y_s|<W/2} A_s e^{jk\left\{\frac{x_o x_s}{z_{os}} + \frac{y_o y_s}{z_{os}}\right\}} dx_s dy_s \; . \tag{4.28}$$

This can be recognised as a 2D Fourier Transform of the aperture function! For a rectangular aperture, as shown in Figure 4.2, with uniform excitation amplitude A_s, this is a 2D sinc() function

$$p_r(\vec{r}_o|\vec{r}_s \in \mathcal{A}, t) \approx A_s \cdot H \cdot W \cdot \frac{e^{j(\omega_0 t - kz_{os})}}{z_{os}} \text{sinc}\left(\frac{kx_o H}{2z_{os}}\right) \cdot \text{sinc}\left(\frac{ky_o W}{2z_{os}}\right) . \tag{4.29}$$

A rule of thumb for the distance from an aperture which represents the start of the far-field (or the Fraunhofer zone)is often quoted as being as at about (aperture width)/(wavelength) of the oscillating signal (for sinusoidal excitation). However, laboratory experience quickly shows this to be a conservative expression. Furthermore, it turns out that under the action of *focusing*, the approximation gains accuracy in the focal zone, i.e., the Fraunhofer approximation turns out to be quite reasonable for estimating the beam patterns of focussed transducers e.g., by placing an acoustic lens in front of the aperture.

The following observations may be made about Equation (4.29):

- the radiation pattern has the appearance of a 2D sinc function in the observation plane.
- the sinc() function implies a concentration of acoustic energy in a direction normal to the aperture plane.
- the main 'lobe' of the beam diverges with distance r_{oa}.
- the concentration of the pressure distribution increases with increasing k. Increasing k is obtained by decreasing the wavelength used for imaging (increasing the excitation frequency, ω_0).
- the concentration of pressure distribution increases with increasing aperture dimensions, W and H.

This purely theoretical treatment of an acoustic aperture is sufficient to illustrate many of the fundamental resolution issues of ultrasonic transducers and imaging.

We provide a simple example of a far-field acoustic distribution pattern based on the Fraunhofer approximation in Figure 4.3. In Figure 4.3, we have used the Fraunhofer approximation to estimate the beam profile of a square transducer, of side 5cm, in a plane 30cm away from the transducer face. The transducer is driven at 3.5MHz. In Figure 4.4, the height, H of the transducer has been reduced to 2.5cm. The corresponding beam profile shows a broadening of shape in the x_o direction relative to the y_o direction.

Another common way of representing 1-D beam profiles i.e., the intensity along a line through the beam axis, along one of the principal axes of alignment) is by a polar plot. An example of a polar plot profile is provided in Figures 4.5 and 4.6, respectively, which are computed from the rectangular aperture transducer, along the y_o and x_o directions respectively. Beam profiles should really be provided in pairs like this.

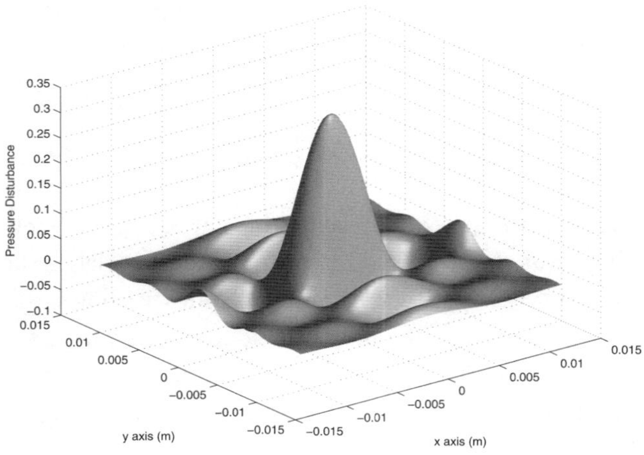

Figure 4.3: Fraunhofer beam pattern from square transducer.

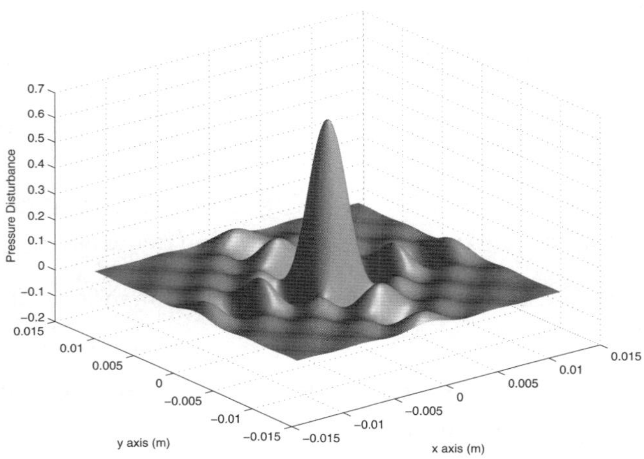

Figure 4.4: Fraunhofer beam pattern from rectangular transducer.

4.2.2 SUMMARY

Using information on the time of flight of an ultrasonic echo, we can deduce the distance of a reflecting boundary from a source. Using a directional source, we can also localise the echoes to some region in front of a transducer (which we have treated as an acoustic aperture). In a later section, we shall discuss the ideal parameters to use for getting the best possible resolution, both in the axial and lateral directions.

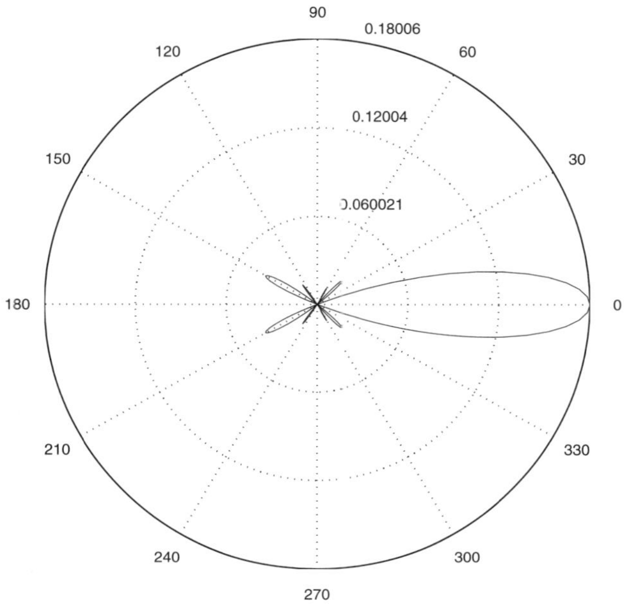

Figure 4.5: Fraunhofer beam pattern profile along x_o direction.

4.3 REAL ACOUSTIC MEDIA

We have, until now, been considering ideal acoustic media, in which there are no inhomogeneities, no support for shear waves, and no tendency to attenuate the energy of an acoustic wave which happens to be propagating through it. The facts of life about acoustic propagation, and, in particular, the propagation of waves in biological media, are far removed from the nonattenuated HPW discussed previously. We can touch on most of these through looking at the attenuating effect of the medium on the propagation of acoustic energy.

4.3.1 ATTENUATION

A number of physical effects contribute to the attenuation of the propagating acoustic wave. In many texts, these mechanisms are split into further categories; they can all be seen as mechanisms which remove energy from a propagating monotonic ultrasonic wave.

Classical Mechanisms

There is a certain time delay in applying a pressure differential and the attainment of what is known as equilibrium condensation (fractional change in density). The wave equation can then be obtained as a lossy Helmholtz equation (Fourier domain solution), and in this form, one finds that the solution

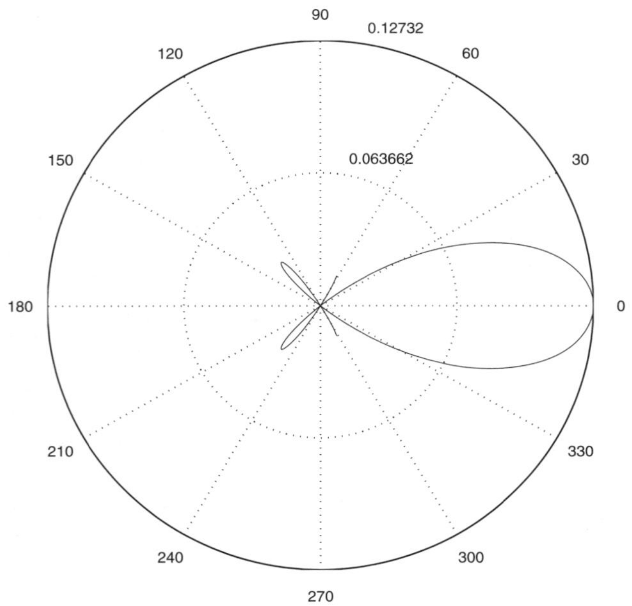

Figure 4.6: Fraunhofer beam pattern profile.

to the wave equation contains an attenuating term, i.e., Equation (4.29) must be written

$$p_r(r, t) = A_1 e^{-\alpha r} e^{j(\omega_0 t - kr)} \tag{4.30}$$

where the term α is real and positive. This shows that the effective amplitude of the wave decreases as it travels through a medium with some nonzero viscosity. These viscous losses will occur in any fluid in which not all adjacent particles are moving together. They can be thought of as being due to friction, and are associated with a bulk rise in temperature of the medium.

In longitudinal waves, there will also be tiny differences in temperature between areas of compression and rarefraction. Under these conditions, realistic media will conduct heat from the region of greater temperature to one of lower, and this is another mechanism whereby energy is lost from wave propagation. Such losses are known as *heat conduction losses*. Thermal and viscous mechanisms of sound absorption together are considered *classical* mechanisms of sound absorption, and they can both be handled by the approach of Equation (4.30).

Molecular absorption processes are, in fact, closely related to the thermal and viscous mechanisms, but these classical mechanisms can be seen as macroscopic properties derived from the underlying microscopic (or molecular) mechanisms. The molecular mechanisms provide certain extra absorptions! For example, certain frequencies of wave propagation will experience bigger absorptions if they happen to correspond to relaxation times of rotational and vibrational molecular states [1]. Molecular mechanisms of this sort cannot be elegantly handled in the classical framework, and are best treated empirically.

Scattering
All media have minor inhomogeneities. If the sizes of these inhomogeneities is very small relative to the wavelength, then the effect on the wave is to scatter a certain portion of the energy into a direction other than that of the principle direction of propagation. So, whilst the energy of wave motion might be maintained and not converted into heat or internal energy, there is a certain redirection of energy away from the forward direction. The fraction of incident energy that gets redirected in such a manner is usually handled by considering the effective collision cross section of the inhomogeneity.

Scattering is also important in understanding the manner in which ultrasound is used for blood velocity estimation, as the returned acoustic energy from blood arises from erythrocytes acting as monopole scatterers.

Non-Linear Effects
Acoustic media are only approximately linear, and there are many nonlinear effects that can occur. For example, oscillatory energy can be frequency shifted under a nonlinear acoustic medium response. These nonlinear effects are difficult to model as a whole, and one must look at specific models under different conditions.

Other Propagation Modes
This can be seen as part of viscous behaviour. Due to viscosity, longitudinal waves can give rise to propagation of transverse waves. This yields a certain amount of "leakage" of the longitudinal wave energy away from the principal wave into transverse (shear) waves. These effects become more significant in stiffer media, such as bone.

The amount of attenuation is usually expressed in terms of the attenuation coefficient, whose values are determined via a largely empirical approach.

4.3.2 EMPIRICAL TREATMENT

The easiest way to determine the attenuation characteristics of a medium is to measure them. The parameterisation of attenuation is then done in a manner similar to that suggested by the classical approach of Equation (4.30). Accordingly, one can define the attenuation coefficient to ultrasound by

$$\alpha = \frac{I}{I dx} \tag{4.31}$$

where dx is a small distance, and so α is a fractional rate of intensity change with distance. One can also write

$$I = I_0 e^{-\alpha x} \tag{4.32}$$

where I_0 is the intensity of a beam incident on a piece of tissue of thickness x, and I is the exit intensity of the beam. This effect excludes reflection. The overall α will be the sum of contributing effects outlined in the previous section.

The units of α are m^{-1}, sometimes referred to as Nepers. Another way of citing α is using the decibel notation. Due to the exponential definition of α, it turns out that the definition of the decibel attenuation coefficient is very simply given by

$$\mu = 4.3\alpha \text{dBm}^{-1} . \tag{4.33}$$

Typical figures for decibel attenuation coefficient, μ, at 1MHz are, blood: 20 dBm^{-1}, muscle: 180 dBm^{-1}, and an average soft tissue value: 70 dBm^{-1}. The parameter α increases as a function of excitation frequency in real biological tissue according to a frequency law anywhere in the range of f to f^4, depending on whether absorption or scattering is the dominant mechanism.

Don't forget that in addition to the two dominant effects of absorption and scattering, we have to remember that at each interface point in the medium, we also have energy losses due to reflection. Remember, too, that in the Fraunhofer zone of a planar acoustic source, one can expect a $1/r^2$ dependency of intensity with distance due to the action of diffraction.

4.4 IDEAL IMAGING PARAMETERS

Having discussed the manner in which we can obtain lateral and axial positional information from an acoustic medium, we ask the following question:

What would be the ideal function to use as source excitation to obtain the best spatial localisation?

We will answer this question as a means of introducing the ideas of resolution in imaging, and illustrating some of the tradeoffs in designing a system for real-world use.

4.4.1 AXIAL RESOLUTION

One test of imaging performance is the ability of a pulse-echo system to distinguish the presence of two successive impedance interfaces i.e., 3 separate layers of differing acoustic parameters) which are closely spaced along a normal to the aperture. See Figure 4.7. We illustrate the observed echo amplitude (shown by the envelope) in Figure 4.8. Note the overlap of the tail of the echo from the first interface with the front end of the echo from the second.

Since the signals are typically oscillatory in nature, it is also possible to have destructive interference between the reflected signals. In the case shown above, a suitable phase shift at the

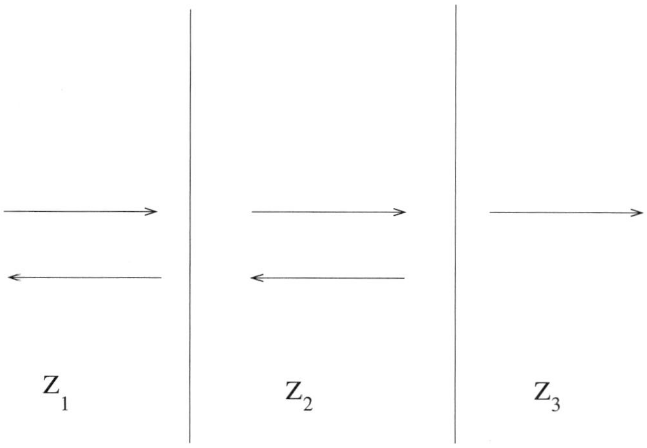

Figure 4.7: Two acoustic boundaries.

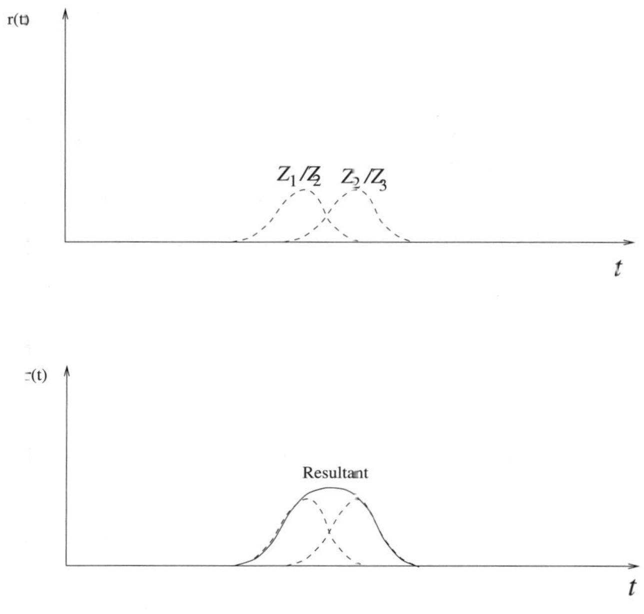

Figure 4.8: Overlapping of echoes.

second boundary could result in a very significantly decreased resultant echo. For this reason, it is best to reduce as much as possible the overlap of echoes from successive boundaries and, therefore, to keep the length of the transmitted pulses as short as possible.

4.4.2 LATERAL RESOLUTION

Consider how we might make the main lobe of the transducer acoustic pressure amplitude distribution narrower. An examination of Equation (4.29) shows that this can be achieved by increasing k. Since we know that k is related to the R.F. frequency of the transmitted pulse f by

$$k = 2\pi f/c \tag{4.34}$$

then it is clear that an increase in frequency will result in a greater of ÿconcentration of energy being radiated about a centred normal to the aperture. This would lead to a better spatial localisation capability in the lateral direction, which we will later identify formally as a better lateral resolution.

If we couple this requirement with the need for short pulses, then one arrives at the use of bursts of high frequency ultrasound as the appropriate source function. The high-frequency content provides the lateral directionality, whilst the short time extent provides the axial localisation capability.

Equation (4.29) also suggests that wider apertures provide better lateral beam localisation. We shall see that there are practical limitations for this, too.

4.4.3 CONSTRAINTS

There are some constraints to these ideal parameters in terms of spatial resolution.

The ability to reliably detect received echoes requires that the signal-to-noise ratio of the system (and its signals) be sufficiently high in order to rise above the noise floor of the instrument and its surroundings. This in turn requires that a certain minimum of acoustic energy must be present in the acoustic echo.

So, reducing the length of the pulses implies reducing the total transmitted energy in each pulse. This, in principle, leads to a reduction in the overall transmitted energy into the deeper regions of the medium, and an overall loss in returned echo energy. There is thus a practical lower limit to the amount of energy, and hence of pulse length, which must be transmitted to achieve a useful echo from a given depth. (There are a few other reasons for wanting to transmit pulses of a certain minimum length, but these normally have to do with Doppler instrumentation). This minimum amount of detected energy is related to the efficiency of the receiving system, the signal-to-noise "floor" level, the display system and viewing conditions, and the experience of the sonographer.

Considering an ideal imaging system also leads to the idea of using a very high R.F. frequency to yield the highest possible lateral resolution. However, increasing the R.F. frequency used for pulse transmission incurs, again, a reduced transmitted energy to deeper regions, due to an increase in the attenuation coefficient, α, with frequency.

Thus, we can identify what the ideal characteristics of an acoustic imaging system might be, but also see that, when looking at real acoustic media, these ideals must quickly give way to choosing parameters for the excitation which are optimal for a given imaging application.

4.4.4 SUMMARY

In the next sections of the course, we shall look at real ultrasonic instrumentation, and its medical applications. We will see how the physical principles that we have outlined in this introductory section can be exploited for anatomical imaging, and velocimetry.

CHAPTER 5

Pulse-Echo Ultrasonic Imaging

5.1 INTRODUCTION

This chapter provides a condensed overview of the principles and applications of pulse-echo ultrasonic imaging (excluding colour-flow mapping). In particular, attention will be placed on the functional aspects of the pulsed ultrasonic imaging system, emphasizing overall concepts rather than specifics.

5.1.1 APPLICATIONS

Pulse-echo ultrasound is currently most widely used in obstetrics, where it affords a safe, noninvasive means of evaluating and following the development of the foetus, and of the detection of such abnormalities as cardiac disorders, encephaly and hydrocephaly. Amniocentesis, for the detection of Down's Syndrome, can be performed under the real-time control of ultrasonic imaging, a fact which has dramatically improved the safety of this technique. However, pulse-echo techniques are becoming increasingly used for vascular imaging (in conjunction with flow imaging through Doppler techniques) and is used in gynaecology, abdominal examination, transrectal imaging of the prostate and ophthalmic examination. Its widespread use and adaptation to imaging such diverse anatomical sites are testimony to its diagnostic potential.

New applications, reviews, and clinical practice can be found in the journal *Ultrasound in Medicine and Biology* Also, see *Journal of Diagnostic Medical Sonography.*

5.1.2 PRINCIPLES OF OPERATION

Unlike Doppler ultrasound velocity measurement, which relies on the scattering of ultrasonic waves from moving erythrocytes, pulse-echo methods rely instead on the reflections which takes place when a forward propagating acoustic pulse encounters a variation in the acoustic impedance of an acoustic medium. Such a variation will give rise to a reflected ultrasonic wave, provided that the variation is sufficiently localized to produce a distinct boundary. A simple example of what is referred to as an A-Mode scan, and the stages of processing required to obtain it, is shown in Figure 5.1. The term "A-mode" is applied to either the second trace or the bottom trace in this figure. After the acoustic pulse is transmitted, the transducer goes into receive mode, and the system collects information on the time of arrival and the intensity of the reflected echoes. The primary method of detecting the echoes is to envelope detect the received R.F. signal, which simply makes the returned signals easier to interpret. A simplified diagram of the processing chain is illustrated in Figure 5.2, *excluding* the time-gain compensation (time-dependent gain) which we shall discuss later on.

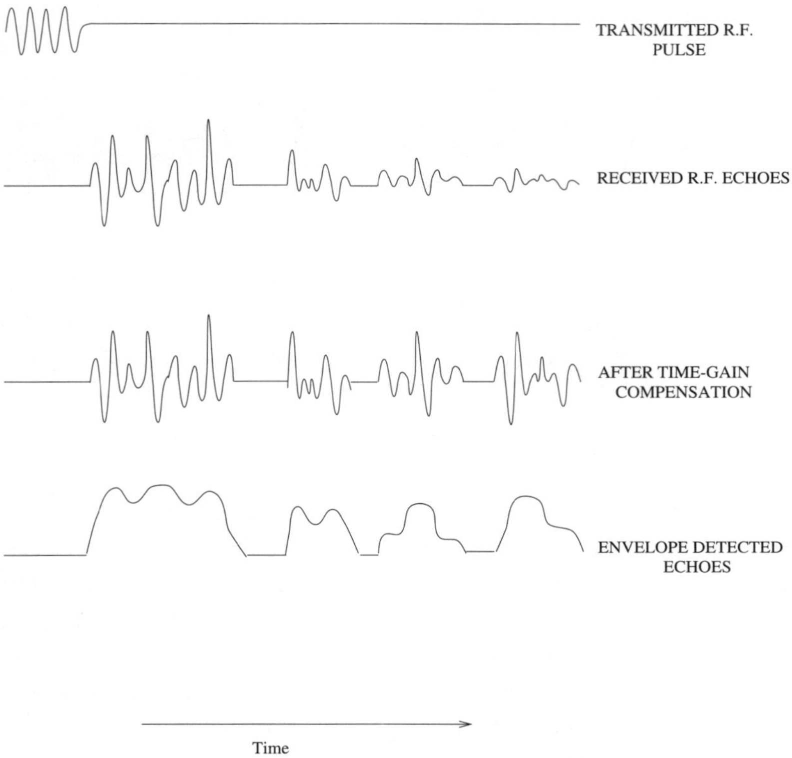

Figure 5.1: A-Mode Pulse-Echo Signals: Raw and Processed.

5.1.3 ACOUSTIC PULSE GENERATION

The acoustic pulse, which is applied to the medium, is typically composed of only a few cycles in length of radio-frequency acoustic excitation, and is most easily generated by applying a short, high amplitude, voltage pulse to the transducer. The natural resonant frequency of the transducer then acts as a filter to produce what is essentially an amplitude modulated sinewave. Transducer design is a critical aspect of an ultrasonic measurement system, and can severely affect the signal to noise ratio and the resolution of the entire system. It is common to have a choice of transducer probe designs available for a particular ultrasound machine, each probe being better suited for imaging different anatomical sites, and quite often at different frequencies. As we shall see soon, transducers usually consist of a number of elements, which may be used for dynamic focusing, application of apodisation, or for beam steering. There will be more on this under the following sections.

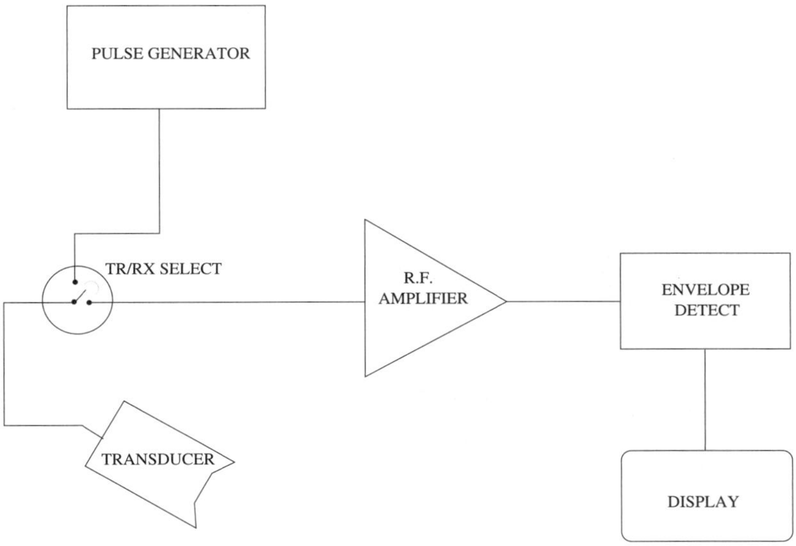

Figure 5.2: Flowchart of stages for Simple A-Mode scanner.

5.1.4 SCANNING GEOMETRIES

A Mode Scanning

The A-Mode signal, which is obtained as a time varying voltage, is essentially a one-dimensional map of the tissue impedance; a large amplitude peak might, for example, represent a muscle/blood-vessel tissue boundary, and the time at which this "echo" is recorded, relative to the time of pulse transmission, provides a measure of the depth at which the boundary occurs in the body. The A-Mode signal is therefore a one-dimensional map of the differential tissue impedance. Localisation of structures in depth can be very good, but one only knows that there is a boundary at a certain depth: the extent of that boundary laterally is not known using only an A-mode system.

From 1D to 2D

In order to obtain a two-dimensional image of differential tissue impedance, we must scan the beam in some fashion. An obvious way of doing this is to collect a number of adjacent A-mode scans, all parallel to each other, but displaced slightly from each other, so that the positions of the transducer form a line. This is depicted in Figure 5.3.

This method of scanning is known as rectilinear B-Mode scanning, and it, and its variations, are one of the most widely used in medical applications of ultrasound. An example B-mode scan (badly blurred by poor quality grabbing, I am afraid) is shown in Figure 5.4.

Conceptually, rectilinear scanning can be achieved very simply by movement of the acoustic aperture back and forth. Consider one drawback of a rectilinear scanning strategy for medical imag-

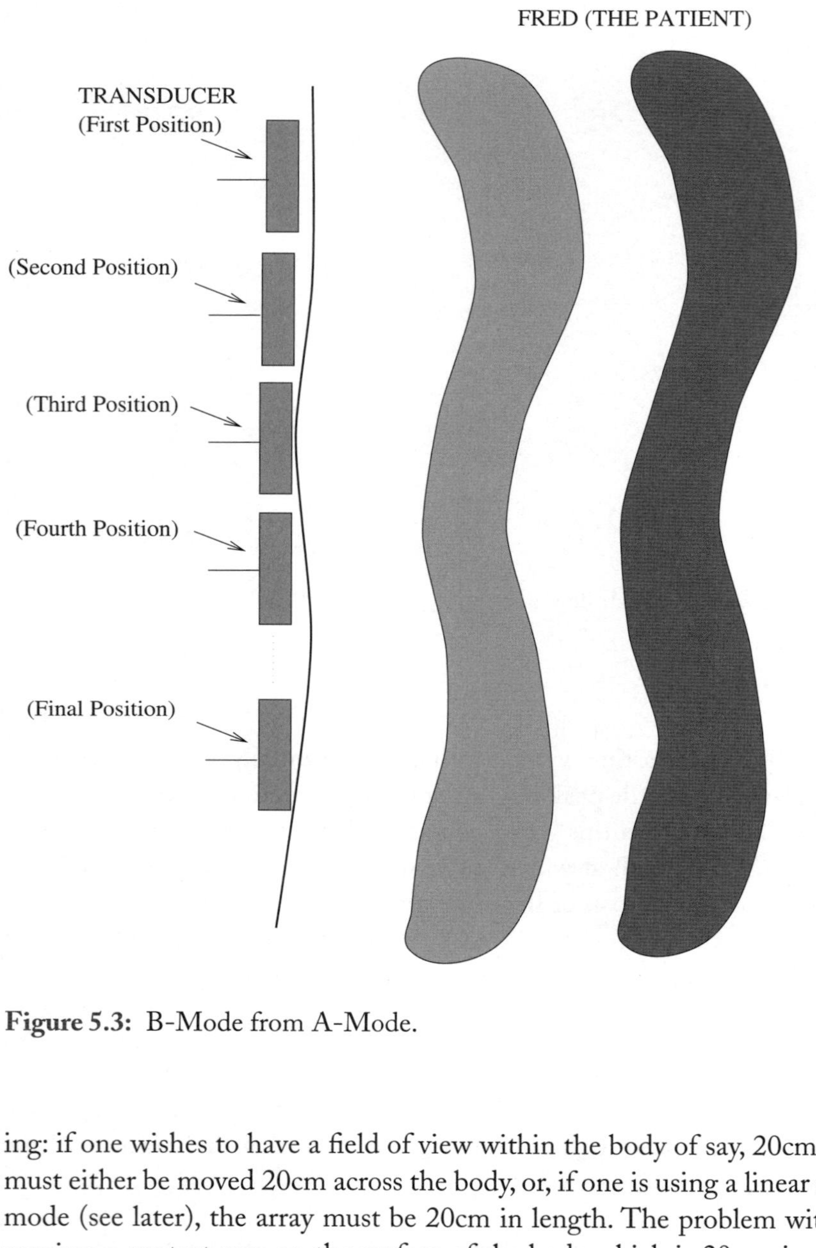

Figure 5.3: B-Mode from A-Mode.

ing: if one wishes to have a field of view within the body of say, 20cm laterally, then the transducer must either be moved 20cm across the body, or, if one is using a linear array transducer in rectilinear mode (see later), the array must be 20cm in length. The problem with this arrangement is that it requires a contact area on the surface of the body which is 20cm, i.e., the same dimensions as the desired lateral field of view. This may not always be possible. Quite often, there will be intervening structures which will distort or attenuate the beam before it gets to the target imaging depth. An extreme example of such a restriction occurs in cardiac imaging, where the ribs can obscure the acoustic aperture. In other cases, the use of a linear array is precluded due to the contours of the body.

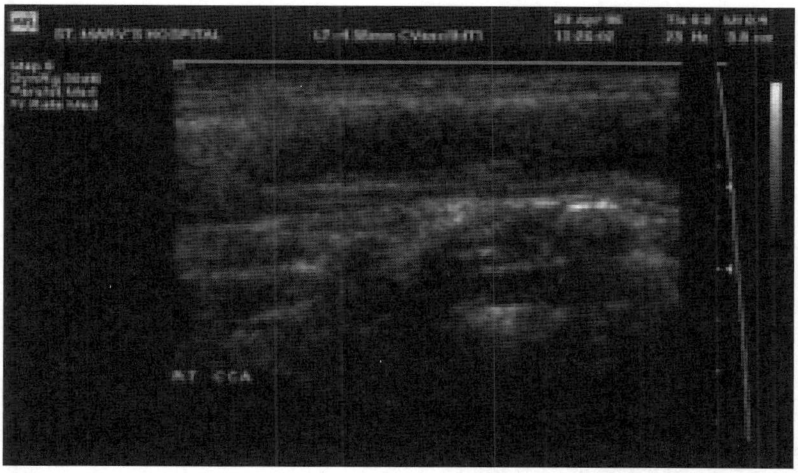

Figure 5.4: Linear B-Mode. Poor quality is due to frame-grabber.

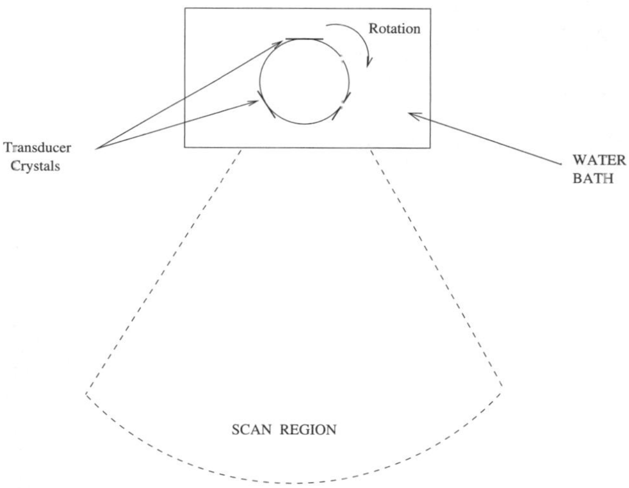

Figure 5.5: B-Mode mechanical sector scanning.

In many instances, it is therefore advantageous to be able to use a *curvilinear* method of scanning. The most widely used curvilinear geometry belongs to that of the sector scan. Here, the beam direction is swept by tilting the aperture plane back and forth. A sector scan therefore has the advantage that the beam aperture does not have to be the same width as the desired field of view. Lateral extent is defined by the depth and angle range of the scanning. A depiction of a (very old) drum-type sector scanner, containing 3 transducers spaced around the circumference of a

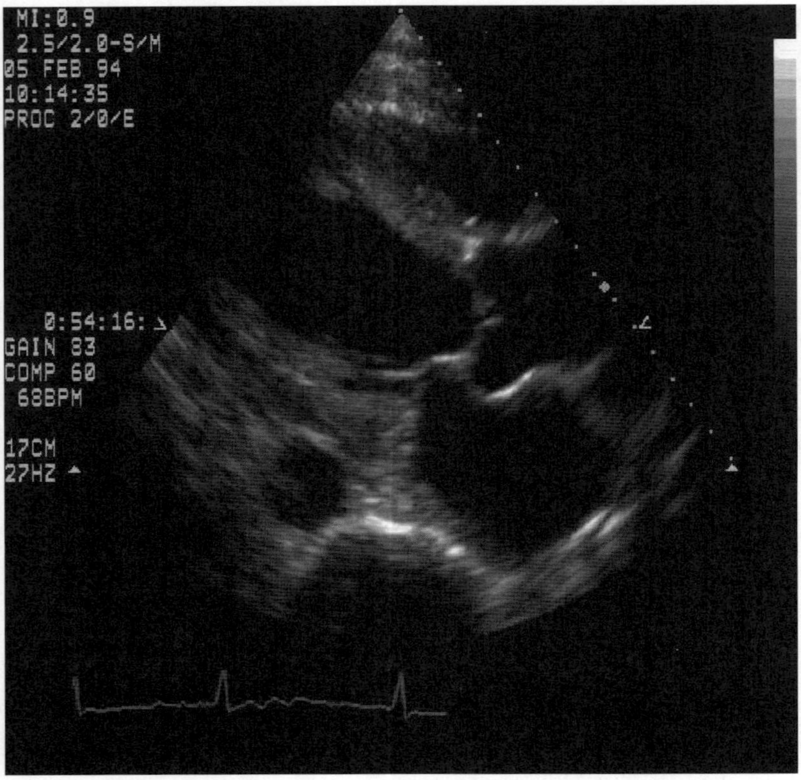

Figure 5.6: B-Mode sector scan cardiac image.

rotating drum, is shown in Figure 5.5. A B-Mode sector image, constructed by electronic scanning, is provided in Figure 5.6.

TM Mode

The TM Mode scan, also known as the M-Mode scan is rather less logical. It consists of a series of A-Mode scans, each acquired from successive pulsings of the transducer. Each A-Mode scan is displayed as an intensity versus depth (or time) graph, and graphs of received A-Mode traces from successive excitations of the media are displayed alongside each other. The scan is depicted in Figure 5.7, and an actual scan is reproduced in Figure 5.8.

The M-Mode scan has proven useful for detailed studies of cardiac and valvular function. To some extent, M-Mode studies are being replaced by real-time B-Mode scans, but M-Mode scans still have an application where a very specific movement along a particular line must be tracked with high precision.

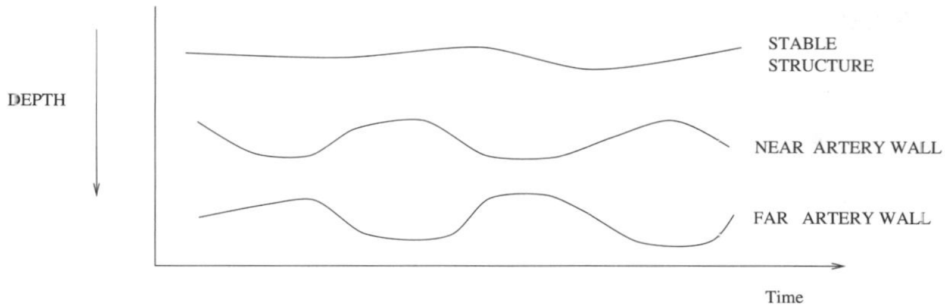

Figure 5.7: Depiction of M-Mode Scan.

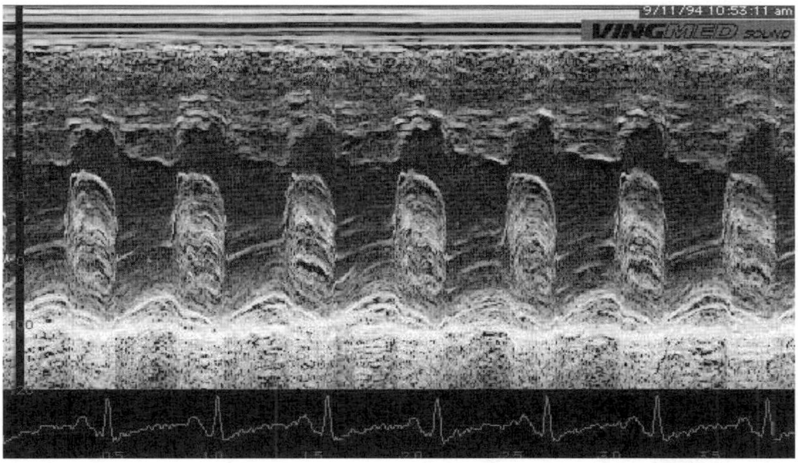

Figure 5.8: M-Mode Scan.

C-Mode

Another possible means of building up a two-dimensional image of differential impedance is less well known in medical application - it is referred to as the C-Mode scan, and is formed by collecting the received A-Mode signal from a particular depth. The probe is scanned in a 2D plane over the skin of the patient. This is clearly time consuming.

C-Mode scans throw away much information by only acquiring data from a particular depth. It is, of course, feasible to record this extra information, so that a 3D map of differential tissue attenuation is obtained. This is known generically as 3D ultrasound. 3D scanners are not real-time (unless one is imaging to a very shallow depth). Also, because of the difficulty in displaying 3D data, clinical implementation of this modality is not widespread, and is very much in its infancy. Note that 3D ultrasound can make use of both curvilinear and rectilinear scanning geometries: a common technique for obtaining 3D scans is to acquire a series of B-Mode scans in roughly parallel planes, and to simultaneously acquire the probe positioning information from a position and angle sensing device. The series of B-mode slices is then interpolated and resampled to yield a 3D volume.

5.1.5 IMPLEMENTATION

We shall now briefly discuss the means whereby the above mentioned methods of scanning are implemented.

A-Mode Scanning

The physical implementation of an A-mode scanning system is a little more complex than the simple example of Figure 5.2. For example, Figure 5.2 is for the case of a single pulse, and so the display of this information would require a sampling oscilloscope. A more realistic diagram for an A-mode system is shown in Figure 5.9. The limiter is used to suppress very high-amplitude echoes in the

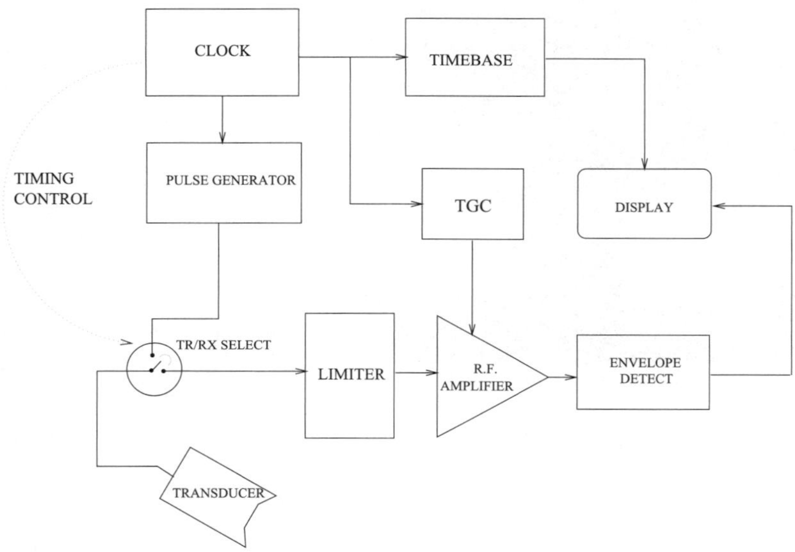

Figure 5.9: A-Mode Scanning.

signal, which typically come from transducer ringing. Time-gain compensation is discussed under B-mode implementations - its function is the same in all ultrasonic scanners.

5.1.6 LINEAR B-MODE

Whilst the physical moving of a single aperture transducer in a linear fashion across the skin of a patient can be used to build up a B-mode scan from A-mode scans, this is a far from satisfactory scanning technique. First of all, as it involves physically moving the acoustic transducer, it is very slow. It is also cumbersome, as one needs to accurately measure the position of the transducer for each A-mode scan-line. For a discussion of the implementation of manual scanning techniques, see [5], [3].

For the past 15 years, linear mechanical scanning has been largely replaced by the linear array transducer (see specification sheet of ATL 5000). This type of scanning essentially consists of the same rectilinear scan method of mechanical movement, but it is achieved by selectively switching on and off groups of transducer elements. Figure 5.10 shows a simple linear transducer array. The entire array is referred to as a transducer, and you can also think of each element of the array as a tiny rectangular acoustic aperture.

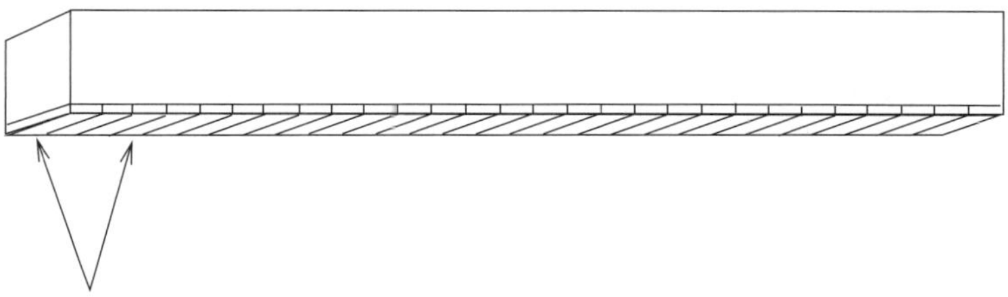

Figure 5.10: Linear array transducer.

A typical figure for the number of elements in a linear array varies from 30 up to over 200, and the typical dimensions of an external (noninvasive) linear probe would be 1.5cm wide by 10cm long. Typical operating frequencies vary from 3.5 to 12.0 MHz. The construction of these arrays is quite difficult. In order to get all the crystals of the array to be similar, they are actually grown from one large crystal, then machined into separate elements. For linear B-mode scanning, the number of elements which simultaneously fire is of the order of 30. The groups of elements fired to make up an A-mode scan can overlap, or be distinct. Overlapping clearly provides a higher density of spatial sampling (closer A-mode scan lines) but is accompanied by an increased scan time, and a higher data rate for demodulation and post processing. Constraints are therefore set by data rate considerations, and permissible scan time. It is conceivable that tissue heating considerations may also be important.

In addition to obtaining a linear sweeping by switching on contiguous groups of elements in sequence, some electronic focusing is often simultaneously applied; this involves employing a technique similar to phased array techniques, in that the each element receives excitation which is slightly phase-shifted with respect to its active neighbour. A parabolic phase shift across elements away from the group centre, for example, will simulate a parabolic (curved) aperture. The technique is known as dynamic focusing, and is quite extensively used with annular transducers arrays.

A great advantage of linear array scanning over mechanical linear scanning is the speed with which the scanning can be achieved. Real-time scanning becomes feasible because of the electronic nature of the scanning process.

In addition to the basic linear array shape shown here, there are also curvilinear arrays, which have a gentle convexity as one moves along the elements from one end to the other (long axis). Also, it is common to introduce some concavity across the width (short axis) of the array, to provide a degree of focus in the image "slice" direction.

B-Mode Sector

B-Mode sector scans are generally considered as presenting a distorted view of anatomical structures. This is not necessarily the case, and the extent of any geometric distortion is dependent on the amount of post-processing and interpolation applied to the received data.

A number of mechanical arrangements may be employed to acquire a curvilinear image. These include a rotating scan head, equipped with a number of transmit/receive probes spaced around its circumference, a rocking arrangement (similar to a car's windscreen wiper) and a rotating acoustic mirror, which is easier to implement. The rotating scan head or mirror are even suitable for real-time application.

As in the case of rectilinear scanning, it is possible to employ electronic scanning of a *linear* transducer array in order to sweep the beam back and forth in a *curvilinear* fashion. The basis of the technique is close to the dynamic focusing mentioned earlier, in that the elements are supplied with different phases, which effectively curves the emergent wavefront from the array. The technique is derived from beam steering techniques in pulsed radar. It is amusing to recall that many researchers in the 1980s were quite sceptical of the efficacy of this technique, due to "errors" introduced into the wavefront as it passed through tissue layers in the body. The argument was that it would not be possible to steer the beam with any degree of precision, since the wavefront would be hopelessly distorted by random phase-shifts experienced in real tissue. The technique of dynamic focusing and beam steering has, however, been since successfully incorporated into products by commercial companies. Moral: Don't believe everything an "expert" tells you!

Compound Scanning

This is a technique of scanning which is a cross between linear and sector scanning, and which has its own advantages and disadvantages. The compound scan is composed by integrating a number of linearly-shifted sector scans. Clearly, the mechanism needed to achieve this with a mechanical scanner is quite intricate. It is more naturally implemented by electronic scanning. Although it is difficult to implement in a real-time system, the compound scan offers improved boundary distinction, reduced speckle (integration of backscatter) and can improve to some extent the lateral resolution of the signal. A compound scan is depicted in Figure 5.11. One final note: purists may wish to consider

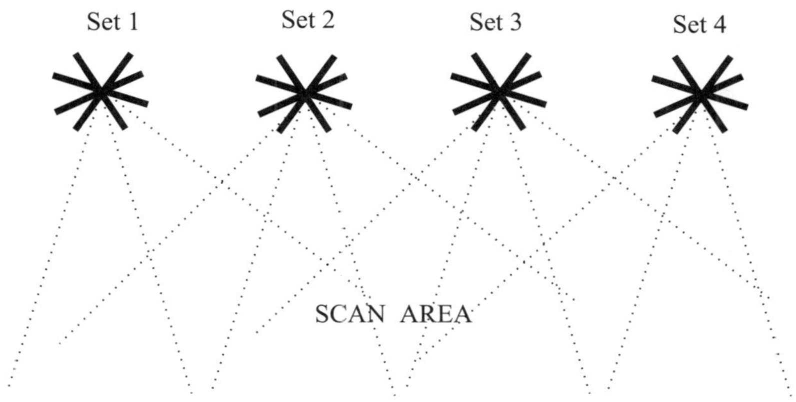

Figure 5.11: Compound Scanning.

compound scanning as being a different scanning geometries. This is a valid argument!

Matrix Scanning

I am not aware of any commercial implementations of matrix scanning. There are difficulties in wiring transducer elements in a matrix, and the large number of R.F. amplifiers necessary to drive all of the array elements. Some matrix transducers are already on the market, but remain less widely in use. Clearly, such a matrix array allows the use of beam steering in two dimensions (rather than in one, as with a linear array). It makes the prospect of capturing 3D volume data easier than the current hybrid methods.

5.1.7 SIGNAL DETECTION

The detection of the reflected echoes from an ultrasonic insonation of a patient is accomplished by a chain of processing stages (See Figure 5.12). The reflected acoustic signal is converted into an electrical signal by the action of the transducer. An R.F. amplifier with a voltage controlled gain is the next stage after detection. The need for a time dependent gain can be explained as follows: as a single burst of ultrasound travels into the body, it is attenuated, via the mechanisms discussed in Chapter 2. This has the effect of reducing the strength of the echoes coming from deeper in the body relative to those from superficial structures. In order to correct for the weakening of the signal with depth, the Time Gain Control (TGC) module increases the gain of the amplification of the A-Mode signal as a function of time (and therefore, depth). Various functions of attenuation vs time can be used; a common choice is a linear ramp. Usually, on the more recent scanners, the user has a choice of preset TGC functions, as well as a piecewise linear interactive control of the gain function at various depths.

Following the TGC R.F. amplification comes envelope detection. This is usually implemented by quadrature demodulation, i.e., by treating the incoming echo as an amplitude modulated sinusoid, and extracting the envelope of this signal by a two-channel mixing process. Low-pass

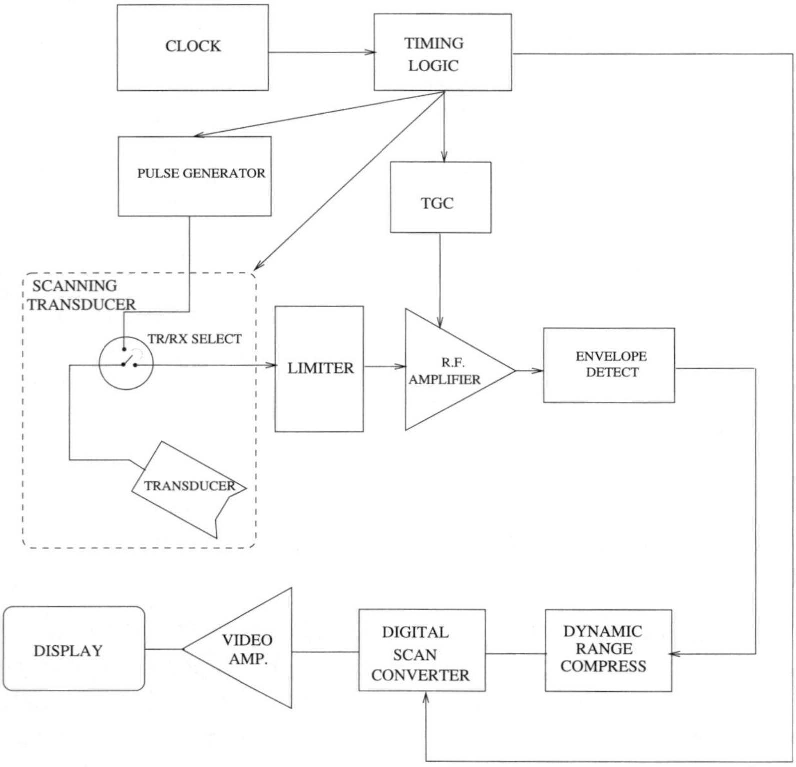

Figure 5.12: B-Mode Imaging.

filtering is also performed in this stage, to remove residual R.F. signal components. The filtering must be carefully controlled; too much, and there will be a risk of losing axial resolution. This is mentioned again under image quality considerations.

The next stage is some form of dynamic range compression of the signal. This is required because the dynamic range of the amplitude of the received echoes is 70 to 80 dB i.e., 20 log (Max. Amplitude / Min. Amplitude)). The TGC module will reduce this somewhat, (to 50 dB) by attenuating the signal from more superficial structures relative to the deeper structures. However, while a typical 8-bit display, capable of grey-scale display has a dynamic range of 48dB, it is very difficult for a user in average room lighting conditions to perceive image details spanning more than about 30dB. For this reason, dynamic range compression is applied to the signal, which effectively reduces the difference in signal level between strongly and weakly reflecting acoustic structures. This operation may be performed by a logarithmic amplifier.

A digital scan converter follows next. The role of the digital scan converter is to take each of the demodulated A-mode scan-lines, and to "stitch" them together into an image. This is a very complex task, and is highly dependent on the scan geometry, the beam patterns, and other factors (such as the video display resolution). A large part of the task of the digital scan converter is to interpolate the pixels between the geometric centres of scan lines. This operation is now performed digitally for a number of reasons:

- The critical step of interpolation can be altered, as necessary, if new scanning from-ends or geometries become available, by a software upgrade, rather than hardware re-design.
- it allows one to easily mix in other information with the ultrasonic image, such as text and graphics,
- it allows digital storage of the reconstructed image.
- Markers and cursors can be more effectively added to the image. These can be customised for different applications.

Other types of enhancement may be applied to the image prior to display on the monitor, outside of the digital scan converter. These include:

- Edge Enhancement, through differentiation of the signal, followed by summation with the original signal.
- Contrast Enhancement. Altering the dynamic range of the image under user control, to improve contrast.

And so, on most modern scanners, a digital post-processing system may be found between the digital scan converter and the video amplifier.

5.2 IMAGE QUALITY

There are a number of issues that need to be considered in assessing image quality of an ultrasound scanner. These include resolution, dynamic range, maximum frame-rates and existence (preferably none!) of artefact.
Unlike the tomographic scanners, there are not many system-induced artefacts on modern ultrasonic scanners. Most of the artefacts listed below are actually due to the physics of the ultrasound beam and its interaction with matter, and so these are all patient and examination dependent, and, in most cases, all ultrasound scanners will suffer from them in a similar manner.

5.2.1 IMAGE ARTEFACT

Acoustic Shadowing

This results when a highly attenuating or reflecting region falls in the path of the beam, effectively creating a "shadow" behind it. If a region which we are interested in visualising lies within this shadow, then the reflected energy from that region which reaches the transducer might be sufficiently low to reduce considerably the contrast detail which we can perceive. Regions which act to produce shadow include bone, kidney stones, some forms of atheromatous plaque and air.

Multiple Reflections
This may be thought of as a type of reverberation which typically occurs between the transducer and a flat, parallel acoustic interface, or between the transducer and some highly reflecting interface. Examples of sites for occurrence include imaging of the abdomen, when reverberation might occur between the transducer and bowel gas, or between the transducer and the bladder. The form that the artefact takes will be multiple echoes, at depths which are multiples of the transducer-reflector transit time (distance/c). Strong reverberation between the transducer and the skin of the patient may be reduced by a well designed probe.

Post-Cystic Enhancement
As the name suggests, a common condition under which this artefact occurs is "behind" a low-attenuation cyst. This causes region immediately behind the cyst to receive a higher ultrasonic intensity than would otherwise be found. This will be contrary to the attenuation assumption made by the TGC module, which will boost the amplitude of the deeper signals as normal. The result of this will be a higher than normal echo intensity behind the cyst.

Finite Beam Width
This is really a problem of lateral resolution; one cannot have a zero-width acoustic beam. In practice, the beam has a finite width, which is a function of depth, and is only minimum at the focus depth of the beam. This implies that the point-spread function of the acoustic imaging system is not constant, but is a function of depth. The result of this is that small objects may appear to be of a different size, depending on their depth.

Refraction
This form of artefact stems from a simple ray physics. If a wavefront of ultrasound crosses a boundary between media of two different acoustic velocities (or impedances), the beam path is bent (refracted) by an angle which is dependent on the ratio of the two impedances. This can cause the same form of error as one observes by placing a straight stick into a pool of water: a lateral and axial displacement of the image.

Registration Error
This is due to the velocity of sound in the tissue being considerably different from that which it is assumed to be, normally 1540 m/s. If this happens, then the depth calibration of the system will be incorrect, and distance measurements become unreliable. An example occurs in the imaging of the eye; the echo of the retina is typically displaced toward the transducer, due to the higher velocity in the lens of the eye than in the aqueous and vitreous humours. For very precise depth determination, this can be a problem.

Sidelobe Artefact
This arises when the side-lobes of the transducer beam pattern are not well suppressed, either in transmission, or by "clever" interpolation. First, imagine a scan line in a region of tissue which contains no reflectors directly in front of the aperture for that line. If a very strong reflector sits on a sidelobe of the beam for that scan line, any reflected energy that is received is interpreted as originating from the main lobe of the transducer. On a B-mode image, this appears as a ghost echo (faint region of brightness) where there is, in fact, no structure. An experienced sonographer learns to interpret such ghost reflections, by noting the presence of any very strong reflectors in adjacent pixels of the image at the same depth, and also by moving the probe around and noting how the appearance of the ghost echo changes. If the suspicious echo "moves" in a certain way with the movement of the probe, the game is up!

5.2.2 RESOLUTION
We begin by emphasising the concept of resolution in an imaging system. Recall that measurements on the ability of an imaging system to resolve two closely spaced objects are somewhat arbitrarily defined. The usual way of presenting the idea of resolution is to illustrate with a graphical example. We assume that two point reflectors are placed in the field of view of the ultrasound scanner. Due to the point spread function of the scanner, the appearance of the reflectors will be broadened into a blob.
If we physically move the reflectors progressively closer together, they will eventually come sufficiently close that it will not be possible to discern that they are distinct reflectors from the image. We then say that this distance is below the resolution limit of the system. If we now move the sources apart until we can just discern from the image that they are indeed two distinct reflections, then we have a well defined measure for the resolution capability of the instrument. This is a very simple definition of resolution, and in practice we must also consider such matters as the relative strength of the sources, the resolution in different directions, and so on. Nevertheless, we can identify the factors which limit the spatial resolution on ultrasound scanners.

In the case of pulse-echo ultrasound (A-Mode, B-Mode, C-Mode) we can identify two primary resolution measurements: the axial resolution and the lateral resolution.

Axial Resolution
This is the ability of the ultrasonic system to resolve two objects which are at different depths in the medium. It is therefore dependent on the ability of the system to recognise two successive echoes as being distinct, rather than being one large echo. This is dependent on

- the width of the transmitted pulse (shorter the better, but must not degrade S/N performance).

- the bandwidth of the transducer (wideband transducers give better axial resolution).

- the bandwidth of the receiver and demodulation processing and, in practice, the gain applied to the received signals.

Lateral Resolution
This is the ability of the system to resolve two small objects placed side by side in the plane perpendicular to the axial direction. It is affected by

- the wavelength of the acoustic wave in the medium (smaller wavelength, better resolution, but higher attenuation).
- by the aperture size of the transducer (larger aperture, better resolution).
- the degree of focusing (if any) used, either electronic/dynamic focusing, an acoustic lens or transducer curvature.
- the depth of the object (diffraction effect).

Slice Thickness
This is not usually considered to be an important issue for 2D B-Mode, but in 3D ultrasound, it is. The main limiting factors are quite similar to the lateral resolution limits:

- The wavelength of the acoustic wave (smaller wavelength is better).
- The aperture size of the transducer (in the direction orthogonal to that for image (pixel) lateral resolution and axial resolution).
- The degree of physical curvature or acoustic lens focusing applied.

5.2.3 FRAME RATE

The frame rate of a B-mode ultrasound scanner is usually limited by the depth in tissue that one wishes to image to, and by the lateral field of view, and density of scan lines. A rough relationship between the maximum attainable frame rate, the number of A-mode scan lines, N_A, and the depth of imaging, d is given by

$$f_{max} = \frac{c}{2dN_A} . \tag{5.1}$$

Remember that the number of scan lines, N_A is the number of A-mode signals used to generate the B-mode image. This is usually well below the pixel width of the image (interpolation!).
Also, note that if dynamic focusing is applied, and a number, N_f, of focusing zones is acquired for each scan line, the maximum frame rate will be reduced by a factor of N_f.

CHAPTER 6

Doppler Velocimetry

6.1 INTRODUCTION

Doppler Ultrasound has been firmly established as a valid clinical technique in the assessment of cardiac performance, evaluation of risk in incidence of cerebrovascular disease and the detection and severity analysis of vessel stenoses. For a brief introduction to the applications of Doppler ultrasound, see [5] and [3]. New clinical applications are constantly being found for the basic Doppler techniques.

- Rijsterborgh and Roelandt, (1987) for a discussion on evaluating the pressure drop across a stenosis by using the modified Bernoulli equation. This may shed some light on the role of haemodynamic insufficiency in the development of neurological symptoms in patients suffering from aortic stenoses.
- Deambriogo and Bernardi, (1986) for applications of Doppler u/s into investigating the occurrence of the vertigo syndrome.
- Jorch et al.,(1986); blood flow in the brains of premature infants with respect to Sudden Infant Death Syndrome (SIDS).

New ways are always being found to exploit subtleties of Doppler ultrasound physics, and the technology is also under constant refinement.

The diagnostically relevant information which is usually sought during Doppler interrogation of the vascular system includes estimates of the peak arterial blood velocities, pulsatility indices, and the detection of the presence of turbulent flow. Pulsatility indices are a measure of vessel compliance, and turbulence can be a measure of vessel occlusion due to the presence of a stenosis. Valvular dysfunction can often be diagnosed using Doppler ultrasound: mitral valve regurgitation can often be detected by observing flow patterns through the valve.

More recently, you will find that most journal reports cover the use of techniques such as *colour flow mapping*. Again, for details of how this is performed, you can have a look at earlier engineering papers (such as by [12] and [11]), which discuss how these systems are constructed.

In what follows, we concentrate on the *basic principles* of measurement which allow blood velocity to be estimated. To move from estimating velocity to colour flow mapping, one has to combine the pulsed Doppler flowmeter with a beaming scanning technique, and some very fast processing in order to generate a colour flow mapping image in which colour-coded velocity maps are superimposed onto the B-mode "reflectance" maps.

6.2 BASIC PHYSICS

A brief discussion of the physics of Doppler ultrasound is essential to an understanding of both the strengths and weaknesses of the technique. We begin by briefly reviewing the basic idea of ultrasound (in a medical context), and provide a number of references which should be helpful in understanding a bit more about the nature of Doppler ultrasound. Let's start with the source of an ultrasonic wave.

During ultrasonic imaging or velocimetry, an ultrasonic transducer is used to mechanically excite the particles of a medium. An ultrasonic transducer is simply a device for converting an electrical signal into a (tiny) mechanical displacement. Transducers normally have a very particular design for a specific application, and so a particular ultrasonic scanner may have two or three probes, all with slightly different constructions, characteristics and shapes for imaging different regions within the body. Remember that current technology employs multi-element transducers, incorporating sophisticated electronic steering. However, to introduce the basics of Doppler instrumentation, we will assume only one- or two-element transducers to begin with.

If, as is typical in an imaging application of ultrasound, one applies a sinusoidal excitation to the transducer, this will result in a simple harmonic wave being generated in the medium. Like all mechanical wave phenomena, this is merely a disturbance of the particles in the medium from their nonexcited positions, and represents an injection of energy into the medium. The acoustic wave also propagates forward into the medium, and this represents a radiation of acoustic energy. More background into acoustic wave propagation may be found in the earlier chapters, or in *Theoretical Acoustics*, by Morse and Ingard [2], or *Fundamentals of Acoustics* by Kinsler et al. [1]

6.2.1 REFLECTION VS SCATTERING

As a wave propagates through an acoustic medium, we know that any interface that it encounters will give rise to a reflected wave. Reflected waves form the basis of echographic Doppler imaging, including A-Mode, B-Mode, M-Mode, and the relatively unknown C-Mode scans. The interfaces that we have dealt with so far are planar interfaces. Just as you can think of a planar wavefront as being produced by a monopole source as the distance from the source tends to infinity, you can think of a planar interface as the surface of a sphere of infinite radius. Imagine this sphere shrinking down to be of finite radius, and then smaller still, to be very tiny in comparison to the wavelength of an incident acoustic wave, and you have what is known as a *scatterer*. The effect of the scatterer on an incident wave is "reflect" a tiny portion of the incident wave energy in all directions. This is known as *scattering*[1]. Scattered waves form the basis of all blood flow measurement.

6.2.2 SCATTERING OF ULTRASOUND BY BLOOD

It has been experimentally verified [13, 14] that the active agents responsible for producing the Doppler signal from moving blood are erythrocytes (red blood cells). These erythrocytes possess a volume of only 80-90μm^3, and as such behave very much as ideal scatterers. Because they are so

[1]Actually, acoustic scatterers come primarily in two types: monopole scatterers, where the wave is scattered equally in all directions, and dipole scatterers, where the wave is scattered primarily into the forward and reverse directions.

tiny, the total amount of scattered energy from each erythrocyte is extremely small. However, they occur in vast numbers, and so the net effect is that they give rise to a measurable scattered wave. If the blood containing the scatterers is moving, then the scattered acoustic wave will contain some velocity information in the form of the Doppler effect.

The precise pressure distribution of the scattered field varies, depending on the random configuration of the "cloud" of scatterers being insonated. As the configuration changes due to the motion of the blood, so too the pressure distribution of the scattered wave will also change. This is a quite separate effect to the Doppler shift of the scattered acoustic signal itself.

6.2.3 DOPPLER EFFECT BASICS

It is rather pointless to repeat the derivation of the Classical Doppler equation here - a high school physics textbook will more than adequately cover this. What I will emphasize is the difference between the classically quoted formulae for the Doppler shift and the particular case of backscattered ultrasound, which we meet in medical systems.

The Classical Doppler shift, is, of course, a change in observed frequency due to the relative motion of source and observer. The frequency which is seen by an observer moving directly away from a monotonically i.e., simple harmonic excitation) excited source is given by

$$f_{s_1} = f_0(1 - v/c) \tag{6.1}$$

where c is the speed of sound in the medium (in this case, blood), v is the velocity of the observer away from the source, and f_0 is the unshifted frequency at which the source is oscillating.

If the motion of the observer is not along a line joining the source and infinity, then Equation (6.1) must to be modified to read

$$f_{s_1} = f_0(1 - v/c)\cos(\theta_1) \tag{6.2}$$

where θ_1 is defined in Figure 6.1.

For the case of a source moving away from the observer, the frequency, f_{s_2} seen by the observer is given by

$$f_{s_2} = \frac{f_0}{1 + v/c}\,. \tag{6.3}$$

Equation (6.3) may be written in terms of a binomial expansion of $(1 + v/c)^{-1}$:

$$f_{s_2} = f_0\{1 - v/c + (v/c)^2 + ...\} \tag{6.4}$$

upon which we invoke the *first order Doppler approximation*. This approximation is based on the assumption that the velocities of blood motion which we are likely to encounter in practise are much smaller than the speed of sound *in vivo*, and leads to the neglection of terms of second order and higher in v/c.

$$f_{s_2} \approx f_0 \left(1 - \frac{v}{c}\right) . \tag{6.5}$$

As in Equation (6.2), we may generalize to the condition in which the source is not moving directly away from the observer:

$$f_{s_2} \approx f_0 \left(1 - \frac{v}{c} \cos(\theta_2\right) . \tag{6.6}$$

We can apply the results from the two source/observer cases which we have considered above to the situation of backscattered ultrasound from a moving scatterer. The particular argument is as follows:

1. Consider two transducers, one transmitting, the other receiving, placed in a configuration as in Figure 6.1.

2. A moving scatterer is represented by P. In the first instance, we consider that Transducer 1 is the source, radiating ultrasound at a frequency of f_0, and that the scatterer P is the observer.

3. If P is moving along the horizontal line through P in the direction shown, then clearly the signal received by the scatterer is at a frequency given by Equation (6.2).

4. If Transducer 2 is now acting as a receiver, with P as the source, then we may employ Equation (6.6) to derive the frequency of the scattered ultrasound received by Transducer 2. However, our "transmitted" ultrasound signal (from the scatterer!) is nothing but the scattered wave from P, which is already Doppler shifted by the scatterer's motion relative to Transducer 1. Therefore, we substitute f_{s_1} for f_0 in Equation (6.2) to obtain

$$f_{s_2} = f_0 \left(1 - \frac{v}{c} \cos(\theta_1)\right) \left(1 - \frac{v}{c} \cos(\theta_2)\right) . \tag{6.7}$$

If we perform the multiplication, and again use the first order Doppler approximation to neglect second order terms, we obtain the result

$$f_{s_2} \approx f_0 \left(1 - \frac{v}{c} (\cos(\theta_1) + \cos(\theta_2))\right) . \tag{6.8}$$

The shift in frequency of the signal is simply

$$\begin{aligned} f_d &= f_{s_2} - f_0 \\ &= -f_0 \tfrac{v}{c} (\cos(\theta_1) + \cos(\theta_2)) . \end{aligned} \tag{6.9}$$

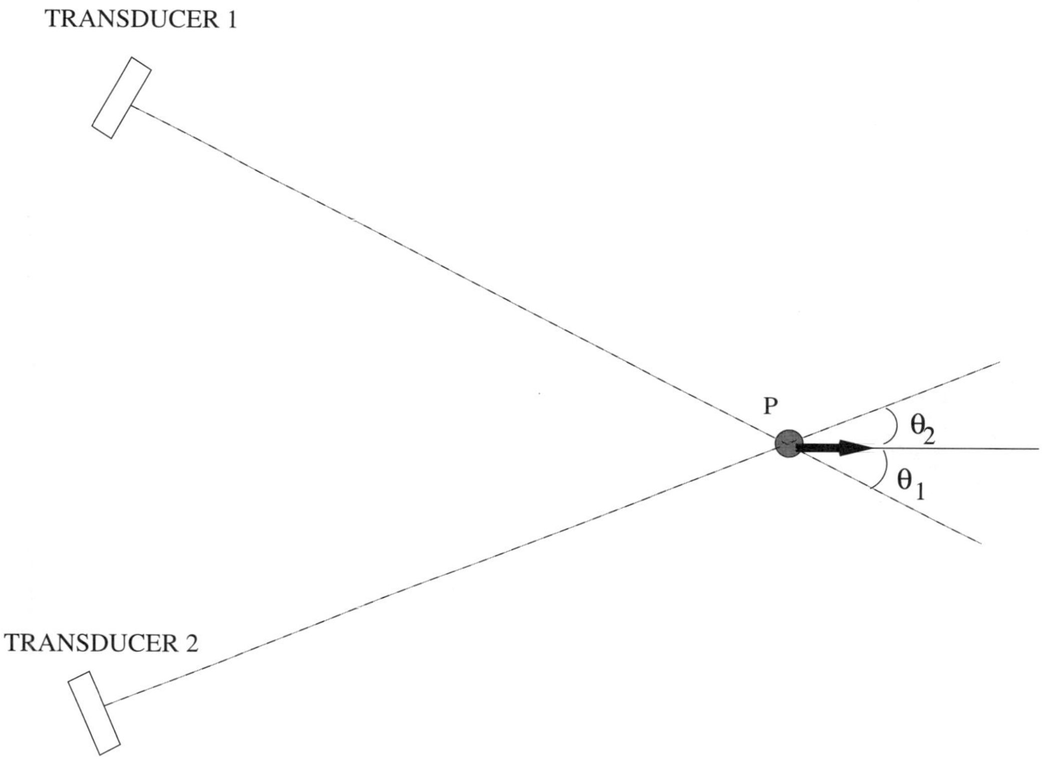

Figure 6.1: Simple Point Observer Doppler Problem.

6.3 THE CONTINUOUS WAVE DOPPLER FLOWMETER

The physical scenario which we have considered, and depict in Figure 6.1, is similar to the situation corresponding to the continuous wave Doppler flowmeter. We illustrate by functional diagram the arrangement of the Continuous Wave (CW) Doppler flowmeter in Figure 6.2.

A brief description of the functional modules is as follows:

- The oscillator provides the excitation signal to the transducer (frequency f_0). Usually, some form of driver is needed to boost the signal level to the transducer, and further electrical impedance matching circuitry may have to be included. A reference R.F. signal is also tapped off, for use by the mixing circuitry.

- One can see from the diagram that both the transmit and receive crystals are mounted in the transducer probe body. This is because transmission and reception are *continuous*.

- The R.F. amplifier boosts the received signal, which includes the Doppler shifted R.F. signal *as well as* reflections of unshifted R.F. from stationary structures in the body.

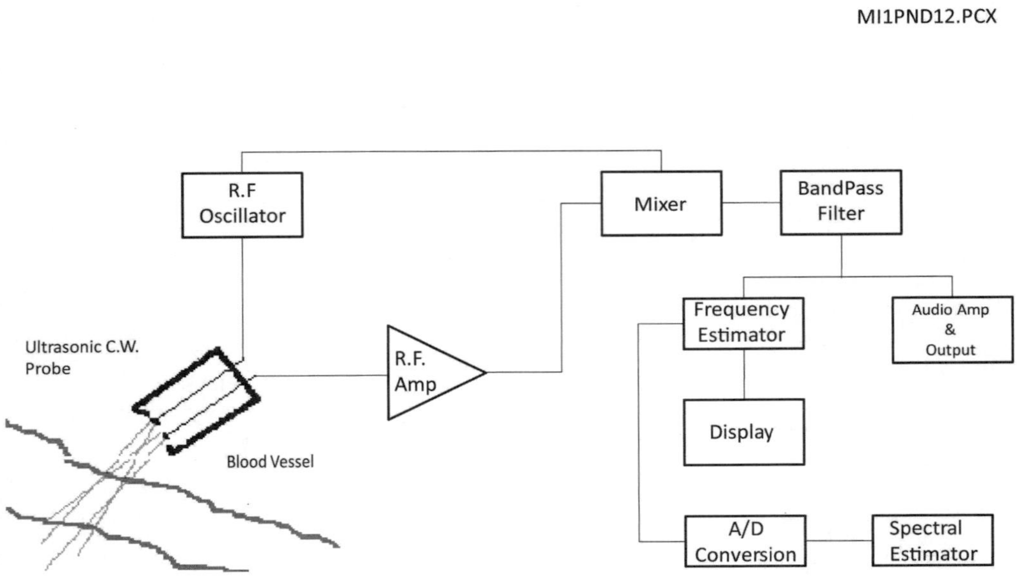

Figure 6.2: Continuous Wave u/s Doppler System. (Adapted from P.J. Fish.)

- The mixer multiplies the received signal with the reference (unshifted) signal obtained directly from the R.F. oscillator. This yields a mixture of signal components, from which the desired Doppler signal must be extracted by bandpass filtering.

- The audio amplifier and the audio output are intended to provide an audible feedback to the operator for positioning the probe to get the best possible signal. Stand-alone CW Doppler systems are often used as an inexpensive way of monitoring blood flow. The audible output is vital to the task of locating a vessel with the acoustic beam if there is no B-mode image to guide the positioning.

- For spectral estimation, older instrumentation relied on zero-crossing counters, or some other frequency counting system, applied to the analogue audio Doppler signal. Modern systems apply specialised or general microprocessors to compute spectral estimates via the Fast Fourier Transform (FFT). This is performed on a digitised version of the Doppler signal.

It is very surprising how much information can be extracted from a Doppler signal by a trained sonographer. Simple CW systems are used routinely in ante-natal scans by midwives to verify the presence of the foetal blood flow (often, the Doppler effect off the moving heart is detected too), and can easily distinguish between the mother's and the foetus's blood flow signals. They are also used during labour to check for signs of foetal distress.

6.3.1 DOPPLER SIGNAL DEMODULATION

If we consider the original signal supplied to the transmitting crystal to be given by

$$s_0(t) = A_0 \cos(2\pi f_0 t) \tag{6.10}$$

then the received signal will be in the form

$$s_r(t) = A_s \cos(2\pi f_0 t) + A_m \cos(2\pi (f_0 + f_d) t) \tag{6.11}$$

where A_s is the amplitude of the unshifted R.F. signal, which is obtained from (i) leakage from the transmitting to the receiving crystal, (ii) reflections of the unshifted ultrasonic wave from stationary tissues in the body[2]. A_m is the amplitude of the Doppler shifted R.F. signal from blood, which is typically 40dB below that of the unshifted signal.

The operation of the mixer may best be described mathematically as it is essentially a multiplication between the total received signal and the reference R.F. signal:

$$s_m(t) = s_r(t) \cdot s_0(t) \ . \tag{6.12}$$

If we apply some trigonometric identities for the multiplication of cosine functions, we can see that the resulting mixed signal will consist of components at twice the R.F. frequency plus the Doppler shift, at the Doppler shift, and a d.c. (zero frequency) component:

$$\begin{aligned} s_m(t) = & \ \{A_s \cos(2\pi f_0 t) + A_m \cos(2\pi (f_0 + f_d) t)\} \cdot A_0 \cos(2\pi f_0 t) \\ = & \ \frac{A_s A_0}{2} \{\cos(2\pi (2 f_0) t) + 1\} + \\ & \ \frac{A_m A_0}{2} \{\cos(2\pi (2 f_0 + f_d) t) + \cos(2\pi f_d t)\} \ . \end{aligned}$$

In addition, imperfections in the mixing process usually lead to some of the original R.F. signal leaking through the output of the mixer.

Because the frequency of the Doppler signal (the rightmost term on the bottom line of Equation (6.13)) is of the order of 1000's of Hertz, while the R.F. signal is of the order of millions of Hertz, we see that a low pass filter will remove those frequency components which we are not interested in. Ideally, we wish to be left only with a sinusoidal signal at the frequency of the Doppler shift.

[2] Strictly speaking, these signals will have very small Doppler shifts on them, since there will always be very small amounts of tissue-transducer relative motion.

6.3.2 REMARKS

We mention a few requirements on the hardware implementation for successful construction of a CW Doppler flowmeter:

1. The R.F. oscillator must be stable, with low drift, despite the fact that a reference signal is taken off it; the Doppler signal received from blood is usually so noisy and of such low power that R.F. drift can easily introduce extra phases throughout the circuit, and lead to a loss of calibration in the later stages of the processing.

2. The R.F. frequency in a CW application is usually selected as a tradeoff between the f_0^n attenuation dependence, and the requirements on the diffraction limited lateral resolution.

3. The transducers for transmission and reception are housed in the same transducer, but must be shielded as much as possible from each other, to reduce the amount of raw, transmitted R.F. signal that leaks through from transmitter to receiver.

4. The mixer is often a balanced, *quadrature* mixer. A quadrature mixer allows positive frequencies to be distinguished from negative frequencies. The output of a quadrature mixer is two channels of Doppler signal, and so all subsequent filtering operations and processing should be two-channel. The final audio output will be a stereo audio pair, with one channel indicating flow towards the transducer, and the other indicating flow away from the transducer.

5. Note that in addition to the low pass filter for removing the R.F. components, a high pass filter is also usually employed to remove signals below about 150 Hz, arising from the high amplitude reflections of stationary structures in the body. This is filter is sometimes known as a *wall-thump* filter.

6.3.3 LIMITATIONS OF THE CW FLOWMETER

- There is no depth information explicitly contained in the signal. Indeed, the only depth localisation arises from the geometry of the two-crystal transducer, which will only pick up flow signals from the "common volume" or "range cell" formed by the overlap of the beam sensitivity patterns of both transducers[3]. Lack of ability to finely localize the origin of the received signal leads to a higher degree of signal clutter from stationary tissue interfaces, and to ambiguity if more than one vessel crosses the range cell. If one is investigating large vessels, with a range of velocities, there is no separation of these Doppler shifts.

- Because of the low amplitude of the scattered signal from blood, the Doppler signal can be quite noisy.

[3]This needs a bit of explaining. A receiving transducer has a *sensitivity pattern*, which is a measure of how the received signal from a transmitting monopole varies as the monopole is placed at different positions in front of the transducer. This is approximately the same as the beam pattern of an identical transducer acting as a transmitter (reciprocity principle).

6.3.4 ATTRIBUTES OF THE CW FLOWMETER

- Cheapest of all Doppler instrumentation.
- Can allow an assessment of pulsatility indices of an artery by the ratio of maximum to minimum velocity computed over the cardiac cycle; also possible by qualitative assessment of the audio signal. Two simple CW probes can also be used to estimate wavespeed in the body - correlated with vessel compliance.
- Can provide an estimate of the volume of blood flow if certain assumptions are made on the blood velocity profile across the vessel in question (Evans, et al., 1989). However, it is possible for ambiguity to creep into the interpretation of these results, due, for example, to geometrical spectral broadening.

6.4 THE PULSED WAVE DOPPLER FLOWMETER

In order to enable a localization of the velocity information using the Doppler effect, the pulsed wave Doppler instrument (or simply Pulsed Doppler), borrowing concepts from radar, has been very successful. Instead of continuously emitting an R.F. wave, the system is designed to emit pulses of short duration, typically containing 10 cycles or so of R.F. signal. The pulse propagates into the medium, giving rise to reflections and backscattered waves.

After sending the pulse, the same transducer is free to act as a receiver, until it is time to transmit the next burst of ultrasound. However, while signals from reflections and scattering of the first pulse begin arriving immediately after the pulse is transmitted, the receiving circuitry only "looks" at the signal after a predetermined length of time known as the range time. This range time defines a certain depth within the region from which we receive signals. By selecting a different range time we can therefore effectively select a region of the medium (in the depth direction) to interrogate. Furthermore, the time over which we "look" at the received signal defines the axial distance over which we are interrogating the medium. This is known as the gating time, and the whole process is referred to as *range-gating*.

In principle, the steps are then:

1. Emit a pulse of acoustic energy with a well-defined central R.F. frequency.
2. Switch the transducer into receive mode. Collect the received information after a time T_R which determines the depth of the range cell, and for a length of time T_G which specifies the axial size of the region over which the information is gathered.
3. Demodulate the signal in the usual way. The only difference is that one effectively has a *sampled* version of the Doppler signal. So, interpolation by low-pass filtering is necessary.
4. Back to step 1. (Repeat)

Note that because only one transducer is used for both transmission and detection, the geometry effectively has Transducer 2 in the same position as Transducer 1 (in Figure 6.1). The Doppler Equation is thus modified to

$$f_d = -\frac{2 f_0 v \cos(\theta)}{c} . \tag{6.13}$$

6.4.1 INSTRUMENTATION

A block diagram of the system is shown below.

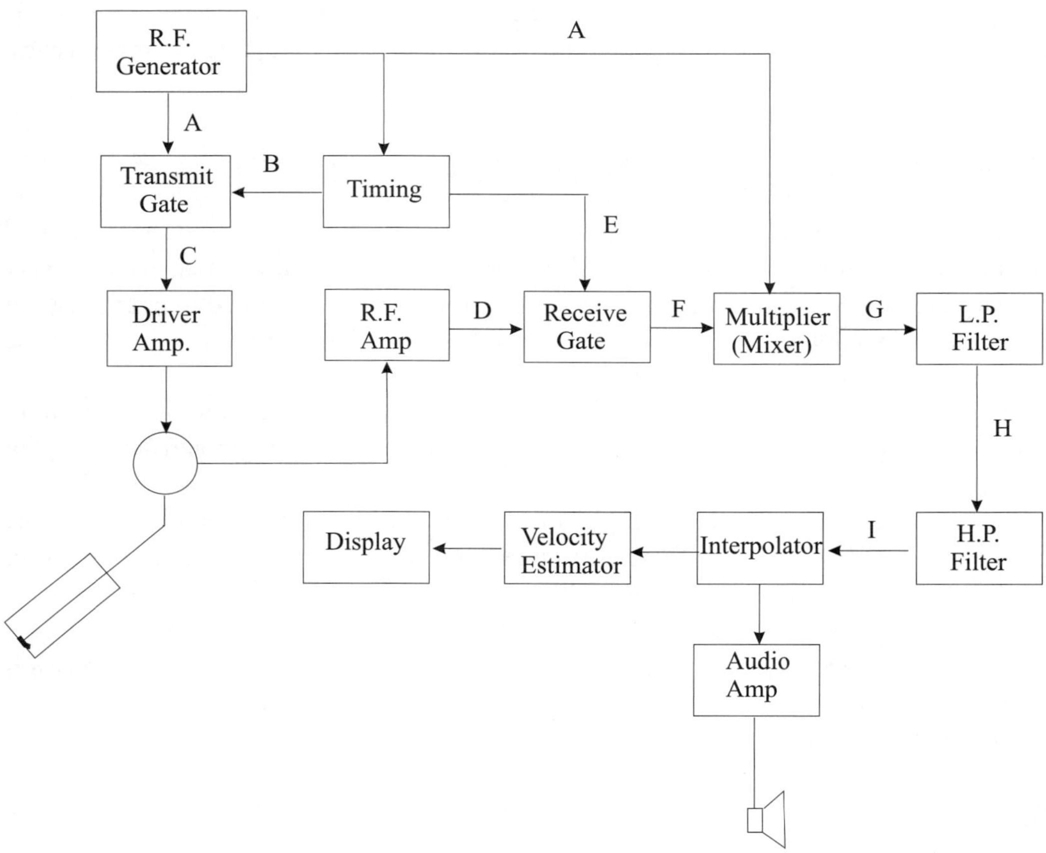

Figure 6.3: Pulsed Doppler System.

The signals at the various points in the circuit are as follows:

(A) This represents the R.F. frequency used in the system, and will be in the range 2-20MHz, depending on the desired axial resolution and on the required depth of interrogation (See notes on resolution in ultrasound).

(B) This is the signal which controls the passage of the R.F. signal through to the driving amplifier and to the transducer, by supplying an "on-off" control signal to the gate. The width of the 'on' times from the timer controls the switching of the gate, and the period between the pulses is the inverse of the *Pulse Repetition Frequency*. Typical values for the P.R.F are 2.5 or 5 kHz.

(C) The signal out of the gate, which is fed into the amplifier is the R.F. pulse train. Note that this may be viewed as a multiplication of the R.F. signal from the master oscillator with the timing signal fed to the gate.

(D) The signal which is received by the transducer acting in pulsed mode includes some left-over ringing from the excitation R.F. train. Note that this signal is very close to an A-mode signal which one sees in conventional pulse-echo imaging.

(E) This timing signal controls both the depth from which signals are collected, and the axial extent of that depth. Clearly, it must be closely coordinated with the timing producing the pulses, in order for the desired ranging effect to be achieved. The repetition frequency must be the same as that of the transmitted R.F. train, but the gate width is generally uncoupled from the transmitted gate width.

(F) This signal resembles the transmitted pulse train, in that it consists of bursts of R.F. Note, however, that it is derived from a certain depth in the medium, and is of very low amplitude. It also contains velocity information in the form of the Doppler shift.

(G) After mixing, we obtain this signal, which contains R.F. noise, low frequency components due to wall motion and the signal we want. This interpretation is extremely important.

(I) The high and low-pass filtering stages are similar in function to those of the C.W. system. Following these stages, we have at (I), a sampled Doppler signal.

(J) The interpolator is simply a low-pass filter, which reconstructs a smooth Doppler signal from the time-samples. Note that if the common volume of a two-crystal probe could extract signals within only the range-cell determined by T_G and T_R of a pulsed Doppler system, then this final interpolated signal would be identical to a CW Doppler signal. Following mixing, we are in effect trying to reconstruct the Doppler signal i.e., a sinusoidal signal at the frequency of the Doppler shift f_d) from a sampled version. Our reconstruction will be in error if the sampling process is aliased.

6.4.2 REMARKS

In terms of instrumentation, we may note the following points:

1. The R.F. oscillator should again be as stable and as free from drift as possible. The oscillator can also be used to provide a reference clock from which the various timing signals are derived. This can be achieved by passing the R.F. signal through a Schmidt trigger, thereby generating

a reference square wave at the frequency of the R.F. oscillator. The timing may then be simply obtained by frequency division of the square wave. Other strategies are now possible, based around computer technology (e.g., a square wave clock provides master timing, from which R.F. oscillations are derived by logic division and filtering).

2. The gating system must have a sufficiently fast response time to support the generation of clean R.F. pulse bursts.
3. It is also possible to use the gated transmission pulse train as the reference signal to the mixer.
4. In practise, most pulsed Doppler systems use a quadrature mixer. This is really two mixers, one fed with the original reference, and the other fed with a the reference signal after a 90 degree phase shift. As with modern CW systems, this results in two signal channels, both of which are low-pass filtered to remove R.F. and P.R.F. components, which are then passed to the frequency detection circuitry as quadrature signals. Quadrature signals allow the resolution of the principle flow vector as being either toward or away from the transducer. There are some other more subtle noise-immunity effects from quadrature signals. More on quadrature detection can be found in Physical Principles of Medical Ultrasound (1986, by Hill (Ed.)).

6.4.3 LIMITATIONS OF THE PULSED DOPPLER VELOCIMETER

- Obtaining a good signal to noise ratio. Factors which tend to reduce the S/N ratio include R.F. interference (for example, radio communication bands transmitting at or about the frequency of the R.F. frequency), electronic noise (usual, random electronic noise!). Also, if attempting to obtain velocity information from a region deep in the medium, then tissue attenuation effects can quickly reduce the power of the backscattered R.F signal.
- Aliasing. When attempting to acquire data from a depth d along the transducer axis, the leading edge of the pulse will take a time to reach the range cell given by

$$T_{forward} = d/c \ . \tag{6.14}$$

Clearly, then, the time for the "round trip" from the transducer to the range cell and back to the transducer is given by

$$T_{total} = 2d/c \ . \tag{6.15}$$

From the timing diagrams for the pulsed Doppler systems, it should be understood that after transmitting one R.F. pulse, it is not possible[4] to transmit a subsequent pulse until the echo from the first has been received. Thus, for a particular depth of interrogation, one cannot exceed a certain pulse repetition frequency, defined by the reciprocal of the total time for a return trip of the ultrasonic pulse from the transducer to the range cell:

$$PRF_{max} = \frac{1}{T_{total}} \ . \tag{6.16}$$

[4] Anything is possible, but somehow, we have to reduce the chance of depth ambiguity, and keep the processing relatively simple.

Equation (6.16) defines the maximum pulse repetition frequency which can be used for obtaining Doppler velocity information from a particular depth in the medium. This has a direct implication for the maximum frequency of the Doppler signal which can be detected at a particular depth. This frequency is given by the sampling theorem as

$$f_d^{max} = \frac{PRF_{max}}{2} .$$

6.5 ROUNDING UP

- The Doppler signal that is output by either a continuous wave or a pulsed Doppler system contains information on the frequency shift induced in the ultrasonic signal by the movement of (usually) blood. This signal, after audio amplification, also provides the characteristic auditory sound that one hears from Doppler instrumentation.
  In order to estimate the blood velocity from this signal by using the equations derived previously, one first needs to *estimate* the frequency of the demodulated Doppler signal. In most cases (in modern systems) the Doppler signal is digitised, and passed to a processing system which applies a Discrete Fourier Transform (usually using the FFT algorithm). One can easily estimate either the peak frequency shift or the mean frequency shift from the output of the DFT. One can even display the *entire* DFT result as a grey-scale trace, which provides a measure not only of the peak frequency shift (and, through the geometry of acquisition, the peak velocity), but also about the *spread* of frequencies, which can be used as a measure of either flow disturbance, or simply a sharp velocity profile.

- As mentioned earlier, the pulsed Doppler system forms the heart of Colour Flow Mapping systems. Older colour flow mapping systems use hardware autocorrelators to extract the mean frequency shift. The speed of current DSP chips means that modern systems can employ FFT algorithms to estimate the frequency shift at each point in the field-of-view for colour-flow mapping.

- One can increase the Doppler signal from blood by using a microbubble contrast agent. The increased backscatter from blood leads to an enhanced Doppler signal from blood, and harmonics produced by the bubbles can also give very interesting information. Whilst micro-bubble contrast agents may be all the rage in the current (medical) scientific literature, remember that this is so because it is an exciting *current* research area: this does not mean that all current Doppler systems are micro-bubble based, nor are they likely to be in the future, since they involve injecting this contrast agent. They will have niche applications.

- Another type of measurement technique is known as *Power Doppler*. As the velocity of blood passing through a range cell of a Doppler instrument increases, the effective number of erythrocytes passing through that range cell also increases. Because of this, there is an increased backscatter from the range cell, *in addition to* the change in Doppler shift frequency. By comparing the level of the backscattered signal over time, one can produce a Power Doppler image. This is good for visualising small blood vessels, or flow in vessels where the Doppler shift might be very small. It does not provide very high-quality information about velocity of flow, because the amount of acoustic signal attenuation will be unpredictable in any given situation; think of it as a comparative imaging technique.

EXAMPLE NUMERICAL WORKING

A Pulsed Doppler ultrasonic probe insonates an artery at an angle of 45 degrees. The range is set to interrogate a region of the flow field in which the average velocity (over time) is 0.15 m/s, and the flow is directed toward the transducer, parallel to the artery. Neglect any local velocity gradients in the range cell. The frequency of the emitted R.F. pulses is 6MHz, and the pulse repetition frequency is 5kHz. The depth of the range cell measured from the transducer face is 6cm.

1. What is the average value of the Doppler shift (in Hz)?

2. During the cardiac cycle of the patient being examined, we note that the maximum frequency of the Doppler shift which we record is 1.5 kHz. What is the absolute value of velocity giving rise to this peak frequency?

3. What is the maximum velocity which we can detect using the parameters specified above, assuming that the PRF is maintained at 5kHz?

4. If we use a Doppler angle of 30 degrees, what is the maximum velocity which we can unambiguously detect?

5. Using a Doppler angle of 60 degrees, we try now to interrogate a jet with a peak velocity of 60cm/s which is 15cm from the transducer face. What happens? Consider the same interrogation using an angle of 30 degrees. Can we, by altering the parameters, obtain a reliable velocity estimate at 30 degrees?

SOLUTION

1. Use (ignoring the sign)

$$f_d = \frac{2 f_0 v \cos(\theta)}{c} \tag{6.17}$$

which gives

$$f_d = \frac{2 \cdot 6000000 \cdot 0.15 \cdot \cos(45°)}{1540}$$

leading to

$$f_d = 826Hz\ .$$

2. Use

$$v = \frac{c \cdot f_d}{2 \cdot f_0 \cos(\theta)} \tag{6.18}$$

to give

$$v_{peak} = 0.27ms^{-1} .$$

3. Because of the requirement to meet the Nyquist rate, the maximum frequency of the Doppler shift which can be unambiguously detected is 5 kHz/2 = 2.5 kHz. Using Equation (6.18), the velocity, v_{max}, corresponding to this shift is 0.45 m/s.

4. As above, using $\theta = 30$ deg, which gives $v_{max} = 0.37ms^{-1}$.

5. Use Equation (6.17) as before, to get f_d= 2.337 kHz. The P.R.F. is 5 kHz, so this figure is just below the Nyquist frequency. If $\theta = 30$ deg, then the Doppler shift will be 4 kHz. Therefore, the pulsed Doppler signal will be aliased, and it will not be possible to unambiguously estimate the velocity. If we were to change the P.R.F. to 8 kHz, we would just be able to correctly sample the Doppler signal. However, the depth of the interrogation is required to be 15cm, so the minimum time for a "round trip" from the transducer to the range cell is given by

$$T_{total} = \frac{2 \cdot 0.15}{1540}$$

which is 194 μs. Therefore, the maximum P.R.F. which we can use is given by

$$PRF_{max} = \frac{1}{T_{total}}$$

or 5.133 kHz. So, we cannot obtain reliable velocity estimates under these conditions, using a Doppler angle of 30 deg.

CHAPTER 7

An Introduction to MRI

7.1 INTRODUCTION

These notes are intended to be used as background reading material; they are intended to constitute a more "gentle" set of MR imaging notes that focus on basic physics, and I hope that they are helpful. They have been produced as a result of comments from many years' worth of students about the poor state of many MR text books (exception is outlined below); one of the aims in this set of notes is to be as consistent and concise as possible in presentation, and to explain some of the core aspects of MR as clearly as possible to engineering students. In doing so, I am undoubtedly going to

- not cover the more modern aspects of MRI instrumentation
- offend the purists by ignoring the many subtleties of the MR phenomenon.

For this I apologise[1].

7.2 BOOKS AND SUGGESTED READING

1. *MRI: From Picture to proton*, by Donald McRobbie, Elizabeth A. Moore, Martin J. Graves, and Martin R. Prince, Cambridge University Press, Second Edition, 2007. Undoubtedly, an excellent book, far and away the best of its kind in MR.

2. *Magnetic Resonance in Medicine: The Basic Textbook of the European Magnetic Resonance Forum*, By Peter A. Rinck, Atle Bjørnerud Published by Blackwell Publishing, 2001 ISBN 0632059869, 9780632059867 245 pages A recognised standard among many MR physicists.

3. *The Physics of Medical Imaging* Ed. S. Webb. IDP 1993 A more general book - good introduction to MR and many other modalities.

4. *MRI: the Basics* If you can stand this riot of diagrams and repetitive text, be my guest. Excellent coverage of some of the more specialised and modern pulse sequences. And, yes, some of the diagrams *are* rather good....

5. *The Physics of Diagnostic Imaging* This has good instrumentation diagrams. Collapses MR physics into very few pages.

[1]For the latter of these "crimes," only grudgingly.

7.3 BASIC PRINCIPLES

Magnetic resonance imaging has seen enormous development over the last 25 years. MRI is noninvasive and does not use ionizing radiation. Whilst in X-Ray CT, the contrast is primarily determined by the atomic number (for equal mass densities) of the constituents of tissue, MRI offers a huge variety of tissue parameters as measures that then become contrast mechanisms, and the expert user has control over these parameters. In the most common form of clinical usage, MR measurements are centred around the hydrogen nucleus, H^+. However, NMR spectroscopy allows details of many other molecular structures.

7.3.1 A BRIEF HISTORY

- 1935: I.I. Rabi uses the magnetic resonance phenomenon to measure spin quantum numbers in the nuclei of particle beams. Wins the 1944 Nobel Prize.
- 1946 Measurement by Bloch *et al.* & Purcell *et al.* in "condensed" (solid) matter. Wins 1952 Nobel Prize.
- 1950 E.L. Hahn proposes the use of the spin-echo sequence for T_2 determination.
- 1954 Carr & Purcell. Spin-echo train.
- 1960's Development of computerized Fourier Transform methods.
- Early 1970's, Damadian suggests that magnetic resonance can be used to image the body, and that MR can be used to visualise tumours.
- 1970s Development of high-field superconducting magnets.
- 1973 Lauterbur, Mansfield & Grannell. Utilization of resonant frequency shift by magnetic field gradient.
- 1977 Damadian *et al.* First whole body image.

7.3.2 MOTION WITHIN THE ATOM

The classical interpretation of the quantum mechanical property of a particle's spin leads to a number of analogies of "spinning" motions present in an atom, amongst which are:

1. Electrons spinning on their own axis;
2. Electrons orbiting the nucleus;
3. The nucleus itself spinning about its own axis.

MRI is based on the third type of motion; the spinning motion of a specific kind of nuclei present in biological tissues. These are called MR active nuclei. MR active nuclei are characterised by their tendency to align their axis of rotation with an applied magnetic field. The physical property responsible for this alignment is a coupling between the magnetic moment of the charged, spinning nucleus and the applied magnetic field. As neutrons are uncharged, they have no magnetic moment. Only certain nuclei have a total magnetic moment, and can interact with an external applied magnetic field. Such a nucleus has a dipole moment **m**, which is given by

$$\mathbf{m} = \gamma_p^r \mathbf{I} \tag{7.1}$$

where **I** is the angular momentum vector (aligned with the spin axis), and γ_p^r is a nucleus-dependent constant, given approximately by $\gamma_p^r = ge/2m$, where e is the charge and m is the mass of the nucleus. g is a "fudge-factor:" for different fundamental particles, it varies. For a nucleus, it is found experimentally to be around 5.56. The mass of a nucleus is provided by the protons and neutrons in it. For a hydrogen atom, the nucleus contains only a proton, with mass 1.672×10^{-27} kg. So, the gyromagnetic ratio, γ_p^r, for a hydrogen nucleus is

$$\begin{aligned} \gamma_p^r &= 5.56 \times \frac{1.609 \times 10^{-19}}{2 \times 1.672 \times 10^{-27}} \\ &= 2.68 \times 10^8 \text{radians s}^{-1}\text{T}^{-1} \end{aligned} \tag{7.2}$$

where the r superscript in γ_p^r denotes that this measurement is a radial frequency measurement: it contains a scaling by 2π. You will also find that there is another, more commonly used value for this ratio, which we will denote γ_p, given by

$$\gamma_p = \frac{\gamma_p^r}{2\pi} \tag{7.3}$$

and which has units of Hz T^{-1}. For a hydrogen nucleus, this is about 42.57×10^6 Hz T^{-1}.

The strength of the magnetic dipole moment is clearly specific to every nucleus and determines its sensitivity to magnetic resonance. Important examples of MR active nuclei, together with their atomic mass numbers, are listed below.

- Hydrogen 1
- Carbon-13 13
- Nitrogen-15 15
- Oxygen-17 17
- Fluorine 19

- Sodium 23
- Phosphorous 31

The number after each element is its atomic mass number, which is equal to the number of protons in their nuclei. You will notice that many of these have odd atomic mass numbers: there is a strong (but imperfect) correlation between the atomic mass numbers (or just mass number) and the quantum property of *nuclear magnetic moment*. Whilst these are all *MR active atoms*, the *hydrogen* nucleus is the primary MR active nucleus used in medical MRI, the hydrogen nucleus contains a single proton (hence, atomic mass number of 1). It is used because it is very abundant in the human body, and because its solitary proton gives it quite a large magnetic moment. When we refer to protons in the context of imaging, we are referring to hydrogen nuclei.

7.3.3 THE BARE NECESSITIES OF THE QM DESCRIPTION

For more on basic quantum mechanics, you can consult texts such as *The Feynman Lectures on Physics* or *Lectures on the Electrical Properties of Materials.*

The quantum mechanical properties of nuclei are related to the classical concept of angular momentum by the relationship

$$|\mathbf{I}| = \hbar[j(j+1)]^{1/2} \tag{7.4}$$

where j is the nuclear spin quantum number, and is a function of the atomic species. The possible components of angular momentum along *any* specified direction[2] are given by the quantity m_l, known as the magnetic quantum number. The possible values of m_l are quite easily determined by

$$m_l = j, j-1, j-2, \dots -j\ . \tag{7.5}$$

For example, for a sodium nucleus, ^{23}Na, the value of j is $\frac{3}{2}$, and so

$$m_l = \frac{3}{2}, \frac{1}{2}, -\frac{1}{2}, -\frac{3}{2} \tag{7.6}$$

i.e., 4 possible spin states[3]. For a hydrogen nucleus, j has the value of $\frac{1}{2}$, leading to

$$m_l = \frac{1}{2}, -\frac{1}{2}\ . \tag{7.7}$$

In the presence of an applied magnetic field $\mathbf{B_0}$, these correspond to the spin up and spin down states. Each of these states has an associated energy which is given by

$$E = -\gamma_p^r \hbar m_l |\mathbf{B_0}| \tag{7.8}$$

[2]Unfortunately, this is very confusing: in quantum mechanics, if one knows the angular momentum of a nucleus along one direction, it imposes uncertainty into the knowledge of angular momentum in other directions. Hence, the oddness of this sentence.

[3]A researcher, Cohen, working with I.I Rabi at Columbia in the 1930's was the first person to measure the spin quantum number for sodium. He recalls travelling early one morning on the New York subway, just after obtaining his experimental result, thinking to himself "I am the only person in the world who knows that the spin quantum number for sodium is 3/2!"

where $\hbar$ is simply $h/2\pi$. The difference in energy between the up and down spin states is given by

$$\Delta E = -\gamma_p^r \hbar |\mathbf{B_0}| \tag{7.9}$$

where the negative sign indicates that the spin down state *opposes* the direction of $\mathbf{B_0}$, and spin up is in the direction of $\mathbf{B_0}$. This equation therefore expresses the energy of spin up relative to spin down. If one has a nucleus in the spin up state (lower energy state), one can flip it into the down state by applying the precise amount of energy ΔE. How can this be done? Consider the absorption of a photon of electromagnetic radiation by a nucleus in the down state. The energy of this photon must be ΔE. Recall the relationship between the frequency of EM radiation and the energy of each photon. We want

$$\begin{aligned} \Delta E &= h f_0 \\ &= \hbar \omega_0 \, . \end{aligned} \tag{7.10}$$

Equations (7.9) and (7.10) leads to the Larmor Equation[4]

$$\omega_0 = -\gamma_p^r |B_0| \, . \tag{7.11}$$

The number of nuclei which flip between low and high energy levels clearly depends on the number of photons injected at the right energy.

Life gets much more complicated, however. The probability functions, used in the Schrödinger wave Equation to calculate the energy states of individual nuclei can interfere. In everyday use, the quantum mechanical description turns out to rather involved to apply easily. There is a way of simplifying matters: if one looks at an *ensemble* of nuclei (that is, observes the behaviour of many of wave functions which are all interfering), one can study the behaviour not of single nuclei, but the averaged effect of many. This leads to the idea of the Net Magnetisation Vector (to be introduced below), which is based on classical physics. It is sometimes common to "cheat," and to apply the classical ideas of precessing nuclei at the level of individual nuclei: we can get away with this in most cases, and this approach has indeed been used in the development of MR physics. For example, Purcell pursued MR measurement from the quantum mechanical viewpoint, whilst Bloch approached it from the classical viewpoint of inductive magnetism. Both were awarded the Nobel prize for the same experimental results[5].

[4]Unfortunately, many texts on MR confuse angular frequency with frequency in Hz. We have solved this problem by making a distinction between γ_p and γ_p^r. You should find all equations in these notes self-consistent. If not, do e-mail.

[5]Purcell received a telegram on the day of the award reading:

> I think it's swell for Ed Purcell
> To share the shock with Felix Bloch.

History does not seem to record the name of the poet.

7.3.4 CLASSICAL DESCRIPTION

A complete explanation of the MR phenomenon needs a quantum mechanical description. However, the QM description needs ridiculously complicated mathematical treatments (the explanation of a transverse component of the Net Magnetisation Vector, for example) for some phenomena. For this reason, the most often used description of MR physics is based on a classical description. Whilst this is unable to completely explain the subtleties of the physics of single nuclei, it is usually sufficient for a population (ensemble) of nuclei spins. We shall only need to resort to the QM description in special circumstances.

In the absence of an applied magnetic field, the magnetic moments of the hydrogen nuclei are randomly oriented. When placed in a strong static external magnetic field, denoted by $\mathbf{B}_0$, however, the magnetic moments of the hydrogen nuclei align themselves with this magnetic field[6]. This alignment is produced as a result of the coupling between the magnetic dipole moment of the spinning nucleus, and an applied external magnetic field. This is expressed in terms of the couple, $\mathbf{C}$, which is a description of the torque exerted on the dipole moment of the nucleus by the presence of $\mathbf{B}_0$.

$$\begin{aligned}\mathbf{C} &= \mathbf{m} \times \mathbf{B}_0 \\ &= \frac{d\mathbf{I}}{dt} .\end{aligned} \tag{7.12}$$

Incorporating the expression for $\mathbf{I}$ yields

$$\frac{d\mathbf{m}}{dt} = \gamma_p^r \mathbf{m} \times \mathbf{B}_0 . \tag{7.13}$$

Equation (7.13) is a differential equation which leads directly to the vector form for the Larmor equation, which indicates the direction of precession relative to the $\mathbf{B}_0$ field:

$$\boldsymbol{\omega}_0 = -\gamma_p^r \mathbf{B}_0 . \tag{7.14}$$

Compare this with the form obtained in Equation (7.11). In terms of classical mechanics, this describes a circular "wobbling" of the nucleus i.e., the magnetic dipole) about the $\mathbf{B}_0$ direction, which is known as *precession*. You can take magnitudes on both sides of this equation, so that, for example, at $|\mathbf{B}_0|$ =1.5 Tesla, $f_0 = 63.855$ MHz $= \omega_0/2\pi$ for a hydrogen nucleus (proton).

7.3.5 ORIENTATION

The vector representation of the dipole moment can be written

$$\mathbf{m} = m_x\mathbf{x} + m_y\mathbf{y} + m_z\mathbf{z} . \tag{7.15}$$

The $\mathbf{B}_0$ direction is usually defined as the z direction in MR imaging. The precessional motion of the magnetic dipole of a single nucleus does not affect the z component of the dipole, m_z, since

[6]The units of $\mathbf{B}_0$ are Tesla, where 1 Tesla =10,000 Gauss.

the axis of precession is in the z direction. However, it turns out that some of the nuclei (those that have enough energy) have a *negative* component m_z. Indeed, when the magnetic field is first applied, there is an approximately equal distribution of parallel and antiparallel alignments of the m_z components of the nuclei. Over the space of several milliseconds, a slight excess of spins with a $+ve$ z component arises. The factors affecting which hydrogen nuclei finally align parallel and which align anti-parallel are determined by the strength of the external magnetic field and the thermal energy of the nuclei. Low thermal energy nuclei do not possess enough energy to oppose the magnetic field in the anti-parallel direction. High thermal energy nuclei, however, do possess enough energy to oppose the field, and as the strength of the magnetic field increases, fewer nuclei have enough energy to do so. The thermal energy of the nucleus is determined by the temperature of the material. The relative distribution of the two directions is given by a Boltzmann distribution

$$\frac{\#\text{ spin up}}{\#\text{ spin down}} = \exp\left(\frac{\Delta E}{kT_s}\right) \tag{7.16}$$

where k is the Boltzman's constant, T_s is the absolute value of the material, and ΔE is the difference in energies between the spin up and spin down states. This energy difference is determined by quantum mechanics in Equation (7.10).

In most situations, this slight excess of "spin up" protons is sufficient to consider the effects of the *net magnetisation vector*, **M**,

$$\mathbf{M} = \sum \mathbf{m} \tag{7.17}$$

i.e., in terms of a summation of all the individual excess spin-up nuclei dipole moments in the sample. **M** is of more practical interest to us, because in principle it represents a macroscopic quantity that we can measure and interact with.

7.3.6 THE NET MAGNETISATION VECTOR

We can consider the components of the net magnetisation vector arising from the precessing spins. Let

$$\mathbf{M} = M_x\mathbf{x} + M_y\mathbf{y} + M_z\mathbf{z} \ . \tag{7.18}$$

The small excess of spins that exists in the direction of the applied magnetic field means that M_z is nonzero. In addition, although the m_x and m_y components represent a circular motion, the fact that all of the nuclei in the material are precessing out of phase with each other means that there is cancellation of the m_x and m_y components over the excess spin-up nuclei in a volume of material placed in a magnetic field; thus $M_x = M_y = 0$.

7.3.7 INTERACTING WITH M

At equilibrium, we have only M_z nonzero. M_z is, however, unchanging. This is not very helpful, as there is nothing to be measured or interact with! We need to get this net magnetisation vector do display some kind of motion which will produce a usable signal. To ensure that we can detect a signal, we want the vector **M** to wriggle, so that it yields a detectable electromagnetic signal. It would also be most helpful if this wriggle was in a direction *normal* to that of the (very strong) static magnetic field, $\mathbf{B_0}$. How do we do this? It would clearly help if we could get the precessing spin-up nuclei into phase.

If we apply an oscillating field, $\mathbf{B_1}$, in the xy plane at the right frequency, we can achieve two things. First, the application of a regular "driving" excitation to the precessing nuclei *at the right frequency* will cause them to spin *in phase*. This will yield a *nonzero* component of $M_{xy} = \sqrt{M_x^2 + M_y^2}$. Secondly, the application of this oscillating $\mathbf{B}_1$ field also implies an injection of energy: a greater proportion of nuclei will be able to flip their m_z components into the antiparallel direction of alignment. This will cause a *decrease* in the value of M_z for the net magnetisation vector. Although the latter of these two effects is best describable in quantum mechanical terms, the net result can be interpreted in classical mechanical terms, through the motion of the **M** vector. The application of the rotating magnetic field is achieved through the generation of a circularly polarised radio-frequency (R.F.) excitation, usually referred to as the *RF pulse*. This pulse *must* be at (or, if it is a wideband pulse, it must have a component at) the Larmor frequency in order for the interaction to occur.

7.3.8 THE MOTION OF M

On the application of this RF pulse (which is often represented by the $\mathbf{B_1}$ field), **M** will experience a torque

$$\mathbf{T} = \mathbf{M} \times \mathbf{B_1} . \tag{7.19}$$

Thus, **M** rotates[7] into the xy plane with angular velocity

$$\omega_1 = -\gamma_p^r \mathbf{B}_1 . \tag{7.20}$$

The angle through which **M** rotates away from the z axis is therefore dependent upon the magnitude and duration of $\mathbf{B_1}$. If the total energy of the supplied pulse (dependent on duration and amplitude) is sufficiently large, then M_z becomes zero, and we are left only with an M_{xy} component in the NMV. This is termed a 90° pulse. If the energy of the RF pulse is sufficient to *reverse* the direction of the NMV, then this is called a 180° pulse.

7.4 RELAXATION PROCESSES

Recall that the application of the $\mathbf{B_1}$ field tilts the net magnetization vector away from the z direction. Thus, M_z decreases from the equilibrium value, M_0. On stopping the application of the

[7] In fact, it is more complex than this: the net magnetisation vector spirals towards the xy plane, due to the combined precessional motions around $\mathbf{B_0}$ and the oscillating $\mathbf{B_1}$. If we express this spiral motion in a frame of reference which is rotating at the Larmor frequency, then the spiral motion is "unseen" and the only motion is the tilting of **M** away from the z direction towards the xy plane. Expressing things in the rotating frame therefore makes the description of the effect of $\mathbf{B_1}$ much easier.

$\mathbf{B_1}$ field, there are two immediate effects. The first is that the "driving" oscillation that keeps all the spin precessions in phase is no longer present. The spins begin to dephase due to two main factors: spin-spin interactions and local inhomogeneities in the $\mathbf{B_0}$ field. The dephasing effect means that the xy component of the net magnetisation vector, described as the *transverse* magnetisation, falls in a manner that is well described by a decaying exponential.

The second effect is that the z component of the magnetisation vector, known as longitudinal magnetisation, begins to grow. This process is well described by a *growing* exponential.

These relaxation processes are simply described by the *Bloch Equations*.

7.4.1 THE BLOCH EQUATIONS

The Bloch equations describe the relaxation processes of the NMV of a sample of nuclei which have limited interactions. In a static field of strength $\mathbf{B_0}$, the z component, of the NMV $\mathbf{M}$ returns to the equilibrium value, M_0 by

$$\frac{dM_z}{dt} = -\frac{M_z - M_0}{T_1} \tag{7.21}$$

where the constant T_1 is known as the longitudinal relaxation time.
Any component of $\mathbf{M}$ that is induced in the xy plane decays after the application of the RF pulse by the expressions

$$\frac{M_x}{dt} = -\frac{M_x}{T_2} \tag{7.22}$$

$$\frac{M_y}{dt} = -\frac{M_y}{T_2} \,. \tag{7.23}$$

The equations given above describe the NMV components in the rotating frame; in the laboratory frame of reference, one must add the precessional angular motion components.

7.4.2 SIGNIFICANCE OF T_1 AND T_2

The longitudinal magnetisation relaxation process, described by the exponential time-constant T_1, is also known as *spin-lattice* relaxation. It is a measure of the transfer of energy from the nuclear spin system to a surrounding lattice. In the absence of paramagnetic nuclei (which have unpaired electrons in their orbits), the main effect is due to the magnetic fields generated by the dipole moments of water molecules; some of these will have xy components that can oscillate at the same frequency as the precessional frequency of the primary spin system we are interested in. They will therefore be able to absorb (or introduce) energy to the spin system, and will therefore affect how quickly energy is transferred from the spin system to the molecular lattice. As a result of this, T_1 values tend to provide a measure of both the water content of tissue, and the degree to which water is bound or absorbed onto surfaces of the tissue. When quoted as a time, it is the time it takes 63 of the longitudinal magnetisation to recover in the tissue. Since the recovery is exponential, one has (if starting from equilibrium)

$$M_z \propto N(H)(1 - e^{-t/T_1}) \tag{7.24}$$

where $N(H)$ is the number of hydrogen atoms, related to what is sometimes described as *proton density*.

T_2 is a complex beast. It is a measure of the rate of decay of the xy component of the magnetisation vector. This happens partly due to loss of the spin-lattice relaxation, which is a relatively slow process described by T_1, but also due to the dephasing of the spins once the RF pulse ($\mathbf{B_1}$) field is turned off. However, the dephasing effect is much more rapid, and this therefore dominates the transverse decay. There is, however, another effect that makes things even more complex, which we shall discuss shortly. For the time being, let us concentrate on what the T_2 relaxation time constant tells us.

The dephasing of spins is caused by the exchange of energy between nuclear spins. This is what leads to the term *spin-spin* relaxation, describing the energy exchange between neighbouring nuclei, leading to a loss of phase. Other effects include dephasing due to very localised precessional fluctuations, governed in turn by the relationship between the nuclei and the local molecular magnetic field. In imaging the body, T_2 dephasing is a highly tissue-dependent process. In most tissue samples, the T_2 time is much faster than the T_1 time; however, if one has some material containing small, rapidly tumbling molecules in a dilute solution, the T_2 time can be of the same order as the T_1 time.

7.4.3 T_2^* VS T_2

The other dephasing effect that occurs in the xy component of the magnetisation vector is due to inhomogeneities in $\mathbf{B_0}$, which can never be perfectly uniform in any given volume element of tissue. This fluctuation in the static magnetic field leads to differences in the precessional frequencies of nuclei in a volume element. This means that in the absence of an applied driving frequency, they will quickly dephase (since they will spin at different rates). The problem here is that *this* dephasing effect is not tissue dependent, but *system* dependent, and of no diagnostic use. For this reason, the exponential decay of the M_{xy} component in response to a *single* RF pulse excitation set to tilt the magnetisation vector into the xy plane is described as being governed by the time constant T_2^*.

7.4.4 SUMMARY OF RELAXATION

So, we have two tissue-dependent relaxation processes that give us different pieces of information. In practice, we use *mainly* three measures to characterise tissue: the proton density ($N(H)$), the longitudinal relaxation time, T_1, and the transverse relaxation time, T_2. In fact, there are numerous practical problems with measuring these. First, the growth of M_z cannot be measured directly; this means that estimating T_1 and $N(H)$ is not entirely straightforward. Secondly, there is the problem that the FID gives us a measure of T_2^* and *not* T_2.

In the 1940's and 1950's, these problems were overcome by the design of *sequences* of radio-frequency pulses i.e., several applications of the $\mathbf{B_1}$ field) in order to solve both of these problems.

7.5 BASIC SEQUENCES

7.5.1 FREE INDUCTION DECAY

Relative to the laboratory frame of reference, the Bloch equations in the M_x and M_y components of the NMV describe an exponentially decaying electromagnetic signal. This is at the same frequency as the precessional motion i.e., the Larmor frequency). This free induction decay signal (FID) is the "raw" signal of MR; it is detected by receiver coils[8]. Before it is used, however, the FID signal it is first multiplied by a quadrature reference signal to bring it into a much lower frequency range. One channel of the resulting complex envelope looks as shown in (idealised form) in the middle trace of Figure 7.1.

There is a fundamental issue here: we have the free-induction decay signal which is measure-able, and which (ideally) can provide us with a measure of the transverse relaxation rate in a sample of homogenous material. How can we measure the *longitudinal* relaxation?

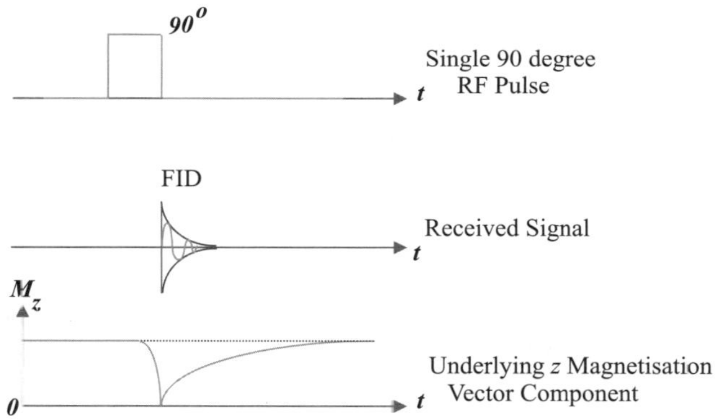

Figure 7.1: Free Induction Decay Signal from a single 90° RF pulse.

7.5.2 PARTIAL SATURATION

If we (could) monitor M_z, we would see a return of magnetisation following a 90° RF pulse. However, we need to go back to the quantum mechanical description of the spin-states in order

[8]The rotating NMV acts as a moving magnet: a coil placed nearby (or surrounding) the volume of tissue in which the NMV is moving will induce a current in the coil.

to see how we could find a way of *indirectly* measuring the T_1 relaxation time. Remember that at equilibrium, there is an excess of spins in the spin-up state. This corresponds to a lower energy state than spin-down. When the RF energy is injected into the system, the proportion of spins in the down state increases. This reduces the number of *excess* spins in the spin up-state. Once the RF field is no longer applied, the photons are re-emitted as a signal, or given up to the surrounding lattice. The number of spins in the up state therefore gradually grows until equilibrium is again reached. Since the NMV strength is directly related to this excess of spin up states this is what is responsible for the growth of the M_z component.

If a *second* 90° RF pulse is applied, then turned off, a second FID signal will be received. The key point, though, is that if the time between the first and second RF pulses is shorter than the recovery time to equilibrium, the second FID signal will be *smaller* in amplitude than the first. For a given time between pulses, the difference in maximum amplitudes between the first and second pulses is dependent on the T_1 time: for a material that has a short T_1 time, the M_z component will have time to return to the saturation value; the second FID will therefore be of similar maximum amplitude to the first FID signal. For a material with a T_1 time much longer than the repetition time, the second FID signal will have a lower amplitude than the first FID signal. Similarly, one can deduce the value of T_1 for a simple homogenous material by applying several pairs of pulses, with different times in between them. The time between successive 90° RF pulses is known as the repetition time or T_R time. If T_R is used as a parameter in a series of FID experiments, one can write

$$f_P(T_R) = M_z(0)\,(1 - \exp(-T_R/T_1)) \tag{7.25}$$

where f_p denotes the peak FID signal level.

7.5.3 SATURATION RECOVERY

If we allow equilibrium to be reached before applying the second RF pulse, then it is clear that, for a given sample, there will be no difference between the FID amplitudes for successive pulses. The amplitude of the FID signal is then dependent on the $M_z(0)$ value: this is dependent on the number nuclei at the Larmor frequency in the sample. Since, for medical applications, we use hydrogen atoms as the MR active nuclei, the maximum FID signal amplitude is proportional to the *density* of protons in the sample.

In practice, there is a controllable delay in time from the application of the RF pulse before we begin to measure the FID. This delay is known as the time to echo, T_E. It is one of the parameters under control of the user (like T_R). For Proton Density (PD) estimation using the saturation recovery sequence, T_E is minimal.

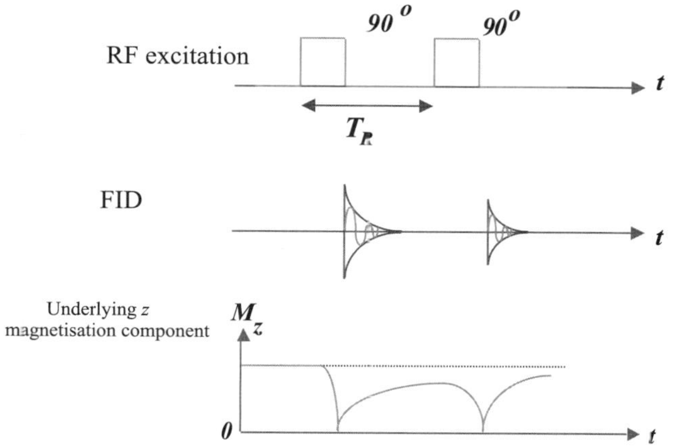

Figure 7.2: One sequence for measuring T_1 is represented by $90° - T_R - 90°$. T_R specifies the delay time between successive 90° RF excitations. Signal intensity is now dependent on the magnitude of the M_z recovery. If we have a long T_1, then recovery is slow, and we can see this with decreased FID amplitude.

7.5.4 INVERSION RECOVERY SEQUENCE

In the *inversion recovery sequence*, a 180° pulse RF pulse is first applied. This flips the NMV so that it lies along the $-ve$ z axis. No measureable signal is produced at this time. Recovery of the M_z vector then proceeds at the usual T_1 recovery rate. The M_z magnetisation vector component is allowed to pass through zero, then to become positive, before a 90° pulse is applied, producing a measureable M_{xy} component. The magnitude of the resulting FID signal clearly depends on the time lag between the 180° and the 90° pulse. We shall consider inversion recovery principle in more detail later on: for the moment, we will just point out that since the M_z component passes through zero, and the time at which it passes through zero is dependent on the T_1 time for a given material, the FID signal for certain types of material (tissue) can be suppressed by applying the 90° at the time of this zero crossing. In samples containing multiple tissues, where one knows what one tissue is, but is unsure about the other, the FID signal can therefore be suppressed for the known tissue.

7.5.5 THE SPIN ECHO SEQUENCE

After a 90° RF pulse, a single FID signal has an exponential envelope on it. The decay of this signal represents a loss of magnetisation from the xy plane due to

- Loss of net signal (T_1 recovery - usually long).

- Loss of phase coherence due to spin-spin interactions (very interesting)with a decay constant of T_2.

- Very slight inhomogeneties in $\mathbf{B_0}$ - This causes slight changes in ω_0 over the volume in which we have the ensemble - leads to a decay constant of T_2^*. This is system dependent and a nuisance.

If we were to spin around at the Larmor frequency with the precessing nuclei, we might observe the pattern of dephasing shown in Figure 7.3. We know how to get at T_1 values, but how do we get at T_2

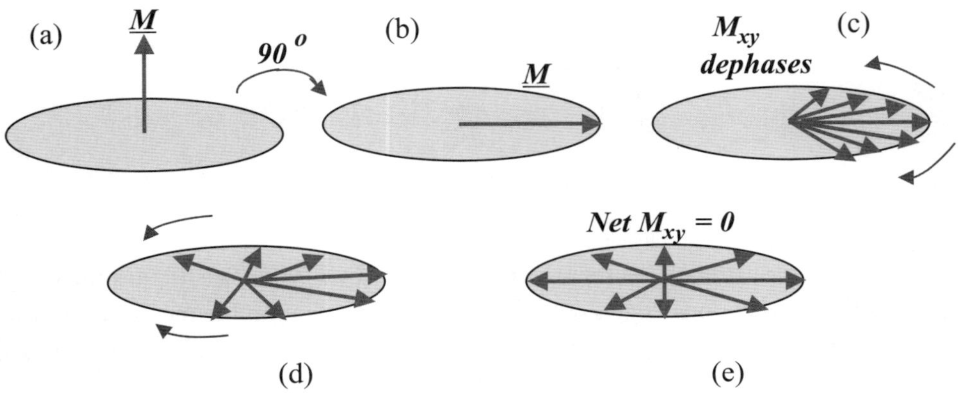

Figure 7.3: After the application of the 90° pulse, the spins are initially all in phase and the moments lie in the xy plane. Once the pulse is off, the spins begin to dephase, because they will have slightly different Larmor frequencies. This dephasing leads to a loss of signal in the xy plane. The time between (b) where the FID magnitude is strongest and (e), where it has decayed to zero, is characterised by T_2^*.

measurements if the FID decay itself has a T_2^* decay? Here is what happened to Hahn, celebrated discoverer of the spin-echo sequence:

> I decided one day – this was during the summer of 1949 – to increase the intensity of the pulses and shorten them to squeeze in more points. After seeing the normal signals on top of the pedestals in sequence, a strange, symmetric signal appeared on the screen with no pedestal under it. I discounted the latter as a false signal from the multivibrator, kicked the apparatus, let go a few obscenities, and was happy to see that the signal finally disappeared. In retrospect, I tremble at the thought that by this particular "ignoration," I could have missed completely the discovery of free precession and spin echoes.

Hahn's discovery of the spin echo sequence allowed the measurement of true T_2 (rather than T_2^*) from what is now referred to as an *echo* signal. How does this work?

Imagine that a group of runners starts off along a straight running track from the same starting line at time $t = 0$. Due to differences in running speeds, the runners would disperse as they ran along the track: the faster runners would move a further distance, the slower a smaller distance in a time interval, T. Now, imagine that at $t = T$, the runners are stopped in their tracks, made to turn around, and run back towards the starting line. At time $t = 2T$, all the runners would be back at the starting line, and approximately together again.

This is pretty much how the spin echo signal works. After applying a 90° pulse, we wait for a period of time given by $T_E/2$. Due to the different precession rates arising from the field inhomogeneities, the spins in a sample volume will begin to go out of phase. A 180° pulse is then applied. This flips all the spins by 180° degrees; the system dependent variation of precession speed resulting in spin dephasing continues, but *in the opposite direction*. This leads to a *rephasing* of the spins in the sample. The maximum point of rephasing is at time T_E. If another 180° pulse is applied at time $3T_E/2$, another echo signal is obtained. If successive 180° pulses are applied, the echo signal is slightly weaker each time: the rate at which the successive echo signals decay in maximum amplitude is determined by the true T_2 time. It's fiendishly clever.

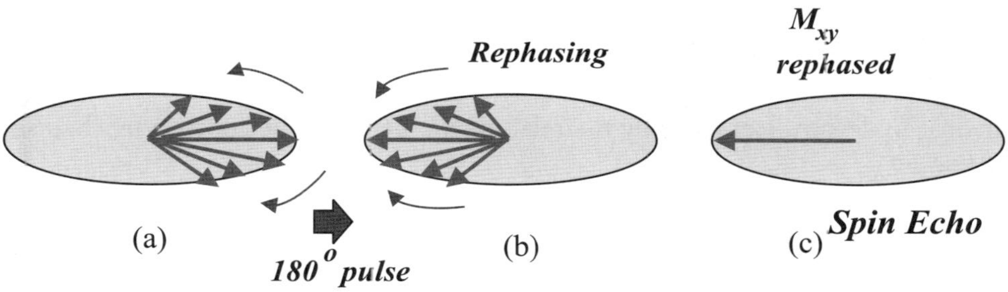

Figure 7.4: Before the application of the 180° pulse, the differing precessional frequencies are acting to dephase the spins (a). This is due to spatial field inhomogeneities. If we flip the spins by 180°, (b) the different precessional frequencies now act to bring the nuclei back into phase again. An echo forms, peaking when all spins are back into phase. Any dephasing at this point is due to spin-spin interactions (c).

A few more points:

- Carr and Purcell subsequently suggested this *spin-echo train* where a single 90° pulse is followed by a series of 180° pulses spaced T_E apart. Since FID's are usually noisy, the use of multiple echoes improves the precision with which one can measure T_2 (see Figure 7.5).
- T_E is known as the *time to echo*; it, together with T_R, are the primary "controls" of MR image contrast.

7.6 CONTRAST

In most cases, the contrast between tissue types in an image will be due to variations of both T_1 and T_2 in the image tissues, an also to differences in proton density, $N(H)$. Let us consider 2 tissues types

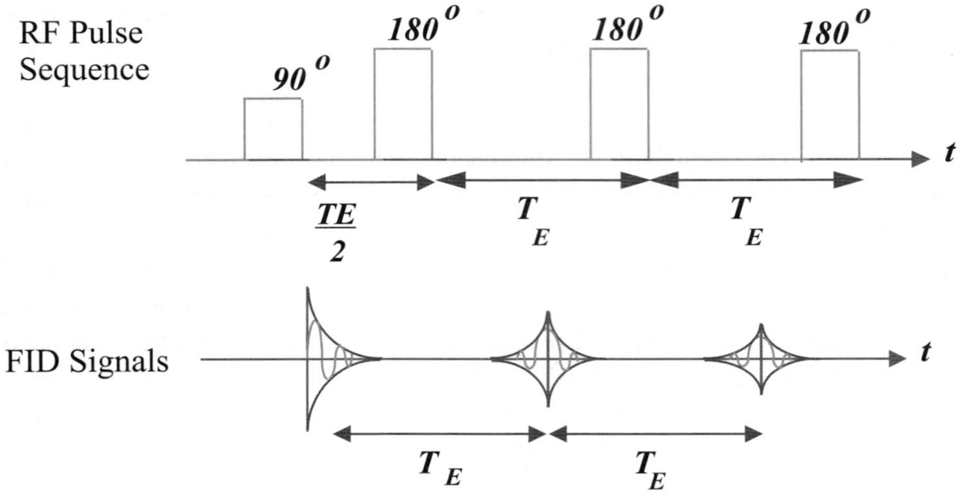

Figure 7.5: Multiple 180° pulses yield multiple echoes, with decaying amplitudes.

a and *b*, of equal hydrogen density. Let us study the relationship between the contrast mechanism and the values of T_E and T_R.

7.6.1 PROTON DENSITY WEIGHTING

Let the following hold

- $T_R >> T_1$
- $T_E << T_2$

Then, neither the relaxation parameters T_1 nor T_2 affect the contrast greatly. Any contrast in the image is due to differing densities, $N(H)$, between tissue types. This is known as *PD weighting* (proton density), because changes in the plateau levels are more significant than differences in the time constants of either relaxation process (vertical lines are T_R, T_E times). We shall see this more explicitly when we look at an example pair of decay curves with different plateaus and relaxation times.

7.6.2 T_2 WEIGHTED

Let the following hold

- $T_R >> T_1$
- $T_E \approx T_2$

This time, differences in the M_{xy} component decay curves are highlighted. Contrast in the image is said to be T_2 *weighted*.

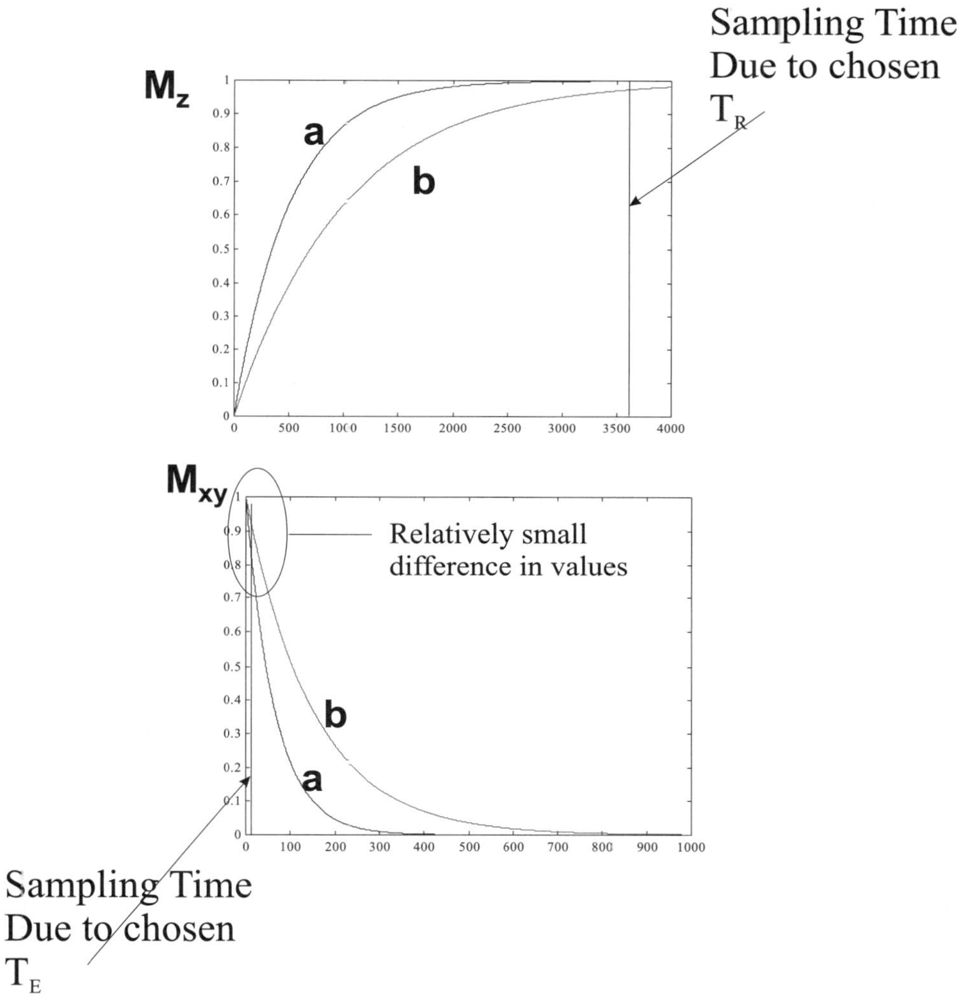

Figure 7.6: PD weighting. Neither the T_1 nor T_2 relaxation curves show great differences *at the points of sampling*. Image intensity is more strongly weighted to the plateau levels (for M_z) or maximum transverse magnetisation M_{xy}. These are affected by *proton density*. Differences in image intensity (contrast) are said to be proton-density weighted.

7.6.3 T_1 WEIGHTED

Let the following hold

- $T_R \approx T_1$
- $T_E << T_2$

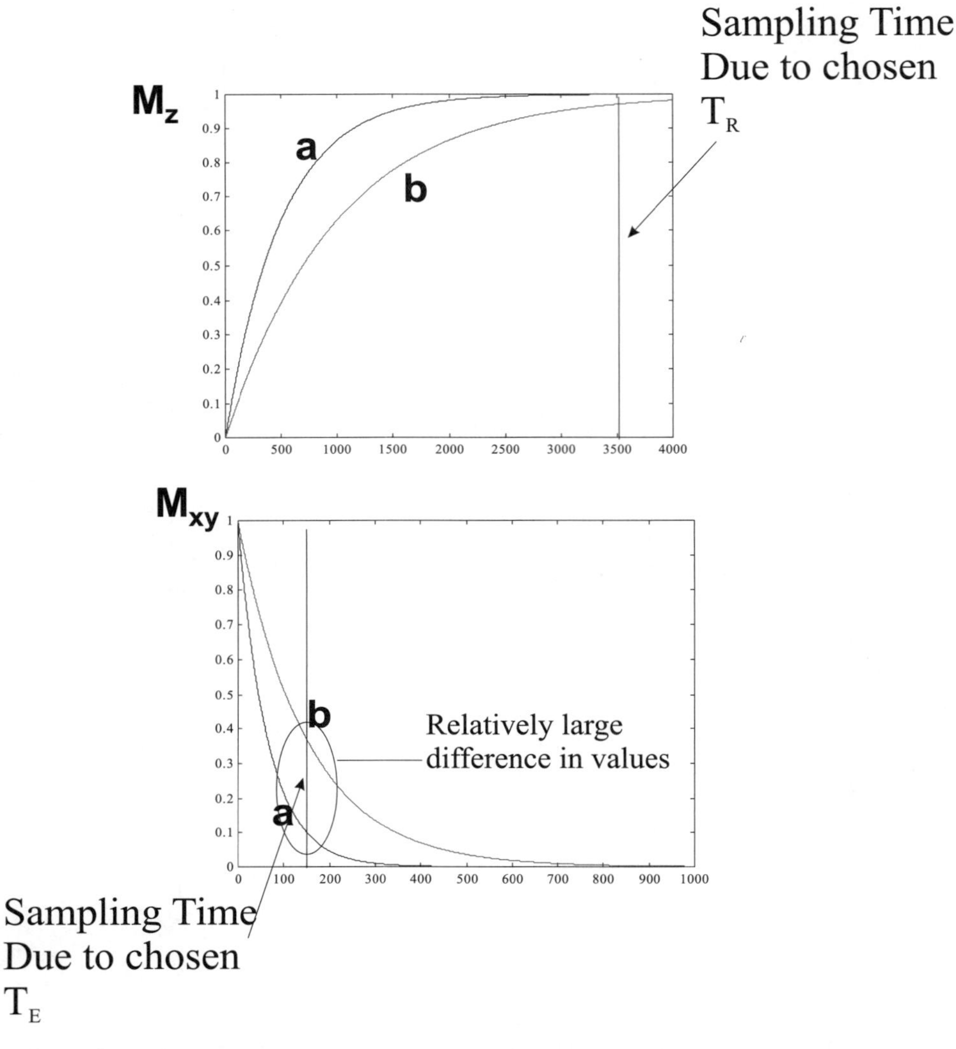

Figure 7.7: T_2 weighting. Any contrast is strongly weighted towards any differences in the M_{xy} decay. This is effectively, therefore, a T_2 weighted measurement.

This time, differences in the M_z component recovery curves are highlighted. Contrast in the image is said to be T_1 *weighted.*

7.6.4 BRAIN TISSUE CONTRAST: EXAMPLE

Now that you have the hang of image contrast, lets look at a more complex example of signal intensities, and the relaxation curves. You can find many more of these examples in *MRI: The Basics*.

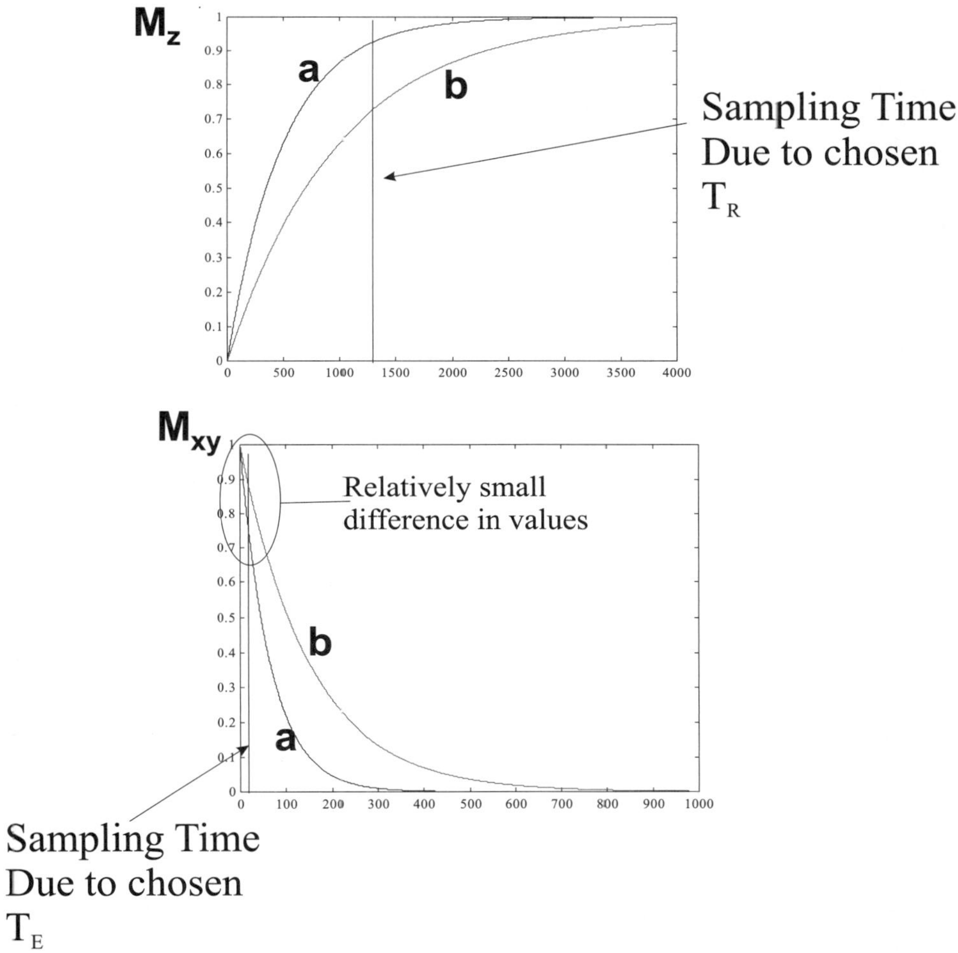

Figure 7.8: T_1 weighting. Any contrast is strongly weighted towards any differences in the M_z decay. This is, therefore, a T_1 weighted measurement.

If we consider white matter and CSF in the brain, these types of tissue would have different plateau levels. The T_R times and T_E times greatly affect the appearance of the image.

7.6.5 SUMMARY

We have now covered the basics of the contrast mechanisms. You can see that T_R and T_E have the ability to subtly alter the relationship in signal between tissues. Compare this with x-ray imaging, where bone almost always appears white, and soft-tissue dark. In MR, the intensity of tissue in the final image depends on the values of T_R and T_E.

What we still have not done is to consider how the signal is localised in MR. This is radically different

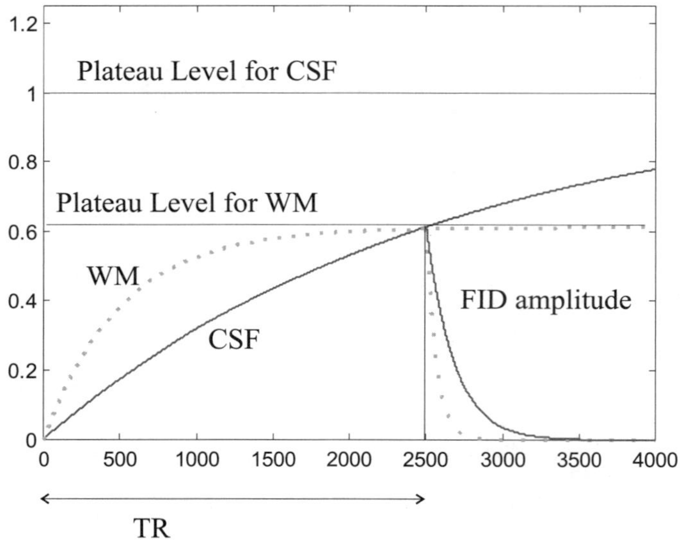

Figure 7.9: Using a T_R at approximately the cross-over point for the M_z curves for white matter and CSF, we see that, for all T_E, CSF is brighter. Maximum weighting for T_2 occurs with $T_E \approx 250ms$.

from the other modalities that we have met so far, and is one of the aspects of MR technology which is truly remarkable.

7.7 WHERE'S THAT ECHO COMING FROM?

In this section, we will introduce spatial localisation in MR. If we have a uniform $\mathbf{B_0}$, then the Larmor frequency will be the same for all regions of tissue in the field. Thus, the RF pulse applied to tilt the magnetisation vector into the transverse plane will be absorbed by all protons in the imaging volume. The raw signal, the FID, will also be obtained from all over the volume placed in the field.

Spatial localisation relies on the relationship between the Larmor frequency and the field strength. Thus, if we have a field strength of 1.5 Teslas, using the appropriate gyromagnetic ratio for hydrogen as 42.58 MHz T^{-1}, the Larmor frequency is given by

$$f_0 = 42.58 \times 10^6 \times 1.5 \tag{7.26}$$

or 63.87MHz. If the field strength varies, then the Larmor frequency varies linearly with it. So, if we impose a magnetic field *gradient* on top of the $\mathbf{B_0}$ field, we can selectively excite a slice of tissue, by altering the frequency of the RF pulse sent into the patient. This is how we *begin* to tackle spatial localisation.

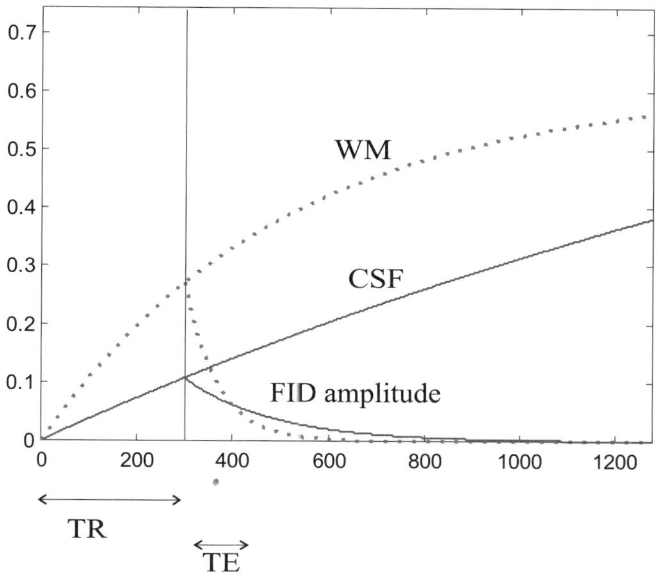

Figure 7.10: Using a very short T_R, the dependency on T_E is now critical: for $T_E < 100ms$, white matter appears brighter; for $T_E > 100ms$, CSF appears brighter.

7.7.1 SLICE SELECTION

Let us suppose that a linear magnetic field gradient $\mathbf{G} = G_x\mathbf{x} + G_y\mathbf{y} + G_z\mathbf{z}$ is superimposed onto the main magnetic field, and that this gradient exists only in the z direction. We can write

$$f(z) = \gamma_p|\mathbf{B_0}| + \gamma_p G_z z \,. \tag{7.27}$$

Usually, the magnetic field is B_0 in the centre of the bore of the scanner, (it is useful, therefore, to define this position as $z = 0$). So, if our 90° RF signal has a central frequency of f_c, then, in principle, only the spins with a Larmor frequency of f_c will be excited. In fact, since we are transmitting a *pulse* of RF excitation, there will be other frequencies present in the pulse too, and so a *slice* of spins at some position along the z axis will be excited. The centre of this slice will correspond to the central frequency of the RF pulse, and will be given by

$$z_c = \frac{f_c/\gamma_p - |\mathbf{B_0}|}{G_z} \,. \tag{7.28}$$

Example. Find the position of the slice excited by a central RF frequency of 63.70MHz when the z field gradient strength is 15mT m^{-1} and the main field strength, $|\mathbf{B_0}|$, is 1.5 T. Using the gyromagnetic ratio for hydrogen as 42.58MHz T^{-1}, we have

$$\begin{aligned} z_c &= \frac{f_c/\gamma_p - |\mathbf{B_0}|}{G_z} \\ &= (63.70/42.58 - 1.5)/0.015\text{m} \\ &= -0.27\,\text{m} \\ &= -27\,\text{cm}\,. \end{aligned} \tag{7.29}$$

Thus, the excited slice is 27 cm down from the centre of the bore. □

The preceding example should make it clear that, for a given main field strength, the slice position is determined by the frequency of the transmitted RF signal, and by the gradient field, G_z. Have a look at Figure 7.11.

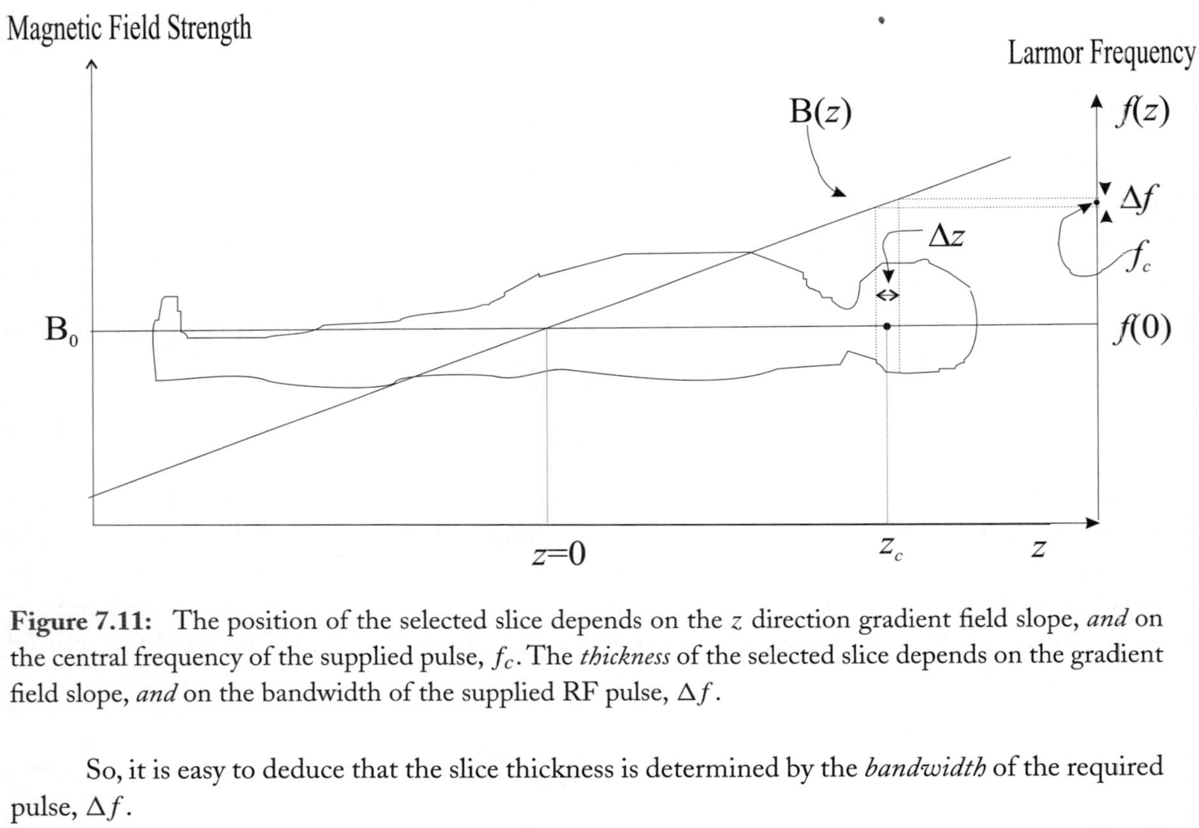

Figure 7.11: The position of the selected slice depends on the z direction gradient field slope, *and* on the central frequency of the supplied pulse, f_c. The *thickness* of the selected slice depends on the gradient field slope, *and* on the bandwidth of the supplied RF pulse, Δf.

So, it is easy to deduce that the slice thickness is determined by the *bandwidth* of the required pulse, Δf.

$$\Delta f = \Delta z G_z \gamma_p\,. \tag{7.30}$$

Example. Find the bandwidth of the RF pulse required to excite a slice 1cm thick when the z field gradient strength is 15mT m^{-1} and the main field strength, $|\mathbf{B_0}|$, is 1.5 T. Using the gyromagnetic ratio for hydrogen as 42.58MHz T^{-1}, we have

$$\begin{aligned}\Delta f &= 42.58 \times 10^6 \times 0.01 \times 0.015\text{Hz} \\ &= 6387\text{Hz}\,. \end{aligned} \tag{7.31}$$

This is a very small bandwidth indeed, relative to the frequency of the signal. This has implications for the design of the RF amplifiers used to provide the excitation pulse. □

A few other points are worthy of mention.

- If one uses an equal weighting of frequencies in the range of $f_c - \frac{\Delta f}{2}$ to $f_c + \frac{\Delta f}{2}$, then this is equivalent to a *rectangular* function in the Fourier domain. This means that the *envelope* of the RF pulse will look like the inverse Fourier transform of a rectangular signal: it will be a $sinc()$ function in time. See Figure 7.12.
- Rectangular slices (as shown and discussed above) are very common nowadays. In the earlier days of MR, Gaussian envelopes on the RF pulses were used, leading to uneven weighting of sensitivity in each selected slice. Since the Fourier Transform of a Gaussian is a Gaussian, the weighting across the slice is also Gaussian in shape, with maximum weighting at the centre of the slice, z_c.
- Traditionally, slice selection is performed in the z direction, yielding an image plane which is normal to the $\mathbf{B_0}$ direction. However, many modern scanners are capable of slice selection in *arbitrary* planes. That is, gradients can be supplied in the x, y and z directions simultaneously, and an RF pulse supplied so that only a slice of spins in an arbitrary plane are excited. Localisation in the plane of the slice is then (again) by phase and frequency encoding.

7.7.2 IN-PLANE LOCALISATION

So, we can perform slice selection as described above. What about localisation *within* the slice plane? We use 2 Fourier encoding techniques.

- Frequency encoding
- Phase encoding.

First, imagine that we have switched on a z gradient field during the application of a shaped RF pulse, and that only the spins in a selected slice of the patient have been excited, and are therefore producing free induction decay signals. Imagine now that we apply a gradient field in the x direction. This will cause a change in precessional frequency across the patient, in the x direction. So, the FID signals that we receive will contain *many* frequency components, with each component

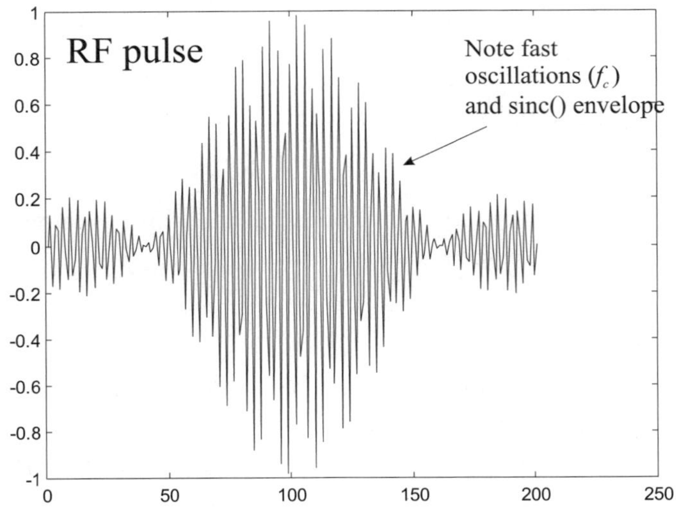

Figure 7.12: Example of a shaped RF pulse. The frequency of the fast oscillations is f_c, and the envelope is a $sinc()$ function. In the presence of a z magnetic field gradient, this will excite only a rectangular slice of tissue in the patient, with thickness and position determined as explained in the text.

corresponding to a different position across the slice in the x direction. As a simple example of this principle, consider the "patient" slice shown below, which contains two test tubes of different materials, displaced at different points in the x direction (Figure 7.13).

If the tubes are at positions x_1 and x_2, then the frequencies of the FID decay will be given by

$$f(x_1) = \gamma_p|\mathbf{B_0}| + \gamma_p G_x x_1 \tag{7.32}$$

and

$$f(x_2) = \gamma_p|\mathbf{B_0}| + \gamma_p G_x x_2 \ . \tag{7.33}$$

If we take the Fourier Transform of the fairly nasty signal shown in Figure 7.13, we will see the separation in the frequency domain, because by the Larmor relationship, frequency is now *directly* related to position! See Figure 7.14.

7.7.3 FREQUENCY ENCODING

The preceding example demonstrates the principle of frequency encoding. However, we remain with a problem. In a more complex field of view i.e., a person!), the detected FID signal in the presence of an x gradient will contain frequency components corresponding to the full range of different positions along the direction of applied gradient. Therefore, for a given position along the

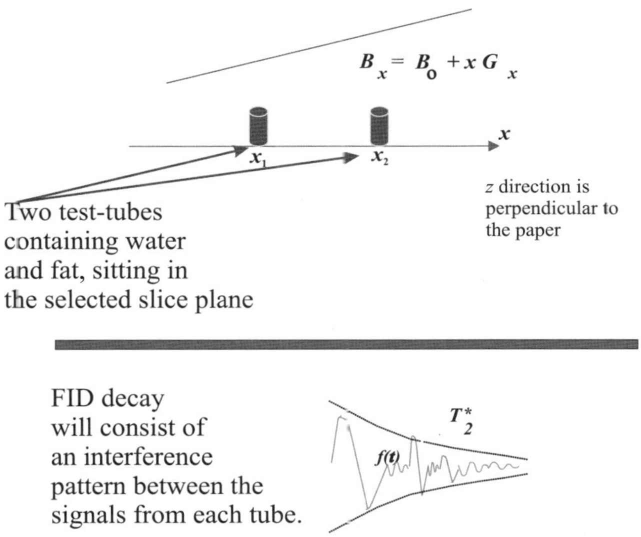

Figure 7.13: Imagine two test tubes containing water and fat placed at two different locations in the x direction, but at the same slice position i.e., same location, z_c). Each tube will produce an FID, but at with a slightly different frequency. These are superimposed in the time-domain, but separated in the frequency domain!

x direction, the Larmor frequencies along the y direction will all be the same! Thus, each frequency component actually contains information on the sum (or integration) of all components in the spaced along the y direction for that particular x position. We need to find a way of also localising in the y direction. One suggestion would be to also apply gradients in the y direction. This is possible, but if one looks into this technique in detail, it becomes clear that one would need to apply a range of combinations of x and y gradients in order to achieve a nonredundant set of measurement data. This would be very time-consuming.

A very elegant solution will present itself by turning to a little Fourier theory. But first, we need to formulate a mathematical model of the signals being detected by the coils, and the demodulation process.

7.7.4 THE SIGNAL DETECTION PROCESS

In order to handle the very high-frequency FID signal, we apply a mixing process, much as we do for ultrasonic signals. The way that this is represented mathematically is via a multiplication of the FID signal by a complex exponential of the form $\exp(-2\pi f_0 t)$. Practically, this is performed by multiplying the incoming FID signal with both sin() and cos() signals at the frequency f_0.

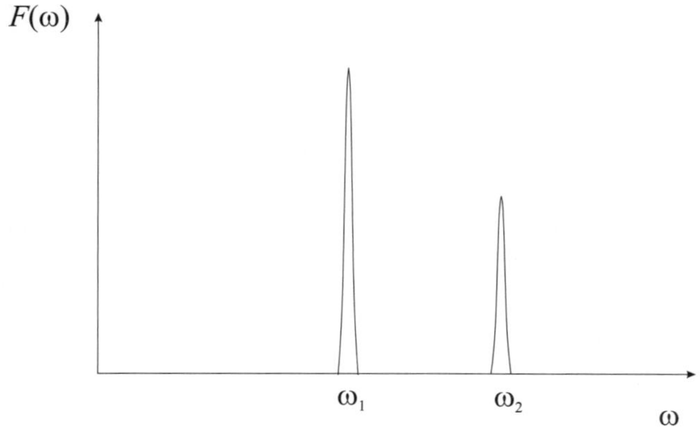

Figure 7.14: The Fourier transform of the FID signal from our tubes will show (if a gradient exists) a separation of the signals in the frequency domain. The heights of the peaks in the frequency domain tell us about (in this case) the proton density in each tube. The shape of each pulse peak in Fourier space is related to the envelope of decay (T_2^*, for a single 90° pulse). Note that $\omega_1 = 2\pi f(x_1)$ and $\omega_2 = 2\pi f(x_2)$.

Now, let us be quite general, and allow the possibility of a distinct frequency for each position in the plane of the slice. Then, a model for the FID signal coming from the entire slice is simply

$$s(t) \propto \int_x \int_y m(x, y) e^{2\pi j\, f(x,y)\, t} dx dy \,. \tag{7.34}$$

Here, the complex exponential term (containing the imaginary number, j), represents the oscillation at the Larmor frequency[9] of the spins in the plane, and $m(x, y)$ represents the decay envelope - for a single 90° degree pulse, this is an exponential decay with time constant T_2^*; for a spin-echo sequence, it is a sampling of the T_2 decay curve, etc. The integration merely shows that in the time domain, all the FID signals over the slice plane are summed together, creating a bit of a mess.

If f_0 corresponds to the Larmor frequency of hydrogen nuclei at the $\mathbf{B_0}$ frequency, then the effect of this multiplication will be to subtract f_0 from the frequency of the FID signal in the slice.

$$s_m(t) \propto \int_x \int_y m(x, y) e^{2\pi j\, (f(x,y) - f_0) t} dx dy \,. \tag{7.35}$$

Let there now be a gradient field in the x and y directions. Let us apply a gradient of magnitude G_x in the x direction of the slice plane during the acquisition of the FID signal. We will now be very cunning: apply the spin-echo pulse sequence to acquire the FID signal, but *before* the appearance

[9] In fact, a single receiver coil produces only a real signal. Sometimes, however, quadrature coils are used, consisting of two receiver coils placed at right angles to each other. Under this situation, a complex FID signal is obtained. Since the complex signal is easier to work with, we will stick with this.

of the spin-echo signal, we apply a *second* gradient of magnitude G_y in the y direction for a brief period of time, τ. We will therefore have (during time, τ):

$$f(x, y) = f_0 + \gamma_p y G_y . \tag{7.36}$$

Over a period of time τ, this will cause the spins in the selected slice to precess at different rates with, again, a linear dependence of rate of spin on position. If one then *switches off* the G_y component, the nuclei in the slice will resume their precession at the original Larmor frequency, but spins will maintain this relative dephasing in the y direction: effectively, what we have done is to create a *phase encoding*[10]. Mathematically, we can represent a phase, $\phi(t)$, as $e^{j\phi(t)}$. Also, recall that frequency is simply the rate of change of phase,

$$f = \frac{1}{2\pi} \frac{\partial \phi(y, t)}{\partial t} . \tag{7.37}$$

Integrating both sides, and starting from zero phase, we can say that

$$\phi(y) = 2\pi \int_0^{\tau} f(y) dt . \tag{7.38}$$

Since f is a constant (with time) over this time interval, the total phase change induced during this period of time is given by (including Equation (7.36))

$$e^{2\pi j(\tau \gamma_p G_y y + f_0 \tau)} . \tag{7.39}$$

The dispersion of the spins in the y direction due to the G_y presence therefore leads to a *relative* dispersion in phase with y given by

$$e^{2\pi j(\tau \gamma_p G_y y)} . \tag{7.40}$$

This phase shift is position dependent, and also depends on G_y and τ[11].

Going back to our expression for the signal, we can include this phase shift into the model. What we shall do to simplify matters further, is to remember that at the time of acquisition of the FID signal, G_y is turned off, and only G_x is nonzero.

$$s_m(t) \propto \int_x \int_y m(x, y) e^{2\pi j(\gamma_p G_x x t + \gamma_p G_y y \tau)} dx dy . \tag{7.41}$$

The first term in the exponential is merely the time varying oscillation, the second term represents a phase shift[12]. If we were to repeat the entire FID measurement many times, but each time using

[10] If we go back to the analogy of our runners, it is as if at the start of the race, all the athletes are running at different rates, and disperse. Suddenly, at a time τ, all of the athletes begin to run at the same speed. They will therefore maintain their relative distances apart. At the finish line, the dispersion is maintained, and depends both on τ, and their relative speed differences at the start of the race.

[11] This is a delicate procedure, and the phase information is fairly temporary, but usable.

[12] Compare with the "standard" form $e^{j(\omega t+\phi)}$ for a complex oscillating sinusoid which includes a phase shift.

a different G_y value, we could write

$$s_m(t, G_y) \propto \int_x \int_y m(x, y) e^{2\pi j(\gamma_p G_x x t + \gamma_p G_y y \tau)} dx dy \,. \tag{7.42}$$

This looks very Fourier like. Indeed, if we write

$$k_x = -\gamma_p G_x t \tag{7.43}$$

and

$$k_y = -\gamma_p G_y \tau \tag{7.44}$$

then we can write

$$s_m(k_x, k_y) \propto \int_x \int_y m(x, y) e^{-2\pi j(k_x x + k_y y)} dx dy \,. \tag{7.45}$$

Where have these k_x and k_y come from? Look again at Equations (7.43) and (7.44): k_x is linearly related to time. k_y is linearly related to G_y. So, as we collect a single FID signal, the time axis represents k_x (with a scaling). Different FID signals, each with a different G_y give us different k_y values. Big deal right? Well, actually, **yes**.

7.7.5 k-SPACE

Equation (7.45) is a 2D Fourier transform of the envelope of the FID signals from each point in the selected slice plane! This is referred to by MR physicists as *k-space*, but it is basically Fourier space.

An explanation of the significance of the k_x and k_y variables is as follows. 2D k-space is just a representation for 2D Fourier space. Basis functions are characterised by a vector, $\mathbf{k}$, with magnitude $|\mathbf{k}|$ oriented at angle, θ. One also has

$$k_x = |\mathbf{k}| \cos(\theta) \tag{7.46}$$

$$k_y = |\mathbf{k}| \sin(\theta) \,. \tag{7.47}$$

The k-numbers of a wave express its relative orientation and frequency. Fourier theory merely says that any two-dimensional function can be constructed by summing up waves with different k numbers, provided that each wave is appropriately scaled in height, and shifted in position.

7.7.6 PRACTICALLY SPEAKING...

In order to obtain $m(x, y)$ from this, we need to apply an inverse Fourier transform to $s_m(k_x, k_y)$. In practice, the demodulated FID signal is *digitised*, and so k_x is a discrete variable. In fact, since k_y is obtained by repeating an excitation with different G_y values, k_y is also discrete. So what we really need is an *inverse 2D discrete Fourier Transform*. We will cover this in detail elsewhere in the

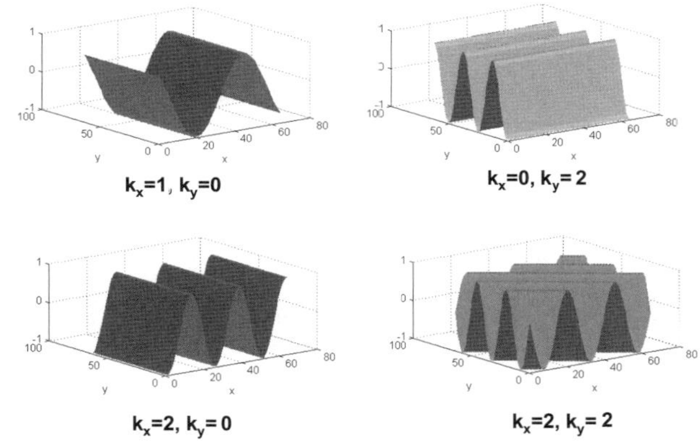

Figure 7.15: k-space axes describe the orientation and frequency of waves as a function of 2D space. This is just a generalisation to 2 dimensions of oscillating sines and cosines - the basis functions of Fourier Theory.

MSc course.

For the time being, let's see how the k space representation is constructed. Let us look at the full representation for the entire pulse sequence. This is a fairly standard diagram, which is intended to indicate the sequence of operations used in building up k-space. Do not be horrified by this; it is really amazingly easy to understand, and just takes a bit of getting used to.

Note, first, that we have adopted the standard notation for these traces[13]. So, $G_{slice} \equiv G_z$, $G_{FE} \equiv$ frequency encoding direction $\equiv G_x$, $G_{PE} \equiv$ phase encoding direction $\equiv G_y$. G_{FE}, driving the G_x field, is on twice: the first dephaser pulse is present to compensate for the phase effects that can accumulate in the x direction during the acquisition of the FID signal, when the Read out gradient is applied: remember, we only want phase differences in the y direction. Note the "ladder" appearance of G_{PE}. This indicates that the entire sequence is repeated many times, each time with a different value of G_{PE}. Although not labelled, the *duration* of the G_{PE} gradient is, of course, τ.

Each time the sequence is run, we get an echo signal, and the G_x gradient is switched on to provide frequency encoding during the read-out. This echo signal is demodulated as described

[13] It is sometimes assumed that, phase encoding is always performed in the y direction, and frequency encoding in the x direction, so that G_{FE} is always G_x, and G_{PE} is always G_y. Modern scanners generally do not have this as a restriction, although it may be adopted as a convention.

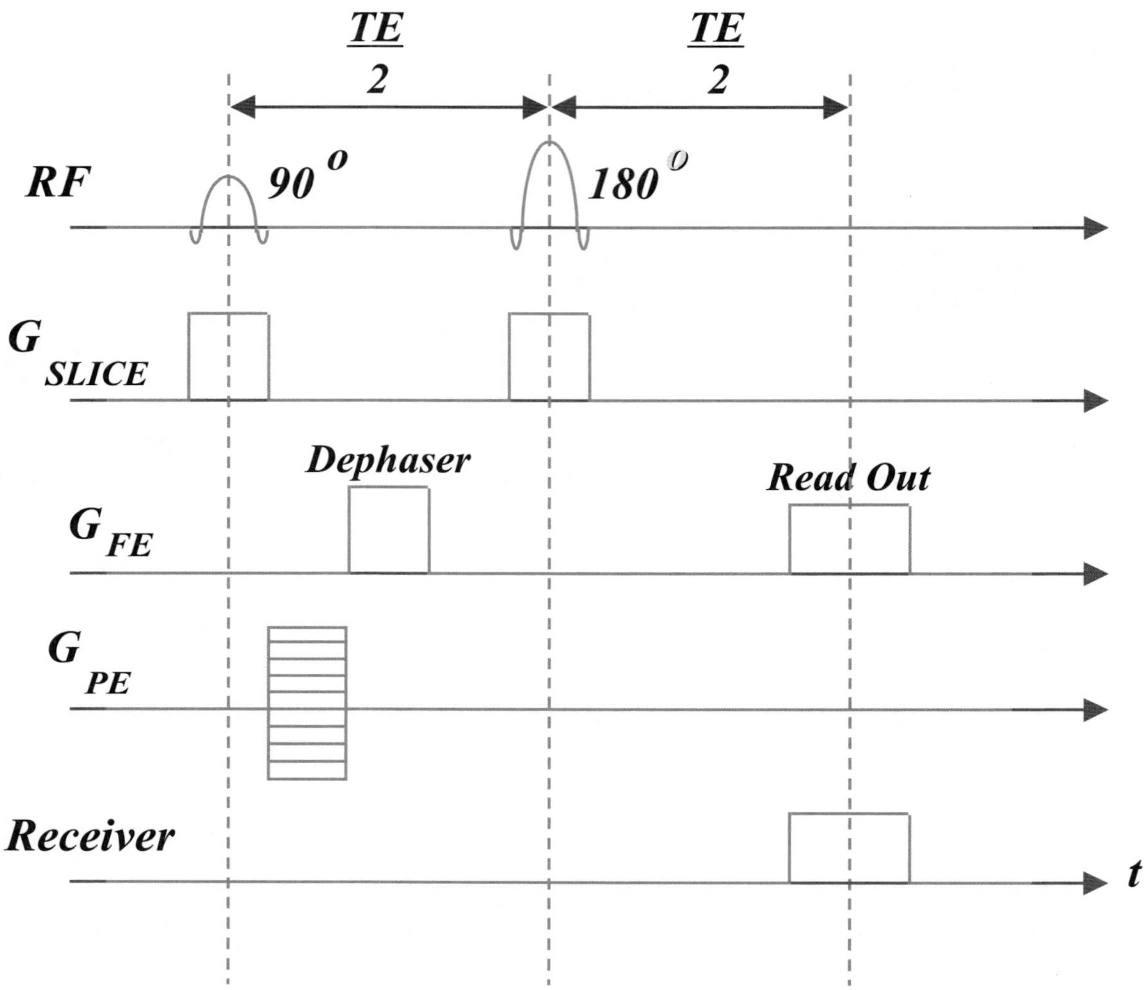

Figure 7.16: A pulse sequence containing the RF excitation, slice selection steps, and frequency and phase encoding steps. $G_{slice} \equiv G_z$, $G_{FE} \equiv$ frequency encoding gradient $\equiv G_x$, $G_{PE} \equiv$ phase encoding gradient $\equiv G_y$.

earlier, is digitised, and then placed into one row of k-space. As we acquire more and more FID signals, each under a different G_{PE} (or G_y), these go into subsequent rows[14] of k-space:

The number of *rows* of k-space corresponds to the number of *phase encoding steps*, or N_{PE}. This has significance for the time taken to acquire an image slice. The number of columns is

[14]Please note that there are different orders of filling up k-space: we have illustrated a simple scheme, in which the columns start from the bottom and move to the top of the matrix. For reasons of time or instrumentation limitations, the rows may be filled in a different order, and the order depends on the way that G_{PE} is varied at each step.

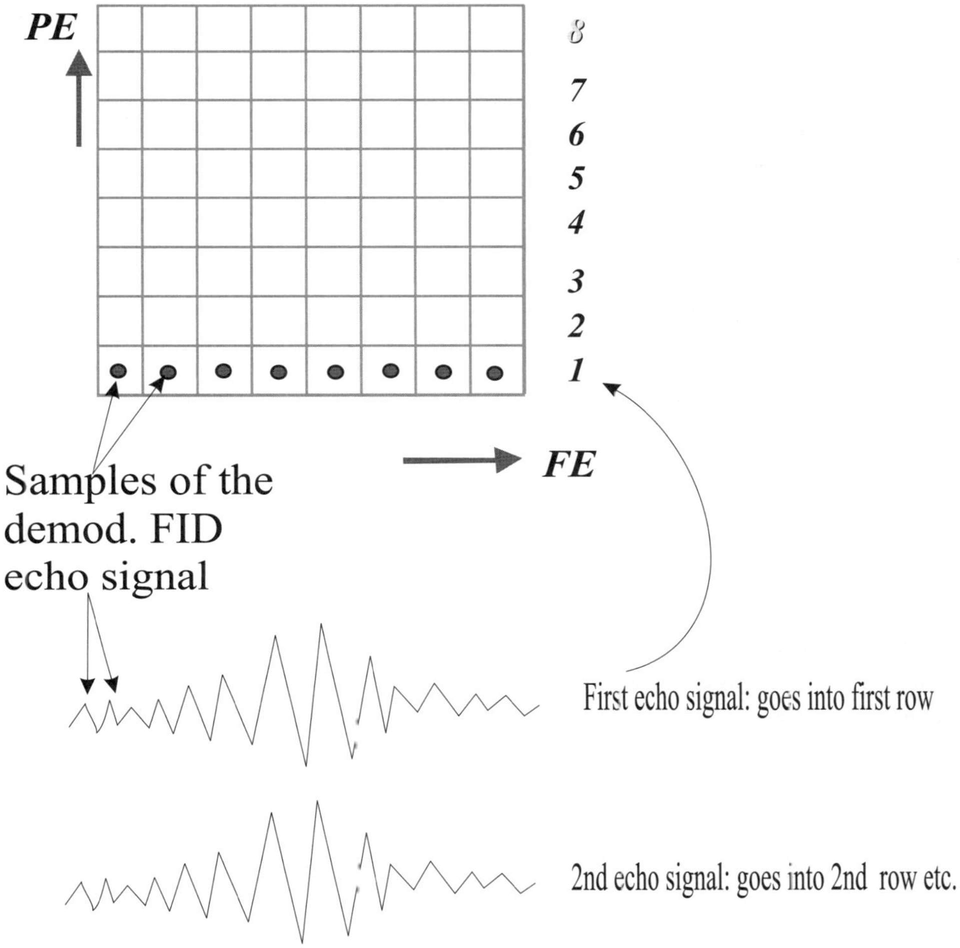

Figure 7.17: An illustration of the filling of a simple 8 × 8 k-space. Usually, k-space is much larger: as many as 128×128 or 256×256 in size.

consequently denoted by N_{FE}, and is related to the speed of digitisation of the FID signal.

However, there is one further complication we should mention. FID signals are quite noisy. The level of noise increases as one tries to get higher resolution measurements. A solution, therefore, is, for each phase encoding "step," to acquire several FID signals. This means that data for *each* k-space row is acquired several times. The number of times of repetition is denoted by either N_{SA}

which stands for "number of signal averages" or, by some manufacturers as, N_{EX}, which stands for number of excitations[15].

Imaging Time

Nowadays, the time to apply the 2D inverse DFT to a k-space matrix is minimal. The real bottleneck in imaging speed is the acquisition of the k-space matrix, whilst keeping the signal-to-noise ratio high enough by performing signal averaging.

A reasonable estimate of single-slice spin-echo imaging time is given by the number of phase-encoding steps, multiplied by the number of signal averages, multiplied by the time between each 90° degree pulse in the spin-echo sequence:

$$T_{slice} \approx N_{SA} \times N_{PE} \times T_R \; . \qquad (7.48)$$

Why do you think that a proton density weighted spin-echo scan will take longer than a T_1 weighted spin-echo scan?

7.8 WRAPPING UP

These notes on MR have just scratched the surface of this mind-bendingly powerful imaging technique. There is much more detail that one would need to cover to understand all aspects of this modality. For example, we have covered (mainly) only a single pulse sequence. There are many more, which achieve different tissue contrast, and which have different types of spatial localisation. In particular, there are many very *fast* image sequences that are becoming increasingly important. Much of these rely on clever ways of filling up k-space, and are well covered in texts such as McRobbie *et al.*, which is also highly recommended for details on MR image quality, and spatial resolution.

There are also safety issues that must be addressed: MR cannot be used on patients with pacemakers, for example, and many patients suffer from vertigo in the closed conditions of the scanner bore. Further safety aspects of this are normally covered in clinically oriented courses on MR imaging.

Instrumentation is an area in its own right, and a significant amount of research and development has been carried out by the major equipment manufacturers to enable todays scanners to perform as well as they do. The instrumentation includes large-current systems for the superconducting magnets and gradient coils, and highly specialised instrumentation for receiver coil interfacing and pickup. Field uniformity is a major technical issue with the main magnetic field in MR, and is quite an art form to get right: field inhomogeneities may be corrected for with either active or passive shimming methods.

[15]Indeed, some MR physicists are known to answer the question "What resolution can this thing (MRI scanner) achieve?" with the answer "How long have you got?"

APPENDIX A

Wave Equations for Ultrasound

A.1 DERIVATION OF THE HWE

To begin with, let us look at a simple frictionless mechanical model which will provide an analogy to the behaviour of an acoustic system. A mass, M, is attached to two springs each of spring constant, k. These springs provide a restoring force when the spring is in tension.

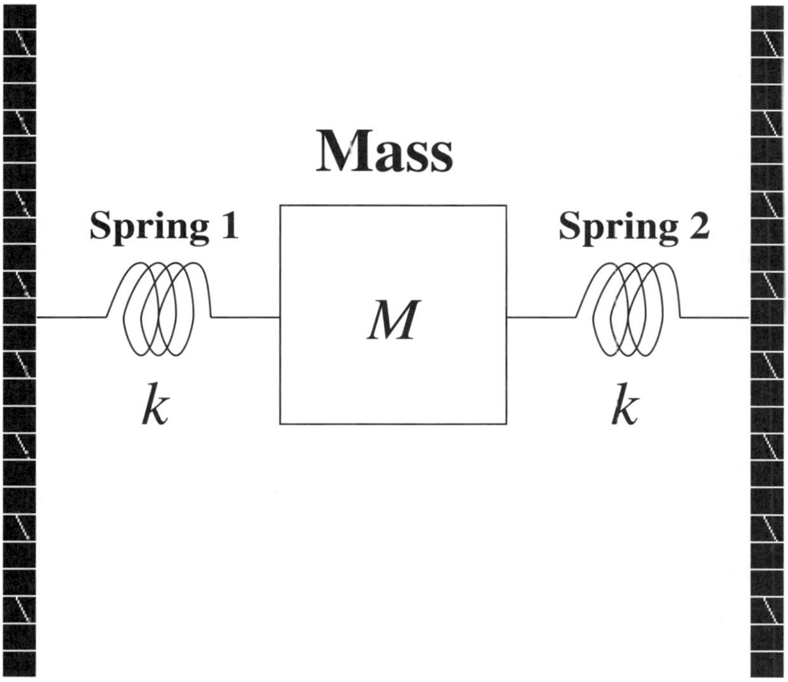

Figure A.1: Simple mass/spring system at rest.

We consider a resting condition where the forces in the springs counteract each other. Now, let us displace the mass by a small amount, x, to the left from its resting position. Spring 1 will exert a decreased force on M, Spring 2 an increased force. The net force applied to the mass under this displacement will thus be $2kx$. The mass is now released, and will therefore start to accelerate under this applied force according to Newton's second law of motion. This acceleration will be in the opposite direction to the displacement x, so that we can write

$$\begin{aligned} a &= -F/M \\ &= -2kx/M \; . \end{aligned} \tag{A.1}$$

However, a is defined as the rate of change of velocity, so

$$\begin{aligned} a &= \frac{dv}{dt} \\ &= \frac{d^2x}{dt^2} \; . \end{aligned} \tag{A.2}$$

So, we have,

$$\frac{d^2x}{dt^2} = -2kx/M \; . \tag{A.3}$$

This can be recognised as having the form of a second order differential equation in the function x with time. The simplest solution to this is the generic cosine solution:

$$x(t) \propto \cos \sqrt{2k/M}t \; . \tag{A.4}$$

However, this solution has not used any formal boundary conditions, so we have only a proportionality. We need to give an "example" correspondence between displacement and time in order to obtain a full solution. Let us say that at $t = 0$, the displacement, $x(0) = A$, where A is a constant. We then have

$$x(t) = A \cos \sqrt{2k/M}t \; . \tag{A.5}$$

Another equally valid *mathematical* solution is

$$x(t) = Ae^{j\sqrt{2k/M}t} \; . \tag{A.6}$$

This solution is known as the *complex phasor representation*. It is often used in describing wave phenomena because it is algebraically simple to work with. Its drawback is that one must remember to take the real or imaginary part of the phasor if one wishes to denote actual spatial distances, which cannot be complex.

Finally, note that we can write

$$\omega_0 = \sqrt{2k/M} \tag{A.7}$$

and we can recognise that the motion of the mass corresponds to *simple harmonic motion* at a single frequency ω_0.

Now, let us consider a more representative example, and see what new pieces of information we need to consider. Consider the series of weights and springs attached to each other, as shown in Figure A.2.

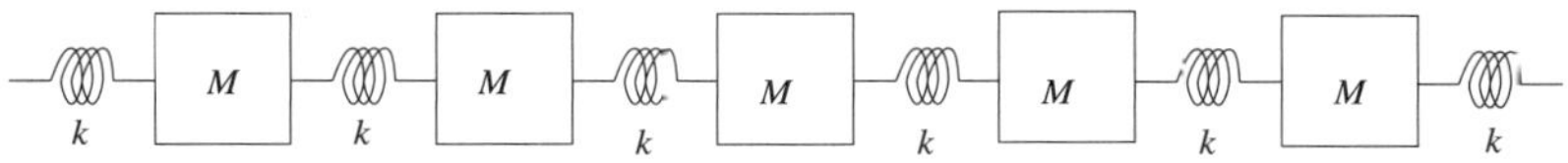

Figure A.2: Complex mass/spring system at rest.

We find that the equations of motion become more complex. The easiest way of handling this problem (and getting the insight we need) is to write a generic set of equations for the forces and motion of the nth mass. The displacement of the nth mass is denoted, $x(n, t)$, and its acceleration,

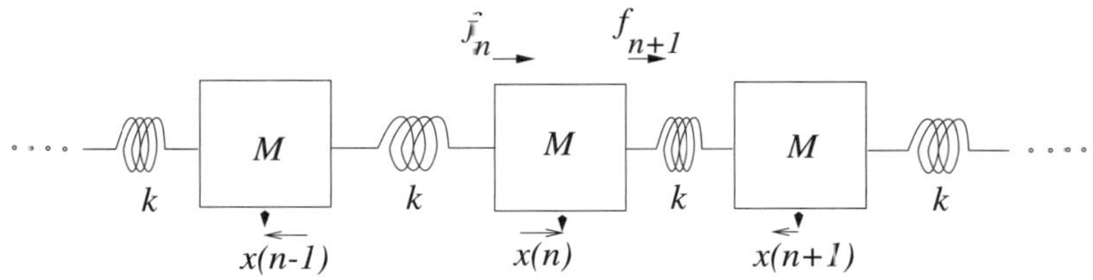

Figure A.3: Parameters in mass/spring system under displacement.

$a(n, t)$. The net force acting on this mass is given by $f_n(t) - f_{n+1}(t)$, the forces shown in the diagram (note that these forces are the forces exerted by the tether points of the springs on either side of the mass). One can immediately write an expression for the acceleration $a(n, t)$,

$$
\begin{aligned}
a(n,t) &= \frac{f_n(t) - f_{n+1}(t)}{M} \\
&= -\frac{k}{M}\{x(n,t) - x(n-1,t) - x(n+1,t) + x(n,t)\} \\
&= -\frac{k}{M}\{2x(n,t) - x(n-1,t) - x(n+1,t)\} \qquad \text{(A.8)}
\end{aligned}
$$

and this must hold for all values of n, except at the ends of the system. This yields a series of coupled differential equations describing the behaviour of the system. A state-space model can be conveniently used for this representation, but this is not necessary for our purposes. We are better off moving straight to the case of the continuous medium. But first, an observation: Looking at Equation (A.8), the left hand side is equivalent to $\partial^2 x(n, t)/\partial t^2$. The right hand side is quite complex, involving the displacements of neighbouring weights. However, a closer look at the r.h.s. will show that this is merely a discrete approximation to a second order derivative if one treats the mass displacements ...$x(n-1), x(n), x(n+1)$... as a discrete sequence. Indeed, as the masses are ordered along the horizontal direction, this can be seen as a discrete second order *spatial derivative* of displacement. This is subject to the condition that the entire system does not "flow", i.e. mass n does not move to the location of mass $n+1$, and that we can indeed refer unambiguously to individual

masses by their equilibrium position. There are analogous conditions to this in considering the equations for a continuous medium.
So, Equation (A.8) is a second order partial derivative equation relating temporal derivatives of displacement to spatial derivatives of displacement. Because the displacements can be related to forces, and the forces to velocities and accelerations, one has a series of equations which are similar to the wave equation in a continuous medium such as a string, (see, for example, [2]).

We will leave the lumped model here, and refer back to it as necessary to explain certain aspects of wave propagation in continuous media.

A.2 THE CONTINUOUS MEDIUM

In order to demonstrate the nature of a propagating disturbance in an acoustic medium, we consider a simple volume element of lossless, homogeneous, isotropic fluid defined as shown in Figure A.4. We shall consider only forces, velocities and accelerations along the x axis of the system (direction of

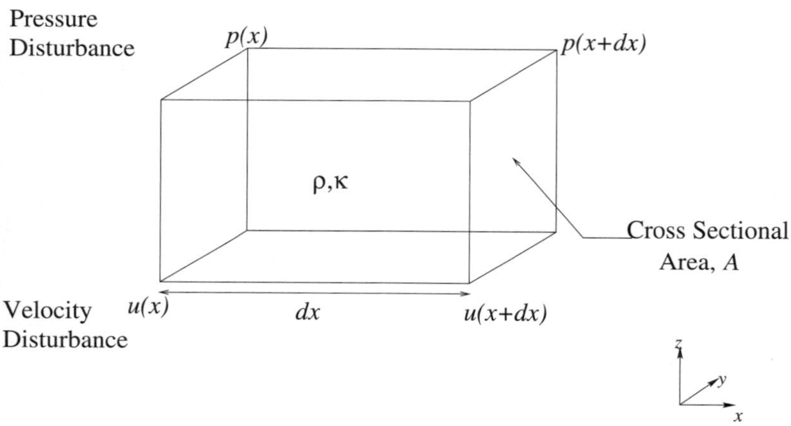

Figure A.4: Diagrammatic representation of a volume element of fluid and physical parameters.

wave propagation for longitudinal waves). In the discrete case, we saw that a force differential across a mass element induced an acceleration. Here, we begin by supposing that there exists a pressure gradient, δp, across the volume element:

$$\delta p = p(x) - p(x + dx) \tag{A.9}$$

so,

$$\delta p = -\frac{\partial p}{\partial x} \cdot dx \, . \tag{A.10}$$

Due to the application of this pressure gradient, we get two effects

- an acceleration of the volume element

• a compression of the fluid in this region

In addition, because we are really applying a pressure *gradient*, we obtain a velocity *distribution* in the x direction. Two equations may be written from the above considerations [1], which are approximately

$$\frac{\partial p}{\partial x} = -\rho \frac{\partial u}{\partial t} \tag{A.11}$$

known as the *linear inviscid force equation*, which is merely a continuum form of Newton's second law of motion, and

$$\kappa \frac{\partial p}{\partial t} = -\frac{\partial u}{\partial x} \tag{A.12}$$

which relates the rate of compression to the local velocity gradient, and is known as the *linearised continuity equation*. ρ is the mass density of the medium, and κ is the adiabatic compressibility of the medium. ρ and κ are clearly analogous to M and $1/k$ in our previous lumped model (why $1/k$?).

If we differentiate Equation (A.11) w.r.t x, and Equation (A.12) w.r.t. t, we obtain

$$\frac{\partial^2 p}{\partial x^2} = -\rho \frac{\partial^2 u}{\partial x \partial t} \tag{A.13}$$

and

$$\kappa \frac{\partial^2 p}{\partial t^2} = -\frac{\partial^2 u}{\partial t \partial x} . \tag{A.14}$$

Provided that we can reverse the order of the derivatives w.r.t. x and t, we can eliminate the expressions in u to obtain the one-dimensional form of the wave equation:

$$\frac{\partial^2 p}{\partial x^2} = \rho\kappa \frac{\partial^2 p}{\partial t^2} . \tag{A.15}$$

A.3 THE 3D ACOUSTIC WAVE EQUATION

The 3D homogeneous wave equation (HWE) can be written as

$$\nabla^2 p(x, y, z, t) - \rho\kappa \frac{\partial^2 p(x, y, z, t)}{\partial t^2} = 0 \tag{A.16}$$

where $p(x, y, z, t)$ is the acoustic pressure disturbance as a function of position and time, and ∇^2 is the Laplacian operator

$$\nabla^2 = \frac{\partial^2}{\partial x^2} + \frac{\partial^2}{\partial y^2} + \frac{\partial^2}{\partial z^2} . \tag{A.17}$$

We shall note that the quantity $\kappa\rho$ could be combined into one constant, but we shall leave things as they are for the moment.

POLAR FORM OF THE HWE

When dealing with wave propagation in a 3D, homogeneous, isotropic medium, it is sensible to look for solutions which are dependent only on Euclidean distance, and not on direction. In this case, a suitable coordinate system (such as a spherical system) is appropriate.

Unfortunately, Equation (A.16) becomes messy when expressed in spherical coordinates [3]:

$$\frac{\partial^2 p_R(r,t)}{\partial r^2} + \frac{2}{r}\frac{\partial p_R(r,t)}{\partial r} - \rho\kappa \frac{\partial^2 p_R(r,t)}{\partial t^2} = 0\,, \tag{A.18}$$

but we can introduce a clever auxiliary function

$$\phi_R(r,t) = r p_R(r,t) \tag{A.19}$$

which allows us to write Equation (A.18) in the one-dimensional form

$$\frac{\partial \phi_R(r,t)}{\partial r^2} - \rho\kappa \frac{\partial^2 \phi_R(r,t)}{\partial t^2} = 0 \tag{A.20}$$

from which a simple expression for $\phi_R(r,t)$ results, based on specified initial conditions (boundary conditions in 4-space). The solution takes the form of d'Alembert's solution:

$$\phi_R(r,t) = f_+(t - r\cdot\sqrt{\rho\kappa}) + f_-(t + r\cdot\sqrt{\rho\kappa}) \tag{A.21}$$

where f_+ is a forward propagating wave function and f_- is a backward propagating wave function. These must be chosen to satisfy the applied boundary conditions (specified as a function of time and three-dimensional space). The following example will show that this solution corresponds to quite a simple result. See also [3].

Example: Solution of the H.W.E. given specified boundary conditions. This is more than an example! It will also provide us with a useful physical scenario which can be extended in a vast variety of ways. The conditions we shall specify will be very simple. As before, we continue to work with an isotropic, homogeneous, lossless medium. We shall additionally specify only

- the pressure disturbance at a single point in the medium
- that the medium be unbounded

We introduce a disturbance (forcing function) at $r = 0$

$$\phi_R(0,t) = \begin{cases} s(t) & \text{for} \quad t \geq 0, \\ 0 & \text{for} \quad t < 0 \end{cases} \tag{A.22}$$

where $s(t)$ is an arbitrary waveform. An example is shown in Figure A.5. Such an excitation is an idealised one, and is referred to as a *monopole source*. The generic solution of Equation (A.21) is first examined. Because the medium is assumed unbounded (limitless), there will be no backward propagating waves, so that the f_- term is zero. The forward propagating wave only is adopted, and at

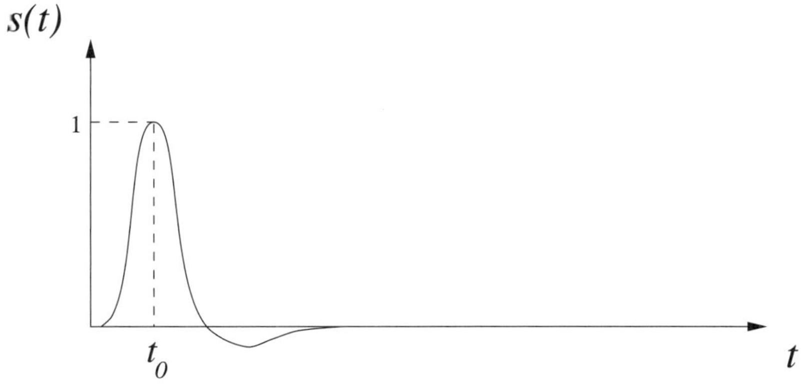

Figure A.5: Plot of pressure disturbance as a function of time at a location defined as r=0 in our 3D acoustic medium.

$r = 0$ the solution must satisfy the forcing function. The solution for $r > 0$ in terms of the auxiliary function is therefore given by

$$\phi_R(r, t) = s(t - r \cdot \sqrt{\rho\kappa}), \qquad r > 0 \tag{A.23}$$

or,

$$p_R(r, t) = s(t - r \cdot \sqrt{\rho\kappa})/r \; . \tag{A.24}$$

This is a very simple result, which has intuitive appeal. The physical implication is as follows: We introduce a localised pressure disturbance at some point $r = 0$ in the medium. The temporal shape of this disturbance, known as a forcing function, is immaterial. The disturbance induces a spherical wave whose characteristic is dependent only on the distance from the source, and the excitation applied at the source, not on direction. At some distance from that point, $r = r_1 > 0$, we would observe that the pressure disturbance as a function of time would be of similar shape to the forcing function, $s(t)$, but scaled in amplitude by $1/r_1$, and shifted along in time by an amount $t_1 = r_1\sqrt{\rho\kappa}$. See Figure A.6.

It is instructive to see how the location of the peak of the pressure disturbance changes with time. The disturbance travels through the medium with uniform speed. To emphasise this, a plot of peak position versus time is shown in Figure A.7.

This demonstrates that the excitation function "propagates" through the acoustic medium, with velocity of wavefront propagation, $c = 1/\sqrt{\rho\kappa}$.

Example: Monopole Source, SHM. We consider an unbounded, isotropic, homogeneous medium once more. We use a forcing function defined by

$$\phi_R(0, t) = A_0 e^{j\omega_0 t} \; . \tag{A.25}$$

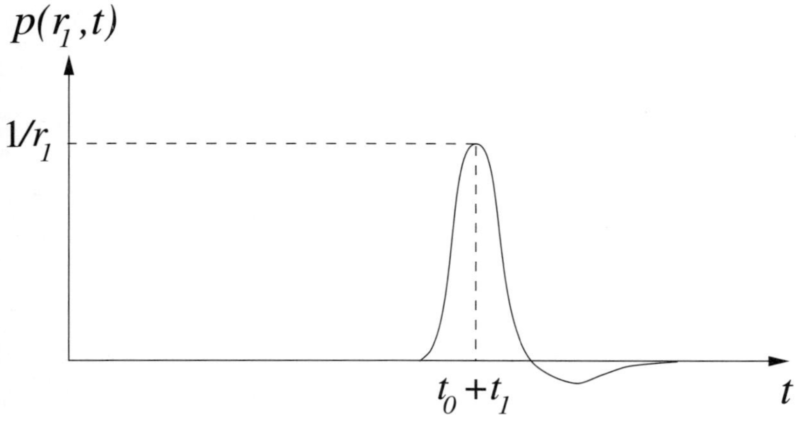

Figure A.6: Plot of pressure disturbance as a function of time at some distance $r = r_1$ away from the origin of the excitation. Note the time axis.

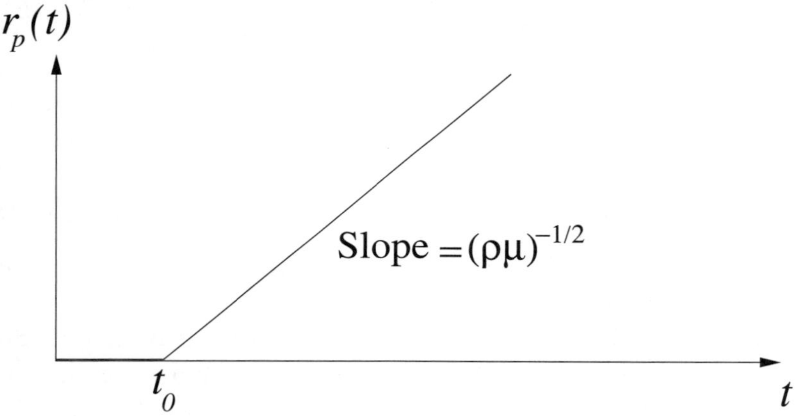

Figure A.7: Plot of disturbance peak location, $r_p(t)$ of pressure disturbance as a function of time.

The solution to the wave equation with this source excitation is

$$p_r(r,t) = A_0 \frac{e^{j\omega_0(t-r/c)}}{r}, \qquad r > 0\,. \tag{A.26}$$

This is the solution to the wave equation for a monopole source oscillating with simple harmonic motion.

Now, consider a point a very large distance away from the "source" at $r = R$

$$p_r(R,t) = A_0 \frac{e^{j\omega_0(t-R/c)}}{R}\,. \tag{A.27}$$

Advance R by an increment δ_R:

$$p_r(R + \delta_R) = A_0 \frac{e^{j\omega_0(t - R/c - \delta_R/c)}}{R + \delta_R} . \tag{A.28}$$

But, it is clear that

$$\frac{1}{R + \delta_R} \approx \frac{1}{R} \tag{A.29}$$

for large R and small δr.

Thus, the rate of change of the amplitude of the spatial oscillations becomes smaller at large distances from the $r = 0$ point, and *the predominant spatial variation is due to the oscillatory (complex exponential) component of pressure*. We may therefore write

$$p_r(r, t) = A_1 e^{j\omega_0(t - r/c)} \tag{A.30}$$

for sufficiently large distances away from the source. A_1 absorbs the A_0/r term. This limiting form (as $r \rightarrow \infty$) of pressure variation is known as a *harmonic plane wave*, and sometimes as an *infinite plane wave*. We will return to this expression to simplify our analytic techniques later in the course. The infinite plane wave is also the basis of the derivation of the first-order Doppler shift equation, and is further used to understand what happens when a planar wavefront approaches a planar boundary.

APPENDIX B

Mathematical Conventions Used

B.1 CONVOLUTION

This operation is denoted by $*$, which is not to be confused with multiplication.

$$f_1(t) * f_2(t) = \int_{-\infty}^{\infty} f_1(\tau) f_2(t-\tau) d\tau \ . \tag{B.1}$$

B.2 SIFTING PROPERTY

Sifting property of the δ function:

$$f(t) * \delta(t) = f(t) \ . \tag{B.2}$$

B.3 FOURIER TRANSFORM

Forward 1D:

$$\begin{aligned} \mathfrak{F}\{f(t)\} &= \int_{-\infty}^{\infty} f(t) e^{-j\omega t} dt \\ &= F(\omega) \ . \end{aligned} \tag{B.3}$$

Inverse 1D:

$$\begin{aligned} \mathfrak{F}^{-1}\{F(\omega)\} &= \frac{1}{2\pi} \int_{-\infty}^{\infty} F(\omega) e^{j\omega t} d\omega \\ &= f(t) \ . \end{aligned} \tag{B.4}$$

Forward 2D:

$$\begin{aligned} \mathfrak{F}\{f(x,y)\} &= \int_{-\infty}^{\infty} \int_{-\infty}^{\infty} f(x,y) e^{-j(ux+vy)} dx dy \\ &= F(u,v) \ . \end{aligned} \tag{B.5}$$

Inverse 2D:

$$\begin{aligned} \mathfrak{F}^{-1}\{F(u,v)\} &= \frac{1}{4\pi^2} \int_{-\infty}^{\infty} \int_{-\infty}^{\infty} F(u,v) e^{j(ux+vy)} du dv \\ &= f(x,y) \ . \end{aligned} \tag{B.6}$$

B.4 POLAR INTEGRALS

$$\int_{-\infty}^{\infty}\int_{-\infty}^{\infty} f(x, y)dxdy = \int_{0}^{\pi}\int_{-\infty}^{\infty} f(r, \theta)|r|drd\theta \ . \tag{B.7}$$

See also [19] for image reconstruction from projections.

Bibliography

[1] L.E. Kinsler, A.R. Frey, A.B. Coppens, and J.V. Sanders, *Fundamentals of Acoustics*, Wiley, 1999.

[2] P.M. Morse and K.U. Ingard, *Theoretical Acoustics*, Princeton University Press, 1986.

[3] S. Webb (Ed.), *The Physics of Medical Imaging*, Taylor and Francis, 1988. DOI: 10.1887/0852743491/b172c16

[4] S. Webb, A review of physical aspects of X-ray transmission computed tomography, *Physical Science, Measurement and Instrumentation, Management and Education, Reviews, IEE Proceedings A*, 134(2), 126–135, 1987. DOI: 10.1049/ip-a-1:19870019

[5] P. Fish, *Physics and Instrumentation of Diagnostic Medical Ultrasound*, Wiley, 1990.

[6] D.J. Dowsett, P.A. Kenny, and R.E. Johnston, *The Physics of Diagnostic Imaging*, Hodder Arnold, 1998.

[7] S. Leeman, V.C. Roberts, and D.A. Seggie, Dirac, Faraday, Born and Medical Imaging, *IEEE Proceedings*, 134(2), 1987. DOI: 10.1049/ip-a-1:19870016

[8] T.S. Curry, J.E. Dowdey, and R.E. Murry, *Christensen's Physics of Diagnostic Radiology*, Lippincott Williams & Wilkins, 4th ed., 1990.

[9] J.T. Bushberg, J.A. Seibert, E.M. Leidholdt, Jr., and J.M. Boone, *The Essential Physics of Medical Imaging*, Lippincott Williams & Wilkins, 2nd ed., 2001.

[10] Y. Bresler and C.J. Skrabacz, Optimal interpolation in helical scan 3D computerized tomography, *Acoustics, Speech, and Signal Processing, 1989. ICASSP-89., 1989 International Conference on*, Vol. 3, 1472–1475, 1989. DOI: 10.1109/ICASSP.1989.266718

[11] W.R. Hedrick and D.L. Hykes, Autocorrelation Detection in Colour Doppler Imaging, *Journal of Diagnostic Medical Sonography*, 11(1), 16–22, 1995. DOI: 10.1177/875647939501100104

[12] C. Kasai and K. Namekawa, Real-Time Two-Dimensional Blood Flow Imaging Using an Autocorrelation Technique, *IEEE 1985 Ultrasonics Symposium*, 953–958, 1985.

[13] K.K. Shung, R.A. Sigelmann, and J.M. Reid, Scattering of Ultrasound by Blood, *Biomedical Engineering, IEEE Transactions on*, Vol. BME-23(6) 460–467, 1976. DOI: 10.1109/TBME.1976.324604

[14] Y.W. Yuan and K.K. Shung, Ultrasonic backscatter from flowing whole blood. I: Dependence on shear rate and hematocrit, *Journal of the Acoustical Society of America*, 84(1), 52–58, 1988. DOI: 10.1121/1.397238

[15] D.H. Evans and W.N. McDicken, *Doppler Ultrasound: Physics, Instrumentation and Clinical Applications*, John Wiley & Sons, 1989.

[16] R.H. Hashemi, W.G. Bradley, and C.J. Lisanti, *MRI: The Basics*, Lippincott Williams & Wilkins, 2004.

[17] R.P. Feynman, *The Feynman Lectures on Physics: Commemorative Issue, Three Volume Set*, Student's Guide edition, Addison Wesley Longman, 1970.

[18] B.M. Moores, R.P. Parker, and B.R. Pullan, *Physical Aspects of Medical Imaging*, John Wiley & Sons, chapters 1, 2, and 6, 1980.

[19] A.K. Jain, *Fundamentals of Digital Image Processing*, chapter "Image Reconstruction From Projections", Prentice Hall, 1992.

[20] E. Kreyszig, *Advanced Engineering Mathematics*, 9th International Edition, John Wiley & Sons, 2005.

Lung Sounds: An Advanced Signal Processing Perspective

Leontios J. Hadjileontiadis
Aristotle University of Thessaloniki

Morgan & Claypool Publishers

ABSTRACT

Lung sounds auscultation is often the first noninvasive resource for detection and discrimination of respiratory pathologies available to the physician through the use of the stethoscope. Hearing interpretation, though, was the only means of appreciation of the lung sounds diagnostic information for many decades. Nevertheless, in recent years, computerized auscultation combined with signal processing techniques has boosted the diagnostic capabilities of lung sounds. The latter were traditionally analyzed and characterized by morphological changes in the time domain using statistical measures, by spectral properties in the frequency domain using simple spectral analysis, or by nonstationary properties in a joint time–frequency domain using short-time Fourier transform. Advanced signal processing techniques, however, have emerged in the last decade, broadening the perspective in lung sounds analysis. The scope of this book is to present up-to-date signal processing techniques that have been applied to the area of lung sound analysis. It starts with a description of the nature of lung sounds and continues with the introduction of new domains in their representation, new denoising techniques, and concludes with some reflective implications, both from engineers' and physicians' perspective. Issues of nonstationarity, nonlinearity, non-Gaussianity, modeling, and classification of lung sounds are addressed with new methodologies, revealing a more realistic approach to their pragmatic nature. Advanced denoising techniques that effectively circumvent the noise presence (e.g., heart sound interference, background noise) in lung sound recordings are described, providing the physician with high-quality auscultative data. The book offers useful information both to engineers and physicians interested in bioacoustics, clearly demonstrating the current trends in lung sound analysis.

KEYWORDS

Lung sounds, advanced signal processing, nonstationarity, nonlinearity, non-Gaussianity, modeling, classification, denoising, heart sound cancellation

Contents

CHAPTER 1

The Nature of Lung Sound Signals

1.1 HISTORICAL OVERVIEW

ὅταν μὲν γὰρ ὁ τῶν πνευμάτων τῷ σώματι ταμίας πλεύμων μὴ καθαρὰς παρέχῃ τὰς διεξόδους ὑπὸ ῥευμάτων φραχθείς, ἔνθα μὲν οὐκ ἰόν, ἔνθα δὲ πλεῖον ἢ τὸ προσῆκον πνεῦμα εἰσιὸν τὰ μὲν οὐ τυγχάνοντα ἀναψυχῆς σήπει, τὰ δὲ τῶν φλεβῶν διαβιαζόμενον καὶ συνεπιστρέφον αὐτὰ τῆκόν τε τὸ σῶμα εἰς τὸ μέσον αὐτοῦ διάφραγμά τ' ἴσχον ἐναπολαμβάνεται, καὶ μυρία δὴ νοσήματα ἐκ τούτων ἀλγεινὰ μετὰ πλήθους ἱδρῶτος πολλάκις ἀπείργασται.

When the lung, which is the dispenser of the air to the body, is obstructed by rheums and its passages are not free, some of them not acting, while through others too much air enters, then the parts which are unrefreshed by air corrode, while in other parts the excess of air forcing its way through the veins distorts them and decomposing the body is enclosed in the midst of it and occupies the midriff thus numberless painful diseases are produced, accompanied by copious sweats.

Plato's *Timaeus*
On Physis: Diseases of the body (84d-e)
Translated by B. Jowett

From the time of ancient Greeks and their doctrine of medical experimentation until at least the 1950s, lung sounds (LS) were considered as the sounds originating from within the thorax and they were justified mainly on the basis of their acoustic impression. For example, writings of the Hippocratic school, in about 400 B.C., describe chest (lung) sounds as splashing, crackling, wheezing, and bubbling sounds emanating from the chest [1]. An important contribution to the qualitative appreciation of LS was the invention of the stethoscope by René Theophil Laennec in 1816. His gadget, which was originally made of wood, replaced the "ear-upon-chest" detection procedure enhancing the emitted LS [2].

Medicine was one of the first fields where conceptual tools of rationality and empiricism were combined with investigation techniques to make the human body an object of knowledge [3].

The use of the stethoscope played a tremendous role in medicine. It was not so much about the actual artifact as the technique that it crystallized, that is, mediated auscultation [4]. Thus, listening became important to the construction of medical knowledge and its application throughout the development of a technique and a technology to go with it, to such an extent that a doctor's hearing tool became the symbol of a profession, even as far back as the 1820s [5].

The development of the stethoscope and mediated auscultation coincided with the development of new theories of sense perception based on a "separation of the senses." To this end, seeing and hearing are to be understood as fundamentally and absolutely different modes of knowing the world, although neither form of knowledge is guaranteed as truth. This was the moment when empiricism collided with subjectivism [6]. If the sensorium was, before this moment, a type of complex whole, it then became an accumulation of parts. Thus, not only vision but hearing became its own, specific object of knowledge over the course of the 19th century, supplemented through technique and technology. From then on, audition became a key modality in perceiving states of patients' bodies.

Attempts for a quantitative approach date to 1930, but the first systematic, quantitative measurement of their characteristics (i.e., amplitude, pitch, duration, and timing in controls and in patients) is attributed to McKusick in 1953 [7]; the door to the acoustic studies in medicine was finally opened.

1.2 MAIN CHARACTERISTICS AND CATEGORIZATION

Sound, in general, consists of audible vibrations transmitted through an elastic solid or a liquid or gas, created by alternating regions of compression and rarefaction of the elastic medium. The density of the medium determines the ease, distance, and speed of sound transmission. The higher the density of the medium, the slower sound travels through it. Sound waves are characterized by the generic properties of waves, which are frequency, wavelength, period, amplitude, intensity, speed, direction, and polarization (for shear waves only).

In LS categorization, the principal characteristics, that is, frequency, intensity, duration, and quality (timbre/texture), are mainly considered. LS are divided into two main categories: normal (NLS) and abnormal (ALS). NLS are certain sounds heard over specific locations of the chest during breathing in healthy subjects. The character of the NLS and the location at which they are heard defines them. Hence, the category of the NLS includes the following types [8]:

- Tracheal LS are heard over the trachea having a high loudness.
- Vesicular LS (VLS) are heard over dependent portions of the chest, not in immediate proximity to the central airways.
- Bronchial LS (BLS) are heard in the immediate vicinity of central airways, but principally over the trachea and larynx.

- Bronchovesicular LS (BVLS), which refers to NLS with a character in between VLS and BLS, are heard at intermediate locations between the lung and the large airways.
- Normal crackles are inspiratory LS heard over the anterior or the posterior lung bases [9, 10].

ALS consist of LS of a BLS or BVLS nature that appear at typical locations (where VLS are the norm). ALS are categorized between continuous adventitious sounds (CAS) and discontinuous adventitious sounds (DAS) [11], and include the following types [12–14]:

- *Wheezes* (WZ) are musical CAS that occur mainly in expiration and invariably associated with airway obstruction, either focal or general, rhonchi, low-pitched sometimes musical CAS that occur predominantly in expiration, associated more with chronic bronchitis and bronchiectasis than with asthma.
- *Stridors* are musical CAS caused by a partial obstruction in a central airway, usually at or near the larynx.
- *Crackles* are discrete, explosive, nonmusical DAS further categorized as:

    - Fine crackles (FC)—high-pitched exclusively inspiratory events that tend to occur in mid-to-late inspiration, repeat in similar patterns over subsequent breaths, and have a quality similar to the sound made by strips of Velcro™ being slowly pulled apart; they result from the explosive reopening of small airways that had closed during the previous expiration.
    - Coarse crackles (CC)—low-pitched sound events found in early inspiration and occasionally in expiration as well, develop from fluid in small airways, are of a "popping" quality, and tend to be less reproducible than the FC from breath to breath.

- Squawks are short, inspiratory wheezes that usually appear in allergic alveolitis and interstitial fibrosis [15], predominantly initiated with a crackle, and caused by the explosive opening and fluttering of the unstable airway that causes the short wheeze.
- Friction rub are DAS localized to the area overlying the involved pleura and occur in inspiration and expiration when roughened pleural surfaces rub together, instead of gliding smoothly and silently.

The variety in the categorization of LS implies changes in the acoustic characteristics either of the source or the transmission path of the LS inside the lungs because of the effect of a certain pulmonary pathology. It is likely that the time and frequency domain characteristics of the LS signals reflect these anatomical changes [16]. In particular, the time domain pattern of the NLS

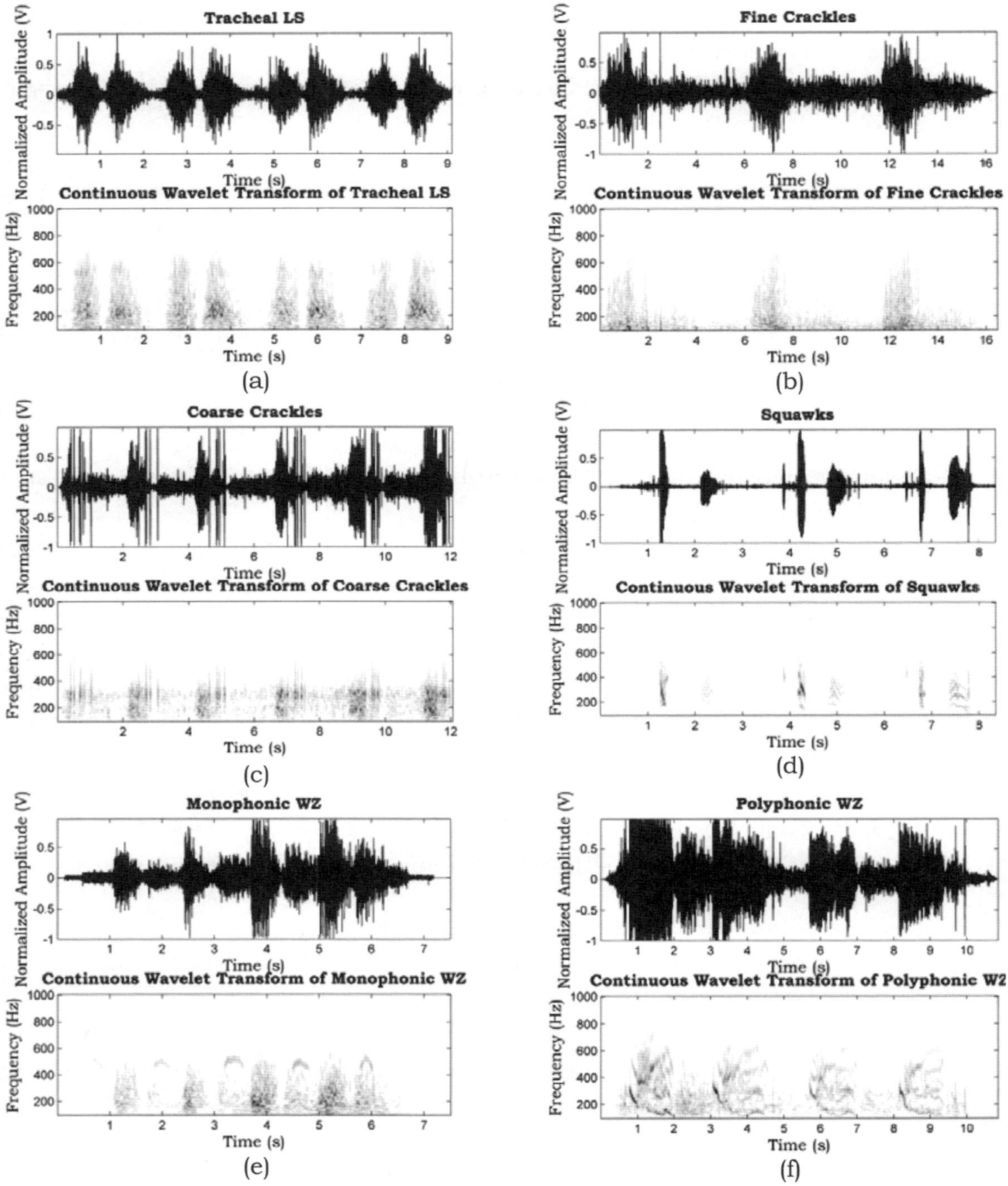

FIGURE 1.1: Typical examples of abnormal lung sounds (ALS). In all subfigures, the upper and lower parts illustrate the time- and time/frequency-domain representation of the ALS signal, respectively. (a) Tracheal LS, (b) fine crackles, (c) coarse crackles, (d) squawks, (e) monophonic wheezes, and (f) polyphonic wheezes.

resembles a noise pattern bound by an envelope, which is a function of the flow rate [16]. Tracheal sounds have higher intensity and a wider frequency band (0–2 kHz) than chest wall sounds (0–600 Hz) and contain more acoustic energy at higher frequencies [17]. The CAS time domain pattern is a periodic wave that may either be sinusoidal or a set of more complex, repetitive sound structures [16]. In the case of WZ, the power spectrum contains several dissonant-like peaks ("polyphonic" WZ) or a single peak ("monophonic" WZ), usually in the frequency band of 200–800 Hz, indicating bronchial obstruction [16]. Crackles have an explosive time-domain pattern, with a rapid onset and short duration [16]. It should be noted that this waveform may be an artifact of high-pass filtering [18]. Their time domain structural characteristics (i.e., a sharp, sudden deflection usually followed by a wave) provide a means for their categorization between FC and CC [19].

Some indicative examples of characteristic ALS are given in Figure 1.1. In this figure, the time domain and time/frequency domain representation (via the continuous wavelet transform; see Section 2.4) of ALS are depicted. These illustrations reveal the characteristics of ALS showing, for example, the explosive character of crackles superimposed on the vesicular sound (see Figure 1.1b, c) or the harmonic spread in the frequency domain when moving from monophonic wheezes to polyphonic ones (see Figure 1.1e, f).

For an extensive description and a variety of examples regarding LS structure and characteristics, the reader should refer to [16].

1.3 RECORDING ISSUES

1.3.1 Standards

LS auscultation is traditionally performed via the stethoscope. Listening to LS by means of a stethoscope involves several physical phenomena, such as vibrations of the chest wall that are converted into pressure variations of the air in the stethoscope, which are then transmitted to the eardrum. However, the stethoscope cannot be used in the quantitative analysis of bioacoustic signals (BAS), because it does not provide any means of signal recording. In addition, it presents a selective behavior in its frequency response instead of a flat one (at least in the area of interest, 40–4000 Hz).

One basic category of recording devices of LS signals refers to microphones. Whatever the type of microphone, it always has a diaphragm, like in the human ear, and the movement of the diaphragm is converted into an electric signal. Two major microphone approaches exist: the "kinematic" approach, which involves the direct recording of chest wall movement ("contact sensor"), and the 'acoustic' approach, which involves the recording of the movement of a diaphragm exposed to the pressure wave induced by the chest wall movement ("air-coupled sensor"). The chest wall movements are so weak that a free-field recording is not possible; it is essential to couple the diaphragm acoustically with the chest wall through a closed-air cavity. Whatever the approach, kinematic or acoustic, vibrations must be converted into electric signals using transduction principles of [20]: electromagnetic induction (movement of a coil in a magnetic field induces an electric current through the coil), condenser principle (changing the distance between the two plates of a charged capacitor induces a voltage fluctuation), and piezoelectric effect (bending of a crystal (rod, foil) for the induction of an electric charge on the surface).

The other category of recording devices includes piezoelectric accelerometers. A piezoelectric accelerometer applies the piezoelectric effect in such a way that the output voltage is proportional to the acceleration of the whole sensor. Early applications used heavyweight sensors with high sensitivity and good signal-to-noise ratio. The disadvantages, because of its heavy mass, are mechanical loading of the surface wall, difficulties with attachment, and a low resonance frequency (well within the band of interest). A piezoelectric accelerometer of very low mass (1 g) has been successfully applied in the past, yet it may be so fragile that routine clinical applications may be difficult [20].

Although both condenser microphones and piezoelectric contact sensors are displacement receivers, the waveforms that they deliver are different because of the coupling differences. Selection criteria of a device should also include size, average lifetime, and maintenance cost. Both sensors have several disadvantages, that is, piezoelectric sensors are very sensitive to movement artifacts (e.g., by the connecting wire, their characteristics depend on the static pressure against the body surface [21] and they are brittle), whereas condenser microphones need mounting elements that change the overall characteristics of the sound transduction [20].

1.3.2 Procedures and Considerations

The most common bandwidth for LS is from 60–100 Hz to 2 kHz when recorded on the chest, and from 60–100 Hz to 4 kHz when recorded over the trachea. In addition, recording of adventitious sounds on the chest requires a bandwidth from 60–100 Hz to 6 kHz [20]. Thus, sampling frequencies of 5–10 kHz are sufficient for the acquisition of all LS signals.

Electret microphones, such as Sony ECM 77, have a flat frequency response from 40 Hz up to 20 kHz in free space and are usually preferred in recording almost all types of LS. For the processing of LS, an analog system is required, which consists of a sensor, an amplifier, and filters that condition the signal before analog-to-digital (A/D) conversion (usually with a 12- or 16-bit resolution). A combination of low-pass filters (LPF) and high-pass filters (HPF) in cascade is usually applied. The purpose of using an HPF is to reduce the heart, muscle, and contact noises. The LPF is needed to eliminate aliasing. The amplifier is used to increase the amplitude of the captured signal so that the full A/D converter range can be optimally used, and sometimes to adjust the impedance of the sensor.

In recent years, some efforts of creating wireless sensors for LS recording have been proposed [22], based on Bluetooth® technology. The implementation of such concept could contribute to the continuous monitoring of LS and their archiving in a patient's database, thus realizing the concept

TABLE 1.1: Summary of recommendations for the case of LS signal acquisition with piezoelectric or condenser sensors [20]

SENSOR SPECIFICATION	
Frequency Response	Flat in the frequency range of the sound. Maximum deviation allowed, 6 dB
Dynamic range	>60 dB
Sensitivity	Must be independent of frequency, static pressure, and sound direction
Signal-to-noise ratio	>60 dB (S = 1 mV Pa^{-1})
Directional characteristics	Omnidirectional
COUPLING	
Piezoelectric contact	Immediate to the skin surface
Condenser air-coupled	Shape: conical; depth: 2.5–5 mm; diameter at skin: 10–25 mm; vented
FIXING METHODS	
Piezoelectric	Adhesive ring
Condenser	Either elastic belt or adhesive ring
NOISE AND INTERFERENCES	
Acoustic	Shielded microphones; protection from mechanical vibrations
Electromagnetic	Shielded twisted pair of coaxial cable
AMPLIFIER	
Frequency response	Constant gain and linear phase in the band of interest
Dynamic range	>60 dB
Noise	Less than that introduced by the sensor

TABLE 1.1: (*cont.*)	
HIGH-PASS FILTERING	
	Cutoff frequency: 60 Hz; roll-off: >18 dB octave^{-1}; phase as linear as possible; minimized ripple
LOW-PASS FILTERING	
	Cutoff frequency: above the higher frequency of the signal; roll-off: >24 dB octave^{-1}; minimized ripple

of "acoustic" file of each patient, something that is currently lacking from most clinical practice. Diffused distribution of many sensors over the patient's back and chest wall has also been proposed as a multichannel LS recording [23]. Based on these simultaneous multisensor recordings (e.g., ≥16 microphones), an acoustic mapping and imaging of the chest is feasible [24]. Although such an approach offers lower resolution compared to computerized tomography, X-ray, or functional magnetic resonance imaging techniques, it provides a new perspective in the exploitation of LS diagnostic information [25].

Table 1.1 tabulates recommendations [20] regarding the acquisition of LS signals.

1.4 TRENDS IN LS ANALYSIS

Although, in practice, LS interpretation is still based on the auscultative findings of the physician, modern computer-based analyses have been proposed as new means to extend LS capabilities by supplying information that is drawn not only directly from sound perception, but also from many other underlying characteristics that foster the diagnostic information of LS. Auscultation, however, remains a widespread technique, because it involves using the stethoscope. To this end, in most cases, the agreement of an automatic method with physician's interpretation is a basic criterion for its acceptance.

Through the years, LS analysis has branched out in various directions; the most popular include respiratory flow estimation [26, 27], heart sound cancellation [28–42], DAS detection and denoising [43–57], nonlinear analysis of LS [58–64], feature extraction, and classification [43, 52, 65–83].

In this book, focus is placed on two main categories of LS analysis from the perspective of advanced signal processing. In particular, the first one refers to the latest domains of LS representation that provide new insight on the structure and character of LS, forming new analysis tools, which can be of great assistance to the physician for a more efficient appreciation of the acquired LS. The second one includes up-to-date denoising techniques that take into account the nature of LS and provide enhanced signal quality, circumventing any noise effects in the interpretation of LS. The objective of presenting such methodologies is not only to reveal the importance of LS as indicators of respiratory health and disease, but also to shed light on the inherent characteristics of the advanced signal processing techniques that manage to adapt to the specific properties of LS, providing a new standpoint in the evaluation of respiratory acoustics.

• • • •

CHAPTER 2

New Domains in LS Presentation

2.1 OVERVIEW

Lung sound (LS) analysis has long existed between two main representations, that is, in time and frequency domain. Although LS time and spectral characteristics proved to be so far quite efficient in the interpretation of LS, they cannot always provide the full image of the diagnostic power of LS. To this end, some new analysis domains have been proposed to define some LS properties seen from different angles, capturing—perhaps, in a more pragmatic way—their behavior. These new representations and features of LS are examined in details in the subsequent sections.

2.2 HIGHER-ORDER STATISTICS (SPECTRA)

2.2.1 Epitomized Rationale

Questions about the Gaussianity, linearity, and minimum phase of the process under study often emerge within a biomedical signal processing framework. In the past, due to the lack of appropriate analytical tools, researches were forced to apply second-order (correlation-based) methods to just about every biomedical signal processing problem, ignoring those important distinguishing features. However, many real-world signals, such as almost all the biomedical ones, are actually non-Gaussian random processes. As a result, distinguishing features, such as non-Gaussianity, nonminimum phase, colored noise, nonlinearity, are important and must be accounted for in a signal processing context. To this end, in the recent decades, a class of tools has been developed to the point of practical application, known as higher-order statistics (spectra) (HOS), which have proven of potential value when dealing with non-Gaussian random processes [84].

HOS, also known as cumulants, are related to and may be expressed in terms of the moments of a random process. In general, the kth-order cumulant of a random process is defined in terms of the process's joint moments of orders up to k [84]. Just as the power spectrum (PS) (the Fourier transform of the autocorrelation) is a useful tool in the signal analysis, so too are the higher-order spectra, also known as polyspectra, which are the associated Fourier transforms of the cumulants. In PS estimation, the process under consideration is treated as a superposition of statistically uncorrelated harmonic components and the distribution of power among them is the estimation outcome.

As such, only linear mechanisms governing the process are investigated because of the suppression of the phase relations among the frequency components [85]. Consequently, the information contained in the PS suffices for the complete statistical description of a Gaussian process of known mean. Looking beyond the PS to obtain information regarding deviations from Gaussianness and presence of nonlinearities, the polyspectra should be used. The most often used ones are the third- and fourth-order spectra, also called *bispectrum* and *trispectrum*, respectively. From this perspective, the PS should be considered, in fact, as a member of the class of higher-order spectra, that is, as a second-order spectrum.

Cumulants not only reveal amplitude information about a process, but also reveal phase information. This is of great importance, because, as is well known, second-order statistics, that is, correlation, are phase-blind. A key characteristic, which differentiates cumulants from correlation, is that cumulants are blind to all types of Gaussian processes, whereas correlation is not. From a practical point of view, this means that when cumulant-based methods are applied to non-Gaussian (or, possibly, nonlinear) processes contaminated by additive Gaussian noise (even colored Gaussian one), they provide an analysis field of automatically improved signal-to-noise ratio (SNR).

Practically speaking, many biomedical signal processes are non-Gaussian, yet corrupted by measurement noise, which can often be realistically described as colored Gaussian process; hence, in these practical applications, the value of the HOS is apparent. Another very attractive property of cumulants is that they enable handlers to work with them as operators. On a practical level, this means that the cumulants of the sum of two statistically independent random processes equals the sum of the cumulants of the individual processes. Higher-order moments, by contrast, do not have that property; hence, they are much less convenient [84].

Based on the aforementioned HOS characteristics, placing LS in the HOS-based domain could reveal their true character, and hence, could further divulge their diagnostic value.

2.2.2 HOS Definitions: Higher-Order Statistics

Given a set of n real random variables $\{x_1, x_2, \ldots, x_n\}$, their *joint cumulants* of order $r = k_1 + k_2 + \cdots + k_n$, are defined as the coefficients in the Taylor expansion (provided it exists) of the second characteristic function $\tilde{\Psi}(\omega_1, \omega_2, \ldots, \omega_n) = \ln\left[E\{\exp(j(\omega_1 x_1 + \omega_2 x_2 + \cdots + \omega_n x_{\mathrm{n}})\}\right]$, about zero [86, 87], that is,

$$\mathrm{Cum}\left[x_1^{k_1}, x_2^{k21}, \ldots, x_n^{k_n}\right] \equiv (-j)^r \frac{\partial^r \tilde{\Psi}(\omega_1, \omega_2, \ldots, \omega_n)}{\partial \omega_1^{k_1} \partial \omega_2^{k_2} \cdots \partial \omega_n^{k_n}} \Big|_{\omega_1 = \omega_2 = \cdots = \omega_n = 0} \cdot \tag{2.1}$$

If $\{X(k)\}$, $k = 0, \pm 1, \pm 2, \ldots$ is a real stationary random process and its moments up to order n exist, then its nth-order cumulants, that is, its nth-order statistics, are given by the following equation:

$$c_n^x(\tau_1, \ldots, \tau_{n-1}) \equiv \mathrm{Cum}\left[X(k), X(k+\tau_1), \ldots, X(k+\tau_{n-1})\right]. \tag{2.2}$$

In the case where $\{X(k)\}$ is Gaussian, its third- and fourth-order cumulants, $c_3^x(\tau_1, \tau_2)$, $c_4^x(\tau_1, \tau_2, \tau_3)$, are identically zero. This can be easily deduced from the rewriting of (2.2) for a real zero-mean stationary random process $\{X(k)\}$ $k = 0, \pm 1, \pm 2, \ldots$, for $n = 3, 4$, in the form of

$$c_n^x(\tau_1, \ldots, \tau_{n-1}) = E\{X(\tau_1)\ldots X(\tau_{n-1})\} - E\{G(\tau_1)\ldots G(\tau_{n-1})\}, \tag{2.3}$$

where $\{G(k)\}$ is a Gaussian random process with the same second-order statistics as $\{X(k)\}$, and $E\{\cdot\}$ denotes the expectation value. What (2.3) clearly states is that cumulants not only display the amount of higher-order correlation, but also provide a measure of a distance of the random process from Gaussianity. Clearly, if $\{X(k)\}$ is Gaussian the cumulants are all zero; this is not only true for $n = 3$ and 4, but for all n.

For n stationary processes $\{X_i(k), i = 1, 2, \ldots, n\}$, the nth-order *cross-cumulant* sequence is defined as:

$$c_n^{x_1,\ldots,x_n}(\tau_1, \ldots, \tau_{n-1}) \equiv \mathrm{Cum}\left[X_1(k), X_2(k+\tau_1), \ldots, X_n(k+\tau_{n-1})\right]. \tag{2.4}$$

The relationship between fourth-order moments and cumulants for zero-mean signals [86] is given by

$$\mathrm{Cum}\left[x_1x_2x_3x_4\right] = E\{x_1x_2x_3x_4\} - E\{x_1x_2\}E\{x_3x_4\} - E\{x_1x_3\}E\{x_2x_4\} - E\{x_1x_4\}E\{x_2x_3\}. \tag{2.5}$$

For a real zero-mean stationary random process $\{X(k)\}$, $k = 0, \pm 1, \pm 2, \ldots$, by setting $\tau_1 = \tau_2 = \tau_3 = 0$ in (2.2), we obtain [86]:

$$\begin{aligned} \gamma_2^x &= c_2^x(0) = E\{X^2(k)\} \text{ (variance)} \\ \gamma_3^x &= c_3^x(0,0) = E\{X^3(k)\} \text{ (skewness)} \\ \gamma_4^x &= c_4^x(0,0,0) = E\{X^4(k)\} - 3\left(\gamma_2^x\right)^2 \text{ (kurtosis).} \end{aligned} \tag{2.6}$$

Skewness shows the degree of asymmetry of a distribution. Kurtosis is a measure of whether the data are peaked or flat relative to a normal distribution. That is, datasets with high kurtosis tend to have a distinct peak near the mean, decline rather rapidly, and have heavy tails. Datasets with low kurtosis tend to have a flat top near the mean rather than a sharp peak; a uniform distribution would be the extreme case.

2.2.3 HOS Definitions: Higher-Order Spectra

Assuming that $c_n^x(\tau_1,\ldots,\tau_{n-1})$ is absolutely summable, the nth-order polyspectrum is defined as the $(n-1)$-dimensional discrete Fourier transform of the nth-order cumulant [84], that is,

$$C_n^x(\omega_1,\ldots,\omega_{n-1}) = \sum_{\tau_1=-\infty}^{\infty} \cdots \sum_{\tau_{n-1}=-\infty}^{\infty} c_n^x(\tau_1,\ldots,\tau_{n-1}) \cdot \exp\left[-j\sum_{i=1}^{n-1}\omega_i\tau_i\right]$$

$$|\omega_i| \le \pi,\ i = 1,2,\ldots,n-1;\ |\omega_1+\omega_2+\cdots+\omega_{n-1}| \le \pi. \tag{2.7}$$

In general, $c_n^x(\omega_1,\ldots,\omega_{n-1})$ is complex for $n > 2$ and thus it can be expressed as

$$C_n^x(\omega_1,\ldots,\omega_{n-1}) = |C_n^x(\omega_1,\ldots,\omega_{n-1})| \exp\left[j\psi_n^x(\omega_1,\ldots,\omega_{n-1})\right]. \tag{2.8}$$

The nth-order polyspectrum is a 2π periodic function. The PS, bispectrum, and trispectrum are special cases of (2.7), that is,

- Power spectrum ($n = 2$):

$$C_2^x(\omega) = \sum_{\tau=-\infty}^{\infty} c_2^x(\tau) \cdot \exp[-j(\omega\tau)],\quad |\omega| \le \pi \tag{2.9}$$

- Bispectrum ($n = 3$):

$$C_3^x(\omega_1,\omega_2) = \sum_{\tau_1=-\infty}^{\infty}\sum_{\tau_2=-\infty}^{\infty} c_3^x(\tau_1,\tau_2) \cdot \exp[-j(\omega_1\tau_1+\omega_2\tau_2)],$$
$$|\omega_1| \le \pi,\quad |\omega_2| \le \pi,\quad |\omega_1+\omega_2| \le \pi \tag{2.10}$$

- Trispectrum ($n = 4$):

$$C_4^x(\omega_1,\omega_2,\omega_3) = \sum_{\tau_1=-\infty}^{\infty}\sum_{\tau_2=-\infty}^{\infty}\sum_{\tau_3=-\infty}^{\infty} c_4^x(\tau_1,\tau_2,\tau_3) \cdot \exp[-j(\omega_1\tau_1+\omega_2\tau_2+\omega_3\tau_3)],$$
$$|\omega_1| \le \pi,\quad |\omega_2| \le \pi,\quad |\omega_3| \le \pi,\quad |\omega_1+\omega_2+\omega_3| \le \pi \tag{2.11}$$

The normalized bispectrum or *bicoherence* (Bic) is defined as [86]:

$$\mathrm{Bic}(\omega_1,\omega_2) = \frac{C_3^x(\omega_1,\omega_2)}{\left[C_2^x(\omega_1)\,C_2^x(\omega_2)\,C_2^x(\omega_1+\omega_2)\right]^{1/2}}, \tag{2.12}$$

where $C_2^x(\omega_i)$, i=1, 2 is the PS defined in (2.9). From (2.12), it is apparent that bicoherence is a function that combines two completely different entities: the bispectrum, $C_3^x(\omega_1, \omega_2)$, and the PS, $C_2^x(\omega)$, of the process. The magnitude of bicoherence

$$\mathrm{BI}(\omega_1, \omega_2) = |\mathrm{Bic}(\omega_1, \omega_2)| \tag{2.13}$$

is called the bicoherence index (BI) and its estimation is not bounded above by 1.0 [88]. Under the Gaussian assumption, the value of the bicoherency is zero. The $\mathrm{BI}(\omega_1, \omega_2)$ quantifies the presence of *quadratic phase coupling* (QPC) between any two frequency components in the process due to their nonlinear interactions. Two frequency components are said to be quadratically phase coupled (with $\mathrm{BI}(\omega_1, \omega_2)$ close to or greater than 1.0) when there exists a third frequency component whose frequency and phase are the sum of the frequencies and phases of the first two components [86]. *Cross-bicoherence* refers to two N-sample processes $\{X(k)\}$, $\{Y(k)\}$, and is defined through the cross-bispectrum, $C_3^{xxy}(\omega_1, \omega_2)$, as

$$\mathrm{CBic}(\omega_1, \omega_2) = \frac{C_3^{xxy}(\omega_1, \omega_2)}{\left[C_2^x(\omega_1) C_2^x(\omega_2) C_2^y(\omega_1 + \omega_2)\right]^{1/2}}. \tag{2.14}$$

The magnitude of cross-bicoherence

$$\mathrm{CBI}(\omega_1, \omega_2) = |\mathrm{CBic}(\omega_1, \omega_2)| \tag{2.15}$$

is referred as *cross-bicoherence index* (CBI), and quantifies the presence of QPC between any two frequency components in the two processes due to their nonlinear interactions.

For a detailed description of HOS properties, symmetries, and estimators, the reader is referred to [86] and references contained therein.

2.2.4 The Parametric Approach

Moving on to a parametric estimation of bispectrum, we consider a real, zero-mean, and stable ARMA(p, q) process $\{X(k)\}$ given by

$$\sum_{i=0}^{p} a(i) X(k-i) = \sum_{j=0}^{q} b(j) W(k-j), \tag{2.16}$$

where $\{W(k)\}$ are independent and identically distributed (i.i.d.) random variables, independent from $\{W(m)\}$ for $m < k$, with zero mean and

$$E\{W(k) W(k+m)\} = Q\delta(m) \tag{2.17}$$

$$E\{W(k)W(k+m)W(k+n)\} = \beta\delta(m,n) \tag{2.18}$$

$\beta\delta(m, n)$ denotes Kronecker's delta function, $Q = E\{W^2(k)\}$ and $\beta = E\{W^3(k)\}$, that is, $\{W(k)\}$ is third-order white. Note that both $\{W(k)\}$ and $\{X(k)\}$ are non-Gaussian. The bispectrum of the system input and output are related by

$$B(\omega_1, \omega_2) = \beta H(\omega_1)H(\omega_2)H^*(\omega_1 + \omega_2), \tag{2.19}$$

where $H(\omega)$ is the frequency response function of the autoregressive moving average (ARMA) system given by

$$H(\omega) = \frac{\sum_{\ell=0}^{q} b(\ell)e^{-j\omega\ell}}{1 + \sum_{\ell=1}^{p} a(\ell)\mathrm{e}^{-j\omega\ell}}. \tag{2.20}$$

2.2.5 LS Quadratic Phase Coupling Detection

QPC refers to peaks at harmonically related positions in the PS. Because the existence of phase-coupling denotes the existence of nonlinearity, which, in turn, in many cases, is related with the existence of pathology, detection of QPC in LS signals using the BI could possibly provide an indicator of anatomical changes of lungs due to pathology.

A nonlinear analysis of musical LS based on BI (see (2.13)) is presented in [59]. The third-order statistics (TOS) and spectra were used both for the detection of QPC between distinct frequency components of musical LS and as a measure of their nonlinear interaction. This harmonic analysis was conducted on preclassified signals (wheezes, ronchi, stridors) selected from international teaching tapes. The derived results have shown a high degree of deviation from Gaussianity along with quadratic self-coupling ($f_2 = 2f_1$; $\phi_2 = 2\phi_1$) within the distinct frequencies of the corresponding PS of musical LS. This is shown in Figure 2.1, where an example of the recorded LS (inspiratory stridor recorded from an infant with croup [59]), the estimated PS, and bispectrum are depicted in Figure 2.1a, b, and c, respectively. From the three distinct peaks seen in the PS (i.e., 410.2, 820.4, 1230.6 Hz), a self-coupling is revealed in the bispectrum domain denoted by the clear peak at (410.2, 410.2) Hz.

Differences in the degree of QPC as well as the nonlinearity and the non-Gaussianity of the processed LS signals could establish a new field in characterization and feature extraction of the pulmonary pathologies associated with musical LS, such as asthma and chronic obstructive pulmonary disease (COPD) [59].

As reported in [52, 58], in the case of monophonic wheezes, the frequency pair of QPC belongs to the low frequency band. The fundamental frequency is revealed in the inspiratory phase

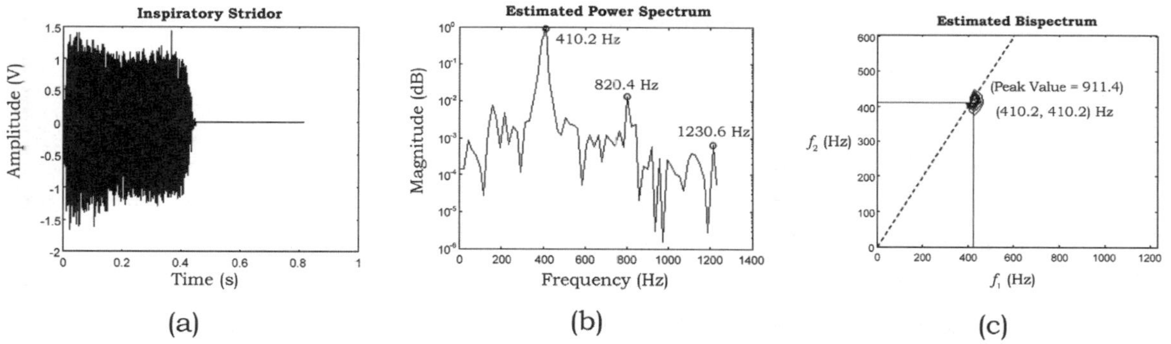

FIGURE 2.1: An example of QPC seen in the case of a recorded inspiratory stridor depicted in (a) from an infant with croup [59], (b) the corresponding PS, (c) the corresponding bispectrum estimated within the principal region in the bifrequency domain. The distinct peak in the bispectrum denotes existence of the QPC and strong deviation from the Gaussian assumption.

of a breath cycle, whereas its first harmonic dominates in the expiratory phase. This fact leads to the conclusion that monophonic wheeze consists of a single note or a single tonality, established by the octaves of the fundamental frequency, due to a nonlinear mechanism. In the case of polyphonic wheezes, the frequency pair of QPC belongs to a higher frequency band than that of monophonic wheezes. In the inspiratory phase, pairs of high frequencies perform QPC, whereas in the expiratory phase, the submultiple frequencies of those in the inspiratory phase perform QPC. The harmonics involved result in a polyphonic chord, because apart from the fundamental frequency (f_0), the octave ($2f_0$) and the fifth of the chord ($3f_0$) perform QPC. This result is consistent with the accepted theory that polyphonic wheezes are made up of several dissonant notes starting and ending simultaneously, such as a chord [12].

In the case of random wheezes, the nonlinearity and non-Gaussianity of the process are evident both in inspiratory and expiratory case. In cases of sibilant (SBR) and sonorous (SNR) ronchi, high- and low-frequency pairs perform QPC, because they correspond to high and low wheezes, respectively. From $(f_1, f_2)|_{expiratory\ SNR} = (1/3)(f_1, f_2)|_{inspiratory\ SBR}$, it can be concluded that sibilant and sonorous ronchi have similar nonlinear production mechanism but different transmission, because high frequencies are only amplified in the case of sibilant ronchi.

In the case of stridors, high frequency pairs, both in inspiration and expiration, perform QPC. The frequencies with QPC corresponding to inspiratory stridors from an infant with croup are higher than those corresponding to inspiratory stridors from a 9-year-old child with croup. As well known, stridor intensity is the only thing that distinguishes a stridor from a monophonic wheeze [12]. From the nonlinear analysis, it can be seen that $(f_1, f_2)_{inspiratory\ MW} = (1/5)\ (f_1, f_2)|_{expiratory\ ST}$ justifying their relationship, but the degree of nonlinearity is much larger in stridor than in monophonic wheeze. This difference can introduce another field of separating the two associated autoregressive musical lung sounds (AMLS).

2.2.6 Autoregressive-HOS Modeling of LS Source and Transmission

Modeling of the system that produces the bioacoustic signals (BAS) contributes to the understanding of their production mechanisms. The manner in which the pathology affects these mechanisms is reflected in the modeling parameters adopted, which is a very important issue in the analysis of BAS, because efficient modeling could lead to an objective description of the changes a disease imposes to the production or transmission path of the BAS.

LS originated inside the airways of the lung are modeled as the input to an all-pole filter, which describes the transmission of LS through the parenchyma and chest wall structure [89]. The output of this filter is considered to be the LS recorded at the chest wall. The recorded LS also contain heart sound interference, the reduction of which is thoroughly addressed in the next chapter. Muscle and skin noise, along with instrumentation noise, are modeled as an additive Gaussian noise. With this model, given a signal sequence of LS at the chest wall, an autoregressive (AR) analysis based on TOS, namely AR-TOS, can be applied to compute the model parameters. Therefore, the source and transmission filter characteristics can be separately estimated, as thoroughly described in [52, 58].

The AR-TOS is described from the following equation:

$$y_n + \sum_{i=1}^{p} a_i y_{n-i} = w_n, \quad a_0 = 1, \tag{2.21}$$

where, y_n represents a pth-order AR process of N samples ($n = 0, \ldots, N - 1$), a_i are the coefficients of the AR model, and w_n are i.i.d., third-order stationary, zero mean, with $E\{w_n^3\} = \beta(0$ and y_n independent of w_l for $n < l$. Because w_n is third-order stationary, y_n is also third-order stationary, assuming it is a stable AR model. For the model of (2.21), we can write the cumulant-based "normal" equations [90]:

$$\sum_{i=0}^{p} a_i R(\tau_1 - i, \tau_2) = 0, \ \tau_1 = 1, \ldots, p \quad \text{and} \quad \tau_2 = -p, \ldots, 0, \tag{2.22}$$

where $R(\tau_1,\tau_2)$ is the third-order cumulant sequence of the AR process. In practice we use sample estimates of the cumulants. Equation (2.22) yields consistent estimates of the AR parameters maintaining the orthogonality of the prediction error sequence to an instrumental process derived from the data [91].

The profound motivations behind the use of the AR-TOS model is the suppression of Gaussian noise, because TOS of Gaussian signals are identically zero. Hence, when the analyzed waveform consists of a non-Gaussian signal in additive symmetric noise (e.g., Gaussian), the parameter estimation of the original signal with TOS takes place in a high-SNR domain, and the whole parametric presentation of the process is more accurate and reliable [86, 92]. The model used for the LS originating inside the airways considers the LS source as the output from an additive combination of three types of noise sequences [17]. The first sequence (periodic impulse) describes the continuous adventitious sound sources, because they have characteristically distinct pitches, and they are produced by periodic oscillations of the air and airway walls [12, 71].

The second sequence (random intermittent impulses) describes the crackle sources, because they are produced by sudden opening/closing of airways or bubbling of air through extraneous liquids in the airways—both phenomena associated with sudden intermittent bursts of sounds energy [12, 71]. Finally, the third sequence (white non-Gaussian noise) describes the breath sound sources, because they are produced by turbulent flow in a large range of airway dimensions [12, 71]. The estimation of the AR-TOS model input (LS source) can be derived from the prediction error via inverse filtering [52, 93].

Two examples of AR-TOS ($p = 2$) modeling (fine crackles from pulmonary fibrosis and squawks from interstitial fibrosis) [58, 94] are shown in Figure 2.2(i) and (ii), respectively. In the case of fine crackles [Figure 2.2(i)], the estimated source waveform [Figure 2.2(i), b] contains impulsive bursts, corresponding to fine crackles [Figure 2.2(i), a]. This could be explained by the associated-with-fine-crackles phenomenon of explosive reopening of small airways that had closed during the previous expiration. The abnormal airway closure that precedes the "crackling" reopening is due to increased lung stiffness [91].

In the case of squawks [Figure 2(ii)], the estimated source [Figure 2.2(ii), b], combines impulsive bursts followed by short, almost exponential, decaying periodic train of impulses. This result is consistent with the accepted theory that squawks are produced by the explosive opening and decaying fluttering of an unstable airway [91]. The transmission filter response in the case of fine crackles [Figure 2.2(i), c] is centered at high frequencies (≈530 Hz), whereas in the case of squawks [Figure 2.2(ii), c], it is centered at lower frequencies (≈380 Hz). From the magnitude of parametric bispectrum of both cases [Figure 2.2(i), d and (ii), d], it can be noticed that pulmonary fibrosis converts the transmission path to a band-pass filter (350–700 Hz), whereas interstitial fibrosis converts it to a rather smoothed low-pass one (cutoff frequency≈480 Hz).

2.3 LOWER-ORDER STATISTICS

2.3.1 Epitomized Rationale

A broad class of non-Gaussian phenomena encountered in practice can be characterized as impulsive. Signals and/or noise in this class tend to produce large-amplitude excursions from the average value more frequently than Gaussian signals. They are more likely to exhibit sharp spikes or occasional bursts of outlying observations than one would expect from normally distributed signals. As a

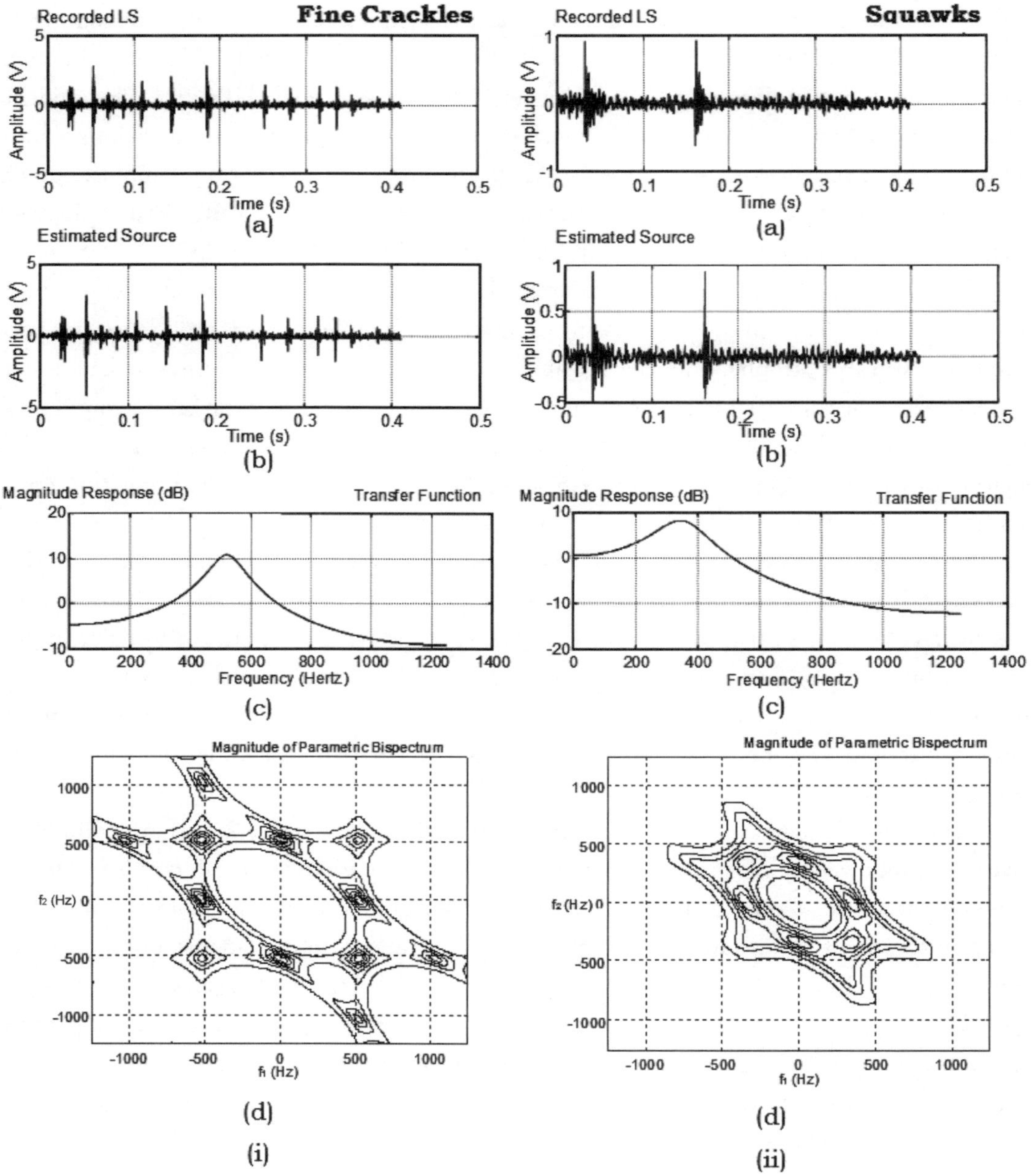

FIGURE 2.2: An example of AR-TOS modeling of LS source and transmission for the case of (i) fine crackles and (ii) squawks [58, 94]. In both columns, (a), (b), (c), and (d) correspond to the recorded LS signal, the AR-TOS-based estimated source, the AR-TOS-based estimated transfer function, and the corresponding parametric bispectrum, respectively. Differences in behavior between the two examined LS in the AR-TOS analysis domain are evident.

result, their density function decay in the tails less rapidly than the Gaussian density function [95]. Among the different types of LS, there are some that seem to belong in the aforementioned class, occurring with a different degree of impulsiveness. Because modeling based on α-stable distribution related to *lower-order statistics* (LOS) is appropriate for enhanced description of impulsive processes [95], its use could provide a useful processing tool for analysis of impulsive LS.

Sustaining the non-Gaussian assumption in the context of analyzing impulsive processes, among the various distribution models that were suggested in the past, the α-stable distribution is the only one that is motivated by the generalized central limit theorem. This theorem states that the limit distribution of the sum of random variables with possibly infinite variances is stable distribution [96]. Stable distributions are defined by the stability property stating that a random variable, X, is stable if and only if the sum of any two independent random variables with the same distribution as X also has the same distribution [96].

Analysis of LS with explosive character, such as crackles, via LOS-based modeling provides an innovating perception in the analysis of impulsive LS, suggesting a new field in their processing for diagnostic feature extraction.

2.3.2 LOS Definitions: α-Stable Distribution

A univariate distribution function $F(x)$ is called α-stable if its characteristic function can be expressed in the following form [95]:

$$\phi(t) = \exp\{ jat - \gamma |t|^{\alpha} [1 + j\beta \text{sign}(t) \omega(t, \alpha)]\}, \tag{2.23}$$

where

$$\omega(t,\alpha) = \begin{cases} \tan\frac{\alpha\pi}{2} & \text{for } \alpha \neq 1 \\ \frac{2}{\pi}\log|t| & \text{for } \alpha = 1, \end{cases} \tag{2.24}$$

$\text{sign}(t) = 1, 0$, and -1 for $t > 0$, $t = 0$, and $t < 0$, respectively, and

$$-\infty < a < \infty, \gamma > 0,\ 0 < \alpha \leq 2,\ -1 \leq \beta \leq 1. \tag{2.25}$$

Thus, the following four parameters can completely determine the α-stable distribution: (1) the location parameter γ, which is the symmetry axis; (2) the scale parameter , also called the *dispersion*, which, in analogy to the variance of the Gaussian distribution, is a measure of the deviation around the mean; (3) the symmetry parameter β, which is the index of skewness; and (4) the characteristic exponent, a, which is a measure of the thickness of the tails of the distribution (a small value of a implies considerable probability mass in the tails of the distribution, whereas a large value of α implies considerable probability mass in the central location of the distribution).

The special cases $\alpha = 2$ and $\alpha = 1$ with $\beta=0$ correspond to the Gaussian distribution and the Cauchy distribution, respectively. When $\beta=0$, the distribution is symmetric about the center α, in which case the distribution is called symmetric α-stable (SαS), and its characteristic function is of the form

$$\phi(t) = \exp\left\{jat - \gamma\,|t|^{\alpha}\right\} \tag{2.26}$$

For SαS distributions, α is the mean when $1 < \alpha < 2$ and the median when $0 < \alpha < 1$.

2.3.3 LOS Definitions: Fractional Lower-Order Moments and Covariation Coefficient

Although the second-order moment of an SαS random variable X with $0 < \alpha < 2$ does not exist, all moments of order less than α do exist and are called the *fractional low-order moments* (FLOM), defined as follows:

$$E\left(|X|^{p}\right) = C(p,\alpha)\gamma^{p/\alpha},\ 0 < p < \alpha,\ a = 0, \tag{2.27}$$

where

$$C(p,\alpha) = \frac{2^{p+1}\Gamma\left(\frac{p+1}{2}\right)\Gamma(-p/\alpha)}{\alpha\sqrt{\pi}\,\Gamma(-p/2)}, \tag{2.28}$$

$E(\cdot)$ denotes the expectation value and $\Gamma(\cdot)$ is the usual gamma function.

The covariation coefficient of jointly SαS random variables X and Y with $\alpha > 1$—which, although it plays the role of correlation coefficient of second-order random variables, can become unbounded—is given by

$$\lambda_{XY} = \frac{E(XY^{\langle p-1\rangle})}{E(|Y|^{p})}, \tag{2.29}$$

for any $1 \le p < \alpha$ For a thorough description of α-stable distributions, the reader is referred to [95].

2.3.4 LOS Analysis Tools: Converging Variance Test

Because the property that differentiates the Gaussian and non-Gaussian distributions is that non-Gaussian stable distributions do not have finite variance, a test based on the convergence of the variance of the population distribution can lead to the adoption of the right assumption. Specifically, if the population distribution $F(x)$ has finite variance, then the *running sample variance* S_n^2, defined as

$$S_n^2 = \frac{1}{n}\sum_{k=1}^{N}(X_k - \bar{X}_n)^2 \quad 1 \leq n \leq N, \tag{2.30}$$

where

$$\bar{X}_n = \frac{1}{n}\sum_{k=1}^{n} X_k, \tag{2.31}$$

and X_k, $k = 1, \ldots, N$ are samples from the distribution, should converge to a finite value [95].

2.3.5 LOS Analysis Tools: Parameter Estimates for SαS Distributions

Because the three parameters (α, a, and γ) determine the SαS distribution, their estimation from the realizations of a symmetric stable random variable could reflect the differences among various types of impulsive LS.

Although estimation of the parameters of a stable distribution is generally severely hampered by the lack of known closed-form density functions (for all but a few members of the stable family), there are several numerical methods that have been suggested in the literature that can be used to perform reliable parameter estimation: fractile method [97], regression method [98], log|SαS| method [95], and negative-order moment method [95]. A short description of the first three methods is included here. For convenience, γ is replaced by a new parameter, c, defined by

$$c = \gamma^{1/\alpha}. \tag{2.32}$$

For the symmetric stable law, the fractile estimate

$$\hat{\upsilon}_\alpha = \frac{\hat{x}_{0.95} - \hat{x}_{0.05}}{\hat{x}_{0.75} - \hat{x}_{0.25}}, \tag{2.33}$$

where $\hat{x}_f$ is the f fractile ($0 < f < 1$), which leads to a consistent estimate of α, $\hat{\alpha}$, using tables, such as those in [97], with matched value of $\hat{\upsilon}$. The location parameter a is estimated as the 75% truncated sample mean of the X_k, ($k = 1, \ldots, N$) observations, $\hat{a}_{75}$, that is, as the arithmetic mean of the middle 75% of the ranked observations. A consistent estimator of c can be found by the following equation:

$$\hat{c} = \frac{\hat{x}_{0.75} - \hat{x}_{0.25}}{\upsilon_c(\hat{\alpha})}, \tag{2.34}$$

using tables, such as those in [97], for calculation of $\hat{\upsilon}_c(\hat{\alpha})$.

The regression method [98] uses the fractile parameter estimates as initial estimates and proceeds iteratively until the change in the estimated values between two successive iterations, *acc*, is sufficient small. The new estimates can be found by solving (via least-squares) the following over-determined system of linear equations:

$$y_l = \mu + \alpha w_l \quad l = 1, \ldots, L, \tag{2.35}$$

where

$$w_l = \log|t|,\ t = (\pi l/25), \tag{2.36}$$

$$y_l = \log\left(-2\log\left|\frac{1}{N}\sum_{k=1}^{N}\exp\left(jtX'_k\right)\right|\right), \tag{2.37}$$

$$X'_k = (X_k - \hat{a}_{75})/\hat{c}_i, \quad k = 1, \ldots, N, \tag{2.38}$$

$$\mu = \log(2c^{\alpha}). \tag{2.39}$$

The log|SαS| method [95] uses the mean, $E(Y)$, and the variance, Var(Y), of a new random variable $Y = \log|X|$, for the estimation of α and γ as follows:

$$\hat{\alpha} = \left(1\Big/\left(\frac{6\mathrm{Var}(Y)}{\pi^2} - 0.5\right)\right)^{1/2} \hat{\gamma} = \exp\left(\hat{\alpha}\left(E(Y) - C_e\left(\frac{1}{\hat{\alpha}} - 1\right)\right)\right), \tag{2.40}$$

where $C_e = 0.57721566\ldots$is the Euler's constant.

2.3.6 LOS Analysis Tools: AR Modeling of SαS Distributions

A pth-order AR SαS process is described by the following equation:

$$X(n) = \sum_{i=1}^{P} \mathrm{a}_i X(n-i) + U(n), \tag{2.41}$$

where $\{U(n)\}$ is a sequence of i.i.d. SαS random variables of characteristic exponent α and dispersion Y_u. Furthermore, the stationary process $X(n)$ is SαS and $X(n)$ and $U(n+j)$ are independent for any $j > 0$. From the several proposed methods of estimating α_i ($i = 1, \ldots, P$) from observations of the output $X(n)$, the least-squares (LS) method [99] and the method that combines FLOM with the Yule–Walker (YW) equation [95] are usually adopted.

In the LS case, the AR model is treated as if it is driven by a second-order process and thus the linear least-squares (LSq) regression can be used. In that case, the LS estimates of α_i $(i \hat{\alpha}\, 1, \ldots, P)$ are the solution of the following equation:

$$\hat{\mathbf{C}}_{\mathrm{LS}}\hat{\mathbf{a}}_{\mathrm{LS}} = \hat{\mathbf{b}}_{\mathrm{LS}} \tag{2.42}$$

where $\hat{\mathbf{C}}_{\mathrm{LS}} = (\hat{\lambda}(i,j))$, $\hat{\mathbf{b}}_{\mathrm{LS}} = (\hat{\lambda}(i))$, and for sufficient large data length, $\hat{\lambda}(i,j) \approx \hat{\lambda}_{\mathrm{LSq}}(1-j)$ and $\hat{\lambda}(i) \approx \hat{\lambda}_{\mathrm{LSq}}(i)$, where $\hat{\lambda}_{\mathrm{LS}}(i)$ is the LSq estimate of the covariation of $X(n+i)$ and $X(n)$.

In the case of YW equation using FLOM, the coefficients of the AR system can be found by solving the following system of YW linear equations:

$$\mathbf{Ca} = \mathbf{b}, \tag{2.43}$$

where $b = [\lambda(1)\lambda(2)\ldots\lambda(P)]^{\mathrm{T}}$, $a = [a_1\ a_2\ \ldots\ a_P]^{\mathrm{T}}$ and

$$\mathbf{C} = \begin{bmatrix} \lambda(0) & \lambda(-1) & \cdots & \lambda(1-P) \\ \lambda(1) & \lambda(0) & \cdots & \lambda(2-P) \\ \vdots & \vdots & \ddots & \\ \lambda(P-1) & \lambda(P-2) & \cdots & \lambda(0) \end{bmatrix}, \tag{2.44}$$

where

$$\lambda(k) = E\left\{X(n)X(n-k)^{\langle p-1\rangle}\right\} \Big/ E\left\{|X(n-k)|^{p}\right\}, \tag{2.45}$$

and $p \geq 1$ is arbitrary.

2.3.7 LOS Analysis of Discontinuous Adventitious Sounds

Following this approach, impulsive LS [i.e., discontinuous adventitious sounds (DAS)] with explosive character can be modeled via LOS [76]. From the estimated parameters of the SαS distribution of the analyzed impulsive LS using the log|SαS| method [95], it is derived that the covariation coefficient $\lambda_{\mathrm{SQ\text{-}FC}}$ calculated for the cases of sound sources of squawks (SQ) and fine crackles (FC) [76] shows an almost 50% correlation between the SQ and FC, confirming the accepted theory that SQ are produced by the explosive opening, due to an FC, and decaying fluttering of an unstable airway [12, 71].

Figure 2.3 depicts the estimated LS source for the case of FC, coarse crackles (CC), and SQ when using the LOS-based AR modeling of (2.40). In fact, by using inverse filtering, an estimation of the input $U(n)$ can be found (Figure 2.3a–c for FC, CC, and SQ, respectively) [76]. Visual inspection of the findings confirms the explosive character of the source sound of impulsive LS, because they are produced by sudden changes in the acoustic properties of the lung.

By estimating the characteristic exponent α for each category (crackles, background noise, and artifacts), using the log|SαS| method [95], a classification criterion could be established based on the estimated α values. This approach has been adopted by Hadjileontiadis et al. [100], who deduced that FC follow approximately a Cauchy distribution (mean $\alpha = 0.92$, which is close to $\alpha = 1.0$ that holds in the case of Cauchy distribution), whereas the CC deviate both from Cauchy and Gaussian distributions (mean $\alpha = 1.33$; the artifacts have high impulsiveness with values of $\alpha \ll 1.0$). These results are illustrated in Figure 2.4. The events that resemble the vesicular sound are seen as background noise and are modeled as Gaussian processes (values of $\alpha > 1.8$ or very close to 2.0, which holds in the case of Gaussian distribution). Using this approach, disputed sounds that could not be classified by the physician (marked as A?) are clearly classified by the proposed method, as shown in Figure 2.4e.

The above examples show that SαS distribution- and LOS-based modeling of impulsive LS provides a measure of DAS impulsiveness, and, at the same time, reveals the underlying relationships between the associated production mechanisms and pathology.

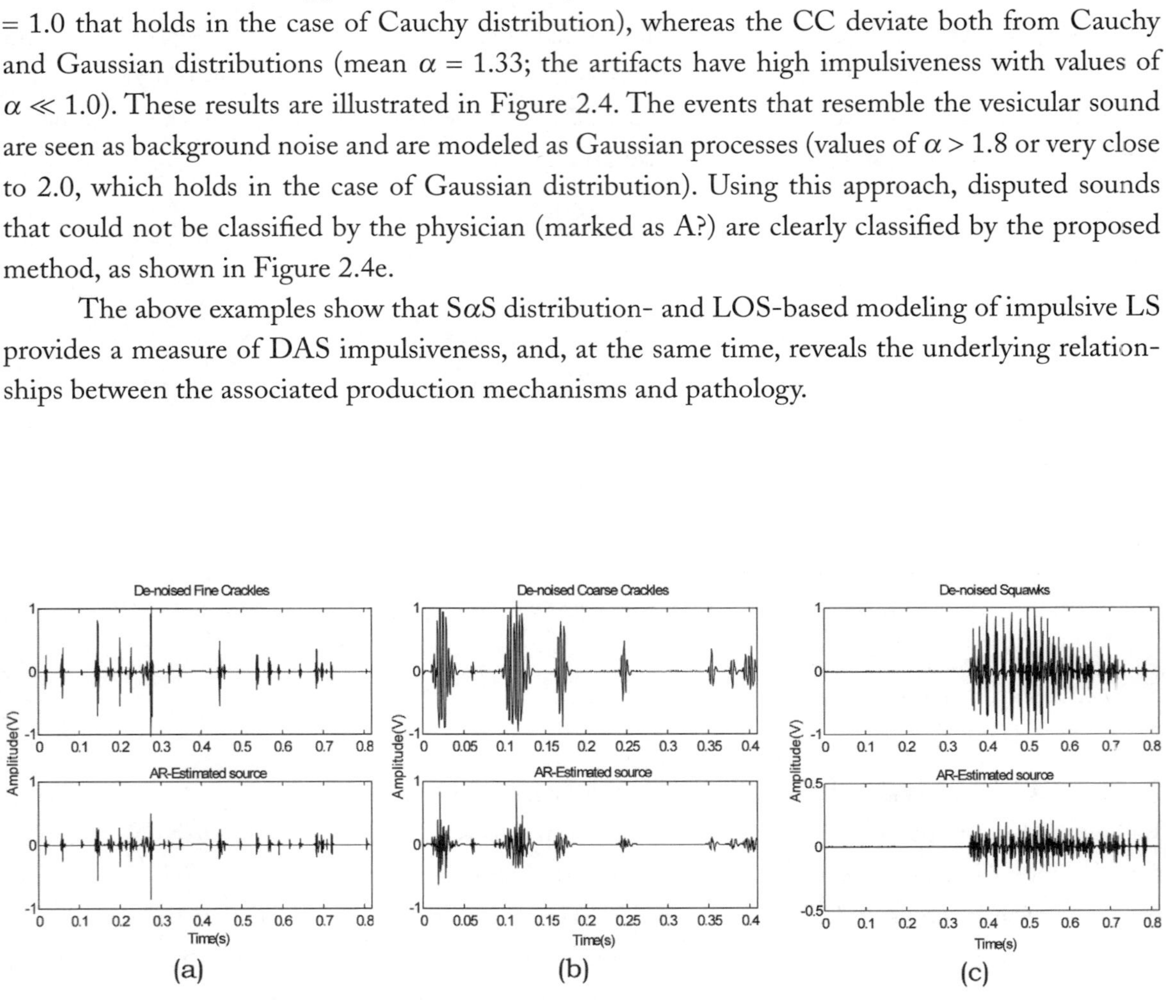

FIGURE 2.3: Estimation of the LS source sound for recorded DAS. In all subfigures, the upper and lower parts denote the recorded DAS (after denoising) and their estimated source using AR-LOS modeling. (a) Fine crackles, (b) coarse crackles, and (c) squawks [76].

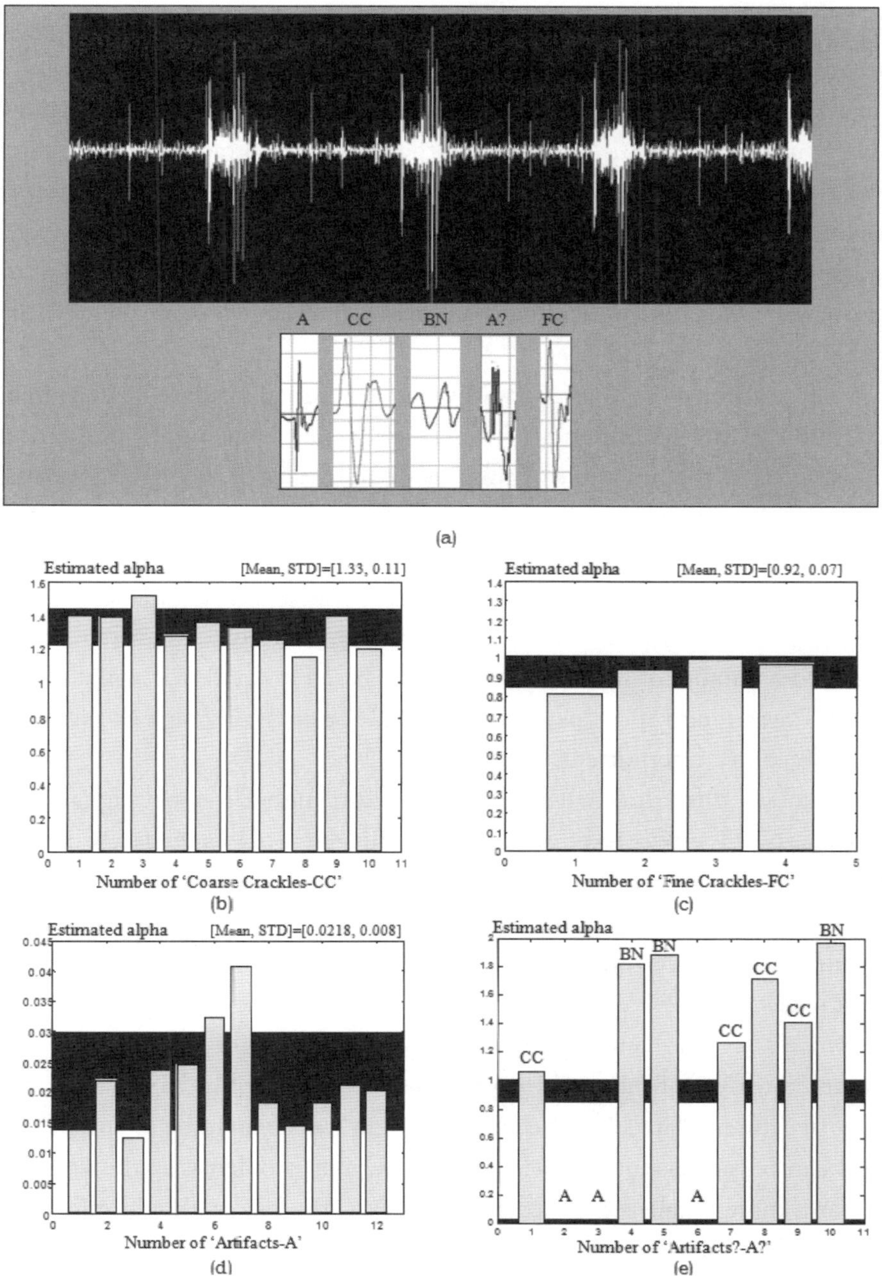

FIGURE 2.4: (a) Analyzed signal [100]. (b–d) Estimated α for crackles and artifacts. (e) Classification results of disputed peaks (A?) using SαS modeling. FC, fine crackles; CC, coarse crackles; A, artifacts; BN, background noise.

2.4 WAVELET ANALYSIS

2.4.1 Epitomized Rationale

The notion of a wavelet (i.e., a small wave) has evolved rapidly from its introduction by Grossmann and Morlet [101] in the mid 1980s as applied to the analysis of properties of seismic and acoustic signals. Nowadays, the family of analyzing functions dubbed wavelets is being increasingly used in problems of pattern recognition; in processing and synthesizing various signals (e.g., speech); in analysis of images of any type (e.g., iris images, X-rays, satellite images, an image of mineral, etc.); for study of turbulent fields, for contraction (compression) of large volumes of information, and in numerous other cases.

The wavelet transform (WT) of a one-dimensional signal involves its decomposition over a basis obtained from a soliton-like function (wavelet), possessing some specific properties, by dilations and translations. Each of the functions of this basis emphasizes both a specific spatial (temporal) frequency and its localization in physical space (time). Thus, unlike the Fourier transform traditionally used in signal analysis, WT offers a two-dimensional expansion of a given one-dimensional signal, with the frequency and the coordinate treated as independent variables. Consequently, the signal could be simultaneously analyzed in physical (time, coordinate) and frequency spaces. This ability has spawned a number of sophisticated wavelet-based methods for signal manipulation and interrogation. WT was found to be particularly useful for analyzing signals that can best be described as aperiodic, noisy, intermittent, and transient [102].

One faces known difficulties when processing short high-frequency signals, or signals with localized frequencies. WT proves to be an extremely efficient tool in adequately decoding such signals because elements of its basis are well localized and possess a moving time–frequency window. It is not a coincidence that many researchers refer to wavelet analysis as a "mathematical microscope," because this term accurately conveys the remarkable capability of the method to offer a good resolution at different scales. This so-called microscope reveals the internal structure of an essentially inhomogeneous process (or field) and exposes its local scaling behavior [103]. In this vein, WT picks out "coherent structures" in a time signal at various scales by shifting the wavelet along the signal, and so coherent structures related to a specific dilation in the signal to be identified.

Wavelets can be successfully applied to solve various problems, as the vast relevant literature justifies. For a more detailed description about WT, the reader should refer to the works of Daubechies [104], Farge [105], and Mallat [106].

2.4.2 Continuous Wavelet Transform

WT analysis uses wavelets to convolve them with the signal under investigation [102]. The WT of a continuous signal with respect to the wavelet function is defined as [102]

$$W_s(a,b) = \int_{-\infty}^{\infty} s(t)\psi_{a,b}^{*}(t)\mathrm{d}t, \tag{2.46}$$

where $s(t)$ is the time-domain signal ($s(t) \in L^2(\Re)$), * is the complex conjugate, $\psi_{a,b}(t)$ is the mother wavelet scaled by a factor a, ($a > 0$), and dilated by a factor b, that is,

$$\psi_{a,b}(t) = \frac{1}{\sqrt{a}}\psi\left(\frac{t-b}{a}\right). \tag{2.47}$$

Substituting (2.47) in (2.46), we get the definition of the continuous wavelet transform (CWT) as

$$W_s(a,b) = \frac{1}{\sqrt{a}}\int_{-\infty}^{\infty} s(t)\psi^{*}\left(\frac{t-b}{a}\right)\mathrm{d}t. \tag{2.48}$$

For $b = 0$, (2.48) takes the form of

$$W_s(a) = \frac{1}{\sqrt{a}}\int_{-\infty}^{\infty} s(t)\psi^{*}\left(\frac{t}{a}\right)\mathrm{d}t. \tag{2.49}$$

If we assume $s(t)$ to be a sine wave (e.g., a musical LS signal, i.e., wheeze), then

$$s(t) = A\sin\left(\omega_s t + \varphi\right), \tag{2.50}$$

where ω_s and ϕ are the angular frequency and phase of the sine wave, respectively, and set $\omega_s = \lambda\omega_c/a$, where ω_c/a is the angular frequency of the analyzing mother wavelet with ω_c being its central angular frequency ($a = 1$), and $\lambda = \omega_s/(\omega_c/a)$, then, assuming $\psi \in \Re$, (2.50) becomes:

$$W_s(a) = \frac{A}{\sqrt{a}}\int_{-\infty}^{\infty} \sin\left(\lambda\omega_c t/a + \varphi\right)\psi\left(\frac{t}{a}\right)\mathrm{d}t, \tag{2.51}$$

which, after using an intermediate parameter of $k = t/a$, it can be simplified in the form

$$W_s(a) = A\sqrt{a}\int_{-\infty}^{\infty} \sin(\lambda\omega_c t + \varphi)\psi(t)\mathrm{d}t. \tag{2.52}$$

Inverse CWT is defined as [102]

$$x(t) = \frac{1}{C_g} \int_{-\infty}^{\infty} \int_{0}^{\infty} W_s(a,b)\psi_{a,b}(t)\frac{\mathrm{d}a\mathrm{d}b}{a^2}, \tag{2.53}$$

where C_g is an admissibility constant [102], and its value depends on the chosen wavelet.

2.4.3 Discrete WT and Multiresolution Representation

By selecting discrete values for a, b parameters, a discrete form of the wavelet of (2.47) is produced. When a power-of-two logarithmic scaling of both the dilation and translation steps is adopted (known as the "dyadic grid arrangement"), a dyadic grid wavelet is formed,

$$\psi_{m,n}(t) = 2^{-m/2}\psi\left(2^{-m}t - n\right), \tag{2.54}$$

where m and n are integers that control the wavelet dilation and translation, respectively. Discrete dyadic grid wavelets are commonly chosen to be orthonomal; they are both orthogonal to each other and normalized to have unit energy. Using the dyadic grid wavelet of (2.54), discrete WT (DWT) can be defined as

$$T_{m,n} = \int_{-\infty}^{\infty} x(t)\psi_{m,n}(t)\mathrm{d}t, \tag{2.55}$$

along with the inverse DWT,

$$x(t) = \sum_{m=-\infty}^{\infty} \sum_{n=-\infty}^{\infty} T_{m,n}\psi_{m,n}(t), \tag{2.56}$$

requiring the summation over all integer m and n.

Orthonormal dyadic discrete wavelets are associated with scaling function and their dilation equations [102]. The scaling function is associated with the smoothing of the signal and is given by

$$\varphi_{m,n}(t) = 2^{-m/2}\varphi(2^{-m}t - n). \tag{2.57}$$

The scaling function is orthogonal to translations of itself, yet not to dilations of itself. The convolution of the scaling function with the signal produces approximation coefficients, that is,

$$S_{m,n} = \int_{-\infty}^{\infty} x(t)\varphi_{m,n}(t)\mathrm{d}t, \tag{2.58}$$

that are simply weighted averages of the continuous signal factored by $2^{\frac{m}{2}}$.

The signal $x(t)$ can be represented using a combined series expansion using both the approximation coefficients and the wavelet (detail) coefficients in the following form:

$$x(t) = \sum_{n=-\infty}^{\infty} S_{m_0,n}\varphi_{m_0,n}(t) + \sum_{m=-\infty}^{m_0}\sum_{n=-\infty}^{\infty} T_{m,n}\psi_{m,n}(t), \tag{2.59}$$

revealing that the original signal can be expressed as a combination of an approximation of itself, at arbitrary scale index m_0, added to a succession of signal details from scales m_0 down to negative infinity. By denoting the signal detail at scale m as

$$d_m(t) = \sum_{n=-\infty}^{\infty} T_{m,n}\psi_{m,n}(t). \tag{2.60}$$

Equation (2.59) can be written in the form of

$$x(t) = x_{m_0}(t) + \sum_{m=-\infty}^{m_0} d_m(t), \tag{2.61}$$

hence,

$$x_{m-1}(t) = x_m(t) + d_m(t). \tag{2.62}$$

Equation (2.62) represents the *multiresolution representation*, because it tells us that if we add the signal detail at an arbitrary scale (index m) to the approximation at that scale we get the signal approximation at an increased resolution, that is, at a smaller scale (index $m-1$) [102].

The above representation can be realized through the concept of vector spaces. For example, in the case of an image analysis, for each vector space, there is another vector space of higher resolution until you get to the final image. Also, each vector space contains all vector spaces that are of lower resolution. The basis of each of these vector spaces is the scale function for the wavelet. For practical purposes, one can think of an image as a vector space—such that V^j would be the perfectly normal image and V^{j-1} would be that image at a lower resolution until you get to V^0, where you just have 1 pixel in the entire image. For each vector space V^j, there is an orthogonal complement called W^j and the basis function for this vector space is the wavelet.

The decomposition of the input signal into approximation and detail space is called *multiresolution approximation* [107], which can be realized using a pair of finite impulse response filters (and their adjoints), which are low pass and high pass, respectively, defining a *multiresolution decomposition–multiresolution reconstruction* scheme (MRD–MRR).

An analytical description and details regarding the implementation of wavelet analysis can be found in [106–111] and in the references contained therein.

2.4.4 Wavelet-based Analysis of LS

Considering the characteristics of various LS signals, a series of research efforts were developed toward the application of WT to LS analysis. In the work of Gross et al. [112], parameters based on multiresolution approximation were able to detect typical pneumonia LS at an early stage of the disease in all 16 examined patients. In particular, using Daubechies-8 coefficient quadratic mirror filters [108], a multiresolution approximation was performed and the bronchial breathing sound pattern was detected by using WT-based ratio R between inspiration and expiration of the frequency band 345–690 Hz (scale 3) described in [113].

Figure 2.5 illustrates the distribution of the WT coefficients at scale 3 during the breathing cycle (Figure 2.5a) for the healthy (Figure 2.5b) and the pneumonia (Figure 2.5c) sites for a patient with bronchial breathing and no crackles. As shown in this figure, WT coefficients in the pneumonia site exhibit reduced amplitude compared to the ones from the healthy site, following a reversed pattern (healthy site: higher amplitude in inspiration and lower in expiration; pneumonia site: lower amplitude in inspiration and higher in expiration).

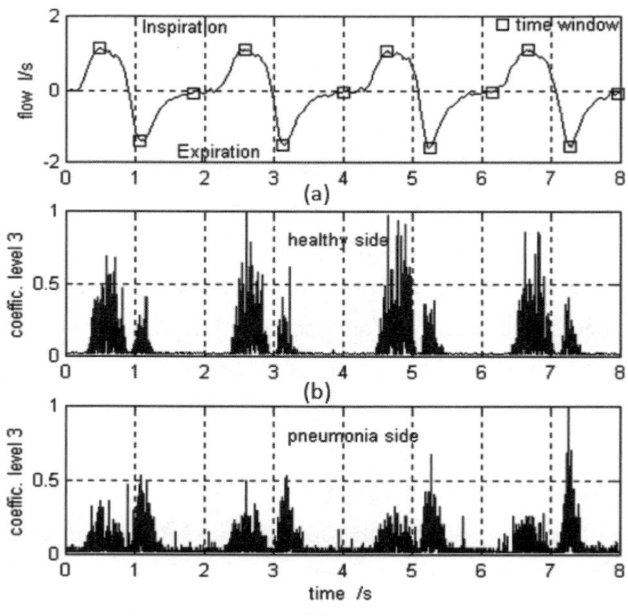

FIGURE 2.5: WT-based analysis of LS from a patient with pneumonia. (a) Breathing pattern, (b) WT coefficients from scale 3 of LS recorded from the healthy site, and (c) WT coefficients from scale 3 of LS recorded from the pneumonia site [112].

The successful WT-based detection of bronchial breathing can be considered as a first step in developing a monitoring system for patients at risk for pneumonia.

Ayari et al. [114] applied the WT transform to normal LS, crackles, and wheezes. In their work, they used wavelets formed by the first derivative of a Gaussian with standard deviation of $\sigma = 32 \times 10^{-5}$ s. Derivatives of Gaussian are most often used to guarantee that all maxima lines propagate up to the finest scale. They used the WT modulus maxima method [115] to characterize the local regularity of LS. The localized singularities correspond to the sharp and slow variations. WT of LS across scales presents local maxima when using a wavelet that is a first derivative of a smoothing function. Consequently, the sharp and slow variations are indicated by the location of the local modulus maxima obtained by one vanishing moment wavelet. Analysis of the behavior of WT modulus maxima across scales permits detection of singularities and estimation of Lipschitz exponents [116] for the sharp and slow variation. Numerical results for the crackle sound showed that singularity at sharp variation was Lipschitz −1. For wheezes, the singularity at sharp variations was Lipschitz 1. This is quite expected, and as wheezing sound seems more like a sinusoidal signal, it appears to be more regular than the normal LS.

WT has also been used in combination with artificial neural networks (ANN) to form an LS classification scheme. In the work of Kandaswamy et al. [117], LS signals were decomposed into the frequency subbands using multiresolution approximation, and a set of statistical features was extracted from the subbands to represent the distribution of wavelet coefficients. An ANN-based system, trained using the resilient back-propagation algorithm (RPROP) [118], was implemented to classify LS into six categories: normal, wheeze, crackle, squawk, stridor, or rhonchus. RPROP is a local adaptive learning scheme, performing supervised batch learning in feed-forward neural networks. The basic principle of RPROP is to eliminate the harmful influence of the size of the partial derivative on the weight step. As a consequence, only the sign of the derivative is considered to indicate the direction of the weight update.

The authors used some statistics from the WT coefficients as features for the ANN-based classification. In particular, they used: (1) the mean of the absolute values of the coefficients in each subband, (2) the average power of the wavelet coefficients in each subband, (3) the standard deviation of the coefficients in each subband, and (4) the ratio of the absolute mean values of adjacent subbands. The first two features represent the frequency distribution of the signal, whereas the next two denote the amount of changes in frequency distribution. These feature vectors, calculated for the frequency scales of 3–7, were used for classification of LS signals by ANN. The authors achieved classification accuracy of >90% when using Daubechies wavelet of order 8 [108].

As is apparent from the examples cited above, WT provides new opportunities in LS representation that allow construction of hybrid analysis tools. The latter stem from the beneficial advantages of WT and, by combining other efficient methods, introduce new approaches in LS analysis and appreciation. This will become clearer in the succeeding section.

2.5 WAVELET-HOS

2.5.1 Epitomized Rationale

Wavelet analysis can be seen as a generalization of the Fourier analysis and in many cases permits a similar interpretation, but amplifies it by adding time resolution—in a more fundamental way than is permitted by the short-time Fourier transform (STFT), because the latter does not remove the objection raised above against Fourier-type methods.

Signals containing coherent couplings have traditionally been analyzed by means of a normalized bispectrum, that is, the bicoherence [see (2.12) and (2.13)]. The bicoherence, when defined properly, quantifies the fraction of power contained in the nonlinearity. However, when the available data have a nonstationary nature, traditional fast Fourier transform (FFT)-based methods [91, 119] may be inadequate due to the inability of the STFT to resolve short-lived transients properly. To circumvent such problem, the wavelet bispectrum (WBS)/bicoherence is proposed [120–122], based on a combination of CWT with HOS.

For the analysis of nonstationary processes, the WBS has two main advantages compared to traditional FFT-based methods. First, because CWT is a timescale representation of a signal, thus, a time axis is introduced in a natural way. Second, wavelets have an inherent constant-Q filtering property, and are consequently well suited for detection of transients.

The wavelet bicoherence (WBC) technique detects QPC while reducing time averages to a minimum, thus permitting short-lived events, pulsed, and intermittency to be resolved [123]. Relatively short data sequences are sufficient to perform an analysis, in contrast to the Fourier bicoherence, which needs long time series to obtain both sufficient frequency resolution and statistics. Estimates of the noise contribution and error level of the WBC provide a criterion for the reliability of the results. A powerful noise reduction is an integral part of the standard technique: as noncoherent contributions are averaged out, weak coherent signals can be detected in very noisy data. Moreover, the wavelet bicoherence is more independent of the frame of reference and may be expected to be more useful in experiments where measurement in the local frame of reference is difficult [120].

2.5.2 CWT-HOS Definitions

Wavelet bispectrum. By analogy to the definition of the bispectrum in Fourier terms [see (2.10)], WBS is defined as [123]

$$B_w(a_1, a_2) = \int_T W_x^*(a, \tau) W_x(a_1, \tau) W_x(a_2, \tau) \, d\tau, \tag{2.63}$$

where $W_x(a_1, \tau)$, $i = 1, 2$ denotes that the CWT defined in (2.46) and the integration is performed over a finite time interval T: $\tau_0 \leq \tau \leq \tau_1$, and a, a_1, a_2 satisfy the following rule:

$$\frac{1}{a} = \frac{1}{a_1} + \frac{1}{a_2}. \tag{2.64}$$

The WBS expresses the amount of QPC in interval T, which occurs between wavelet components of scale lengths a_1, a_2, and a of $x(t)$ such that the sum rule of (2.64) is satisfied. By interpreting the scales as inverse frequencies, $\omega = 2\pi/a$, the WBS can be interpreted as the coupling between wavelets of frequencies that satisfy $\omega = \omega_1 + \omega_2$, within the frequency resolution.

Wavelet bicoherence. Similarly to the definition of bicoherence [see (2.12)], WBC can be defined as the normalized WBS, that is,

$$b_w(a_1, a_2) = \frac{B_w(a_1, a_2)}{\left\{\left[\int_T |W_x(a_1, \tau) W_x(a_2, \tau)|^2 d\tau\right]\left[\int_T |W_x(a, \tau)|^2 d\tau\right]\right\}^{1/2}}, \tag{2.65}$$

in which magnitude $|b_w(a_1, a_2)|$, namely, wavelet bicoherence index, can attain values between 0 and 1. For ease of interpretation, the *squared WBC* plotted in the (ω_1, ω_2) plane, that is, $|b_w(\omega_1, \omega_2)|^2$, is preferred. Because of the symmetries in the definition and the limitation set by the Nyquist frequency ω_s [124], the estimation of WBC in the whole bifrequency plane can be based on its values in the principal region $\{\Delta: \omega_1, \omega_2 \leq \omega_1, \omega_1 + \omega_2 \leq \omega_s\}$.

Summed wavelet bicoherence. For comparing cases computed under the same numerical conditions, the summed wavelet bicoherence (SWBC) could be introduced as

$$b_w^2(\omega) = \sum_\Delta b_w^2(\omega_1, \omega_2). \tag{2.66}$$

In general, the numerical values of SWBC depend on the chosen calculation grid; thus, they basically provide qualitative summarization of the underlying information.

As van Milligen et al. [123] note, the use of nonorthogonal wavelets, such as Morlet, results in nonstatistically independent wavelet coefficients. This introduces a statistical noise level in the estimation of WBC, with an upper bound given by [123]

$$N_{\mathrm{b}}(\omega_1, \omega_2) \approx \left[\frac{\pi}{\min(|\omega_1|, |\omega_2|, |\omega_1 + \omega_2|)} \frac{1}{T} \right]^{1/2}. \tag{2.67}$$

From (2.67), it can be easily observed that the statistical noise affects the low frequencies of WBC, whereas at higher frequencies it drops rapidly with T. This reveals the ability of WBC to serve as a noise filter for coherent signals.

Evolutionary wavelet bicoherence. Because the WBC defined in (2.65) refers to a certain time interval T, its value is made to correspond to the center of this interval, that is, $t_0 = T/2$. Consequently, evolutionary WBC (EWBC) can be defined as

$$\begin{aligned} &\mathbf{b}_{\mathrm{w}}(\omega_1, \omega_2, t) = \left\{ b_{\mathrm{w}}(\omega_1, \omega_2) \,|_{t = t_0 + k\Delta T_1} \right\}, \; k = 0, 1, 2, 3, \ldots; \\ &(2\pi/\omega_{\mathrm{s}}) \leq \Delta T_1 \wedge k\Delta T_1 \leq T_{\mathrm{total}} - 2t_0, \end{aligned} \tag{2.68}$$

where T_{total} is the total time duration of the analyzed signal $x(t)$.

When using EWBC, the evolution of the nonlinearities across time can be represented, within a time-resolution controlled by the selection of the ΔT_1 value.

2.5.3 Wheeze Analysis With CWT-HOS

Because of the nature of wheezes, the notion of QPC detection examined in Section 2.2 is expanded by using wavelet bispectrum and wavelet bicoherence as a means to track and quantify the evolution of the nonlinear characteristics of wheezes within the breathing cycle. To this end, the combination of WT with third-order statistics/spectra introduces the nonlinear analysis of wheezes in the time–bifrequency domain. This was investigated by Taplidou and Hadjileontiadis [125], who analyzed breath sound signals from asthmatic patients by using CWT-HOS-based parameters.

Figure 2.6a depicts one breathing cycle of a breath sound signal along with the normalized airflow (superimposed with a dotted line) recorded from an asthmatic patient. The positive and negative airflow values indicate inspiratory and expiratory phases, respectively. As Figure 2.6a shows, the expiratory phase is prolonged compared to the inspiratory one; this is due to the existence of asthma. Similarly, the breath sound signal exhibits a profound high amplitude section (~1.2–2.8 s) corresponding to inspiratory wheeze, whereas an extended expiratory wheeze (~3.7–6.3 s) with decaying amplitude dominates the expiratory phase.

Figure 2.6b shows the representation of the signal in the time–frequency domain via the CWT of the signal. In this figure, the harmonic character of wheezes (both inspiratory and expiratory) is apparent. It is clear that there are coexisting distinct spectral peaks within the area of 150–500 Hz that emerge during the appearance of wheezes, revealing their polyphonic character (such as a chord). In addition, a frequency sweep from low to high frequencies and vice versa can

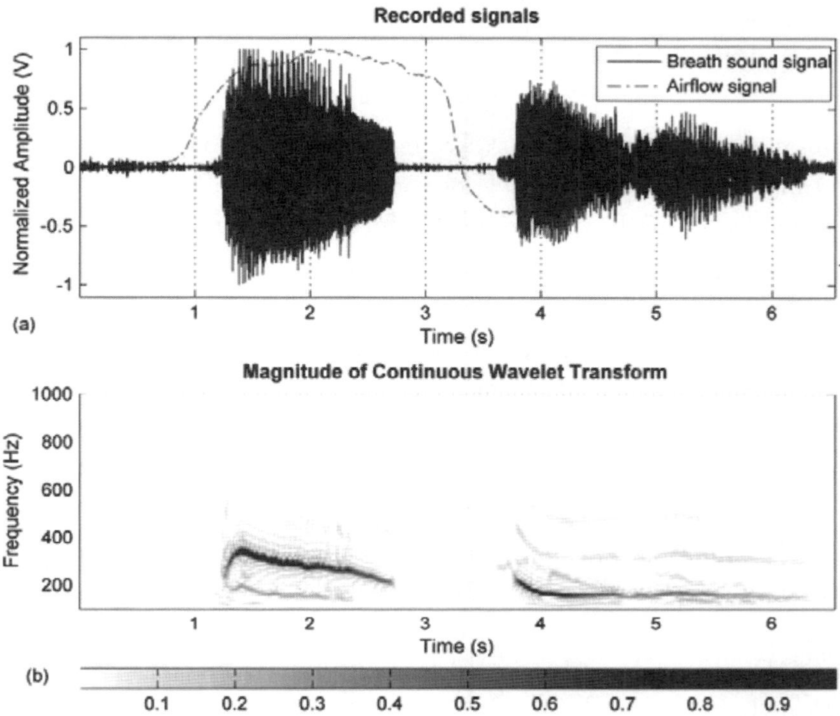

FIGURE 2.6: (a) One breathing cycle of a breath sound recording from an asthmatic patient with two dominant wheezes (one inspiratory and one expiratory). (b) CWT of the analyzed signal [125].

be noticed, mainly at the beginning and the end of wheezes, because of the increase or decrease in airflow signal, respectively. Moreover, the spectral peak within 200–400 Hz that sustains its high amplitude during the inspiratory wheeze subsides in the expiratory one, where the spectral peak about 200 Hz seems to be the most evident one.

The variation in time–frequency characteristics of wheezes justifies the necessity of using wavelet-based HOS to reveal their nonlinearities.

The ability of WBS and WBC to capture the existence of nonlinearities in wheezes is shown in Figure 2.7. In particular, the estimated magnitude of WBS (top), WBC (middle), and squared WBC (bottom) are illustrated in Figure 2.7(i) and (ii), for a time section without wheeze (2.75–3.75 s) and with wheeze (3.75–4.75 s), respectively, corresponding to sections from the breath sound signal depicted in Figure 2.6a. From the comparison of the estimated WBS in Figure 2.7(i) and (ii), it is apparent that the WBS of the section without wheeze [Figure 2.7(i), top] is spread in the area of low frequencies (100–150 Hz) and does not exhibit any distinct peak at a higher-frequency range; furthermore, its values are significantly lower compared to those of the section with wheeze

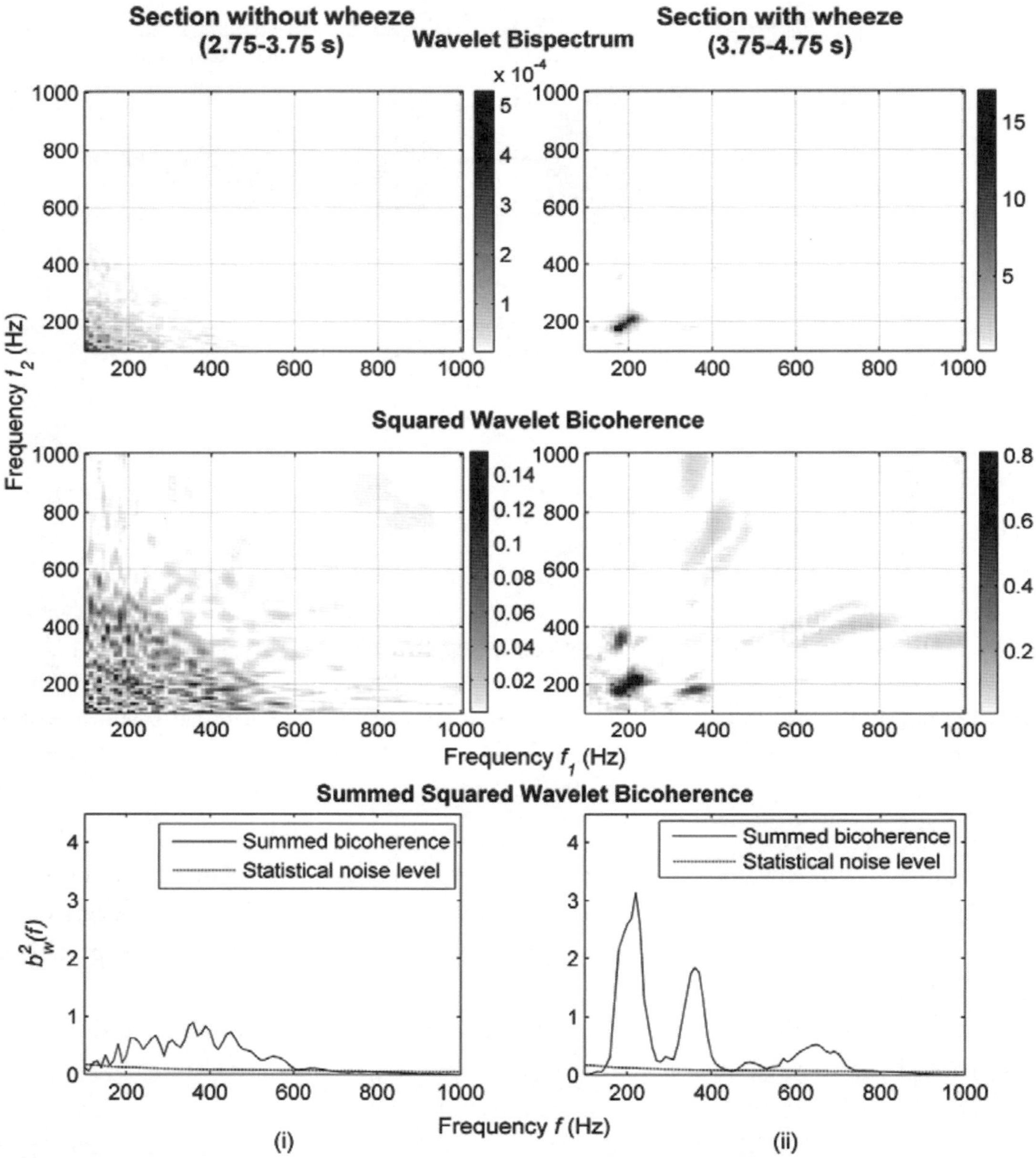

FIGURE 2.7: (i) Section without wheeze (2.75–3.75 s). (ii) Section with wheeze (3.75–3.75 s), with reference to the recorded LS signal depicted in Figure 2.6a. In both cases, WBS (top), squared WBC (middle), and SWBC along with statistical noise level (······) (bottom) are depicted accordingly [125].

[Figure 2.7(ii), top]. In the latter, the WBS reveals a concentration of its values around a peak located approximately at $(f_1, f_2) \approx (200, 200)$ Hz. This implies that a possible quadrature self-phase-coupling exists at $(f_1, f_2) \approx (200, 200)$ Hz related to the frequency located at $f_3 = f_1 + f_2 \approx 400$ Hz. Clearly, the inspection of CWT in Figure 2.6b at the selected section justifies the existence of f_3.

Looking at the estimated magnitude of the squared WBC for both cases [Figure 2.7(i) and (ii), middle], a more enhanced resolution of the bifrequency content is provided. In Figure 2.7(i) middle, a wider spread is noticed, spanning from 100 to 600 Hz, whereas in Figure 2.7(ii) middle, additional distinct peaks appear at higher frequencies. Comparing the values of squared WBC in Figure 2.7(i) and (ii) middle, it can be seen that they are much lower in the time section that is without wheeze (maximum value of 0.15) than those of the time section that contains wheeze (maximum value of 0.8). This shows that practically there is no QPC in the WBC content corresponding to the breath sound signal that does not contain wheeze; in contrast, a strong QPC exists between the harmonics revealed in the WBC content corresponding to the breath sound signal that contains wheeze.

The SWBC, estimated for both cases, is shown in Figure 2.7(i) and (ii) bottom, respectively. The SWBC is plotted along with the upper bound of the statistical noise level (dotted line) defined in (2.67), which in all cases is significantly lower than the estimated SWBC. From the two sub-figures, it is apparent that the SWBC of the breath sound without wheeze [Figure 2.7(i), bottom] sustains a low value at a frequency range of 100–600 Hz, whereas the SWBC of the breath sound with wheeze [Figure 2.7(ii), bottom] demonstrates high peaks of about 200 and 350 Hz and a lower one of about 650 Hz, showing a clear difference from the previous case.

Figure 2.8 illustrates a graphical representation of the ESWBC for whole breath sound signal shown in Figure 2.6a. In this way, the evolution of the nonlinearities in breath sound signal that contains wheezes is demonstrated. From this figure, it is apparent that QPC occurs in both breathing phases only at the time instances where the wheezes exist. Moreover, the main frequency pair with QPC [$(f_1, f_2) \approx (200, 200)$ Hz] is sustained both in inspiratory and expiratory wheezes; however, additional pairs at higher frequencies with QPC emerge during the expiratory wheeze. This relates to the pathology of asthma, because it affects the expiratory phase more than the inspiratory one of asthmatic patients [126].

The CWT-HOS-based analysis of wheezes presented here could be expanded to different types of wheezes as a means to characterize them according to the nonlinear properties they exhibit as they evolve within the breathing cycle. This would shed light on the differences between different categories of wheezes (e.g., monophonic vs. polyphonic) and different pathologies (e.g., COPD and asthma). In either case, CWT-HOS provides an efficient hybrid analysis tool that allows enhance-

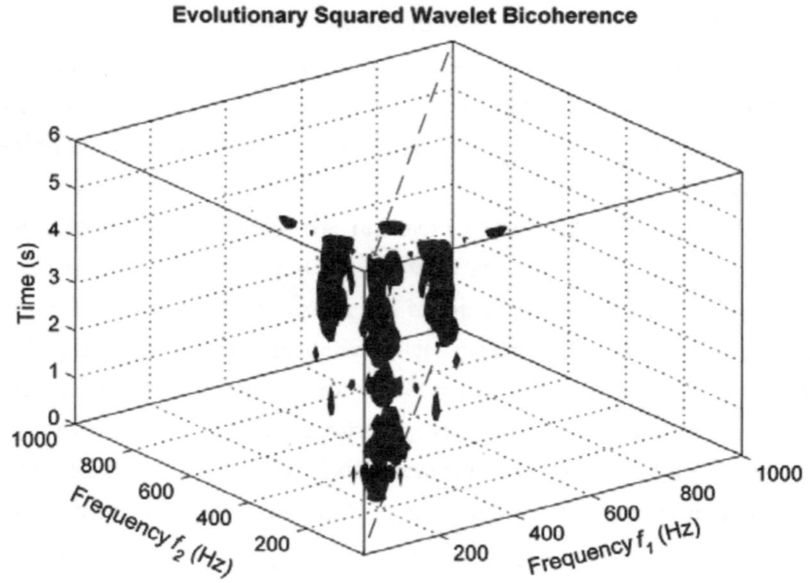

FIGURE 2.8: ESWBC with reference to the recorded LS signal depicted in Figure 2.6a. The time axis includes one breathing cycle [125].

ment of significant details in the LS interpretation, which could not be perceived by auscultation solely.

2.6 HIGHER-ORDER CROSSINGS

2.6.1 Epitomized Rationale

The LS analysis domains that we have presented so far refer to transforms of the initial form of the signal, that is, its time-domain representation, to a new one (e.g., HOS, LOS, CWT domains). In this section, time characteristics of LS would be considered as the main focus. Taking into consideration that almost all observed time series are oscillatory, displaying local and global up and down movements as time progresses, proper time series analysis could "capture" these oscillatory patterns and use them as a new domain of representation. This is done here by using the concept of higher-order crossings (HOC) [127]. HOC analysis of time series is a way to study oscillation of stochastic processes combinatorially by counting and provides an alternative to commonly used spectral methods.

Time ordering makes it possible to apply to a time series linear filtering, one of the most indigenous elements of time series analysis. This filtering, in fact, changes the time series oscillations and this can be reflected in the change in the zero-crossing count effected by the filtering procedure. The application of a specific sequence of filters (or family of filters) to a time series forms the corresponding sequence (or family) of zero-crossing counts (namely, HOC) and, thus, provides a summary of the oscillation "history" observed in the time series and its filtered versions.

Following the properties of HOC estimated from different LS signals, new discrimination features could be established that provide successful differentiation among LS with similar acoustic behavior.

2.6.2 HOC Definitions

For a finite, wide-sense stationary, zero-mean series $\{Z_t\}$, $t = 1, \ldots, N$, its oscillatory pattern about level zero can be expressed through the zero-crossing count. By assuming the iterative procedure of applying a filter to the time series, and counting the number of zero crossings in the filtered time series, applying yet another filter to the original time series, and again observing the resulting zero crossings and so on (filtering and counting), the resulting zero-crossing counts, that is, HOC, could be produced [127]. When a specific sequence of filters is applied to a time series, the corresponding sequence of zero-crossing counts is obtained, resulting in the so-called HOC sequence. Many different types of HOC sequences can be constructed by appropriate filter design, according to the desired spectral and discrimination analysis (DA).

Let ∇ be the backward difference operator defined by

$$\nabla Z_t \equiv Z_t - Z_{t-1}. \tag{2.69}$$

The difference operator ∇ is a high-pass filter. If we define the following sequence of high-pass filters

$$\Im_k \equiv \nabla^{k-1},\ k = 1, 2, 3, \ldots, \tag{2.70}$$

with $\Im_1 \equiv \nabla^0$ being the identity filter, and with a transfer function given by

$$H(\omega; k) = (1 - \exp(-j\omega))^{k-1}, \tag{2.71}$$

we can estimate the corresponding HOC, namely, simple HOC [127], by

$$D_k = \mathrm{NZC}\{\Im_k(Z_t)\},\ k = 1, 2, 3, \ldots;\ t = 1, \ldots, N, \tag{2.72}$$

where NZC$\{\cdot\}$ denotes the estimation of the number of zero crossings and

$$\nabla^{k-1} Z_t = \sum_{j=1}^{k} \binom{k-1}{j-1} (-1)^{j-1} Z_{t-j+1} \text{ with } \binom{k-1}{j-1} = \frac{(k-1)!}{(j-1)!(k-j)!}. \tag{2.73}$$

In practice, we only have finite time series and lose an observation with each difference. Hence, to avoid this effect we must index the data by moving to the right, that is, for the evaluation of k simple HOC, the index $t = 1$ should be given to the kth or a later observation. For the estimation of the number of zero crossings in (2.72), a binary time series $X_t(k)$ is initially constructed given by

$$X_t(k) = \begin{cases} 1 & \text{if } \Im_k(Z_t) \geq 0 \\ 0 & \text{if } \Im_k(Z_t) < 0 \end{cases}, \quad k = 1, 2, 3, \ldots; \quad t = 1, \ldots, N, \tag{2.74}$$

and the desired simple HOC are then estimated by counting symbol changes in $X_1(k),\ldots,X_N(k)$, that is,

$$D_k = \sum_{t=2}^{N} [X_t(k) - X_{t-1}(k)]^2. \tag{2.75}$$

From (6), it follows that if $X_{t-1}(k) \neq X_t(k)$, then $X_t(k+1) = X_t(k)$. Thus, every symbol change in $\{X_t(k)\}$ leads to at least one symbol change in $\{X_t(k+1)\}$. Consequently, we should have $D_{k+1} \geq D_k$, a fact that reveals the monotonic character of simple HOC. In finite data records, what we really have is the sure inequality $D_{k+1} \geq D_k - 1$ [127]. In addition, as k increases, the discrimination power of simple HOC diminishes, because different processes yield almost the same D_k [127]. A value of $K = 20$ usually suffices for effective discrimination purposes [127].

2.6.3 HOC Discrimination Tools

HOC-based discrimination between signals consists of the process of determining from HOC sequences their degree of similarity. Sometimes, instead of comparing directly the two signals, it is useful to measure their distance from a given reference signal. In many respects, white Gaussian noise is the most convenient, for which the expected HOC are given by [127]

$$E\{D_k\} = (N-1)\left\{\frac{1}{2} + \frac{1}{\pi}\sin^{-1}\left(\frac{k-1}{k}\right)\right\}. \tag{2.76}$$

The limits for which D_k falls within with a probability of approximately 95% for each k are [127]

$$\pm 1.96(N-1)^{1/2}\left\{\frac{1}{4} - \left[\frac{1}{\pi}\sin^{-1}\left(\frac{k-1}{k}\right)\right]^2\right\}^{1/2}. \tag{2.77}$$

With (2.76) and (2.77), a *white noise test* (WNT) could be constructed to examine whether a time series oscillates as Gaussian white noise. Based on WNT, the hypothesis of white noise should be rejected when at least one D_k, $k = 1, 2, \ldots, K$, for some K, falls outside the limits of (2.76) [127].

Using WNT, each time series is tested against the white Gaussian noise. By superimposing the results from both tests in the HOC domain, the two time series are actually compared to each other. This indirect comparison could reveal the optimum order, k_{op}, of HOC that results in the maximum discrimination of the two time series, corresponding to the maximum distance between their HOC sequences.

Based on the results of WNT, scatter plots of HOC of k_{op} order versus HOC of other (usually neighboring) orders can reveal the presence of two classes (clusters). If the classes do not overlap in the scatter plots, linear discrimination could be achieved and a simple decision rule of the form

$$aD_{k_{op}\pm n} + bD_{k_{op}} + c = 0,\ n = 1 \text{ or } 2 \text{ or } 3;\ a,b,c = \text{constants}, \tag{2.78}$$

could be constructed in deciding upon class membership. This also means that the HOC sequences of the two classes do not overlap in the WNT, at least for one value of k. Otherwise, linear discrimination is not feasible and the classes overlap in the scatter plots.

To examine the possibility of potential overlapping between the two groups, the mean (MN) and the standard deviation (σ) of the distance from cluster center, alongside the standard error of the MN (s), both for the two classes in the HOC scatter plot, are estimated. Consequently, the borderline of the two clusters is defined as the perpendicular line at the $\mathbf{C}(l_1,l_2)$ point of their center distance, with the $\mathbf{C}$ vector satisfying the following equations:

$$\frac{\text{dis}(\mathbf{C},\mathbf{c}_1)}{\text{MN}_1 + \sigma_1 + s_1} = \frac{\text{dis}(\mathbf{C},\mathbf{c}_2)}{\text{MN}_2 + \sigma_2 + s_2} \tag{2.79}$$

$$\text{dis}\,(\mathbf{C},\mathbf{c}_1) + \text{dis}\,(\mathbf{C},\mathbf{c}_2) = \text{dis}\,(\mathbf{c}_1,\mathbf{c}_2)\,, \tag{2.80}$$

where dis($\mathbf{a}_1$,$\mathbf{a}_2$) denotes the Euclidean distance between vectors $\mathbf{a}_1$ and $\mathbf{a}_2$, and $\mathbf{c}_i$, MN$_i$, σ_i, s_i, $i = 1, 2$, denote the estimated center vector, the mean, the standard deviation, and the standard error of the mean for the two clusters, respectively.

For an extended coverage of the HOC-based analysis of time series, the reader is encouraged to consult the book by Kedem [127] and its references.

2.6.4 HOC Analysis of DAS

HOC have been used to perform a discrimination analysis of DAS, especially FC, CC, and SQ [128]. In fact, HOC assesses the changes in oscillatory pattern of DAS according to their type, by estimating the relevant HOC sequence. Based on the connection between filtering and zero crossings, the proposed method constitutes a possible tool for DAS discrimination analysis (namely,

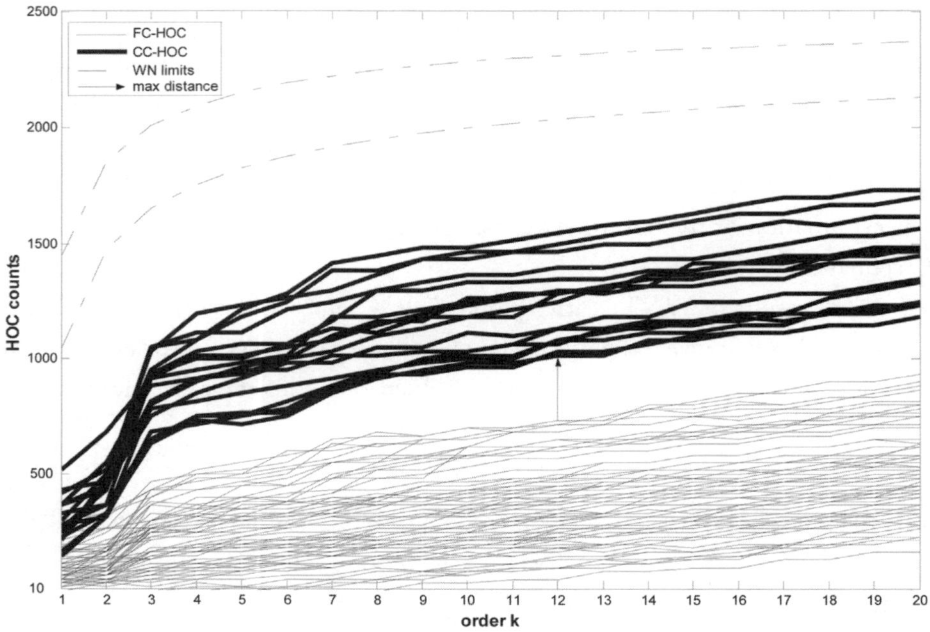

(a)

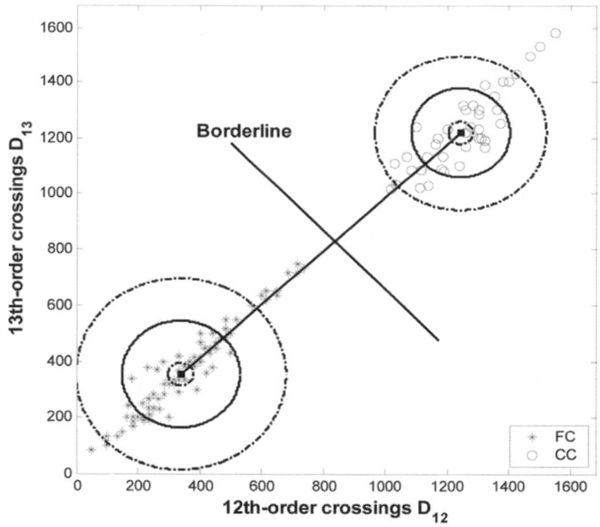

(b)

FIGURE 2.9: (a) WNT for FC and CC with k order ranging from 1 to 20. (b) Scatter plot of 12th-order (D_{12}) vs. 13th-order (D_{13}) crossings of FC (stars) and CC (circles). The ■ symbol denotes the center of each group (FC and CC), whereas the borderline is the estimated linear border that clearly discriminates the two groups. The solid and dash-dotted circles correspond to MN and $[\pm(\sigma + s)]$ of the distance from cluster center, respectively, both for FC and CC [128].

HOC-DA), which is quite attractive because of its simplicity in the evaluation of the HOC sequence and its ability to provide simple discrimination criteria between the examined classes, when it is feasible.

An example of HOC-DA is given in Figure 2.9a, which illustrates the results from WNT for FC and CC. The estimated HOC sequences of FC (plain solid lines) and CC (bold solid lines) are directly compared to the white Gaussian noise limits (dash-dotted lines) and indirectly to each other. The HOC order k ranged from 1 to 20 and its optimum value was found $k_{\text{op}}^{\text{FC-CC}}$ = 2 because there, the distance (denoted by an arrow) between the topmost HOC line of FC and the bottommost HOC line of CC is maximized. This implies that scatter plots of the $k_{\text{op}}^{\text{FC-CC}}$-order crossings could provide a clear (linear) discrimination between FC and CC. This is true from the HOC scatter plot of Figure 2.9b, where the HOC pair (D_{12}, D_{13}) from the time series of FC (stars) and CC (circles) is depicted. Clearly, the two processes give rise to distinctly different clusters of pairs (D_{12}, D_{13}), which can be linearly discriminated. In this scatter plot, the coordinates of the two cluster centers were found equal to c_1^{FC} (338.8, 354.5) and c_1^{CC} (1242.6, 1218.7) with a distance of $\text{dis}(c_1^{\text{FC}}, c_1^{\text{CC}}) = 1251.2$, and are denoted as solid squares connected with solid line, respectively. Furthermore, the estimated $\text{MN}_1^{\text{FC}} = 190.7$; $\sigma_1^{\text{FC}} = 133.9$; $s_1^{\text{FC}} = 15.6$ and $\text{MN}_1^{\text{CC}} = 158.9$; $\sigma_1^{\text{CC}} = 103.2$; $s_1^{\text{CC}} = 16.1$ are also depicted as solid circles (MN) and dash-dotted ones [$\pm(\sigma + s)$], both for FC and CC clusters. The $\mathbf{C}^{\text{FC-CC}}$ vector was derived by (2.79) and (2.80), using as $\mathbf{c}_1 = \mathbf{c}_1^{\text{FC}}$, $\mathbf{c}_2 = \mathbf{c}_1^{\text{CC}}$, $\text{MN}_1 = \text{MN}_1^{\text{FC}}$, $\sigma_1 = \sigma_1^{\text{FC}}$, $s_1 = s_1^{\text{FC}}$, $\text{MN}_2 = \text{MN}_1^{\text{CC}}$, $\sigma_2 = \sigma_1^{\text{CC}}$, $s_2 = s_1^{\text{CC}}$ and its coordinates were found equal to $\mathbf{C}^{\text{FC-CC}}(836.1, 830)$, defining the borderline shown in Figure 2.9b.

The above results show that FC and CC are linearly discriminated and justify the selection of "FC range" and "CC range" that provide discrimination with an accuracy of 100% (SE = 100%, SP = 100%) between FC and CC in the HOC domain as

$$r_{100\%}^{\text{FC-CC}} = \left\{[D_{12}(40 \div 800), D_{13}(40 \div 800)]^{\text{FC}},\ [D_{12}(1000 \div 1600), D_{13}(1000 \div 1600)]^{\text{CC}}\right\}$$

Further examples of the application of the HOC-DA to {FC, SQ} and {CC, SQ} DAS pairs are included in [128]. The described analysis show the ability of the HOC-based discrimination tool to efficiently reveal the differences in the oscillatory patterns of the DAS types that occur when higher (>1) order crossings are used and sets a different perspective in the time-domain analysis of LS.

2.7 EMPIRICAL MODE DECOMPOSITION

2.7.1 Epitomized Rationale

The HOC analysis presented in the previous section deals mainly with (in a broad sense) stationary signals. To accommodate the nonstationarity of the processes, the zero-crossing rate could be estimated either within a moving data window or adaptively, giving more weight to more current data as a means for summarizing a time dependent oscillation pattern. Nevertheless, a more efficient approach lies in the field of empirical mode decomposition (EMD), proposed by Huang et al. [129] in 1998. EMD decomposes the signal into components with well-defined instantaneous frequency. Each characteristic oscillatory mode extracted, namely, intrinsic mode function (IMF), is symmetric and has a unique local frequency, and different IMFs do not exhibit the same frequency at the same time [129].

The EMD method is necessary to deal with both nonstationary and nonlinear data and, contrary to almost all the previous methods, EMD is intuitive—that is, the basis of the expansion is generated in a direct, a posteriori, and adaptive manner, derived from the data [129]. The main idea behind EMD is that all data consist of different simple intrinsic modes of oscillations, represented by the IMFs. An IMF represents a simple oscillatory mode as a counterpart of the simple harmonic function, yet it allows amplitude and frequency modulation; thus, it is much more general. The EMD method considers the signals at their local oscillation scale, subtracts the faster oscillation, and iterates the residual.

EMD can also be viewed as an expansion of the data in terms of IMFs. Then, these IMFs, based on and derived from the data, can serve as the basis of that expansion, which can be linear or nonlinear as dictated by the data, and it is complete and almost orthogonal. Most important of all, it is adaptive, and, therefore, highly efficient.

Representation of LS in the EMD domain reveals their intrinsic characteristics, because these are reflected in the IMFs, and provides a new analysis space where detailed exploitation of LS diagnostic information is feasible.

2.7.2 EMD Description

EMD is formed through the estimation of IMFs. According to Huang et al. [129], an IMF satisfies two conditions:

- In the whole dataset, the number of extrema and the number of zero crossings must either equal or differ at most by one.
- At any point, the mean value of the envelope defined by the local maxima and the envelope defined by the local minima is zero.

By virtue of the IMF definition, the EMD procedure for a given signal $x(t)$ can be summarized as follows [129]:

- Identify the successive extrema of $x(t)$ based on the sign alterations across the derivative of $x(t)$.

- Extract the upper and lower envelopes by interpolation, that is, the local maxima (minima) are connected by a cubic spline interpolation to produce the upper (lower) envelope. These envelopes should cover all the data between them.
- Compute the average of upper and lower envelopes, $m_1(t)$.
- Calculate the first component $h_1(t) = x(t) - m_1(t)$.
- Ideally, $h_1(t)$ should be an IMF. In reality, however, overshoots and undershoots are common, which also generate new extrema and sift or exaggerate the existing ones [129]. To correct this, the sifting process has to be repeated as many times as is required to reduce the extracted signal as an IMF. To this end, treat $h_1(t)$ as a new set of data, and repeat first to fourth steps up to k times (e.g., $k = 7$) until $h_{1k}(t)$ becomes a true IMF. Then set $c_1(t) = h_{1k}(t)$. Overall, $c_1(t)$ should contain the finest scale or the shortest period component of the signal.
- Obtain the residue $r_1(t) = x(t) - c_1(t)$
- Treat $r_1(t)$ as a new set of data and repeat first to sixth steps up to N times until the residue $r_N(t)$ becomes a constant, a monotonic function, or a function with only one cycle from which no more IMFs can be extracted. Note that even for data with zero mean, $r_N(t)$ still can differ from zero.
- Finally,

$$x(t) = \sum_{i=1}^{N} c_i(t) + r_N(t), \tag{2.81}$$

where $c_i(t)$ is the ith IMF and $r_N(t)$ is the final residue.

The above procedure results in a decomposition of the data into N-empirical modes, and a residue $r_N(t)$, which can be either a monotonic function or a single cycle. It is noteworthy that, in order to apply the EMD method, there is no need for a mean or zero reference; EMD only needs the locations of the local extrema to generate the zero reference for each component (except for the residue) through the shifting process.

A useful characteristic of (2.81) is the potential of filtering that it provides. Indeed, using the IMF components, a time-space filtering can be devised simply by selecting a specific range of them in the reconstruction procedure [e.g., in (2.81), for high-pass filtering: $i = 1{:}k$, $k < N$, or for band pass one: $i = b{:}k$, $1 < b$, $k < N$). This time-space filtering has the advantage that its results preserve the full nonlinearity and nonstationarity in the physical space [130].

2.7.3 EMD Considerations and Extensions

As described above, the process of finding IMFs is indeed like sifting: to separate the finest local mode from the data first based only on the characteristic timescale. The sifting process, however, has two effects: (1) to eliminate riding waves and (b) to smooth uneven amplitudes. Although the first condition is absolutely necessary for the instantaneous frequency to be meaningful, the second condition is also necessary in case the neighboring wave amplitudes have too large a disparity. Unfortunately, the second effect, when carried to the extreme, could obliterate the physically meaningful amplitude fluctuations. Therefore, the sifting process should be applied with care, because carrying the process to an extreme could make the resulting IMF a pure frequency-modulated signal of constant amplitude.

To guarantee that the IMF components retain enough physical sense of both amplitude and frequency modulations, we have to determine a criterion for the sifting process to stop. This can be accomplished by limiting the size of the standard deviation, SD, computed from the two consecutive sifting results [131, 132]. Historically, two different criteria have been proposed. The first one was introduced by Huang et al. [129], determined by using a Cauchy type of convergence test. Specifically, the test requires the normalized squared difference between two successive sifting operations defined as

$$\mathrm{SD}_k = \frac{\sum_{t=0}^{T} \left| h_{1(k-1)}(t) - h_{1k}(t) \right|^2}{\sum_{t=0}^{T} h_{1(k-1)}^2(t)}, \tag{2.82}$$

to be small. If this squared difference SD_k is smaller than a predetermined value, the sifting process will be stopped. This definition seems to be rigorous, but it is very difficult to implement in practice. This criterion does not depend on the definition of the IMFs. The squared difference might be small, but nothing guarantees that the function will have the same numbers of zero crossings and extrema, for example.

These shortcomings prompted Huang et al. [133, 134] to propose a second criterion based on the agreement of the number of zero crossings and extrema. Specifically, an S number is preselected. The sifting process will stop only after S consecutive times, when the numbers of zero crossings and extrema stay the same and are equal or differ at most by 1. This second choice has its own difficulty: how to select the S number. Obviously, any selection is ad hoc, and a rigorous justification is needed. In a recent study of this open-ended sifting, Huang et al. [134] used the many possible choices of S numbers to form an ensemble of IMF sets, from which an ensemble mean and confidence were derived. Furthermore, through comparisons of the individual sets with the mean, Huang et al. [134] established an empirical guide. For the optimal siftings, the range of S number should be set between 4 and 8.

Ensemble EMD. One of the possible drawbacks of EMD is the occasional appearance of *mode mixing*, which is defined as a single IMF either consisting of signals of widely disparate scales, or a signal of a similar scale residing in different IMF components. Mode mixing is a consequence of signal intermittency. Recently, a new sifting approach was proposed, Ensemble EMD (EEMD) [135–137], which successfully deals with the mode mixing problem. This new approach consists of sifting an ensemble of white noise-added signal and treats the mean as the final true result. Finite, not infinitesimal, amplitude white noise is necessary to force the ensemble to exhaust all possible solutions in the sifting process, thus making the different scale signals to collate in the proper IMFs dictated by the dyadic filter banks. Because EMD is a time-space analysis method, the white noise is averaged out with sufficient number of trials; the only persistent part survives the averaging process is the signal, which is then treated as the true and more physical meaningful answer. The effect of the added white noise is to provide a uniform reference frame in the time–frequency space; therefore, the added noise collates the portion of the signal of comparable scale in one IMF. With this ensemble mean, one can separate scales naturally without any a priori subjective criterion selection as in the intermittence test for the original EMD algorithm. This new approach uses the full advantage of the statistical characteristics of white noise to perturb the signal in its true solution neighborhood, and to cancel itself out after serving its purpose; therefore, it represents a substantial improvement over the original EMD.

Complementary ensemble EMD. An extension to the EEMD was just recently proposed by Yeh et al. [138], with the purpose of removing the residue of added white noises. They modified the EEMD method by using the complementary sets of added white noises to remove residue of added white noises. This modified EEMD, complementary EEMD (CEEMD), was also developed as a new technique to overcome intermittence, hence, mode mixing.

CEEMD extends the concept of EEMD method by generating two sets of averaged IMFs, that is, averaged IMFs with positive and negative residues of added white noises. In CEEMD, a number (e.g., 20) of arbitrary white noises are chosen for added white noises, because a single white noise cannot solve all intermittent signals. Two mixtures of the original signal and added white noise are then produced by the following equation:

$$\begin{bmatrix} M_1 \\ M_2 \end{bmatrix} = \begin{bmatrix} 1 & 1 \\ 1 & -1 \end{bmatrix} \begin{bmatrix} S \\ N \end{bmatrix} \tag{2.83}$$

where S is the original signal, N is the added white noise, M_1 is the positive mixture, and M_2 is the negative mixture.

The averaged IMFs decomposed from positive mixtures are the averaged IMFs with positive residue of added white noises. Similarly, the averaged IMFs with negative residue of added white noises are obtained. Then, the averaged IMFs without added residue of white noises can be derived by the following equation

$$\begin{bmatrix} \mathrm{IMF}_S \\ \mathrm{IMF}_N \end{bmatrix} = \mathrm{inv}\begin{bmatrix} 1 & 1 \\ 1 & -1 \end{bmatrix} \begin{bmatrix} \mathrm{IMF}_\mathrm{p} \\ \mathrm{IMF}_\mathrm{n} \end{bmatrix}, \tag{2.84}$$

where IMF_p is the averaged IMF with positive residue of added white noises; IMF_n is the averaged IMF with negative residue of added white noises; IMF_S is the averaged IMF without the residue of added white noises; IMF_N is the part of averaged IMF contributed by the added white noises. The averaged IMF without the residue of added white noises is the final result of CEEMD [138].

For the signal reconstruction process, only significant IMFs are used, defined through Monte Carlo verification. In particular, for a normalized white noise, the distribution of energy density-correspondent averaged period of IMFs should locate around the diagonal of Monte Carlo plot [135, 139]. For a normalized signal with significant components, the relationship between the energy density and averaged period of a significant IMF should locate on the area above the diagonal. Hence, based on this, significant IMFs are identified [138].

2.7.4 EMD Crackles Analysis

EMD was applied to describe crackles behavior [140] according to the characteristics of the corresponding IMFs. The main findings of the study [140] were:

- Oscillatory information of crackles embedded in basic inspiratory noise are distributed on several IMFs, starting FC information at IMF_2 and CC at IMF_3, yet the IMF that contains the ending information is not clear.
- Oscillatory information of basic inspiratory sounds starts at IMF_3, but again, the last IMF containing basic respiratory sound information is not clear.
- Depending on the SNR, FC are easier to discriminate from basic respiratory sound than CC.
- Overlapped crackles appear even at IMF_1, but individual components are not differentiated.
- Combined but time-separated FC and CC have no effect on the oscillatory information for individual events.

As Villalobos et al. [140] have stated, the EMD technique improved the visual identification of crackles embedded in respiratory sound. Although crackles and respiratory sound were mixed, the authors were able to identify their respective oscillations at different IMFs. For example, FC were more easily observed in IMF_2, whereas CC were associated to oscillations in IMF_3, probably due to their spectral content differences. Such distinction, however, depends on the intensity of the background noise and how this noise changes the crackle morphology. Because crackles information is propagated to more than one IMF, combining information from different IMFs could help to identify even low amplitude FC and CC and discard artifacts. Therefore, EMD analysis may assist, in addition to crackles counting, in determining the crackles type. EMD analysis is advantageous because it generates oscillations that are associated with crackle locations, and it could separate individual FC that overlapped by a small amount.

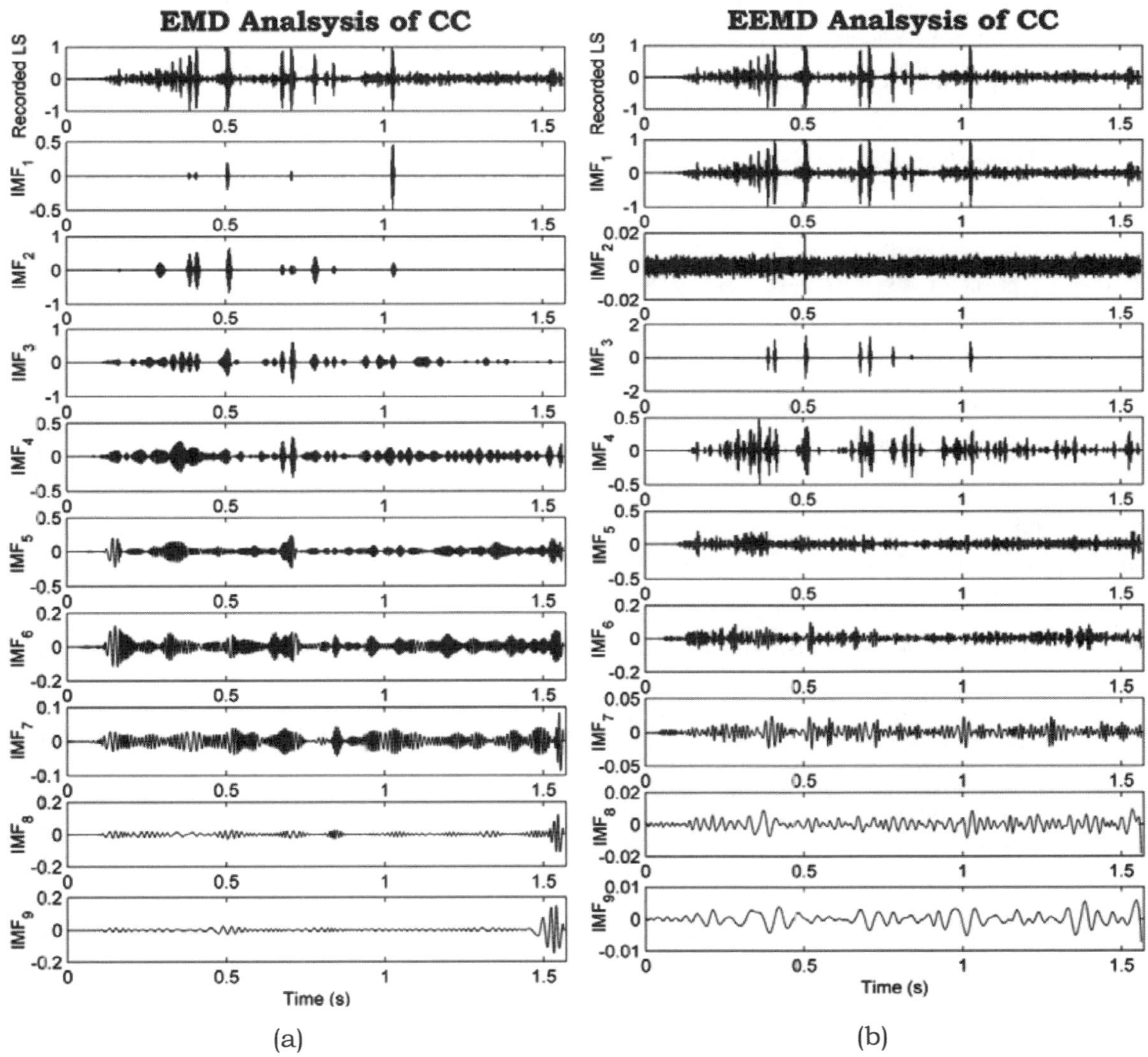

FIGURE 2.10: Application of (a) EMD and (b) EEMD to LS containing CC superimposed on vesicular LS (see first part of Figure 1.1c). In both cases, the recorded LS signal along with the first nine IMFs is depicted, respectively.

An example of the application of EMD and EEMD to recorded CC is illustrated in Figure 2.10a and b, respectively. In particular, the recorded LS signal along with the first nine IMFs is depicted, showing the distribution of the vesicular sound and CC waveforms across the decomposition levels. When comparing the morphology of the IMFs between the two decompositions, some differences can be identified. In particular, IMF_1 clearly differs from the original signal in the case of EMD (Figure 2.10a), whereas in the case of EEMD (Figure 2.10b) it exhibits strong similarity with the original signal. CC are more isolated in the first four EMD-IMFs, whereas CC dominate in the third and fourth EEMD-IMFs. Clearly, EEMD-IMF_2 corresponds to the additive noise used in the EEMD realization. In general, EEMD-IMFs seem to better represent CC than the EMD-IMFs, despite the domination of noise in some IMFs, because EEMD reduces the mode mixing effect (e.g., at IMF_3 and IMF_4 level).

EMD analysis has also been used in a DAS denoising process, as will be thoroughly described in the next chapter.

2.8 FRACTAL DIMENSION–LACUNARITY ANALYSIS

2.8.1 Epitomized Rationale

Fractal dimension. The term "fractal dimension" (FD) can more generally refer to any of the dimensions commonly used for fractals characterization (e.g., capacity dimension, correlation dimension, information dimension, Lyapunov dimension, Minkowski–Bouligand dimension) [141]. In other words, FD is a measure of how "complicated" a self-similar figure is. To this end, FD can be considered as a relative measure of the number of basic building blocks that form a pattern [142]. Consequently, FD could reflect the signal complexity in the time domain. This complexity could vary with sudden occurrence of transient signals, such as explosive LS.

As Gnitecki and Moussavi [62] note, LS will possess valid FD values, based on their morphological properties. Most obviously, the signals do not self-cross. In physical systems, the property of self-similarity in pure fractal objects is not strict but is probabilistic, and there are minimum and maximum scaling limits [143]. In reality, an object occurring in nature, such as a physiological signal, that exactly duplicates itself over several scales does not exist. Thus, the fractality of LS is in the self-affine sense [62]. Furthermore, LSs exhibit clear quasiperiodicity because they emerge from natural biological processes, such as breathing, which implies that they are not purely random [143]. Consequently, FD analysis can shed light on LS from a functional point of view.

Lacunarity. Laennec [144], in his attempt to develop a metalanguage of sound (as reported in his magnum opus *Treatise*), defined a set of descriptions for the shape and texture of sounds that was independent of subjective experience (i.e., independently verifiable). By using the texture of LS in this codification, Laennec showed the importance of this sound property in the hydraulic hermeneutics of mediated auscultation. Following this pathway, LS texture could be approached through the concept of *lacunarity*. The latter was originally developed to describe a property of fractals [145–148], and to discriminate textures and natural surfaces that share the same FD.

Gefen et al. [147] define lacunarity as the deviation of a fractal from translational invariance. Translational invariance can also be a property of nonfractal sets [148], and it is highly scale-dependent; sets that are heterogeneous at small scales can be quite homogeneous when examined at larger scales or vice versa [149]. From this perspective, lacunarity can be considered a scale-dependent measure of heterogeneity or texture of an object, whether or not it is fractal [148]. Lacunarity features can be used to assess the largeness of gaps or holes of one- (signals) or two-dimensional sets (images). It can describe the complex intermingling of the shapes and the distribution of gaps within a set. A set with low lacunarity is homogeneous and transitional invariant, whereas one with high lacunarity has gaps distributed across a broad range of sizes [150].

2.8.2 FD Estimation

There are two principal approaches to estimate the FD of a time series, one that operates directly on the waveform and one that operates in a reconstructed state space [151]. In the first approach, FD is derived from an operation directly on the signal and not on any state space. This means that the data series does not have to be embedded into higher dimensional space for the FD estimation; hence, this type of FD estimation has fast computational implementation. In the second approach, the waveform is regarded as a planer set in R^2, where it is considered a geometric object. A typical time series has a dimension somewhere in between 1 and 2, as it is more complicated than a line but never covers the whole two-dimensional space.

There are several estimation techniques of FD [152], a few of which are *box counting dimension* (based on counting how many d-dimensional hypercubes with side length ε are needed to cover an attractor), *information dimension* (extended version of the box counting dimension by weighting its count by measuring how much of the attractor is contained within each hypercube), *correlation dimension* (a measure of how the number of neighbors in the reconstructed state space varies with decreasing neighborhood sizes ε), *Katz FD* [153], *Sevcik FD* [154], *variance FD* [143]. The latter three are described more analytically below.

Katz FD. According to Katz [153], the FD of a curve defined by a sequence of N points is estimated by

$$\mathrm{FD}^{\mathrm{K}} = \frac{\log_{10}(n_{\mathrm{s}})}{\log_{10}\left(\dfrac{d}{L_{\mathrm{c}}}\right) + \log_{10}(n_s)}, \tag{2.85}$$

where K denotes Katz's definition of FD; L_c is the total length of the curve, realized as the sum of distances between successive points, that is,

$$L_c = \sum_{i=1}^{N-1} \text{dist}(i, i+1), \tag{2.86}$$

where dist(i,j) is the distance between the i and j points of the curve; d is the diameter estimated as

$$d = \max[\text{dist}(i, j)]; \; i \neq j; \; i, j \in [1, N]; \tag{2.87}$$

for curves that do not cross themselves, usually, d diameter is estimated as the distance between the first point of the sequence and the point of the sequence that provides the farthest distance [153], that is,

$$d = \max[\text{dist}(1, i)]; \; i \in [2, N]; \tag{2.88}$$

and n_s is the number of steps in the curve, defined as

$$n_s = \frac{L_c}{\bar{a}}, \tag{2.89}$$

where $\bar{a}$ denotes the average step, that is, the average distance between successive points. In this way, a general unit or "yardstick" is formed that eliminates the dependence of the FD estimates derived by (2.85) on the measurement units used [153].

Sevcik FD. Sevcik [154] uses the following definition for the FD of an N-sample curve:

$$\text{FD}^{\text{S}} = 1 + \frac{\ln(L_c)}{\ln[2 \cdot (N-1)]}, \tag{2.90}$$

where S denotes Sevcik's definition of FD and L_c is defined as in (2.86). Before applying (2.90), Sevcik proposes, for convenience, linear transformations of the waveform, in order to transform it into a normalized space where all axes are equal [154]. He proposes normalization of every point in the abscissa as:

$$x_i^* = \frac{x_i}{x_{\max}}, \; i = 1, \ldots, N, \tag{2.91}$$

where x_i, $i = 1, \ldots, N$, are the original values of the abscissa and $x_{\max} = \max(x_i)$, and ordinate normalization as:

$$y_i^* = \frac{y_i - y_{\min}}{y_{\max} - y_{\min}}, \; i = 1, \ldots, \text{N}, \tag{2.92}$$

where y_i, $i = 1, \ldots, N$, are the original values of the ordinate, $y_{\max} = \max(y_i)$ and $y_{\min} = \min(y_i)$.

Variance FD. The variance FD (VFD) of a $s(t)$ signal (i.e., LS) is estimated via the Hurst exponent given by

$$H = \lim_{\Delta t \to 0} \left[\frac{\frac{1}{2} \log\left\{ \operatorname{var}\left[(\Delta s)_{\Delta t} \right] \right\}}{\log(\Delta t)} \right], \tag{2.93}$$

which is a measure of signal smoothness. In (2.92), $(\Delta s)_{\Delta t}$ denotes $s(t_2) - s(t_1)$ and $\Delta t = |t_2 - t_1|$. Using (2.92), VFD is given by

$$\text{VFD} = E + 1 - H, \tag{2.94}$$

where E corresponds to the Euclidean dimension, which equals one for a 1-D time series; hence, (2.93) can be rewritten as

$$\text{VFD} = 2 - H. \tag{2.95}$$

In fact, (2.94) implies a power-law relationship between the variance of the amplitude increment of $s(t)$, which is produced by a dynamical process, over a time increment Δt. This power-law is of the form

$$\operatorname{var}\left[(\Delta s)_{\Delta t} \right] \propto \Delta t^{2H}. \tag{2.96}$$

The choice of the time increment depends on the aim of the application; unit time increment is preferred when separating a signal from noise, whereas a dyadic is preferred for separating different components within the signal [143].

2.8.3 Lacunarity Estimation

There are many algorithms to define, quantify, and calculate lacunarity [145, 147]. In recent years, the most popular methods all seem to be based on the use of the intuitively clear and computationally simple "gliding box" method of Allain and Cloitre [148]. Stemming from the latter, Plotnick et al. [149] extended the concept of lacunarity to describe the spatial distribution of a real dataset even if it is not an ideal fractal set. Furthermore, Dong [155] and Du and Yeo [150] proposed two lacunarity estimating methods for image texture analysis and segmentation, respectively. The "gliding box" method of Allain and Cloitre [148] consists of the following main steps.

Consider a set of data with length M; at first, a gliding box of length $r(\leq M)$ is placed at the origin of the set and the number of occupied sites within the box, that is, locations with existence of information, termed as "box mass" s, are counted. Next, the box is moved to the right one space along the set and the box mass is counted again. This process is repeated over the entire set, producing a frequency distribution of the box masses $n(s,r)$, which is then converted into probability distribution $Q(s,r)$ by dividing by the total number of boxes $N(r)$ of size r. Then, the first and second moments of $Q(s,r)$ are estimated by

$$Z_1 = \sum_s sQ(s,r), \tag{2.97}$$

$$Z_2 = \sum_s s^2 Q(s,r). \tag{2.98}$$

Using (2.97) and (2.98), the lacunarity for size r is defined as

$$\Lambda(r) = \frac{Z_2^2}{Z_1^2}. \tag{2.99}$$

The calculation of (2.99) is repeated over a range of box sizes, ranging from $r = 1$ to some fraction of the length of the analyzed data M. Considering the form of (2.97) and (2.98), it is clear that they can be rewritten in the form of

$$Z_1 = \overline{s(r)}, \tag{2.100}$$

$$Z_2 = \sigma_s^2(r) + \left(\overline{s(r)}\right)^2, \tag{2.101}$$

with $\overline{s(r)}$ and σ_s^2 denoting the mean and the variance of the number of sites per box, respectively. Consequently, (2.99) could be written as

$$\Lambda(r) = \frac{\sigma_s^2(r)}{\left(\overline{s(r)}\right)^2} + 1. \tag{2.102}$$

Based on (2.102), it is clear that the lacunarity statistic is a dimensionless representation of the variance to mean ratio. The representation of lacunarity in (2.102) reveals its texture relationship. In particular, sparse regions have low means of the box mass; in other words, they appear to be less homogeneous than dense regions. Consequently, they will have higher lacunarity values than dense regions while counted with the same box size. For the same region, large boxes will generally be more homogeneous than small ones, so the lacunarity value of the former case is lower than that of the latter. Before the size of the boxes reaches the scale of randomness of the dataset, which is the scale at which gaps in the dataset appear to be almost homogeneous, significant variation in lacunarity values can be observed. When the variance of the gaps between occupied cells approaches zero, lacunarity values approach one [150].

Original applications of lacunarity analysis have used binary datasets only [147, 148]. However, as Plotnick et al. [149] denote, the use of quantitative (instead of binary) data is analogous to beginning the analysis at a coarser level of resolution. Lacunarity can thus be calculated by using the sum (or integral) of the distribution in a box of size r.

2.8.4 FD Analysis of LS

An analysis of LS with FD is presented in the work of Gnitecki and Moussavi [62], who examined the fractality of LS from normal subjects reflected in the estimated values of FD using FD^K, FD^S, and VFD, defined in (2.85), (2.90), and (2.95), respectively. LS were sequestered corresponding to 85–100% of the maximum flow per inspiratory breath. Three sliding window sizes were used for each FD calculation across the signals (50, 100, and 200 ms, respectively), shifted forward by 25 ms.

In [62], apparent similarities, in terms of both morphology and overall variability, between three scalings of LS were found. Overall, both VFD and FD^K showed an increased with airflow, whereas FD^S appeared to be unaffected by change in flow and hence in LS intensity. From the characteristics of the FD analysis of the examined normal LS, a fractal character was identified; this finding, however, was not justified for the case of abnormal LS.

FD was also used in the analysis of DAS for their detection and isolation from the vesicular sound; this is thoroughly described in the next chapter.

2.8.5 Lacunarity Analysis of DAS

The texture-based approach of LS using lacunarity analysis is a fairly new concept. Application of lacunarity analysis to DAS classification has been recently examined in [83]. In this work, denoised signals from each DAS category, that is, FC, CC, and SQ, formed data series that correspond to the three textures, respectively, as illustrated in Figure 2.11(i). Thirty-two sections of the same size W from every texture (data series) were randomly chosen. Next, from these 32 sections, 24 sections (75%) were used to estimate the lacunarity value of the texture, that is, these 24 sections were used to train a classifier. The remaining eight sections (25%) were used to test the classification performance. This sequence was repeated 200 times, and classification accuracy was averaged out. By repeating the classification experiment many times with different (random) training and test sections, robust estimate of the classifier performance and the texture (i.e., DBS) description ability of the proposed lacunarity-based approach were obtained. The classification process used in [83] was the discriminant analysis [156, 157]. The classification was based on the Mahalanobis method, which uses Mahalanobis distances with stratify covariance estimates [156, 157].

Figure 2.11(ii) depicts the estimated normalized lacunarity $\Lambda(r)$ [see (2.99)] for the FC–CC–SQ comparison group for the training and testing dataset, respectively, versus the box size r when

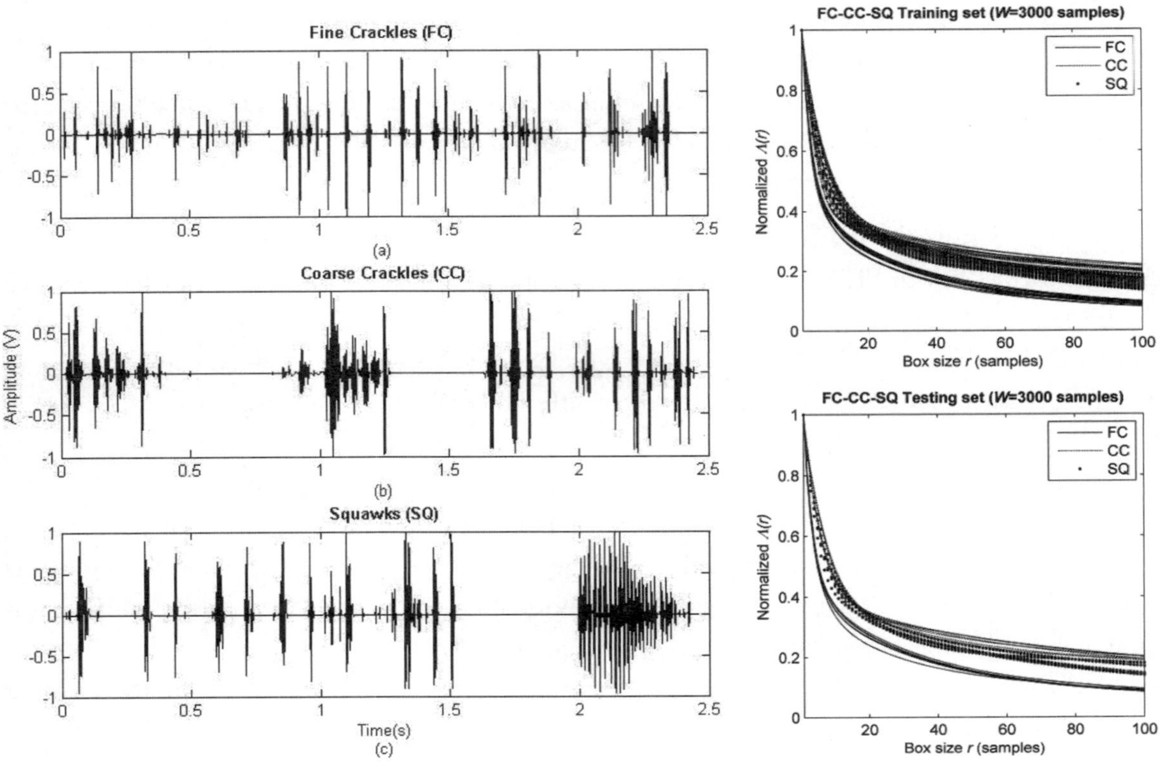

(i) (ii)

FIGURE 2.11: Lacunarity analysis of DAS. (i) The three "texture-series" corresponding to FC, CC, and SQ. (ii) Estimated normalized lacunarity $\Lambda(r)$ for the training (top) and testing (bottom) set with $W = 3000$ samples ($\Lambda(r)$ that corresponds to FC, CC, and SQ is plotted with solid, dashed, and dotted line, respectively) [83].

$W = 3000$ samples (sampling at 2.5 kHz). In Figure 2.11(ii), it is clear that curves are grouped in a manner that makes texture separation feasible. In particular, both in the training and testing sets, FC-related lacunarity curves are grouped in a markedly distant manner from the lacunarity curves related to other two textures (CC and SQ), almost independently of the box size, whereas CC- and SQ-related lacunarity curves show a more neighborly behavior that makes their separation dependent on the value of box size r.

Results shown in Figure 2.11(ii) indicate that the behavior of the normalized $\Lambda(r)$ resembles that of the hyperbola. In particular, during the transition from CC to SQ and further to FC cases, the concavity of the hyperbola function changes, corresponding to a transition from higher to lower lacunarity, respectively. This change in the concavity of the hyperbola function seems to be the basic parameter that contributes to the successful categorization among DBS types using lacunarity analysis. The higher values of lacunarity observed in CC compared to those in FC are due to the difference in their reproduction. CC behave more randomly than FC over subsequent breaths [12]; hence, they create sparser regions [see Figure 2.11(i), b] that have low means of the box mass, reduce their homogeneity, and result in higher lacunarity values than those for the FC regions, while counted with the same box size. An intermediate situation is seen in the case of SQ, where the combination of FC with a short wheeze results in less sparse regions than CC, yet more than FC [see Figure 2.11(i), c), creating smaller lacunarity values than CC, but higher than those in FC [see Figure 2.11(ii)].

When applied to two datasets of DAS, lacunarity analysis exhibited high classification performance, yielding results with a combined mean classification accuracy of more than 99.6% for all comparison groups (FC–CC, FC–SQ, CC–SQ, FC–CC–SQ) [83]. The randomized training and testing sections used in the evaluation of lacunarity analysis augurs a promising performance of LAC analysis under different case scenarios. The simplicity of LAC analysis facilitates the customization of its computational environment under a real-time context, resulting in an easily implemented and user-friendly processing tool of DAS.

• • • •

CHAPTER 3
Denoising Techniques

3.1 OVERVIEW

Noise contamination of the signal of interest is an ordinary effect in signals acquired from pragmatic systems, such as the human body. When focusing on lung sounds (LS), this is clearly evident, as different types of noises coexist with the useful LS; heart sounds, respiratory muscle sounds, friction rubs, ambient noise sources are some primary examples. Because physicians include LS interpretation, alongside other symptomatic indicators, in their process of diagnosing a respiratory pathology, denoising of useful information could contribute to more accurate and objective appreciation of LS information.

In this chapter, a series of denoising techniques applied to LS enhancement are described as indicative examples. It should be noted that the concept of denoising adapts to the specific problem examined, as, for instance, DAS denoising refers to their detection and extraction from the vesicular sound (VS), whereas heart sound cancellation focuses on leaving most unaffected the remaining LS after the efficient removal of heart sounds.

The techniques described here refer directly or indirectly to the new domains of LS representation presented in the previous chapter; hence, the latter could serve as a background basis for the methodologies adopted in the denoising techniques presented in this chapter.

3.2 WAVELET-BASED DENOISING

Two LS denoising techniques based on wavelets are presented: continuous wavelet transform-wheeze episode detector (CWT-WED) [158] and wavelet transform-based stationary–nonstationary (WTST-NST) filter [46]. CWT-WED refers to the use of CWT in wheeze detection and extraction from breathing sound, whereas the WTST-NST filter provides efficient separation of discontinuous adventitious sounds (DAS) from VS (background-type noise).

3.2.1 CWT-WED

In the CWT-WED algorithm, the focus is placed on the automatic location and identification of wheezing episodes during breathing. In particular, CWT (see (2.47)) is used to form a wheezing episode detector taking into account the beneficial properties of CWT representation over the Fourier one, because, unlike the latter, CWT provides useful information on the coherent nature of localized features within the signal, such as wheezing episodes.

CWT-WED is structured in a twofold manner. In particular, it not only focuses in identifying the true location of wheezes within the breathing cycle, it also aims at separating the wheeze signal from the acquired lung sound. In this way, the diagnostic character of the wheeze is unveiled, because its contamination due to the superimposition of the breathing sound is circumvented. As a result, CWT-WED acts both as an identification and denoising tool.

Wheeze identification. The N-sample acquired lung sound signal, $x(k)$, can be considered as the sum of an envelop "confined" sinusoidal signal of interest, $s(t)$, that is, wheezes, and the breathing sound, $n(k)$ (such as background noise); hence, it can be written as

$$x(k) = s(k) + n(k), \; k = 1, \ldots, N. \tag{3.1}$$

CWT is realized using the real Morlet wavelet [102], defined as

$$\psi(t) = \sqrt{\pi f_{\mathrm{b}}} \exp\left(-t^2/f_{\mathrm{b}}\right) \cos\left(2\pi f_0 t\right), \tag{3.2}$$

where f_{b} is the bandwidth parameter that controls the Gaussian envelope, which "confines" the sinusoidal waveform with a center frequency of f_0. By taking the square of the modulus of the resulted CWT, the scalogram is estimated as

$$S_{\mathrm{c}}(a,b) = |W_{\mathrm{s}}(a,b)|^2 \tag{3.3}$$

which is analogous to the spectrogram [102], and highlights the location and scale of dominant energetic features within the signal. The increased coherency between the signal of interest, $s(t)$ (i.e., wheezes), and the analyzing wavelet, unlike the breathing sound, $n(k)$w, is reflected in the scalogram by presenting distinct isolated peaks. Consequently, unlike the spectrogram, the scalogram provides direct information about the time and scale (frequency) location of the wheeze.

Wheeze denoising. The denoising performance of CWT-WED is achieved by applying *scale-dependent thresholding* [102]. This helps to reduce the high- and low-frequency noise components in the reconstructed signal, because the inverse CWT (ICWT) of (2.51) is performed over a range of scales, that is, $a^{\circ} < a < a^{\bullet}$, where a° and $a^{\bullet}$ denote the cutoff scales for the high and low frequencies, respectively.

Due to the localization of CWT both in time and scale (frequency), the reconstructed signal resembles the output of an adaptive filtering (AF), retaining the desired structural characteristics only at the true locations of the signal of interest $s(t)$.

Figure 3.1 depicts an example of the application of CWT-WED to LS recordings, presenting the wheeze identification part of CWT-WED. The top subfigure shows the recorder signal (normalized amplitude) along with the superimposition of the normalized airway flow signal. This contributes to the identification of the inspiratory (negative flow) and expiratory phase (positive flow) of the breathing cycle and the characterization of wheezes as inspiratory or expiratory, respectively. The middle subfigure depicts the corresponding spectrogram, estimated using a 256-point Hanning windowed signal with 50% overlapping and a discrete Fourier transform length of 2048 points. The bottom subfigure illustrates the scalogram (see (3.3)) from the estimated CWT of the analyzed LS signals. Figure 3.1 shows that the scalogram (bottom), unlike the spectrogram (middle), provides a timescale (frequency) representation of the wheezes that eliminates most of the

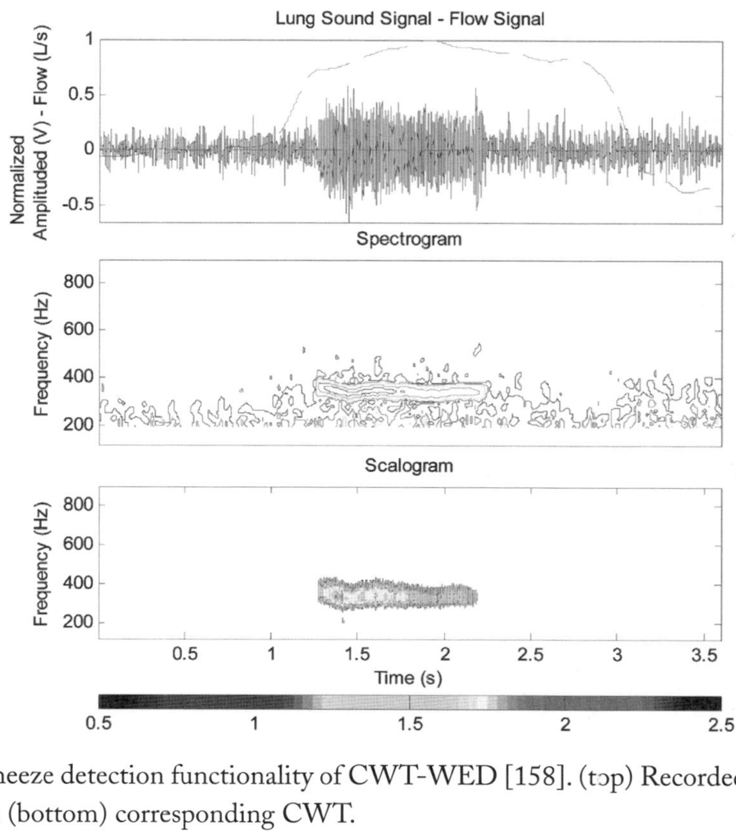

FIGURE 3.1: Wheeze detection functionality of CWT-WED [158]. (top) Recorded LS; (middle) corresponding STFT; (bottom) corresponding CWT.

undesired information related to background noise. As a result, the time–frequency characteristics of wheezing episodes are clearly revealed and identified.

Figure 3.2a depicts the CWT-WED wheeze denoising functionality. In particular, the top, middle, and bottom subfigures present the recorded lung sound (zoomed version of the recording in Figure 3.1), estimated wheeze, and breathing sound (background noise), respectively. In this figure, the adaptive character of CWT-WED in isolating the appropriate portion of the signal that corresponds to the wheeze location in time is evident. This is further justified by the corresponding power spectrum (PS) of the three signals included in Figure 3.2a, as shown in Figure 3.2b. From the latter, the dominant frequency of the wheeze that is evident in the PS of the LS recording (solid line) totally coincides with the main content of the PS of the estimated wheeze (dash-dotted line). Clearly, the PS of the remaining breathing sound (dashed line) follows the morphology of the original signal in the area outside the location of the dominant frequency peak of the wheeze. This proves the efficiency of CWT-WED not only in locating but also in extracting the signal of interest from the background noise.

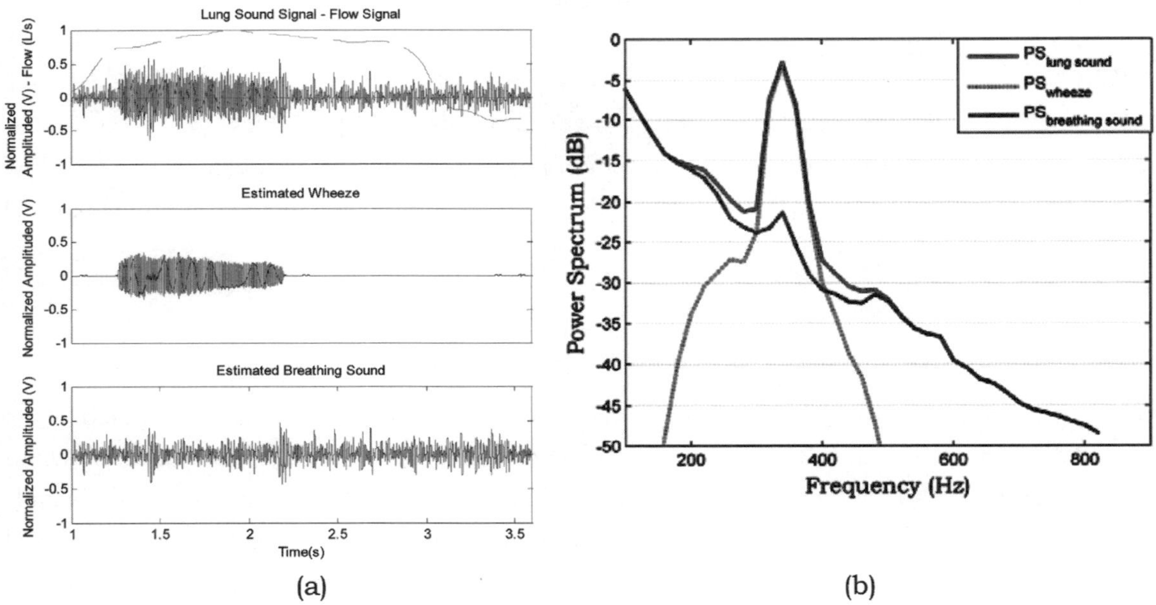

FIGURE 3.2: Wheeze denoising functionality of CWT-WED [158]. (a) The top part denotes the recorded LS; the middle part illustrates the corresponding estimated wheeze episode; the bottom part shows the corresponding breathing sound (background noise), (b) Estimated PS corresponding to the three signals in panel (a); wheeze denoising with CWT-WED is clearly reflected in the frequency domain.

3.2.2 WTST-NST Filter

Automated DAS detection and isolation from VS is a difficult task, due to DAS characteristics (Section 1.2). To address this problem, the nonstationarity of DAS must be taken into account. Consequently, the use of high-pass filtering (HPF) fails to separate the nonstationary sounds, destroying the waveforms. Furthermore, a level slicer cannot overcome the small amplitude of fine crackles (FC). Application of time-expanded waveform analysis in crackle time domain analysis [43, 159] results in separation; it is, however, time consuming and has large interobserver variability. Nonlinear processing [44] yields more accurate results, but requires empirical definition of the set of parameters of its stationary–nonstationary (ST–NST) filter.

As discussed in Section 2.4, WT provides a new perspective in the analysis of LS, because it can decompose them into multiscale details, describing their power at each scale and position [108]. By applying a threshold-based criterion at each scale, a filtering scheme which weights WT coefficients according to signal structure can be composed. Separation of signal from "noise" can be achieved through an iterative reconstruction–decomposition process, based on the derived weighted WT coefficients at each iteration.

The aforementioned concept was introduced in [46], where the implementation of a WTST–NST filter for the separation of DAS (nonstationary waves) from vesicular ones (stationary waves) is described. WTST-NST is a wavelet domain filtering technique, based on the fact that explosive peaks (DAS) have large components over many wavelet scales, whereas "noisy" background (VS) dies out swiftly with increasing scale. This fact allows characterization of the wavelet transform coefficients with respect to their amplitude; the most significant coefficients at each scale, with amplitude exceeding some threshold, correspond to DAS, whereas the rest correspond to VS. Consequently, a wavelet domain separation of wavelet transform coefficients corresponding to DAS and VS, respectively, can offer a time domain separation of DAS from VS using an iterative multiresolution decomposition–multiresolution reconstruction (MRD–MRR) scheme (Section 2.4). Although VS are characterized as "noisy" background, it is better to view them as an N-sample signal being incoherent relative to a basis of waveforms—that is, it does not correlate well with the waveforms of the basis, that is, its entropy is of the same order of magnitude as $\log N - \varepsilon$, with small ε [160]. From this notion, the separation of DAS from VS becomes a matter of coherent structure extraction from the breath sounds.

Figure 3.3 illustrates an example of the application of the WTST-NST filter to recorded LS that contain coarse crackles (CC). In Figure 3.3, the top subplot illustrates the original recorded LS sounds; the nonstationary character of CC is clearly evident. The result of the application of the WTST-NST filter to the recorded LS is depicted in the middle subplot of Figure 3.3, whereas the estimated VS signal (serving as background noise) is shown in the bottom subplot of Figure 3.3.

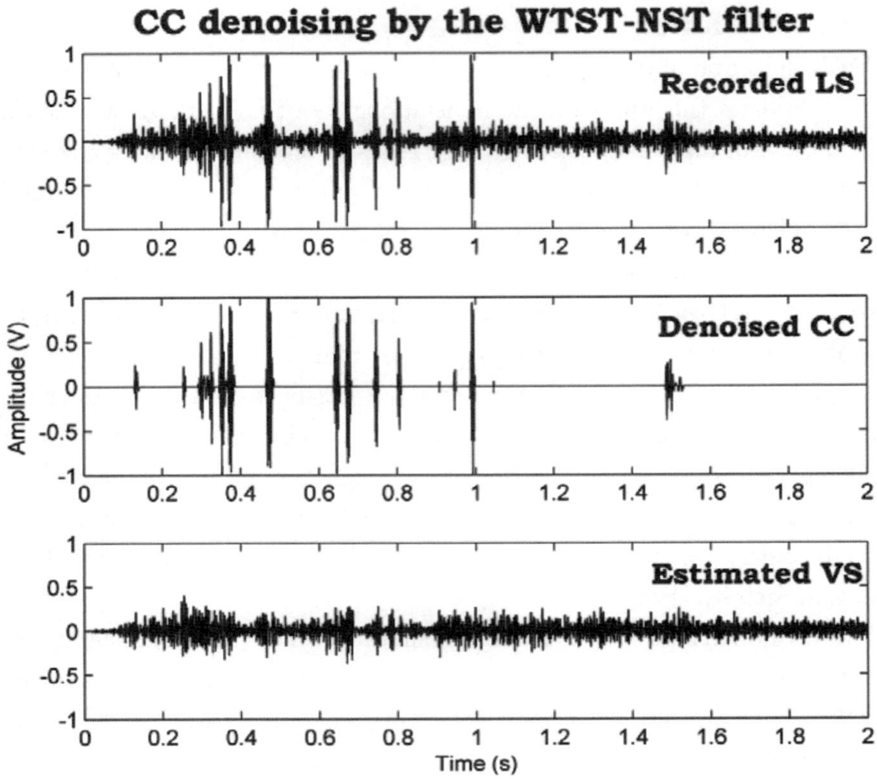

FIGURE 3.3: An example of the application of the WTST-NST filter [46] to recorded LS that contain CC. (top) Original recorded LS; (middle) denoised CC; (bottom) estimated VS (background noise-like). Results shown in this figure illustrate the efficient performance of the WTST-NST filter to identify and extract DAS from VS.

When comparing the output of the WTST-NST filter with the original signal, the ability of the WTST-NST filter to efficiently identify and extract DAS from VS is realized. Extended implementation details and further experimental results can be found in [46].

3.3 KURTOSIS-BASED EXTRACTOR

Higher-order statistics (HOS)-based analysis of LS has proved to be highly efficient as a new representation domain for the extraction of LS characteristics (Section 2.2). In this vein, HOS has also been used as basis for the construction of a denoising tool, especially for DAS. This was introduced in [56], where an iterative kurtosis-based detector (IKD) was formed.

The IKD scheme is based on the estimation of the kurtosis [see (2.6)] within a sliding window along the LS recordings. Because kurtosis is (theoretically) zero for Gaussian signals [86], such as VS (DAS background noise), significant deviations from this value can be attributed to the presence of non-Gaussian signals, such as DAS. This deviation from zero value can be used in forming a criterion for identifying the presence of nonstationary transient signals (i.e., DAS). The IKD segments the signal into two regions: one that consists of DAS plus VS, and one that consists only of VS. Unlike IKD, a detector based on variance changes fails in cases where Gaussian noise is present.

In general, for an N-sample sequence $\mathbf{x} = \{x(k): k = 1, 2, \ldots, N\}$, such as real observations, the estimate of the normalized kurtosis [see (2.6)] (the superscript x is omitted for simplicity) is calculated as [161]

$$\widehat{\gamma}_4 = \frac{\sum_{k=1}^{N} [x(k) - \widehat{m}]^4}{(N-1)\widehat{\sigma}^4} - 3, \tag{3.4}$$

where $\widehat{m}$ and $\widehat{\sigma}$ denote the estimates of the mean and standard deviation of $\mathbf{x}$, respectively.

Kurtosis is a measure of the heaviness of the tails in the probability density function of $\mathbf{x}$ [86], which changes due to nonstationarity. In particular, outliers of abrupt changes in $\mathbf{x}$ sequence affect the tails of the probability density function. Hence, estimation of kurtosis via (3.4) could be used to establish an effective statistical test to identify abrupt changes in signals, such as those that characterize the existence of DAS in LS recordings.

The IKD scheme is implemented in an iterative form. In particular, along the given sound recording, $\mathbf{x}$, an M-sample sliding window ($M \ll N$) is defined, with one sample difference between successive windows. Over each sound segment obtained from the sliding window, the kurtosis $\widehat{\gamma}_4$ is estimated using (3.4). Afterwards, the derived time series $\widehat{\gamma}_4 = \{\widehat{\gamma}_{4,j} : j = 1,2,\ldots,N - \mathrm{M} + 1\}$ is demeaned and its absolute value is obtained, defining the time series $\mathbf{K} = \{K_j\}$. The time series $\mathbf{K}$ is then compared to a threshold, which is proportional to the standard deviation of $\mathbf{K}$ normalized by the standard deviation of the estimated VS that initially coincides with the sound recording $\mathbf{x}$. This normalization results in the adjustment of the threshold according to the mean power of the estimated VS and it tunes the threshold to efficient value as far as the detectability of the algorithm is concerned. The parts of the signal whose corresponding K_j value exceeds the threshold are assigned as signal DAS; the rest of the signal $\mathbf{x}$ is considered as the estimate of the VS.

However, by this approach, it is possible to have significant DAS portions that are not recognized as signal of interest and have been left in the VS part. Thus, the aforementioned procedure is iteratively applied as an effort to extract new DAS portions from the current VS signal. The iterative procedure is terminated when the mean power of the current DAS portion that is added to the previously estimated ones is smaller than a preselected power level $\varepsilon(\varepsilon \ll 1)$. The final DAS signal is the sum of all DAS portions estimated at all previous iterations, whereas the VS estimated at the last iteration qualifies as the final VS signal [56].

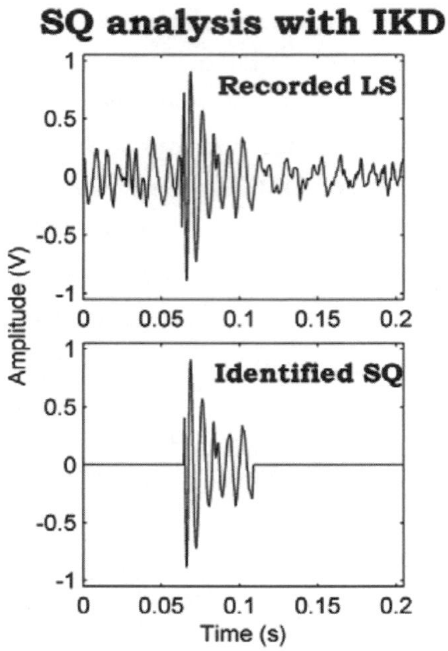

FIGURE 3.4: An example of the application of IKD [56] to recorded LS that contain SQ. (top) Original recorded LS; (bottom) detected SQ. Results shown in this figure illustrate the efficient performance of the IKD filter to detect DAS in LS recordings.

Figure 3.4 illustrates an example of the implementation of IKD to LS recordings that includes a squawk. As can be seen from this figure, IKD keeps the part of the signal that includes the DAS only, and outputs zeroes at places where only VS signal exists. Nevertheless, as already noted, the VS signal that coexists in the DAS portion is also extracted in the IKD output, meaning that, unlike the WTST-NST filter presented in Section 3.2, IKD efficiently allocates the portions of the LS recordings that include DAS and extracts them from the original signal, yet without separating DAS from VS in the extracted segments. However, as Figure 3.4 shows, IKD preserves the characteristics of squawks (SQ) in its output, because the two parts of SQ, that is, initial FC followed by a short wheeze, are clearly identified.

For more details about the performance of the IKD scheme, the reader is referred to [56].

3.4 FRACTAL DIMENSION-BASED DETECTOR

Apart from the fractal analysis of LS presented in Section 2.8, a fractal dimension (FD)-based detector is presented here, and applied to DAS detection within the LS recordings. In particular, the FD-based detector (FDD) introduced in [53] assesses the complexity of sound recordings in the time domain in order to efficiently detect the time location and duration of the nonstationary DAS in LS data.

The FDD scheme is based on the FD of the LS recordings. As discussed in Section 2.8, FD could be used as a measure of signal complexity in the time domain. This complexity could vary with different pathophysiological conditions. Consequently, by means of FD the complexity variations are linked with the changes over time of sound recordings, providing a fast computational tool that tracks the nonstationarities, that is, the location of the explosive DAS.

The FDD scheme initially employs an M-sample sliding window, which is shifted along the N-sample section of the LS recordings, with a 99% percentage of overlap, obtaining point-to-point values of the estimated FD. Over each sound segment obtained from the sliding window, the FD is computed using Katz's algorithm [see (2.82)], and the value of the estimated FD^K is assigned to the midpoint of the sliding window. When FD^K is estimated, an FD-peak peeling algorithm (FD-PPA) [53] is used to automatically identify the location and duration of the explosive DAS through the FD peaks. The FD-PPA iteratively "peels" the estimated FD^K signal, gradually gathering those parts that construct its peaks. In this way, even small-amplitude FD^K peaks are accurately identified. In the unlikely case with merged DAS peaks, which have not been successfully separated by the first pass of the FD-PPA, the latter is applied again locally; hence, with one or two passes of the FD-PPA, the time location and duration of all explosive DAS are accurately and automatically identified [53].

Figure 3.5 shows an example of the application of FDD to LS recordings that include FC. In this figure, the top subfigure illustrates the original LS recordings with FC superimposed on the VS

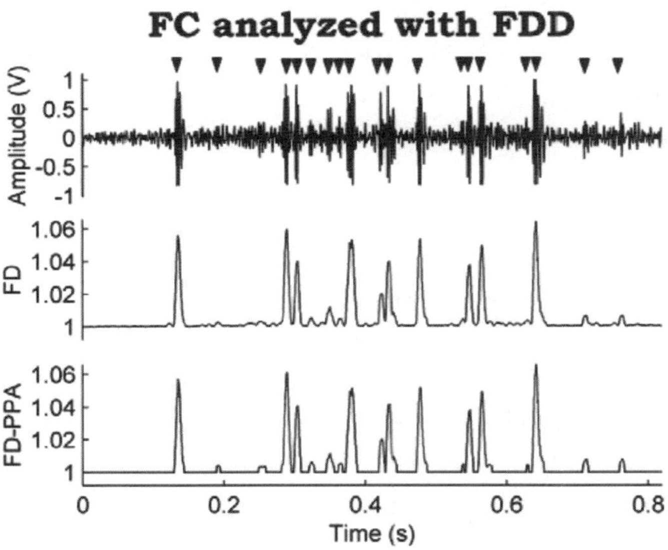

FIGURE 3.5: An example of the application of FDD [53] to recorded LS that contain FC (denoted by arrowheads). (top) Original recorded LS; (middle) estimated FD^K; (bottom) output of the FD-PPA corresponding to valid FC peaks.

sound. The middle subfigure denotes the estimated FD^K, whereas the bottom subfigure shows the output of the FD-PPA applied to the estimated FD^K, keeping only the FD^K peaks that correspond to valid FC ones. As shown in Figure 3.5, variation in the amplitude of FC peaks does not affect the performance of the FDD scheme, because it efficiently detects all FC—despite their high, medium, or low amplitude—compared to the VS amplitude level.

The FDD technique has the potential advantage of data volume reduction. This could be realized by posing criteria (e.g., threshold values) to the estimated FD^K values, which could help an expert system in deciding whether to store or discard the incoming signal included in the analysis window. This is of great importance when LS are continuously recorded for long-term analysis.

More details about the FDD scheme could be found in [53].

3.5 WAVELET-FRACTAL DIMENSION-BASED DENOISING

In Sections 3.2 and 3.4, denoising techniques based on WT and FD, respectively, were presented. WT-FD filter, a combinatory approach that merges the advantages of WT and FD analysis, was introduced [54, 55] in order to achieve superior performance in the enhancement of DAS. In particular, the efficiency of the FDD to accurately locate DAS in LS recordings (Section 3.4) is transferred to the WT domain as a means that values the WT coefficients according to their significance in the signal structure. Consequently, an FD-based thresholding procedure is formed, resulting, through an iterative reconstruction–decomposition process, in efficient separation of the desired signal (i.e., DAS) from the undesired VS (background noise).

The main thrust of the WT-FD filter is the more enhanced selection of WT coefficients that correspond to DAS during the MRD–MRR procedure (Section 2.4), compared to that of the WTST-NST filter (Section 3.2). The WT-FD filter overcomes the requirement of the WTST-NST filter for empirical setting of a multiplicative parameter in the definition of its threshold [46], providing a different perspective in the categorization of the WT coefficients by using FD analysis to construct an efficient method of thresholding. As explained in Section 3.4, FDD enables the detection of DAS, but not their extraction from background noise. In the WT-FD filter, the detection ability of the FDD scheme is transferred from the time domain to the WT domain. In this manner, both detection and extraction of DAS signals from background noise are achieved, by reconstructing the signal of interest (DAS) through the WT coefficients efficiently selected by the FDD.

Using FD-based thresholding, the WT-FD filter acts *only* at the true locations of the desired signal presence, keeping the rest of the input signal unchanged. Consequently, similar to the WTST-NST filter (Section 3.2), the WT-FD filter also performs as an *adaptive* noise removal tool for the analysis of LS.

Figure 3.6 illustrates an example of the application of the WT-FD filter to LS recordings containing CC. As shown in this figure, the efficiency of the FDD scheme to identify the useful

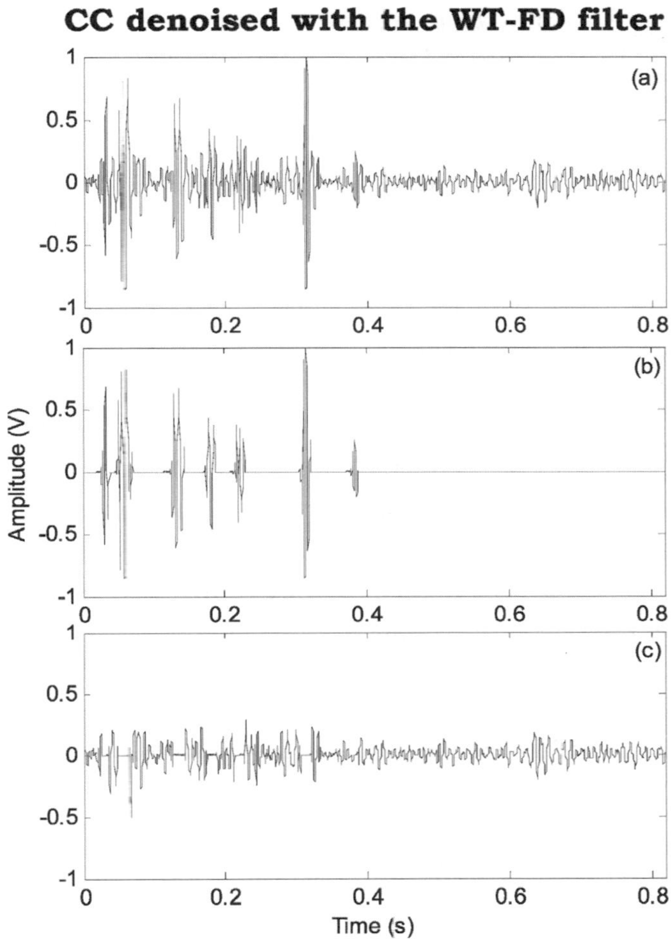

FIGURE 3.6: An example of the application of the WT-FD filter [54, 55] to recorded LS that contain CC. (a) Original recorded LS; (b) denoised CC; (c) estimated VS (background noise-like).

WT coefficients that correspond to valid DAS peaks has resulted in accurate extraction of CC (Figure 3.6b) from the original LS recordings (Figure 3.6a). The remaining signal (Figure 3.6c) qualifies as VS and/or background noise. A thorough description of the structure and performance of the WT-FD filter can be found in [54, 55].

3.6 EMPIRICAL MODE DECOMPOSITION-FRACTAL DIMENSION-BASED DENOISING

An extension of the combinatory approach used in the WT-FD filter to the field of empirical mode decomposition (EMD; Section 2.7) is introduced in [57]. Instead of WT, the EMD is used to decompose the sound signal into components with well-defined instantaneous frequency. In this manner, the oscillatory characteristics of DAS are reflected to the intrinsic mode function (IMF) analysis domain. Consequently, by applying FD analysis in the latter, the important (high FD value) and unimportant (low FD value) portions of IMFs can be identified, corresponding to DAS and background noise, respectively. Because none of the signal is lost in EMD, the sum of the selected portions of IMFs per category gives back the denoised DAS and the background noise, accordingly.

The aforementioned approach forms a new denoising tool, the EMD-FD filter, which adaptively captures the nonstationarity of ELS and successfully extracts them from the background noise. According to the desired accuracy in its output, the EMD-FD filter is formed in two ways: in a noniterative and an iterative way.

In the noniterative realization of the EM-FD filter, an EMD process is applied to the M-sample LS signal $x(k)$, $k = 1, 2, \dots, M$, decomposing it into N IMFs, that is, $c_i(k)$, $k = 1, 2, \dots, N$, and the residue $r_N(k)$, $k = 1, 2, \dots, M$, following the procedure described in EMD analysis (Section 2.7). From the N estimated IMFs, the first L ones are selected according to an energy criterion [57]. The FDD scheme (Section 3.4) is then applied to the L selected IMFs, and for each one, the valid peaks of the corresponding FD sequences, $\mathrm{FD}_j^{\mathrm{K}}(k)$, $j = 1, 2, \dots, L$, $k = 1, 2, \dots, M$, are estimated, that is, $\mathrm{FDPP}_j^{\mathrm{K}}(k)$, $j = 1, 2, \dots, L$, $k = 1, 2, \dots, M$, picked by using the FD-PPA scheme within the FDD (Section 3.4). After the generation of the $\mathrm{FDPP}_j^{\mathrm{K}}(k)$ sequence, two binary thresholds are constructed, signal binary threshold $\mathrm{SBTH}_j^{\mathrm{K}}(k)$ and noise binary threshold $\mathrm{NBTH}_j^{\mathrm{K}}(k)$, defined as follows:

$$\mathrm{SBTH}_j(k) = \begin{cases} 1 & \mathrm{FDPP}_j(k) \neq 1 \\ 0 & \mathrm{FDPP}_j(k) = 1 \end{cases}, \tag{3.5}$$

$$\mathrm{NBTH}_j(k) = \left[1 - \mathrm{SBTH}_j(k)\right], j = 1,2,\dots, L,\ k = 1,2,\dots, M. \tag{3.6}$$

When the binary thresholds of (3.5) and (3.6) are constructed, they are both multiplied with the corresponding IMFs. In this way, by means of $\mathrm{SBTH}_j(k)$, the portions of IMFs that are related to the desired signal (DAS), $c_j^{\mathrm{DAS}}(k)$, $j = 1,2,\dots,L$, $k = 1,2,\dots,M$, are kept, whereas using the $\mathrm{NBTH}_j(k)$, those related to the background noise, $c_j^{\mathrm{BN}}(k)$, $j = 1,2,\dots,L$, $k = 1,2,\dots,M$, are obtained. By combining all coherent parts from L IMFs, the estimated denoised DAS signal is produced as

$$x^{\mathrm{DAS}}(k) = \sum_{j=1}^{L} c_j^{\mathrm{DAS}}(k),\ k = 1,2,\dots, M, \tag{3.7}$$

whereas the noncoherent part (background noise) is estimated as the summation of the remains, that is,

$$x^{\mathrm{BN}}(k) = \sum_{j=1}^{L} c_j^{\mathrm{BN}}(k) + \sum_{i=L+1}^{N} c_i(k) + r_N(k),\ k = 1,2,\ldots,\ M. \tag{3.8}$$

The noniterative structure presented above can be further expanded to an iterative one, when the $x^{\mathrm{DAS}}(k)$ output [see (3.7)] is not of the desired accuracy. This means that some portions of the DAS signal still exist in $x^{\mathrm{BN}}(k)$ [see (3.8)]. This is probably due to the amplitude-overshadowing of some DAS peaks over some others, yet still valid. In other words, some high-amplitude FD peaks that correspond to high-amplitude DAS peaks have led to an underestimation of the FD peaks that correspond to the lower-amplitude DAS peaks, resulting in a misidentification by the FDD scheme. To circumvent this problem and to provide refinement alternatives within the performance of the EMD-FD filter, an iterative structure of it was formed [57]. In this EMD-FD realization, the $x^{\mathrm{BN}}(k)$ derived by (3.8) is used iteratively as an input signal to the EMD-FD analysis. At each iteration, an energy-based stopping criterion is applied [57], and its convergence to ε, where e is a small positive number ($o \ll \varepsilon \ll 1$) that corresponds to the desired accuracy in the refinement procedure, is tested. When convergence is achieved and the stopping criterion is met after the Cth iteration, the final output of the EMD-FD filter—denoised DAS, $x_{it}{}^{\mathrm{DAS}}(k)$, and background noise, $x_{it}{}^{\mathrm{BN}}(k)$—are given as

$$x_{it}^{\mathrm{DAS}}(k) = \sum_{l=1}^{C} x_l^{\mathrm{DAS}}(k) \quad \text{and} \quad x_{it}^{\mathrm{BN}}(k) = x_C^{\mathrm{BN}}(k),\ k = 1,2,\ldots,\ M, \tag{3.9}$$

where $x_l{}^{\mathrm{DAS}}(k)$ corresponds to the estimated denoised DAS signal [similar to (3.7)] at the lth iteration.

From the procedure above, it is clear that the denoised DAS signal is formed bit by bit, adding all the estimated DAS parts derived at each iteration. In addition, the background noise estimated at the Cth iteration qualifies as the final version of the background noise, because further analysis contributes nothing more to the refinement procedure.

Figure 3.7 depicts an example of the application of the EMD-FD filter to LS recordings that include two SQ (denoted with arrowheads). As can be deduced from this figure, denoised SQ (Figure 3.7b) are clearly separated from the original LS recordings (Figure 3.7a), whereas the remaining signal (Figure 3.7c) resembles the characteristics of the background noise (e.g., compare it with portions of LS recordings in Figure 3.7a outside the regions of the SQ existence). Moreover, in the denoised version of SQ (Figure 3.7b), it is clear that the EMD-FD filter sustains the morphology of the SQ. This is important, because SQ include both an FC and a short wheeze [12]; hence, not only their initial part corresponding to FC but the part with the fluctuation corresponding to the short wheeze should be identified and separated from the background noise. This is clearly achieved

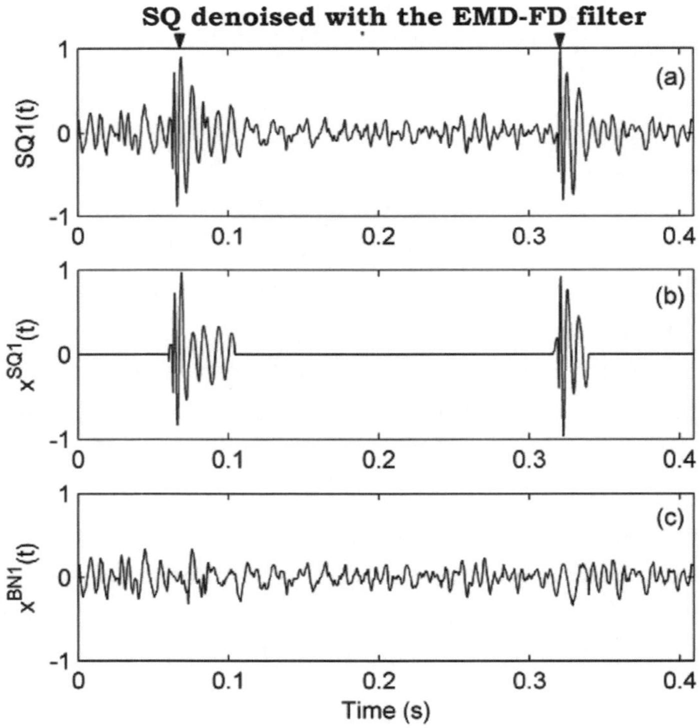

FIGURE 3.7: An example of the application of EMD-FD filter [57] to recorded LS that contain SQ. (a) Original recorded LS; (b) denoised SQ; (c) estimated background noise.

by the EMD-FD filter, as it can be deduced from Figure 3.7b, justifying the adaptive performance of the EMD-FD filter.

For an analytical description of the EMD-FD filter the reader should refer to [57].

3.7 HEART SOUND CANCELLATION

Heart sounds produce an incessant noise during LS recordings that causes an intrusive, quasi-periodic interference that influences the clinical auscultative interpretation of LS. The introduction of pseudo-periodicities, the masking of the relevant signal and the modification of the energy distribution in the spectrum of LS, due to heart sounds [30], require an effective reduction of heart sounds from the contaminated LS signal, to yield a successful LS classification.

HPF and AF are the two main approaches for heart noise reduction. Although HPF (with a cutoff frequency varying from 50 to 150 Hz) is effective in heart sound reduction [17, 162], it degrades the respectively overlapped frequency region of breath sounds and fails to track the changing signal characteristics. The AF technique overcomes these limitations, because it is based on a gradual reduction of the mean square error between the primary input signal (contaminated LS) and a recorded or artificially produced reference signal, highly correlated to the noise component of the input signal (heart sounds) [30, 34, 163]. This approach was initially proposed by Iyer et al. [30], who used AF with a noise reference signal derived from a modified electrocardiogram signal. A modification to this method, proposed by Kompis and Russi [34], combined the advantages of AF with the convenience of using only a single microphone input. Unfortunately, this approach has resulted in a moderate heart sound reduction by 24%–49%. Stemming from these two initial efforts, more efficient heart sound reduction techniques have been developed over the years, using advanced signal processing methodologies. These can be further categorized into those that analyze the entire LS record and reduce heart sounds effect, and those that remove heart sounds included in portions of the LS record and then estimate the signal in the gaps [25]. Several indicative examples of heart sound cancellers (HSCs) from both categories are described in the succeeding subsections.

3.7.1 HOS-Based HSC

HOS have been used in the construction of an HSC in order to adaptively reduce the heart sound noise from LS recordings. In particular, an adaptive noise canceller based on fourth-order statistics (ANC-FOS) was introduced in [42], as a means to form an efficient HSC.

ANC-FOS has two main parts; the first one analyzes the incoming breath signal $x(k)$ to generate the reference signal $z(k)$ for the adaptive filter, whereas the second one computes the FOS of $x(k)$ and $z(k)$, forms the adaptive filter, and yields the final output. To generate the reference signal, the real location of the heart sound in the incoming breath signal must be detected. To this end, in the first part of the ANC-FOS $x(k)$ is analyzed and searched for the true locations of heart sound noise, based on amplitude, distance, and noise-reduction percentage criteria [42]. Its output, $z(k)$, is a localized signal, with precise tracking of the first and second heartbeats, highly correlated with heart sounds, with no extra recording requirement as in Iyer et al.'s method [30]. The second part of ANC-FOS includes the adaptive section of the algorithm and starts with the computation of the fourth-order cumulants of $z(k)$, $c_4^{zzzz}(\tau_1,\tau_2,\tau_3)$, and the fourth-order cross-cumulants of $x(k)$ and $z(k)$, $c_4^{zzzz}(\tau_1,\tau_2,\tau_3)$ (Section 2.2), assuming that they exist and they are not identically zero. These FOS are then used to form the adaptive-filter update equation, given by

$$H_{\mathrm{f}}(k+1) = H_{\mathrm{f}}(k) - \mu(k)\nabla(k), \tag{3.10}$$

with

$$H_{\mathrm{f}} = [h(0), h(1), \ldots, h(N-1)]^T, \tag{3.11}$$

and

$$\nabla \equiv 2\left(c_4^{zzzzT} c_4^{zzzz} H_{\mathrm{f}} - c_4^{zzzzT} c_4^{xzzz}\right), \tag{3.12}$$

where $\boldsymbol{h}(i)$, $i = 0, \ldots, N-1$, denotes the adaptive filter coefficients (N taps) and $m(k)$ corresponds to the step size [42]; bold letters indicate the use of vectors.

Given that the FOS of Gaussian processes are identically zero (Section 2.2), the update equation (3.10) of the adaptive filter consists only of the fourth-order cumulants of the incoming and

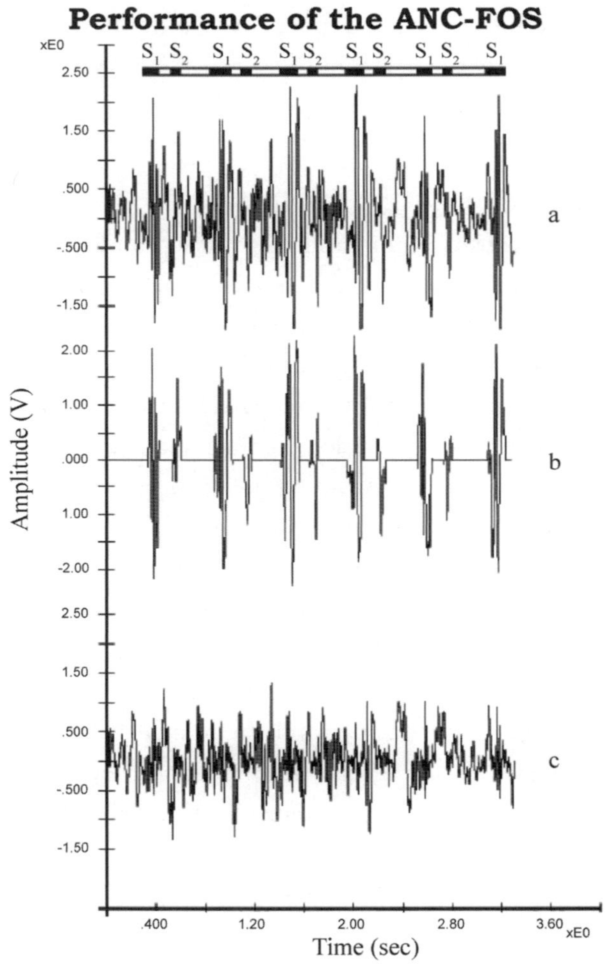

FIGURE 3.8: An example of the application of ANC-FOS [42] to recorded LS, severely contaminated with heart sound noise. (a) Original recorded LS; (b) created noise-reference signal $z(k)$; (c) denoised LS. S_1 and S_2 indicate the first and second heart sounds, respectively.

reference signals and is not affected by Gaussian uncorrelated noise. In addition, ANC-FOS has a localized effect on the input signal at locations pointed out from the reference signal, unlike HPF that deteriorates the whole signal due to elimination of low frequencies.

Figure 3.8 illustrates an example of heart sound cancellation in LS recordings during normal breathing when using the ANC-FOS scheme. When comparing Figure 3.8a and b, it is realized that the reference signal $z(k)$ is highly correlated to the heart sounds existing in the initial LS recording (Figure 3.8a) and exists at the time instances of the heart sounds only. Clearly, in the denoised output of the ANC-FOS (Figure 3.8c), the heart sound interference has been removed, yet without destroying the coexisting LS.

Further details about the ANC-FOS can be found in [42].

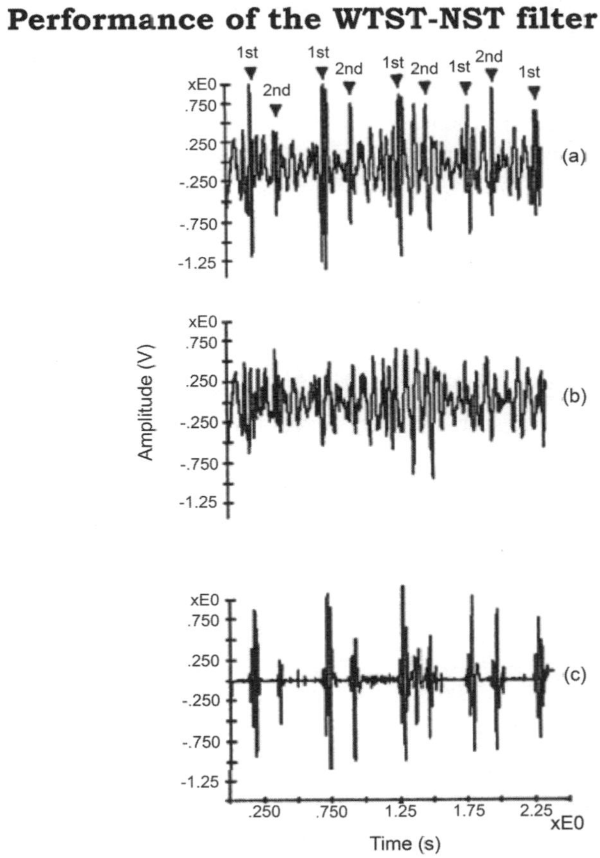

FIGURE 3.9: An example of the application of WTST-NST as HSC [28] to recorded LS, severely contaminated with heart sound noise. (a) Original recorded LS; (b) denoised LS; (c) estimated heart sound interference. Arrowheads denote the first and second heart sounds.

3.7.2 WT-Based HSC

A WT-based adaptive denoising of LS contaminated by heart sound noise is presented in [28]. In fact, this approach refers to the application of the WTST-NST filter (Section 3.2) to the problem of heart sound cancellation. The adaptive separation of signal from "noise," without requiring any reference signal, can be achieved through the iterative reconstruction–decomposition process embedded in the WTST-NST filter, based on the derived weighted WT coefficients at each iteration.

In the realization of the WTST-NST filter as HSC, the DAS, which were originally the signal of interest, are replaced by the heart sound peaks, whereas the originally estimated background noise qualifies here as the desired signal of interest, that is, the denoised LS.

Figure 3.9 presents an example of the performance of the WTST-NST filter as HSC, when applied to severely contaminated LS. As Figure 3.9c shows, when compared to Figure 3.9a, the heart peak locations in input and in heart sound WTST-NST output are synchronized, without using any reference signal. Furthermore, when comparing Figure 3.9a and b, it can be seen that the signal included between the heart sounds remains unchanged. Consequently, the WTST–NST filter when used as an HSC has a localized effect on the input signal, at true locations of heart sound presence without requiring any reference signal.

More details about the WTST-NST filter acting as a HSC can be found in [28].

3.7.3 Recursive Least Squares-Based HSC

Recursive least squares (RLS) have been proposed for linear AF of heart sound noise by Gnitecki et al. [35]. They used a bandpass filtered version of the recorded LS signal to initially detect the presence of heart sound peaks (following Moiussavi et al. [164]) and then they replaced all the components of the bandpass filtered original signal by zero, except for the heart sound that included parts. The resulting impulse-like signal was used as the noise-reference signal for the RLS-based HSC. The latter tries to find the component in the main signal that matches with the noise reference and performs filter coefficient optimization using RLS. The adapted-to-noise output is then subtracted from the original LS recordings, resulting in the denoised version of LS; more details can be found in [35].

3.7.4 Time–Frequency-Based HSC

The use of short-time Fourier transform (STFT) as a means to construct an HSC was presented by Pourazad et al. [37]. In fact, the main concept was to use a time-varying filtering based on STFT to remove heart sound segments from LS records and then by using simple interpolation to fill the gaps in the time–frequency plane. The denoised LS are then produced via the inverse STFT; more details can be found in [37].

An extension to the above-mentioned approach was introduced in [39, 40], where heart sound segments are detected by multiscale product and removed from the WT coefficients at every scale. Removal of heart sound segments is followed by an estimation of the missing data via linear prediction using autoregressive (AR) or moving average (MA) models.

As reported in [39, 40], the choice of AR or MA modeling depends on the location of the heart sound segment within the breathing cycle; for heart sound segments close to the onset of inspiration or expiration, MA modeling is preferred, because most of the information about the samples in the gap is in the next lung sound segment; otherwise, AR modeling is preferred. For a more complete description, the reader is referred to [39, 40].

3.7.5 Recurrent Time Statistics and Nonlinear Prediction-Based HSC

Another HSC approach, which is based on a recurrence time statistic combined with nonlinear prediction, was introduced by Ahlstrom et al. [41]. They explored the sensitivity of a recurrence time statistic to changes in a reconstructed state space (multivariate vector space). Signal segments that are found to contain heart sounds are removed, and the arising missing parts are replaced with predicted LS by using a nonlinear forward and backward prediction scheme, dividing the missing segment in two equal halves.

The prediction operates in the reconstructed state space and uses an iterated integrated nearest trajectory algorithm. Similar trajectories in state space share the same waveform characteristics in time domain, and a way of predicting missing data is thus to mimic the evolution of neighboring trajectories. Six LS segments surrounding the heart sound segment were used to reconstruct the state space, and five nearest neighbors were used in the prediction.

The predicted reconstructed data are then brought back into the time domain as the first coordinate of the embedding vector. The authors report auditory high-quality results, with successful removal of heart sounds from the LS recordings, yet with some shifting effect to higher frequencies in the denoised LS. Analytical description of the performance of the recurrent time statistics and nonlinear prediction-based HSC can be found in [41].

3.8 OTHER DENOISING APPROACHES

Apart from the aforementioned denoising/HSC techniques, a few other techniques have been reported in the literature; they are briefly described here for the sake of completeness.

3.8.1 Fuzzy Logic-Based Denoising

Throughout the years, a whole family of LS denoising techniques has been based on the use of fuzzy logic, aiming at real-time analysis of LS. In particular, the fuzzy rule-based stationary–

nonstationary (FST-NST) filter has been presented by Tolias et al. [47], in an attempt to construct an enhanced real-time separation scheme of DAS (nonstationary waves) from VS (stationary waves). The FST-NST filter consists of two adaptive neurofuzzy inference systems [165] that operate in parallel, for the estimation of the stationary and the nonstationary waves for crackles. An alternative approach for real-time separation of DAS has been suggested by Mastorocostas et al. [48] with the orthogonal least squares method fuzzy filter (OLS-FF). This filter also consists of two fuzzy inference systems that operate in parallel. A complete model-building process is also proposed, based on the orthogonal least squares method. The OLS-FF performs separation of crackles and squawks.

The above-mentioned neurofuzzy models are static models, which attempt to capture the plant dynamics by feeding the networks with delayed values of the input signal. However, for identification of temporal processes, dynamic models have been used, exhibiting some type of memory. The recurrent models are able to learn the system dynamics without assuming much knowledge about the structure of the system under consideration. In this vein, a filter that was based on the dynamic fuzzy neural network (DFNN) [166] was constructed for the separation of all types of DAS [167]. DFNN is a recurrent model and is capable of effectively identifying the dynamics of a process. A recurrent neural filter has been reported in [168], based on the block-diagonal recurrent neural network (BDRNN) [169], which is a simplified form of the fully recurrent network, with no interlinks among neurons in the hidden layer. As shown in [168], due to its internal dynamics, the BDRNN filter is capable of performing efficient real-time separation of DAS from VS.

3.8.2 Independent Component Analysis-Based HSC

The recovery of independent source signals, when a linear mixture of signals is only available with unknown sources and mixing processes, is often addressed by using independent component analysis (ICA). The latter finds a linear coordinate system where the recovered signals are statistically independent [170]. For cases where signals are not linearly mixed but convolved, application of ICA to the spectrograms of the mixed signals (at least two versions are required) provides a satisfactory solution, assuming that the source signals are originally statistically independent and at least one of them is Gaussian. An ICA-based HSC was proposed in [36], where two recordings of LS contaminated with heart sounds were assumed as the mixed signals. By combining the proper independent components from each frequency derived after the ICA, the spectrograms of the separated signals were produced. Using inverse STFT (ISTFT), the separated signals were reconstructed back into the time domain. As Porazad et al. report [36], the proposed HSC method reduces the effect of heart sound noise, yet a weakened heart sound could still be heard in the resultant signal along with a noticeable acoustic change in the LS portions free of heart sounds.

3.8.3 Variance Fractal Dimension-Based Heart Sound Localization

A variance fractal dimension (VFD)-based (Section 2.8) heart sound localization technique was proposed in [171]. The VFD trajectory was created by calculating the VFD of the recorded LS in a moving window of 100 ms with an increment size of 25 ms. The VFD trajectory was then compared with an adaptive threshold (tuned for every subject); the segments above the threshold were considered as those that contain heart sounds. This approach exhibits increased computational burden and its accuracy depends on the window size of VFD calculation [171].

3.8.4 Entropy-Based Heart Sound Localization

A Shannon entropy-based heart sound localization technique was proposed in [27]. This method uses the statistical information that the entropy provides to differentiate between LS segments with and without the presence of heart sounds. In fact, due to changes in the probability density function, the entropy of the LS segments with heart sounds is greater than that of the segments without heart sounds. Mean plus standard deviation of the estimated entropy was used a threshold for the detection of heart sound locations; the entropy of the latter takes values that exceed the threshold. Yadollahi and Moussavi state [27], this approach exhibits high heart sound detectability, high resolution in boundary locations of the heart sound, and robustness to variations between subjects.

For an analytical comparison among ICA-based HSC, and VFD- and entropy-based heart sound localization methods, the reader is referred to the work of Moussavi [25, Section 5.3].

• • • •

CHAPTER 4

Reflective Implications

4.1 FROM AN ENGINEER'S VIEWPOINT

The new analysis domains and the denoising methodologies presented in the previous two chapters, respectively, show the potentiality of the advanced signal processing techniques to offer a wide range of new opportunities in lung sound (LS) analysis. Apparently, this evolution of capabilities is related to a series of parallel developments, such as technology growth, introduction of new theories, better understanding of the mechanisms of biological function, gradual integration of the biomedical engineering research findings in clinical practice. Therefore, the development of technologies related to medical care, health, and welfare are considered extremely crucial. In addition to expectations for the development of medical technologies arising from needs of the global economy, there are also increasing expectations for the creation of new medical technologies stemming from trends in development and practical application of the life sciences. As a result, developments in science and technology focusing on humans have become one of the mainstream trends.

Progress in biomedical engineering in the 20th century has greatly transformed the concept of health care itself, by dramatically transfiguring it from a form of classical medical care to that of leading edge medical care assisted by instrumentation and imaging. Furthermore, social demands to enhance and maintain a high quality of medical treatment, health, and welfare in this aging society are behind the anticipation for further developments in this field in the future. Development of biomedical instrumentation technology to measure the physical and chemical information and signals emitted by the living body in a minimally (or non-)invasive and continuous manner, was responsible for the establishment of the concept of patient monitoring within clinical medicine. It thus contributed greatly to the enhancement of the quality of medical care. Nevertheless, such perspective still lacks in the area of bioacoustics. Except in a few rare cases [23, 24, 172], continuous monitoring of LS (and bioacoustic signals, in general, that is, heart sounds, bowel sounds, joint sounds) is totally absent from clinical practice. Physicians use their temporal memory to store "acoustic impression" of their patients, which is quickly banished by a new one as the next patient comes in. This reality shows that although auscultation is one of the oldest diagnostic procedures still in practice, it has not followed the same rhythm of technology-based enhancement. The biomedical engineer who deals with bioacoustics finds her-/himself in an ambiguous situation. On one hand, (s)he finds difficulties in gathering systematic bioacoustic experimental data, due to the lack of everyday recordings in hospitals. On the other hand, the slow development curve of bioacoustics analysis itself provides an open field for further research and exploitation. The latter could be successfully realized through an interdisciplinary approach, that is, by combining new signal processing and modeling techniques with the interpretation of bioacoustic data by physicians. Results amassed from the interdisciplinary field of engineering and medicine are beneficial in the maintenance of life, in particular, human life, and improvement of the environment.

The previous chapters made it clear that the intrinsic characteristics of LS, such as embedded periodicity (musical LS), nonlinearity, nonstationarity, and non-Gaussianity, require appropriate techniques that efficiently deal with LS properties in a straightforward manner. The combination of processing schemes with appropriate modeling could result in a more effective performance. This could be further enhanced if knowledge from biomechanical dynamic models of the cardiovascular system is transferred to the accompanying filed of sound modeling. Integration between signal processing and such modeling would be capable of directly attributing pathophysiological meaning to parameters obtained from signal processing algorithms.

4.2 FROM A PHYSICIAN'S VIEWPOINT

Sir A. C. Doyle (the writer of *Adventures of Sherlock Holmes* stories) in his *Scandal in Bohemia* offers a glimpse of the medical customs of the late 1800s; he wrote:

> As to your practice, if a gentleman walks into my rooms, smelling of iodoform, with a black mark of nitrate of silver upon his right forefinger, and a bulge on the side of his top-hat to show where he has secreted his stethoscope, I must be dull indeed if I do not pronounce him to be an active member of the medical profession.

Note that men's clothing of the day did not have pockets large enough for the Laennec stethoscope; therefore, it was unscrewed and carried in the physician's top hat [173]. This quotation reveals two things: the role of the stethoscope as a physician's hallmark and the necessity of its portability.

Modern-day physicians still sustain such relationship with the stethoscope. The only changes are the developed form of the stethoscope and its placement around physician's neck or in the pocket of their white robe; combined with the latter, the stethoscope is still the doctor's badge. From this viewpoint, efforts toward an intelligent auscultation system seem like intrusions that could well be viewed as altering the physician's image, devaluating his/her expertise and undermining his/her intellectual contribution to understanding and treating diseases. Actually, the locus of control of medical decision making is a key factor in examining the effects of technology-based assistive tools on a physician's behavior. This suggests that increased physician reliance on computerized auscultation, which may shift control of decision making away from the individual physician, may further depersonalize the doctor–patient relationship. Adding to this, computer-aided decisions are frequently based on rules and criteria embedded in the programs unknown to physicians. Moreover, these decision rules become relatively immune to change. Furthermore, computerized decision support systems are based almost entirely on quantifiable data. Nonquantifiable information that cannot easily be processed by the computer is largely ignored. Intelligent computerized auscultation, however, could offer high-quality auscultative data and indicators that carry important diagnostic information, helping the physician to come up with a more reliable decision about the patient's pathology.

Toward an intelligent auscultation system, considerations such as sensor (or multiple sensors) type, graphical display, built-in memory, wireless data transmission, and adaptive noise reduction should be taken into account; most importantly, all these should be fitted into a small, lightweight, portable, handheld device (a Pocket PC-like). Through synchronization, handheld data can be backed up and managed on a server-like PC (e.g., running database/data mining applications), further increasing the ways in which information can be gathered and distributed. Having high-quality auscultative information readily available at the point of care can be extremely useful in view of the growing amount of medical information available, the increased expectations to follow guideline, and formulary restrictions, and the time limitations placed on physicians. Handheld-type stethoscopes can improve accuracy, help avoid medical errors, and achieve better patient outcomes. To attain higher interfacing of handheld-type stethoscopes with hospital and clinic systems both directly and wirelessly, measures will have to be instituted to assure adequate security and compliance with new, more stringent regulations about medical record privacy and access.

Clinical validation, real-time implementation, generalization, noise robustness, and appropriate training are some of the key issues that need to be considered for a realistic scenario of intelligent auscultation systems. Although the signal processing methods described in this book do not account for *all* these issues, they create a bed-set for more pragmatic exploitation of lung sounds and contribute toward the enhancement of their diagnostic value. The ample space for new discoveries and optimization is the driving force that keeps bioacoustic research alive; yet, physicians' support and collaboration are the catalyst for the success of such endeavor.

• • • •

References

[1] V.S. McKusick, *Cardiovascular Sound in Health and Disease*, Williams & Wilkins, Baltimore, MD, p. 3 (1958).

[2] J. Rapoport, Laennec and the discovery of auscultation, *Isr. J. Med.*, **22**, 597–601 (1986).

[3] M. Foucault, *The Birth of the Clinic: An Archaeology of Medical Perception* (A. M. Sheridan Smith, Trans.), Pantheon Books, New York, NY (1973).

[4] J. Sterne, Mediate auscultation, the stethoscope, and the 'autopsy of the living': medicine's acoustic culture, *J. Med. Humanit.*, **22**(2), 115–136 (2001).

[5] A.B. Davis, *Medicine and its Technology: An Introduction to the History of Medical Instrumentation*, Greenwood Press, Westport, CT, pp. 88–89 (1981).

[6] J. Crary, *Techniques of the Observer: On Vision and Modernity in the Nineteenth Century*, MIT Press, Cambridge, MA, pp. 89–90 (1990).

[7] V.S. McKusick, *Cardiovascular Sound in Health and Disease*, Williams & Wilkins, Baltimore, MD, p. 13 (1958).

[8] S.S. Kraman, Vesicular (normal) lung sounds: how are they made, where do they come from and what do they mean?, *Semin. Respir. Med.*, **6**, 183–191 (1985).

[9] R.E. Thacker and S.S. Kraman, The prevalence of auscultatory crackles in subjects without lung disease, *Chest*, **81**(6), 672–674 (1982).

[10] P. Workum, S.K. Holford, E.A. Delbono, and R.L.H. Murphy, The prevalence and character of crackles (rales) in young women without significant lung disease, *Am. Rev. Respir. Dis.*, **126**(5), 921–923 (1982).

[11] A.J. Robertson, Rales, ronchi, and Laennec, *Lancet*, **1**, 417–423 (1957).

[12] S.S. Kraman, *Lung Sounds: An Introduction to the Interpretation of Auscultatory Findings*, Amer. College of Chest Physicians, Northbrook, IL, workbook (1983).

[13] R.L.H. Murphy, Discontinuous adventitious lung sounds, *Semin. Respir. Med.*, **6**, 210–219 (1985).

[14] D.W. Cugell, Lung sound nomenclature, *Am. Rev. Respir. Dis.*, **136**, 1016 (1987).

[15] J.E. Earis, K. Marsh, M.G. Rearson, and C.M. Ogilvie, The inspiratory squawk in extrinsic allergic alveolitis and other pulmonary fibroses, *Thorax*, **37**(12), 923–936 (1982).

[16] N. Gavriely and D.W. Cugell, *Breath Sounds Methodology*, CRC Press, Boca Raton, FL, p. 2 (1995).

[17] N. Gavriely, Y. Palti, and G. Alroy, Spectral characteristics of normal breath sounds, *J. Appl. Physiol.*, **50**, 307–314 (1981).

[18] T. Katila, P. Piirila, K. Kallio, E. Paajanen, T. Rosqvist, and A.R.A. Sovijarvi, Original waveform of lung sound crackles: a case study of the effect of high-pass filtration, *J. Appl. Physiol.*, **71**(6), 2173–2177 (1991).

[19] R.L.H. Murphy, S.K. Holford, and W.C. Knowler, Visual lung-sound characterization by time-expanded wave-form analysis, *N. Engl. J. Med.*, **296**, 968–971 (1978).

[20] L. Vannuccini, J.E. Earis, P. Helistö, B.M.G. Cheetham, M. Rossi, A.R.A. Sovijärvi, and J. Vanderschoot, Capturing and preprocessing of respiratory sounds, *Eur. Respir. Rev.*, **10**(77), 616–620 (2000).

[21] M.J. Mussell, The need for standards in recording and analysing respiratory sounds, *Med. Biol. Eng. Comput.*, **30**, 129–139 (1992).

[22] L. J. Hadjileontiadis, BlueSteth: a Bluetooth-based stethoscope for healthcare and educational applications, *IEEE Computer Society, CSIDC 2002 Final Report*, pp. 1–30 (2002).

[23] R.L.H. Murphy, Localization of chest sounds with 3D display and lung sound mapping, U.S. Patent 5,844,997, Dec. 1, 1998.

[24] M. Kompis, H. Pasterkamp, and G. Wodicka, Acoustic imaging of the human chest, *Chest*, **120**(4), 1309–1321 (2001).

[25] Z. Moussavi, *Fundamentals of Respiratory Sounds and Analysis*, 1st edn, Morgan & Claypool, San Rafael, CA (2007).

[26] Z. Moussavi, M.T. Leopando, H. Paterkamp, and G. Rempel, Computerized acoustical respiratory phase detection without airflow measurement, *Med. Biol. Eng. Comput.*, **38**(2), 198–203 (2000).

[27] A. Yadollahi and Z. Moussavi, A robust method for estimating respiratory flow using tracheal sound entropy, *IEEE Trans. Biomed. Eng.*, **53**(4), 662–668 (2006).

[28] L.J. Hadjileontiadis and S.M. Panas, A wavelet-based reduction of heart sound noise from lung sounds, *Int. J. Med. Inform.*, **52**, 183–190 (1998).

[29] I. Hossain and Z. Moussavi, An overview of heart-noise reduction of lung sound using wavelet transform based filter, *Proc. 25th IEEE Eng. Med. Biol. Soc. (EMBS)*, 458–461 (2003).

[30] V.K. Iyer, P.A. Ramamoorthy, H. Fan, and Y. Ploysongsang, Reduction of heart sounds from lung sounds by adaptive filtering, *IEEE Trans. Biomed. Eng.*, **33**(12), 1141–1148 (1986).

[31] L. Yip and Y.T. Zhang, Reduction of heart sounds from lung sound recordings by automated gain control and adaptive filtering techniques, *Proc. 23rd IEEE Eng. Med. Biol. Soc. (EMBS)*, 2154–2156 (2001).

[32] L. Guangbin, C. Shaoqin, Z. Jingming, C. Jinzhi, and W. Shengju, The development of a portable breath sound analysis system, *Proc. 14th IEEE Eng. Med. Biol. Soc. (EMBS)*, 2582–2583 (1992).

[33] L. Yang-Sheng, L. Wen-Hui, and Q. Guang-Xia, Removal of the heart sound noise from the breath sound, *Proc. 10th IEEE Eng. Med. Biol. Soc. (EMBS)*, 175–176 (1988).

[34] M. Kompis and E. Russi, Adaptive heart-noise reduction of lung sounds recorded by a single microphone, *Proc. 14th IEEE Eng. Med. Biol. Soc. (EMBS)*, 691–692 (1992).

[35] J. Gnitecki, I. Hossain, Z. Moussavi, and H. Pasterkamp, Qualitative and quantitative evaluation of heart sound reduction from lung sound recordings, *IEEE Trans. Biomed. Eng.*, **52**(10), 1788–1792 (2005).

[36] M.T. Pourazad, Z. Moussavi, F. Farahmand, and R. Ward, Heart sounds separation from lung sounds using independent component analysis, *Proc. 27th IEEE Eng. Med. Biol. Soc. (EMBS)*, 2736–2739 (2005).

[37] M.T. Pourazad, Z. Moussavi, and G. Thomas, Heart sound cancellation from lung sound recording using adaptive threshold and 2D interpolation in time–frequency domain, *Proc. 25th IEEE Eng. Med. Biol. Soc. (EMBS)*, 2586–2589 (2003).

[38] M.T. Pourazad, Z. Moussavi, and G. Thomas, Heart sound cancellation from lung sound recording using time–frequency filtering, *J. Med. Biol. Eng.*, **44**(3), 216–225 (2006).

[39] Z. Moussavi, D. Floras, and G. Thomas, Heart sound cancellation based on multiscale products and linear prediction, *Proc. 26th IEEE Eng. Med. Biol. Soc. (EMBS)*, 3840–3843 (2004).

[40] D. Floras, Z. Moussavi, and G. Thomas, Heart sound cancellation based on multiscale product and linear prediction, *IEEE Trans. Biomed. Eng.*, **54**(2), 234–243 (2007).

[41] C. Ahlstrom, O. Liljefeldt, P. Hult, and P. Ask, Heart sound cancellation from lung sound recordings using recurrence time statistics and nonlinear prediction, *IEEE Signal Process. Lett.*, **12**(12), 812–815 (2005).

[42] L.J. Hadjileontiadis and S.M. Panas, Adaptive reduction of heart sounds from lung sounds using fourth-order statistics, *IEEE Trans. Biomed. Eng.*, **44**(7), 642–648 (1997).

[43] R.L.H. Murphy, S.K. Holford, and W.C. Knowler, Visual lung sound characterization by time-expanded waveform analysis, *N. Engl. J. Med.*, **296**, 968–971 (1977).

[44] M. Ono, K. Arakawa, M. Mori, T. Sugimoto, and H. Harashima, Separation of fine crackles from vesicular sounds by a nonlinear digital filter, *IEEE Trans. Biomed. Eng.*, **36**(2), 286–291 (1989).

[45] L.J. Hadjileontiadis and S.M. Panas, Nonlinear separation of crackles and squawks from vesicular sounds using third-order statistics, *Proc. IEEE 18th EMBS Conf. (EMBS)*, **5**, 2217–2219 (1996).

[46] L.J. Hadjileontiadis and S. M. Panas, Separation of discontinuous adventitious sounds from vesicular sounds using a wavelet-based filter, *IEEE Trans. Biomed. Eng.*, **44**(12), 1269–1281 (1997).

[47] Y.A. Tolias, L.J. Hadjileontiadis, and S.M. Panas, A fuzzy rule-based system for real-time separation of crackles from vesicular sounds, *Proc. 19th IEEE Eng. Med. Biol. Soc. (EMBS)*, 1115–1118 (1997).

[48] P.M. Mastorocostas, Y.A. Tolias, J.B. Theocharis, L.J. Hadjileontiadis, and S.M. Panas, An orthogonal least squares-based fuzzy filter for real time analysis of lung sounds, *IEEE Trans. Biomed. Eng.*, **47**(9), 1165–1176 (1997).

[49] L.J. Hadjileontiadis and S.M. Panas, Enhanced separation of crackles and squawks from vesicular sounds using nonlinear filtering with third-order statistics, *J. Tenn. Acad. Sci.*, **73**(1–2), 47–52 (1998).

[50] Y.A. Tolias, L.J. Hadjileontiadis, and S.M. Panas, Real-time separation of discontinuous adventitious sounds from vesicular sounds using a fuzzy rule-based filter, *IEEE Trans. Inf. Technol. Biomed.*, **2**(3), 204–215 (1998).

[51] L.J. Hadjileontiadis, D.A. Patakas, N.J. Margaris, and S. M. Panas, Separation of crackles and squawks from vesicular sounds using a wavelet-based filtering technique, *COMPEL*, **17**(5/6), 649–657 (1998), MCB University Press.

[52] L.J. Hadjileontiadis, Y.A. Tolias, and S.M. Panas, Intelligent system modeling of bioacoustic signals using advanced signal processing techniques, in: C.T. Leondes (ed.), *Intelligent Systems: Technology and Applications*, Vol. 3, CRC Press, Boca Raton, FL, pp. 103–156 (2002).

[53] L.J. Hadjileontiadis and I.T. Rekanos, Detection of explosive lung and bowel sounds by means of fractal dimension, *IEEE Signal Process. Lett.*, **10**(10), 311–314 (2003).

[54] L.J. Hadjileontiadis, Wavelet-based enhancement of lung and bowel sounds using fractal dimension thresholding—Part I: Methodology, *IEEE Trans. Biomed. Eng.*, **52**(6), 1143–1148 (2005).

[55] L.J. Hadjileontiadis, Wavelet-based enhancement of lung and bowel sounds using fractal dimension thresholding—Part II: Application results, *IEEE Trans. Biomed. Eng.*, **52**(6), 1050–1064 (2005).

[56] I.T. Rekanos and L.J. Hadjileontiadis, An iterative kurtosis-based technique for the detection of nonstationary bioacoustic signals, *Signal Process.*, **86**, 3787–3795 (2006).

[57] L.J. Hadjileontiadis, Empirical mode decomposition and fractal dimension filter: a novel technique for denoising explosive lung sounds, *IEEE Eng. Med. Biol. Mag.*, **26**(1), 30–39 (2007), Art. No. 15.

[58] L.J. Hadjileontiadis and S.M. Panas, Autoregressive modeling of lung sounds using higher-order statistics: estimation of source and transmission, *Proc. IEEE Signal Processing Workshop on Higher-Order Statistics (SPW-HOS)*, 4–8 (1997).

[59] L.J. Hadjileontiadis and S.M. Panas, Nonlinear analysis of musical lung sounds using the bicoherence index, *Proc. 19th IEEE Eng. Med. Biol. Soc. (EMBS)*, 1126–1129 (1997).

[60] J. Gnitecki, Z. Moussavi, and H. Pasterkamp, Classification of lung sounds during bronchial provocation using waveform fractal dimensions, *Proc. 26th IEEE Eng. Med. Biol. Soc. (EMBS)*, 3844–3847 (2004).

[61] E. Conte, A. Vena, A. Federici, R. Giuliani, and J.P. Zbilut, A brief note on possible detection of physiological singularities in respiratory dynamics by recurrence quantification analysis of lung sounds, *J. Chaos Solitons Fractals*, **21**, 869–877 (2004).

[62] J. Gnitecki and Z. Moussavi, The fractality of lung sounds: a comparison of three waveform fractal dimension algorithms, *J. Chaos Solitons Fractals*, **26**(4), 1065–1072 (2005).

[63] J. Gnitecki, Z. Moussavi, and H. Pasterkamp, Geometrical and dynamical state space parameters of lung sounds, *5th Int. Workshop on Biosignal Interpretation (BSI)*, pp. 113–116 (2005).

[64] C. Ahlstrom, A. Johansson, P. Hult, and P. Ask, Chaotic dynamics of respiratory sounds, *J. Chaos Solitons Fractals*, **29**, 1054–1069 (2006).

[65] M. Mori, K. Kinoshita, H. Morinari, T. Shiraishi, S. Koike, and S. Murao, Waveform and spectral analysis of crackles, *Thorax*, **35**(11), 843–850 (1980).

[66] M. Matsuzaki, Polarity of crackle waveforms: a new index for crackle differentiation, *Hokkaido Igaku Zasshi*, **60**(1), 104–113 (1985).

[67] J. Hoevers and R.G. Loudon, Measuring crackles, *Chest*, **98**(5), 1240–1243 (1990).

[68] A. Cohen, Signal processing methods for upper airway and pulmonary dysfunction diagnosis, *IEEE Eng. Med. Biol. Mag.*, **9**(1), 72–75 (1990).

[69] M. Munakata, H. Ukita, I. Doi, Y. Ohtsuka, Y. Masaki, Y. Homma, and Y. Kawakami, Spectral and waveform characteristics of fine and coarse crackles, *Thorax*, **46**(9), 651–657 (1991).

[70] N. Al Jarad, S.W. Davies, R. Logan-Sinclair, and R. M. Rudd, Lung crackle characteristics in patients with asbestosis, asbestos-related pleural disease and left ventricular failure using a time-expanded waveform analysis—a comparative study, *Respir. Med.*, **88**(1), 37–46 (1994).

[71] N. Gavriely and D.W. Cugell, *Breath Sounds Methodology*, CRC Press, Boca Raton, FL (1995).

[72] A.R. Sovijarvi, P. Helistö, L.P. Malmberg, K. Kallio, E. Paajanen, A. Saarinen, P. Lipponen, S. Haltsonen, L. Pekkanen, P. Piirila, L. Naveri, and T. Katila, A new versatile PC-based

lung sound analyzer with automatic crackle analysis (HeLSA); repeatability of spectral parameters and sound amplitude in healthy subjects, *Technol. Health Care*, **6**(1), 11–22 (1998).

[73] T. Kalayci, G. Celebi, Y. Ozturk, and M. Ozhan, Automatic detection and classification of crackles by using a neural network, *Proc. 14th IEEE Eng. Med. Biol. Soc. (EMBS)*, **6**, 2580–2581 (1992).

[74] B. Sankur, Y.R. Kahya, E.C. Güler, and T.S. Engin, Comparison of AR-based algorithms for respiratory sounds classification respiratory disease diagnosis using lung sounds, *Comput. Biol. Med.*, **24**(1), 67–76 (1994).

[75] L. Pesu, P. Helistö, E. Ademovic, J.C. Pesquet, A. Saarinen, and A.R. Sovijarvi, Classification of respiratory sounds based on wavelet packet decomposition and learning vector quantization, *Technol. Health Care*, **6**(1), 65–74 (1998).

[76] L.J. Hadjileontiadis and S.M. Panas, On modeling impulsive bioacoustic signals with symmetric alpha-stable distributions: application in discontinuous adventitious lung sounds and explosive bowel sounds, *Proc. 20th IEEE Eng. Med. Biol. Soc. (EMBS)*, **1**, 13–16 (1998).

[77] Y.P. Kahya and C.A. Yilmaz, Modeling of respiratory crackles, *Proc. 22nd IEEE Eng. Med. Biol. Soc. (EMBS)*, **1**, 632–634 (2000).

[78] L.J. Hadjileontiadis, Discrimination analysis of discontinuous breath sounds using higher-order crossings, *Med. Biol. Eng. Comput.*, **41**(4), 445–455 (2003).

[79] A. Kandaswamy, C. Sathish Kumar, R.P. Ramanathan, S. Jayaraman, and N. Malmurugan, Neural classification of lung sounds using wavelet coefficients, *Comput. Biol. Med.*, **34**, 523–537 (2004).

[80] C.A. Yilmaz and Y. P. Kahya, Modeling of pulmonary crackles using wavelet networks, *Proc. 27th IEEE Eng. Med. Biol. Soc. (EMBS)*, 7560–7563 (2005).

[81] C.A. Yilmaz and Y.P. Kahya, Multi-channel classification of respiratory sounds, *Proc. 28th IEEE Eng. Med. Biol. Soc. (EMBS)*, **1**, 2864–2867 (2006).

[82] O. Taketoshi, S. Hayaru, and K. Shoji, Discrimination of lung sounds using a statistics of waveform intervals, *IPSJ SIG Technical Reports*, **2006**(68(MPS-60)), 1–4 (2006).

[83] L.J. Hadjileontiadis, A texture-based classification of crackles and squawks using lacunarity, *IEEE Trans. Biomed. Eng.*, submitted (2008).

[84] J.M. Mendel, Tutorial on higher-order statistics (spectra) in signal processing and system theory: theoretical results and some applications, *Proc. IEEE*, **79**(3), 278–305 (1991).

[85] P.J. Huber, B. Kleiner, T. Gasser, and G. Dumermuth, Statistical methods for investigating phase relations in stationary stochastic processes, *IEEE Trans. Audio Electroacoust.*, **19**, 78–86 (1971).

[86] C.L. Nikias and A.P. Petropulu, *Higher-Order Spectra Analysis: A Nonlinear Signal Processing Framework*, Prentice-Hall, Englewood Cliffs, NJ (1993).

[87] J.G. Proakis, C.M. Rader, F. Ling, and C.L. Nikias, *Advanced Digital Signal Processing*, Macmillan, New York (1992).

[88] S. Elgar and R.T. Guza, Statistics of bicoherence, *IEEE Trans. Acoust. Speech Signal. Process.*, **36**(10), 1667–1668 (1988).

[89] V.K. Iyer, P.A. Ramamoorthy, and Y. Ploysongsang, Autoregressive modeling of lung sounds: characterization of source and transmission, *IEEE Trans. Biomed. Eng.*, **36**(11), 1133–1137 (1989).

[90] M.R. Raghuveer and C.L. Nikias, Bispectrum estimation: a parametric approach, *IEEE Trans. Acoust., Speech Signal Process.*, **33**(4), 1213–1230 (1985).

[91] C.L. Nikias and M.R. Raghuveer, Bispectrum estimation: a digital signal processing framework, *Proc. IEEE*, **75**(7), 869–891 (1987).

[92] A. Swami, J.M. Mendel, and C.L. Nikias, *Higher-Order Spectral Analysis Toolbox*, 3rd edn, The Mathworks, Natick, MA, chap. 1 (1998).

[93] L.J. Hadjileontiadis, Analysis and processing of lung sounds using higher-order statistics-spectra and wavelet transform, Ph.D. dissertation, Aristotle University of Thessaloniki, Thessaloniki, Greece, pp. 139–175 (1997).

[94] L.J. Hadjileontiadis and S.M. Panas, Higher-order statistics: a robust vehicle for diagnostic assessment and characterisation of lung sounds, *Technol. Health Care*, **5**(5), 359–374 (1997).

[95] C.L. Nikias and M. Shao, *Signal Processing with Alpha-Stable Distributions and Applications*, Wiley & Sons, Inc., New York, USA (1995).

[96] W. Feller, *An introduction to probability theory and its applications*, Vol. II, Wiley & Sons, Inc., New York (1966).

[97] J.H. McCulloch, Simple consistent estimators of stable distribution parameters, *Commun. Stat. Simul. Comput*, **15**(4), 1109–1136 (1986).

[98] I.A. Koutrouvelis, An iterative procedure for the estimation of the parameters of stable laws, *Commun. Stat. Simul. Comput.*, **10**(1), 17–28 (1981).

[99] M. Kanter and W.L. Steiger, Regression and autoregression with infinite variance, *Adv. Appl. Probab.*, **6**, 768–783 (1974).

[100] L.J. Hadjileontiadis, A.J. Giannakidis, and S.M. Panas, a-Stable modeling: a novel tool for classifying crackles and artifacts, in *Proc. 25th Int. Lung Sounds Association Conference (ILSA)*, H. Pasterkamp, ed., Chicago, IL (2000).

[101] A. Grossmann and J. Morlet, Decomposition of Hardy functions into square integrable wavelets of constant shape, *SIAM J. Math. Anal*, **15**, 723–736 (1984).

[102] P.S. Addison, *The Illustrated Wavelet Transform Handbook: Introductory Theory and Applications in Science, Engineering, Medicine and Finance*, Institute of Physics (IOP) Publishing, Bristol, UK (2002).

[103] N.M. Astaf'eva, Wavelet analysis: basic theory and some applications, *Physics-Uspekhi*, **39**(11), 1085–1108 (1996).

[104] I. Daubechies, *Ten Lectures on Wavelets* (CBMS Lecture Notes Series), SIAM, Philadelphia, PA (1991).

[105] M. Farge, Wavelet transforms and their applications to turbulence, *Annu. Rev. Fluid Mech.*, **24**, 395–457 (1992).

[106] S.G. Mallat, *A Wavelet Tour of Signal Processing*, Academic Press, San Diego (1998).

[107] S.G. Mallat, A theory for multiresolution signal decomposition: the wavelet representation, *IEEE Trans. Pattern Anal. Machine Intell.*, **11**(7), 674–693 (1989).

[108] I. Daubechies, Orthonormal bases of compactly supported wavelets, *Commun. Pure Appl. Math.*, **41**, 909–996 (1988).

[109] A. Cohen and J. Kovačević, Wavelets: the mathematical background, *Proc. IEEE*, **84**(4), 514–522 (1996).

[110] M. Vetterli and J. Kovačević, *Wavelets and Subband Coding*, chap. 4, Prentice-Hall, Englewood Cliffs, NJ, pp. 201–298 (1995).

[111] S.G. Mallat, Special issue on wavelets, *Proc. IEEE*, **84**(4), 507–686 (1996).

[112] V. Gross, T. Penzel, L.J. Hadjileontiadis, U. Koehler, and C. Vogelmeier, Electronic auscultation based on wavelet transformation in clinical use, *Proc. 24th IEEE Eng. Med. Biol. Soc. (EMBS)*, pp. 1531–1532 (2002).

[113] V. Gross, T. Penzel, P. Fachinger, M. Fröhlich, J. Sulzer, and P. von Wichert, A simple method for detecting pneumonia with using wavelet-transformation, *Proc. Med. Biol. Eng. Comput. (EMBEC)*, **37**(Suppl. 2), 536–537 (1999).

[114] F. Ayari, A.T. Alouani, and M. Ksouri, Wavelets: an efficient tool for lung sounds analysis, in *Proc. IEEE Computer Systems and Applications (AICCSA)*, pp. 875–878 (2008).

[115] L. Ke and W. Houjun, A novel wavelet transform modulus maxima based method of measuring Lipschitz exponent, *Proc. International Conference Communications, Circuits and Systems (ICCCAS)*, pp. 628–632 (2007).

[116] S.G. Mallat and W. L. Hwang, Singularity detection and processing with wavelets, *IEEE Trans. Inf. Theory*, **38**(2), 617–643 (1992).

[117] A. Kandaswamy, C.S. Kumarb, R.P. Ramanathan, S. Jayaraman, and N. Malmurugan, Neural classification of lung sounds using wavelet coefficients, *Comput. Biol. Med.*, **34**, 523–537 (2004).

[118] M. Riedmiller and H. Braun, A direct adaptive method for faster backpropagation learning: the RPROP algorithm, in H. Ruspini (ed.), *Proc. IEEE Int. Conf. Neural Networks (ICNN)*, pp. 586–591 (1993).

[119] Y. Birkelund and A. Hanssen, Multitaper Estimators for Bispectra, *Proc. IEEE SP Workshop on Higher-Order (SPW-HOS)*, pp. 207–213 (1999).

[120] B.Ph. van Milligen, E. Sánchez, T. Estrada, C. Hidalgo, B. Brañas, B. Carreras, and L. Garcia, Wavelet bicoherence: a new turbulence analysis tool, *Phys. Plasmas*, **2**(8), 3017–3032 (1995).

[121] T. Dudok de Wit and V.V. Krasnosel'skikh, Wavelet bicoherence analysis of strong plasma turbulence at the earth's quasiparallel bow shock, *Phys. Plasmas*, **2**(11), 4307–4311 (1995).

[122] Y. Larsen and A. Hanssen, Wavelet-polyspectra: analysis of non-stationary and non-Gaussian/non-linear signals, *Proc. IEEE Workshop on Statistical Signal and Array Processing (WSSAP)*, pp. 539–543 (2000).

[123] B.Ph. van Milligen, C. Hidalgo, and E. Sánchez, Nonlinear phenomena and intermittency in plasma turbulence, *Phys. Rev. Lett.*, **74**(3), 395–398 (1995).

[124] Y.C. Kim and E.J. Powers, Digital bispectral analysis of self-excited fluctuation spectra, *Phys. Fluids*, **21**, 1452–1453 (1978).

[125] S.A. Taplidou and L.J. Hadjileontiadis, Nonlinear analysis of wheezes using wavelet bicoherence, *Comput. Biol. Med.*, **37**, 563–570 (2007).

[126] J.R. Duke Jr., J.T. Good Jr., L.D. Hudson, T.M. Hyers, M.D. Iseman, D.D. Mergenthaler, J.F. Murray, T.L. Petty, and D.R. Rollins, Frontline assessment of common pulmonary presentations, in: J.F. Murray, L.D. Hudson, T.L., and Petty (eds.), *A Monograph for Primary Care Physicians*, Snowdrift Pulmonary Conference, Inc., Denver, CO, 2000. Available at: http://www.lungcancerfrontiers.org/pdf-books/asmnt_cmn_pulmryPrsntn.pdf.

[127] B. Kedem, *Time Series Analysis by Higher-Order Crossings*, IEEE Press, Piscataway, NJ (1994).

[128] L.J. Hadjileontiadis, Discrimination analysis of discontinuous breath sounds using higher-order crossings, *IEEE Med. Biol. Eng. Comput.*, **41**(4), 445–455 (2003).

[129] N.E. Huang, Z. Shen, S.R. Long, M.L. Wu, H.H. Shih, Q. Zheng, N.C. Yen, C.C. Tung, and H.H. Liu, The empirical mode decomposition and Hilbert spectrum for nonlinear and nonstationary time series analysis, *Proc. R. Soc. Lond. A*, **454**(1971), 903–995 (1998).

[130] P. Gloersen and N.E. Huang, Comparison of interannual intrinsic modes in hemispheric sea ice covers and other geophysical parameters, *IEEE Trans. Geosci. Remote Sens.*, **41**(5), 1062–1074 (2003).

[131] N.E. Huang and S.S.P. Shen, *Hilbert–Huang Transform and Its Applications*, World Scientific, Singapore (2005).

[132] N.E. Huang and N.O. Attoh-Okine, *The Hilbert–Huang Transform in Engineering*, CRC Press (Taylor & Francis Group), Boca Raton, FL (2006).

[133] N.E. Huang, Z. Shen, and S.R. Long, A new view of nonlinear water waves—the Hilbert spectrum, *Annu. Rev. Fluid Mech.*, **31**, 417–457 (1999).

[134] N.E. Huang, M.L. Wu, S.R. Long, S.S. Shen, W.D. Qu, P. Gloersen, and K.L. Fan, A confidence limit for the empirical mode decomposition and Hilbert spectral analysis, *Proc. R. Soc. Lond.*, **459A**, 2317–2345 (2003).

[135] Z. Wu and N.E. Huang, A study of the characteristics of white noise using the empirical mode decomposition method, *Proc. R. Soc. Lond.*, **460A**, 1597–1611 (2004).

[136] Z. Wu and N.E. Huang, Ensemble Empirical Mode Decomposition: a noise-assisted data analysis method, *COLA Technical Report 193*, 2005. Available at: ftp://grads.iges.org/pub/ctr/ctr_193.pdf.

[137] Z. Wu and N.E. Huang, Ensemble Empirical Mode Decomposition: a noise-assisted data analysis method, *Adv. Adapt. Data Anal.*, 2008 (in press).

[138] J.R. Yeh, T.Y. Lin, J.S. Shieh, Y. Chen, N.E. Huang, Z. Wu, and C.K. Peng, Investigating complex patterns of blocked intestinal artery blood pressure signals by empirical mode decomposition and linguistic analysis, *J. Phys.: Conf. Ser. 96*, 1–7 (2008), doi:10.1088/1742-6596/96/1/012153.

[139] K.T. Coughlin and K.K. Tung, 11-Year solar cycle in the stratosphere extracted by the empirical mode decomposition method, *Adv. Space Res.*, **34**(2), 323–329 (2004).

[140] S.C. Villalobos, R.G. Camarena, G.C. Lem, and T. A. Corrales, Crackle sounds analysis by empirical mode decomposition: nonlinear and nonstationary signal analysis for distinction of crackles in lung sounds, *IEEE Eng. Med. Biol. Mag.*, **26**(1), 40–47 (2007), Art. No. 15.

[141] S.N. Rasband, Fractal dimension, in *Chaotic Dynamics of Nonlinear Systems*, , chap. 4, Wiley-Interscience, New York, pp. 71–83 (1997).

[142] R. Esteller, G. Vachtsevanos, J. Echauz, T. Henry, P. Pennell, C. Epstein, R. Bakay, C. Bowen, and B. Litt, Fractal dimension characterizes seizure onset in epileptic patients, *Proc. IEEE Int. Conf. Acoust. Speech Signal Process. (ICASPP)*, **4**, 2343–2346, Phoenix, AZ (1999).

[143] W. Kinsner, Batch and real-time computation of a fractal dimension based on variance of a time series, *Technical Report, DEL94-6*, Dept of Electrical & Computer Eng, University of Manitoba, June 1994.

[144] R.T.H. Laennec, *A Treatise on the Diseases of the Chest and on Mediate Auscultation*, 3rd edn. (J. Forbes, Trans.), Samuel Wood and Sons, and Collins and Hannay, New York, NY (1830).

[145] B.B. Mandelbrot, *The Fractal Geometry of Nature*, Freeman, New York, NY (1983).

[146] B. Lin and Z.R. Yang, A suggested lacunarity expression for Sierpinski carpets, *J. Phys. A*, **19**(2), L49–L52 (1986).

[147] Y. Gefen, Y. Meir, B.B. Mandelbrot, and A. Aharony, Geometric implementation of hypercubic lattices with noninteger dimensionality by use of low lacunarity fractal lattices, *Phys. Rev. Lett.*, **50**(3), 145–148 (1983).

[148] C. Allain and M. Cloitre, Characterizing the lacunarity of random and deterministic fractal sets, *Phys. Rev. A*, **44**(6), 3552–3558 (1991).

[149] R.E. Plotnick, R.H. Gardner, W.W. Hargrove, K. Prestegaard, and M. Perlmutter, Lacunarity analysis: a general technique for the analysis of spatial patterns, *Phys. Rev. E*, **53**(5), 5461–5468 (1996).

[150] G. Du and T.S. Yeo, A novel lacunarity estimation method applied to SAR image segmentation, *IEEE Trans. Geosci. Remote Sens.*, **40**(12), 2687–2691 (2002).

[151] R. Esteller, G. Vachtsevanos, J. Echauz, and B. Litt, A comparison of waveform fractal dimension algorithms, *IEEE Trans. Circ. Syst.*, **48**, 177–183 (2001).

[152] P.S. Addison, *Fractals and Chaos: An Illustrated Course*, Taylor & Francis, Gloucester, UK (1997).

[153] M. Katz, Fractals and the analysis of waveforms, *Comput. Biol. Med.*, **18**(3), 145–156 (1988).

[154] C. Sevcik, A procedure to estimate the fractal dimension of waveforms, *Complexity International [Online]*, Vol. 5, 1998. Available at: http://journal-ci.csse.monash.edu.au/ci/vol05/sevcik/sevcik.html.

[155] P. Dong, Test of a new lacunarity estimation method for image texture analysis, *Int. J. Remote Sens.*, **21**(17), 3369–3373 (2000).

[156] W.J. Krzanowski, *Principles of Multivariate Analysis: A User's Perspective (Oxford Statistical Science Series)*, Oxford University Press, New York, NY, Rev Sub edition (2000).

[157] G.A.F. Seber, *Multivariate Observations (Wiley Series in Probability and Statistics)*. John Wiley & Sons, Inc., San Francisco, CA (2004).

[158] S.A. Taplidou, L.J. Hadjileontiadis, V. Gross, T. Penzel, and S.M. Panas, WED: an efficient wheezing-episode detector based on breath sounds spectrogram analysis, *Proc. 25th IEEE Eng. Med. Biol. Soc. (EMBS)*, pp. 2531–2534 (2003).

[159] R. Loudon and R.L.H. Murphy, Jr., Lung sounds, *Am. Rev. Respir. Dis.*, **130**, 663–673 (1984).

[160] R.R. Coifman and M.V. Wickerhauser, Adapted waveform 'denoising' for medical signals and images, *IEEE Eng. Med., Biol. Mag.*, **14**(5), 578–586 (1995).

[161] P.J. Bickel and K.A. Doksum, *Mathematical Statistics*, Holden-Day, San Francisco, CA (1977).

[162] G. Charbonneau, J.L. Racineux, M. Sudraud, and E. Tuchais, An accurate recording system and its use in breath sounds spectral analysis, *J. Appl. Physiol.*, **55**(4), 1120–1127 (1983).

[163] B. Widrow, J.R. Glover, J.M. McCool, J. Kaunitz, C.S. Williams, R.H. Hearn, I.R. Zeidler, E. Dong, and R.C. Goodlin, Adaptive noise cancelling: principles and applications, *Proc. IEEE*, **63**(12), 1692–1716 (1975).

[164] Z. Moussavi, R.M. Rangayyan, G.D. Bell, C.B. Frank, and K.O. Ladly, Screening and adaptive segmentation of vibroarthrographic signals, *IEEE Trans. Biomed. Eng.*, **43**(1), 15–23 (1996).

[165] J.S.R. Jang, ANFIS: adaptive network-based fuzzy inference system, *IEEE Trans. Syst. Man Cybern.*, **23**(3), 665–685 (1993).

[166] P.A. Mastorocostas and J.B. Theocharis, A recurrent fuzzy-neural model for dynamic system identification, *IEEE Trans. Syst. Man Cybern.—Part B*, **32**(2), 176–190 (2002).

[167] P.A. Mastorocostas and J.B. Theocharis, A recurrent fuzzy-neural filter for real-time separation of lung sounds, *Proc. IEEE Int. Joint Conf. Neural Networks (IJCNN)*, **5**, 3023–3028 (2005).

[168] P.A. Mastorocostas and J.B. Theocharis, A stable learning method for block-diagonal recurrent neural networks: application to the analysis of lung sounds, *IEEE Trans. Syst. Man Cybern.—Part B*, **36**, 242–254 (2006).

[169] S.C. Sivakumar, W. Robertson, and W.J. Phillips, On-line stabilization of block-diagonal recurrent neural networks, *IEEE Trans. Neural Networks*, **10**, 167–175 (1999).

[170] T.W. Lee, *Independent Component Analysis: Theory and Applications*, 2nd edn, chaps. 2 and 4, Kluwer, Dordrecht (2000).

[171] J. Gnitecki and Z. Moussavi, Variance fractal dimension trajectory as a tool for heart sound localization in lung sounds recordings, *Proc. 25th IEEE Eng. Med. Biol. Soc. (EMBS)*, 2420–2423 (2003).

[172] F. Javed, P.A. Venkatachalam, and A.F. Hani, Knowledge based system with embedded intelligent heart sound analyser for diagnosing cardiovascular disorders, *J. Med. Eng. Technol.*, **31**(5), 341–350 (2007).

[173] L.A. Geddes, Birth of the stethoscope, *IEEE Eng. Med. Biol. Mag.*, **24**(1), 84–86 (2005).

Author Biography

Leontios J. Hadjileontiadis was born in Kastoria, Greece, in 1966. He received his B.S. degree in Electrical Engineering in 1989 and Ph.D. degree in Electrical and Computer Engineering in 1997, both from the Aristotle University of Thessaloniki, Thessaloniki, Greece. In December 1999, he began his stint as a faculty member in the Department of Electrical and Computer Engineering, AUTh, where he is currently an assistant professor, working on lung sounds, heart sounds, bowel sounds, electrocardiogram data compression, seismic data analysis, and crack detection in the Signal Processing and Biomedical Technology Unit of the university's Telecommunications Laboratory. His research interests are centered on higher-order statistics, α-stable distributions, higher-order zero crossings, wavelets, polyspectra, fractals, neurofuzzy modeling for medical, mobile, and digital signal processing applications. Dr. Hadjileontiadis is a member of the Technical Chamber of Greece, the IEEE, the Higher-Order Statistics Society, the International Lung Sounds Association (Secretary), and the American College of Chest Physicians. He won 2nd place in the Best Paper Competition of the 9th Panhellenic Medical Conference on Thorax Diseases '97, Thessaloniki. He was also an open finalist at the Student paper Competition (Whitaker Foundation) of the IEEE EMBS '97, Chicago, IL; a finalist at the Student Paper Competition (in memory of Dick Poortvliet) of the MEDICON '98, Lemesos, Cyprus; the recipient of the Young Scientist Award of the 24th International Lung Sounds Conference '99, Marburg, Germany; and the recipient of the 2nd Innovation Award, Research Committee, AUTh, 2008. In 2004, 2005, and 2007, he organized and served as a mentor to three 5-student teams that have won international awards at the Imagine Cup Competition (Microsoft) in Sao Paulo (2004), Yokohama (2005), and Seoul (2007), respectively, with projects relating to assistive technology for disabled people (blind, deaf, autistic kids). Dr. Hadjileontiadis also holds a Ph.D. degree in music composition (University of York, UK, 2004) and he is currently a Professor in composition at the State Conservatory of Thessaloniki, Greece.

An Outline of Informational Genetics

Gérard Battail
École Nationale Supérieure des Télécommunications (retired), Paris

MORGAN & CLAYPOOL PUBLISHERS

ABSTRACT

Heredity performs literal communication of immensely long genomes through immensely long time intervals. Genomes nevertheless incur sporadic errors referred to as mutations which have significant and often dramatic effects, after a time interval as short as a human life. How can faithfulness at a very large timescale and unfaithfulness at a very short one be conciliated?

The engineering problem of literal communication has been completely solved during the second half of the XX-th century. Originating in 1948 from Claude Shannon's seminal work, information theory provided means to measure information quantities and proved that *reliable* communication is possible through an *unreliable* channel (by means left unspecified) up to a sharp limit referred to as its capacity, beyond which communication becomes impossible. The quest for engineering means of reliable communication, named *error-correcting codes*, did not succeed in closely approaching capacity until 1993 when Claude Berrou and Alain Glavieux invented turbocodes. By now, the electronic devices which invaded our daily lives (e.g., CD, DVD, mobile phone, digital television) could not work without highly efficient error-correcting codes. Reliable communication through unreliable channels up to the limit of what is theoretically possible has become a practical reality: an outstanding achievement, however little publicized.

As an engineering problem that nature solved aeons ago, heredity is relevant to information theory. The capacity of DNA is easily shown to vanish exponentially fast, which entails that error-correcting codes must be used to regenerate genomes so as to faithfully transmit the hereditary message. Moreover, assuming that such codes exist explains basic and conspicuous features of the living world, e.g., the existence of discrete species and their hierarchical taxonomy, the necessity of successive generations and even the trend of evolution towards increasingly complex beings.

Providing geneticists with an introduction to information theory and error-correcting codes as necessary tools of hereditary communication is the primary goal of this book. Some biological consequences of their use are also discussed, and guesses about hypothesized genomic codes are presented. Another goal is prompting communication engineers to get interested in genetics and biology, thereby broadening their horizon far beyond the technological field, and learning from the most outstanding engineer: Nature.

KEYWORDS

error-correcting codes, information theory, molecular genetics, nested codes, soft codes.

Contents

Foreword

The scientific and engineering literature uses an impersonal style, where the author traditionally refers to himself or herself as *we*. Science and engineering strive for objectivity, so subjectivity should be avoided in this literature as somewhat indecorous. I depart from this tradition in this Foreword but also in Part I, entitled "An Informal Overview," because its content depends to a large extent on my personal experience as a communication engineer. Moreover, an informal style is perhaps the best way to introduce this lecture so as to motivate its reading.

Until my retirement in 1997, I taught information theory and error-correcting codes for almost 25 years at the *École nationale supérieure des Télécommunications* in Paris, France. Then, as a hobby, I contemplated applications of these topics to sciences of nature. The present lecture is devoted to expounding some results of this research as regards genetics.

Molecular genetics is based on the postulate that DNA is a long-term memory which faithfully conserves the genetic information during time intervals at the geological scale. The trouble is that DNA is subject to sporadic errors, or mutations, which have dramatic effects on the living beings, like aging or cancer, at the much shorter timescale of individual lifetimes. The existence of such mutations blatantly contradicts the assumed long-term conservation of the genetic message borne by DNA. Indeed, DNA may be interpreted as a channel for communicating through time. Information theory associates with any channel a fundamental quantity which measures its ability to convey information, its *capacity* . The capacity actually sets an impassable limit for any communication using the channel. One easily computes the DNA capacity as a function of time, showing beyond any doubt that it vanishes as time increases unless the mutation rate is exactly zero. The mechanism alleged by geneticists, i.e., copying it using the Watson-Crick template-replication mechanism, thus fails to account for the faithful communication of the genetic message through time.

On the other hand, communication engineers developed very powerful tools, referred to as error-correcting codes, to cope with errors occurring in communication channels. They are by now widely used in objects of daily life like computer memories, mobile phones, CD, DVD, digital television, etc. They enable *reliable* communication through *unreliable* channels, what information theory proves to be possible however paradoxical it may look (and the engineering practice plainly confirms). Assuming that such tools are actually implemented in genomes is the only way to account for the faithful conservation of the genetic message. Extremely inventive, nature has widely predated and often surpassed human engineers, which makes this necessary assumption highly plausible. Most geneticists are not aware of the extreme difficulty of conserving genomes during geological times and ignore that communication engineers know how it can be performed. Information theory shows that error-correcting codes are actually the only possible solution to this crucial but overlooked biological problem. It turns out, unfortunately, that both the theoretical possibility of 'error-free' communication and the practical means to implement it are almost unknown, except of course by communication engineers.

Let me illustrate the situation with a fable. Imagine a world where physicists and biologists are strictly separated. The lack of communication between both communities would possibly result in biologists explaining facts of the living world by means that physicists know to be impossible. For instance, imagine that biologists explain the flight of flies, bats, and birds by assuming that wings cancel gravity (anti-gravity is an old trick of science fiction). Establishing a communication between both communities would sooner or later ruin the biologists' explanation of animal flight. Now replace physicists by information-theorists, experts in animal flight by molecular geneticists, and their assumed separation by the ever-increasing specialization of scientific disciplines and their recourse to technical jargons. The situation is comparable and even worse. First of all, nobody ever proved that anti-gravity is physically impossible, while the channel capacity of information theory is a mathematically proven impassable limit. Second, and more importantly, animal flight is not such a central point while the conservation of genomes is the cornerstone of biology as a whole: indeed, the very existence of the living world crucially depends on it. Fortunately, however, information theory does not only prove that merely copying the genomes cannot ensure their conservation but also tells what means can perform it, namely, error-correcting codes.

I do not mean that biologists failed to realize that information is a crucial component of life phenomena. Indeed, many biologists clearly understood its importance, especially in heredity. Maynard Smith quotes Weissmann as having first realized it was so. For instance, he wrote [58]: '[Weissmann] understands that heredity is about the transmission of information, and not just matter and energy.' Although Maynard Smith endorsed this opinion, he quickly discarded information theory, arguing that it could be of no use in biology because it ignores semantics [60, 73]. This is a fundamental misunderstanding. Ignoring semantics is not a weakness of information theory, but maybe its strongest methodological position. As far as literal communication is concerned (as performed by telecommunication means), semantics is indeed not relevant: the meaning of a message does no affect its transportation, and a messenger who carries it has not to know about its content. Rejecting information theory for this reason is just *lâcher la proie pour l'ombre* (literally, letting go the prey for the shadow), i.e., giving up a coherent, complete and quantitative science of communication for no science at all. Despite the efforts of Yockey who pleaded for the use of information theory in biology [79, 80], the position of Maynard Smith has been widely adopted by biologists. (As requiring no educational effort, it is also the laziest one.) Then 'biological information' became a fuzzy, undefined concept except that, in a purely negative way, it was claimed to differ from engineering information. The word 'information' has become ubiquitous in the biological literature but most often used with a loose meaning. In my opinion, *a priori* rejecting information theory is as poorly justified and as harmful to the progress of biology as has been in its time the now obsolete concept of an irreducible 'organic chemistry.'

This lecture is intended for both geneticists and communication engineers. The impossibility of achieving faithful genetic transmission by mere copying is easily understood by the latter who know that the channel capacity is a strictly impassable limit and who can easily check that the capacity of DNA, considered as a channel for communicating through time, is actually a vanishing function of time. However, a vast majority of geneticists have no knowledge at all of information theory so an argument based on the decrease of DNA capacity is mere dead letter for them. Before I can convince them that mere copying is not the solution but that error-correcting codes are necessarily involved, I have to discuss information theory up to the point where they fully understand the meaning of capacity and the operation of error-correcting codes. A large part of this lecture must thus be devoted to this end. Most geneticists would have thrown out the book well before this point! Although communication engineers will accept my argument, they will probably feel uncomfortable with molecular genetics, so I must present some sketchy discussion of this topic which, as always in biological matters, is quite complex in its details. Intending this lecture for both geneticists and communication engineers, I must thus spend many pages discussing matters which reveal to be strongly connected to each other only at the end of the book.

This is why Part I is a summary of the lecture content in an informal style serving as a common ground for both biologists and engineers, hopefully overcoming their different cultural habits and methods. Much like the characters of a novel are progressively revealed as the reader proceeds, it does not comply with the rule of defining all the scientific words and concepts at their first appearance. Part II brings the precision and details which Part I lacks. Part III states, in the usual scientific style, the hypotheses that information theory shows to be necessary in order to explain the genome conservation, and discusses their consequences on the living world and its evolution. Many of their actual features find here an explanation, and answers are proposed to debated questions. I refer to the mandatory conditions stated above as hypotheses, although the word 'hypothesis' is too weak and has a speculative connotation, because I am yet unable to exhibit the details of genomic error-correcting codes. Furthermore, the means which implement them and achieve genome conservation remain to be identified. Making them fully explicit is a formidable task which exceeds by far the means available to me. It requires a collaboration of communication engineers and practicing geneticists which remains to be set up.

A collaboration of engineers and geneticists needs of course a common language. Engineers developed information theory as the science of communication engineering. As a mathematical science, it needed a precisely defined vocabulary. Biologists then borrowed this vocabulary to communication engineering, but used it loosely, in a descriptive rather than mathematical manner. For instance, the word 'code' belongs to the vocabulary of information theory since its very beginning, i.e., Shannon's paper [72], but geneticists began using the phrase 'genetic code' in the 1960s. What 'code' means for them is rather different from the information-theoretic meaning of the word, of which they were apparently unaware. I'll thus need to be very fussy in regards to definitions and vocabulary appropriateness. This runs the risk of looking pedantic, but it is the price to be paid for avoiding ambiguity.

Again for avoiding ambiguity, I'll name 'shannon' the binary unity of information quantity, at variance with the current information-theoretic literature which, following Shannon himself [72], names it 'bit', an acronym for 'binary digit.' However, a binary digit does not always bear information, and when it does the information quantity it bears is not always this unity. A unit and a digit are moreover definitely different entities, so using different words for naming them is better. Naming 'shannon' the unit of information quantity moreover made me free to use 'bit' as a convenient abbreviation for 'binary digit', especially in Chs. 5 and 6.

No special effort has been made to avoid redundancy. Some of the main ideas are developed in different chapters of this lecture, hopefully shedding more light on the subject when dealt with in different perspectives. Just like redundancy is the key for eliminating errors in communication engineering, I believe it is necessary in the semantic field for an efficient didactic communication.

Attempts were made earlier to apply information theory to biology. Undeniably, the great pioneer in this field is Yockey [79, 80]. Besides their scientific content, his books castigate the 'apartheid in the olive grove of academe' which prevents a majority of geneticists to realize that molecular biology actually needs information theory. I fully concur with this criticism. Not only do biologists ignore potentially relevant and useful tools, but it is a pity that ideas that information theory shows to be wrong are still taken as true by many of them and taught in textbooks.

As far as I know, the first attempt to identify error-correcting codes in genetics was made in 1981 by Forsdyke [34], who later gathered in *Evolutionary Bioinformatics* [37] the fruit of his rich experience. Many topics dealt with in this excellent book illustrate, in biological rather than engineering terms, the concept of 'soft code' I discuss in this lecture (Ch. 10). I read it too late to take its content into account here, but I'll hopefully exploit it in further works as will researchers wishing to work along the same line as me.

I would like to also mention that the book by Barbieri, *The Organic Codes* [5], proposes a picture of the living world rather similar to mine through a very different approach. The 'codes' of the title are not error-correcting codes *stricto sensu*, but they may well be interpreted in terms of 'soft codes,' a concept to be discussed in Ch. 10 below. Moreover, they assume the structure of 'nested codes,' another concept I think is necessary for explaining the permanency of the older information, which is discussed in Ch. 9.

Thanks are due to Daniel Mange of the *Ecole polytechnique fédérale de Lausanne* in Switzerland who gave me the opportunity to meet Marcello Barbieri of the University of Ferrara, Italy. I found in him an enthusiastic and friendly interlocutor who generously invited me several times to discuss my ideas both orally and in writing; I am of course deeply indebted to him. I also sincerely thank other people who manifested interest in my work and gave me opportunities to discuss the ideas contained in this lecture or with whom I exchanged ideas, especially Claude Berrou, John Enderle, Paddy Farrell, Donald R. Forsdyke, Jean-Michel Labouygues, Vittorio Luzzati, Elebeoba E. May, Olgica Milenkovic, Mark E. Samuels, and Hubert P. Yockey.

Gérard Battail
École Nationale Supérieure des Télécommunications (retired), Paris
September 2008

Part I

An Informal Overview

CHAPTER 1

Introduction

If you want to understand life, don't think about vibrant, throbbing gels and oozes, think about information technology.

Richard Dawkins, *The Blind Watchmaker*

1.1 GENETICS AND COMMUNICATION ENGINEERING

Although at first sight they may look unrelated, genetics and communication technology are both concerned with *transfer of information*. Besides the obvious fact that the latter is a matter of human engineering while the former is concerned with natural processes, they mainly differ in not operating in the same dimension: the latter is intended to communicate messages in space, from one place to another, while the former communicates hereditary messages in time, from one instant to another.

The outstanding success of communication technology relies of course on tremendous progresses in physical hardware but also, much less visibly, on the development of very powerful conceptual tools, consistently gathered in *information theory* [72]. Although these tools were originally intended for communication through space, they are comprehensive enough to apply as well to communication through time. The central question to which this lecture proposes an answer is:

Can the theoretical framework elaborated by human engineers shed light on the natural processes of hereditary communication?

The answer is yes, beyond any expectation: the perspective provided by information theory deeply renews the vision we can have of the living world. It definitely shows that the template-replication paradigm now in force does not account for the faithful conservation of genomes. This negative result would suffice to call for a deep reappraisal of the fundamentals of genetics, hence of biology as a whole.

I can thus claim that conventional biology gives an inadequate answer to the crucial question of genome conservation. Since the discovery of its double-helix structure, the replication of DNA using one of its strands as a template has become the paradigm of genome conservation and it has remained almost unquestioned. The starting point of the research reported in this lecture is its questioning, from an outer perspective, that of communication engineering. As defined by information theory, the capacity of a channel is the quantity which measures its ability to convey information. Considering DNA as a channel for communicating through time, it is a trivial matter to compute its capacity or, more precisely, a simple upper bound on it. It turns out that it vanishes exponentially fast due to the cumulative effect of casual errors. The memory of heredity—DNA—is thus not permanent as believed but actually ephemeral at the geological timescale. Without further means to ensure the permanence of the genomic message, DNA replication cannot account for its faithful communication. Genome conservation is not the rule and error is not the exception.

On the positive side, information theory asserts that faithful genome conservation can actually be secured provided very efficient *error-correcting codes* are used. One *must* thus assume that it is so. This assumption will be referred to as the *main hypothesis*. (I already explained in the Foreword why I use the word 'hypothesis' for referring to a mandatory condition, despite its weakness and speculative connotation.) Accounting for the faithful conservation of very old information leads, furthermore, to assume as a subsidiary hypothesis that the genomic error-correcting code takes the form of a layered structure—referred to as 'nested codes' or 'nested system'—which combines several codes which successively appeared in time: new information has been appended to a previously encoded message, resulting in a sequence which has been further encoded, and this process has been repeated several times in the past. Information is thus encoded by a number of component codes, the larger, the older it is. Being encoded by more numerous component codes, old information is better protected against errors than the more recent.

Although they properly recognized the importance of permanence in the evolutive success of a species, it seems that biologists did not realize how difficult it is to maintain the information integrity through the ages. Dawkins wrote about the needed accuracy of replication [29, pp. 16–17]: 'We do not know how accurately the original replicator molecules made their copies. Their modern descendants, the DNA molecules, are astonishingly faithful compared with the most high-fidelity human copying process [...]'. Dawkins expresses his astonishment as regards such a high reliability but he *does not question about the means to obtain it* nor the consequences which may result from their use. It will be shown below that high reliability cannot actually result from mere copying, but that intrinsic properties of the genome must make it resilient to casual errors.

For ensuring the genome permanence, the template-replication of DNA actually needs to be complemented with error-correcting means. Genomic messages must be endowed with error-correcting properties, and their conservation can only result from a dynamic process involving these intrinsic properties. Moreover, the genome regeneration based on them must be performed after a short enough time interval, in order to avoid the channel capacity becoming too small or, in terms of error correction, that the cumulated number of errors in the genome exceeds its intrinsic error-correcting ability.

Once it is clear that the conservation of genomes can result from a dynamic process which succeeds only when stringent conditions are met, one may expect these conditions to have a great impact on the living world as a whole, due to the central role that genomes play in it. Indeed, although no explicit description of the assumed genomic code can yet be given, the main and subsidiary hypotheses just stated provide explanations to many basic features of the living world yet unexplained, some of them because biologists did not think they need an explanation and preferred to deal with them as basic postulates. These hypotheses also hint at answers to some debated questions. Moreover, no irreducible contradictions between them and established biological facts were found although, of course, some points will need further elaboration.

The ideas to be developed below have their seeds in a number of papers and symposium presentations, among which [13, 15, 16, 17, 18, 19].

1.2 SEEING HEREDITY AS A COMMUNICATION PROCESS EXPLAINS FEATURES OF THE LIVING WORLD

1.2.1 MAIN AND SUBSIDIARY HYPOTHESES

For emphasis, we state again here the two 'hypotheses' on which this lecture is based. Most of Part III will be devoted to develop the consequences of these hypotheses and to compare them with features of the actual living world.

Main hypothesis: Any genome actually belongs to an efficient error-correcting code, to be referred to below as *genomic code*. Its necessity stems from the need of a faithful conservation of the genetic information.

Subsidiary hypothesis: The genomic error-correcting code postulated by the main hypothesis actually takes the form of a layered structure, to be referred to as *nested system*, which combines several codes which successively appeared in time. This hypothesis is actually needed in order to account for the faithful conservation of very old genetic information.

Chapters 3 and 6 deal with error-correcting codes. Precisely describing the nested system will be deferred to Ch. 9. Although several codes are involved in the nested system postulated by the subsidiary hypothesis, the phrase 'genomic code' will be used to collectively designate the set of codes which, combined into the nested system, contribute to making the genome resilient to errors. An error-correcting code will often be restricted to sequences which satisfy some set of constraints. Then the word 'code' will be used in two different but related meanings: to designate either this set of sequences, or the set of constraints they satisfy. Doing so hopefully avoids lengthy periphrases, the context being most often sufficient to avoid any ambiguity.

Anticipating on what follows, the picture of the living world which derives from the above hypotheses is now roughly outlined.

1.2.2 A STATIC VIEW OF THE LIVING WORLD: SPECIES AND TAXONOMY

An error-correcting code is a collection of symbol sequences, referred to as its *words*, which are as different as possible from each other. A very convenient geometrical language, which moreover has a rigorous mathematical meaning, interprets these codewords as points of an abstract space having as many dimensions as the words have symbols. Each coordinate of these points is associated with a symbol at a precise location in the words and assumes numerical values which unambiguously represent the alphabet symbols. The distance between two points in this space is moreover defined as the number of locations where their coordinates differ (i.e., the number of symbols where the corresponding words differ). It is referred to as the *Hamming distance* and will be used throughout in the rest of the book. The messages to be communicated are assumed to belong to a given finite set. A set of words, as many as the possible messages, is used for communicating them. It is referred to as a *code*. Each word of this code represents one of the messages according to a one-to-one correspondence. Communicating a message consists of transmitting the codeword which corresponds to it. The symbol errors which affect the received word tend to diminish its distance with respect to other codewords but, if the errors are not too many, the received word remains closer to the actually transmitted word than to any other codeword. Then the actually transmitted word can be unambiguously determined. To be efficient, an error-correcting code should thus have its words as distant as possible from each other. The smallest distance between any two different words of an error-correcting code is referred to as its *minimum distance*.

In genetics, the symbols are *nucleic bases*, also referred to as *nucleotides*, the succession of which along the DNA molecule constitutes the genomic message. Assuming that a genomic error-correcting code exists implies that genomes are neatly separated, hence the very existence of discrete species is a mere consequence of the main hypothesis. Moreover, the subsidiary hypothesis implies a distance hierarchy within the layers of the nested system: the words which belong to the central (older) layers are more distant from each other than those of the peripheral (more recent) ones. Not only the biological fact that distinct species exist is explained, but also the very possibility of their ordering according to a hierarchical taxonomy.

It has been stated already that the genome regeneration based on its error-correcting properties must be performed after a short enough time interval. That life proceeds by successive generations is a trivial biological fact, but it receives here an explanation which greatly enhances its significance. 'Successive generations' must now be understood as 'successive *re*generations,' meaning that each new living being starts with an exactly regenerated genome, except for the very few ones which suffered a regeneration error.

1.2.3 A DYNAMIC VIEW OF THE LIVING WORLD: EVOLUTION

An error-correcting code endows a genome with resilience to casual errors, hence making its exact recovery possible despite the errors which may affect the nucleotides. It works only within certain limits beyond which another word, at a distance of at least the minimum distance of the code, results from attempting to recover the transmitted word. This very unusual event transforms a genome into a neatly different one, hence originating in a new species. Let us now elaborate on the dynamic view of the living world which results from this remark.

The engineering function of decoding aims at recovering the codeword which has been transmitted in order to represent some message, given the corresponding received sequence of symbols, i.e., the transmitted codeword affected by errors. Its first step is intended to identify the actually transmitted codeword, and in the genomic context it will be referred to as *regeneration*. The second step, i.e., recovering the encoded message, is trivial once the first one is performed since a one-to-one correspondence between the possible messages and the codewords has been assumed. Regeneration is made possible because distances separate the codewords. The optimum regeneration rule consists of choosing the word of the code the closest to the received word. If the symbol errors are too many, it may occur that the word chosen according to this rule is not the correct one and, in the presence of casual errors, such a regeneration error occurs with nonzero probability. Contrary to intuition, the probability of a regeneration error can be made arbitrarily small by *increasing* the length of the codewords (also referred to as the length of the code) and using a properly designed code, but only provided the average quantity of information borne by each symbol of the codeword is smaller than a fundamental quantity, referred to as the *capacity* of the channel, which depends on the symbol error probability. If this inequality is not satisfied, a regeneration error occurs almost certainly, whatever the code and means intended for regenerating the transmitted codeword.

In normal conditions, the time interval between regenerations is as short as to make the probability of regeneration failure extremely small. It is necessarily so to account for the stability of species which we observe at the scale of individual lifetimes. As an extremely infrequent event which changes a genome into another neatly different one, we may think of a regeneration error as the possible origin of a new species. The main hypothesis thus simply leads to a saltationist vision of biological evolution. Moreover, the subsidiary hypothesis hints at a phyletic tree which coincides with the taxonomic hierarchy since the higher in this hierarchy is a component code, the larger its minimum distance, hence the more infrequent is a regeneration error at its hierarchical level.

Assuming that species originate in regeneration errors also suggests their contingency, since regeneration errors are chance events. It should, however, not be concluded that the genomic messages are purely random: on the contrary, they actually obey the constraints which define the genomic error-correcting code. The wrong recovery of a codeword is a chance event which concerns the word *as a whole*; each of its symbols is not chosen at random. This remark is important since it refutes arguments from the 'intelligent design' upholders on the improbability of errors simultaneously affecting two different nucleotides. The probability of two simultaneous errors equals the square of the probability of error in a single symbol (and thus is much smaller) only if the symbol errors are independent events. Since the error event consists of choosing a codeword as a whole, the symbol errors at two specified locations are no longer independent events so the argument does not hold.

Species do not only originate in regeneration errors. Transpositions, chromosome rearrangements, and integration of genetic material of outer origin, especially viral, are already known mechanisms which deeply modify genomes. The setting up of a new code in the nested system assumed according to the subsidiary hypothesis necessarily results from an event of this kind. A better knowledge of the genomic error-correcting codes will be necessary in order to understand their possible connection with such phenomena.

The main hypothesis suffices to explain the yet poorly understood trend of evolution towards increasing *complexity*. It has been stated above that the probability of a regeneration error can be made arbitrarily small by increasing the length of the codewords, although the longer the codewords, the larger the average number of errors which affect them. The ability to survive, thanks to a low probability of regeneration errors, is an evolutive advantage, just as the ability of fast reproduction. The ability of fast reproduction is limited by the genome length, which should be as large as to specify the replication machinery (if we except viruses which use the cellular machinery of their hosts for their own reproduction), so there is a lower limit to this length. On the other hand, there is no limit to the evolutive advantage which can result from the lengthening of genomes: the average genome lifetime which results from using an error-correcting code varies as the reciprocal of the probability of a regeneration error, as shown in Ch. 8. Lengthening the genome can result in an arbitrarily small regeneration error probability, hence there is no limit to the increase in average lifetime it can provide. The subsidiary hypothesis of a nested system moreover implies that lengthening the genome by introducing one more coding layer enhances the code minimum distance, hence diminishes the regeneration error probability of the already present genomic information.

1.3 REGENERATION VERSUS REPLICATION

Replication and regeneration are deeply different functions. Their confusion would make this lecture incomprehensible. Let us summarize their difference before we proceed further.

Replication makes a copy of the original genome as faithful as possible, thus ideally identical to it. Since copying an already erroneous genome just keeps the errors it bears, it does not faithfully ensure the long-term conservation of genomes.

Regeneration is intended to rewrite the genomic message in such a way that:

— the rewritten message strictly satisfies the constraints of the genomic code; and

— is the closest to the received genomic message possibly affected by errors.

Then, the casual errors which may affect the genome are corrected, except if the number of these errors exceeds the error-correcting ability of the code. For a properly designed code and a short enough interval between regenerations, the probability of a regeneration error is extremely small. The crucial importance of conserving genomes necessarily led natural selection to a good enough code and a short enough time interval. Regeneration in itself does not suffice to ensure the survival of a species which needs that the number of its members increases, so a regenerated genome must be replicated.

Notice that, contrary to replication, regeneration does not result in a message faithful to the original one. However, it is faithful to the genomic code seen as a set of constraints. The genomic code appears here as more important than the individual genomes the conservation of which it ensures. Implementing regeneration, unlike replication, is costly in terms of processing complexity.

The above topics will be dealt with at greater length in the following. Before I can proceed, I must roughly describe molecular genetics, and briefly expound information theory and error-correcting codes. These overviews are somewhat lengthy but I have been unable to make them shorter without hampering their intelligibility.

CHAPTER 2

A Brief Overview of Molecular Genetics

2.1 DNA STRUCTURE AND REPLICATION

Excellent books popularize molecular genetics, among which I particularly appreciated those by Dawkins [29, 30] for my own initiation to the topic (my background is in engineering). They were especially helpful to me as they provided a neatly defined model.

Everybody by now knows that the agent of genetic communication is DeoxyriboNucleic Acid (DNA). It is a very long unidimensional polymer which assumes the form of a double helix. Each strand of the double helix consists of a regular succession of deoxyribose molecules alternating with phosphate groups. Deoxyribose, a sugar, is tied to the phosphate groups by covalent bounds. The deoxyribose–phosphate chain has a mechanical role but does not convey information (it is often metaphorically referred to as the DNA 'backbone'). One of the four 'nucleic bases' (or 'nucleotides')—adenine (**A**), thymine (**T**), guanine (**G**), and cytosine (**C**)—is covalently bound to each sugar molecule of the deoxyribose–phosphate chain. The succession of nucleotides according to a well-defined direction along the backbone of one strand constitutes a message written using the four-letter alphabet {**A**, **T**, **G**, **C**}. The genetic information is borne by this message where the nucleotides can *a priori* assume any order, so a string of n nucleotides can bear in principle 4^n different messages. The other strand in the double helix structure bears a similar message which is entirely determined once the message borne by the first strand is given. In front of each nucleotide **A** of one strand, **T** is found on the other one; and **C** in front of **G**. In other words, the nucleotides appear in *complementary pairs*, **A**–**T** and **G**–**C**. The DNA molecule then appears as a (twisted) ladder, with the molecular 'backbones' as uprights and the pairs of complementary nucleotides as rungs. DNA as described is the material support of heredity for all living things (if we except some viruses where this function is assumed by ribonucleic acid, RNA, another unidimensional polymer whose chemical structure and properties are similar to DNA). DNA is combined with proteins in large structures named 'chromosomes.' In prokaryotes, i.e., unicellular beings having no nucleus, the total DNA content is referred to as its 'genome'. In eukaryotic cells which, contrary to prokaryotes, possess a nucleus separated from the remainder of the cell, the genome is the total DNA content which is enclosed within the nucleus, but some extra DNA is also found in mitochondria or blastocytes, which are organelles of the cell outside the nucleus. (They are believed to be former prokaryotes symbiotically integrated into the eukaryotic cell.) In a multicellular organism, the nucleus of each cell contains the same genome.

The nucleotides of the pair **A–T** are tied together by two hydrogen bounds, while three hydrogen bounds tie together those of the pair **G–C**. Hydrogen bounds are rather weak chemical bounds (much weaker than covalent bounds), so the double helix can fairly easily be separated into its two strands, each bearing a message complementary to that of the other one. DNA replication entirely relies on this property. Splitting the DNA molecule into its two strands, endowing each nucleotide of one strand with its complementary and constructing another 'backbone' as a support for the sequence of newly appended complementary nucleotides eventually results in two identical copies of the initial double-strand molecule. In the following, we shall refer to this replication mode as the 'template-replication paradigm' and question its ability to faithfully conserve the genetic message.

2.2 DNA DIRECTS THE CONSTRUCTION OF A PHENOTYPE

The message borne by the DNA molecule directs the construction of a phenotype. The best-known step in this process is how it directs the synthesis of the main constituents of living matter, complex molecules named *proteins*.

Some parts of the genome, referred to as 'genes,' control the synthesis of proteins. Proteins are made of a properly folded chain of amino-acids referred to as *polypeptidic chain*. An amino-acid is made of a carbon atom to which are attached a hydrogen atom, an amino group (NH_2), a carboxy group (COOH), and a specific side chain. The amino-acids of a polypeptidic chain are linked together by binding the carboxy group of an amino-acid and the amino group of another one, with elimination of a water molecule. Proteins are typically made of a few hundred amino-acids, each of them taken from a given set of as few as 20 amino-acids which differ from each other only as regards their side chain. The side chains of these amino-acids have various chemical compositions (some include a sulphur atom) and various properties, some of them being hydrophilic and others hydrophobic, and others bearing positive or negative electric charges.

The polypeptidic chain is determined by the sequence of nucleic bases of the gene. Each of its amino-acids is specified by a triplet of nucleic bases referred to as a *codon*. The order of the successive codons in the DNA strand determines the order of the corresponding amino-acids in the polypeptidic chain. The rule which tells which amino-acid corresponds to each codon is referred to as the '*genetic code*'. (We shall always use quotes when referring to it, in order to avoid any confusion with the genomic error-correcting code the existence of which we are led to assume.) After being properly folded, polypeptidic chains become proteins, which are both constituents of the living matter and actors of life processes, especially as enzymes, i.e., catalysts for the chemical reactions which take place inside the cell. An upstream 'promoter sequence' warns that a sequence of codons to be translated into amino-acids will be found. Then reading a single well-defined codon both starts the process and 'codes' for the amino-acid methionine, while reading later any of three stop codons stops it. The starting codon also determines the reading frame, i.e., what groups of three nucleotides should be dealt with as codons.

The genes constitute a part of the content of genomes which widely varies depending on the species, from almost its totality for bacteria down to a few percent, e.g., in humans, or even less. Much of the function of DNA outside the genes remains poorly understood but it is probably much more important than yet believed. In any case, DNA directs the construction of a phenotype. In turn, the phenotype hosts the genome in which it originates and its membranes shield it against mechanical and chemical aggressions.

2.3 FROM DNA TO PROTEIN, AND FROM A GENOME TO A PHENOTYPE

The path from codons in DNA to the corresponding amino-acids is not straightforward. To perform the synthesis of a protein, the DNA string of a gene is first copied into an RNA molecule referred to as *messenger RNA* (mRNA). This step is referred to as *transcription*. Then, the messenger RNA is processed so as to build the chain of amino-acids which will eventually become a protein. Each mRNA triplet which corresponds to a codon in DNA generates a short RNA molecule named *transfer RNA* (tRNA) which binds itself to the amino-acid which corresponds to this particular codon according to the genetic 'code'. The transfer RNAs are processed by an organelle of the cell named *ribosome* which joins together the successive amino-acids brought by transfer RNAs, forming the polypeptidic chain to be later folded into a protein. All this process is referred to as *translation*.

The paragraph above roughly describes the mechanism by which prokaryotic cells produce proteins (details of the actual process have been omitted). The mechanism is significantly more complicated in eukaryotes. In this case, the DNA string of a gene is first copied into a *premessenger RNA* which is later spliced: some of its parts named *introns* are removed from it, while the remaining parts, named *exons*, are joined together to constitute the messenger RNA which actually directs the synthesis of the polypeptidic chain. Introns and exons are reliably recognized by the molecular machinery which processes the genome but how they are recognized remains poorly understood.

In eukaryotes, only a small fraction of the DNA is actually used for directing the synthesis of proteins (it is estimated to about 1.3% of the human genome). Some of the remainder has regulatory functions, but the actual function of most of the genome remains unknown. This is why much of the genome has been dubbed 'junk DNA' and the DNA which actually directs the synthesis of proteins as 'coding DNA' (the word 'coding' referring here to the so-called genetic 'code'). The 'junk DNA' was believed to have no usefulness at all, being often interpreted as fossil or parasitic. I'll strongly suggest below that all the DNA of a genome contributes to its own conservation, hence that parasitic DNA cannot have but a short life. Recent researches show that, although a small fraction of DNA encurs both transcription and translation, hence participates in the synthesis of proteins, much of the DNA is transcribed into short RNA sequences the function of which is still poorly understood.

In multicellular organisms, although (almost) each cell possesses the genome in its entirety, only certain genes are expressed, i.e., used in order to direct protein syntheses, the other ones being 'repressed.' The genes which are expressed depend on the kind of the cell which, being differentiated, is now specialized. The gene expression is repressed by RNA molecules or proteins binding themselves at some particular places in the genome thus hampering certain regions in it to be transcribed or translated. Moreover, the expression or repression of genes depends on the particular history of the cell to which it belongs.

2.4 GENOMES ARE VERY LONG

Genomes are *very* long, and, moreover, their length widely depends on the species. The genome lengths range from thousands of base pairs for viruses up to hundreds of billions base pairs for certain animals and plants. The length of the human genome, for instance, is about 3.2 billion base pairs. The number of *a priori* possible genomes of a given length n is $4^n = 10^{n \log_{10} 4} \approx 10^{0.6n}$. Even for the shortest genomes, this number is so large that it defies imagination. For comparison purpose, 4^{133} equals about 10^{80}, the estimated number of atoms in the visible universe. The number of genomes actually associated with living beings is therefore only a tiny fraction of the total number of possible genomes. In the parlance of information theory, genomes are thus extremely *redundant*. We shall see in Sec. 3.5 and Ch. 6 that redundancy is a necessary property of error-correcting codes.

Not only all genomes are very long, but the range of their lengths is extremely large. Interestingly, the number of genes is much less variable from a specias to another (if we except viruses) since it hardly exceeds an order of magnitude. This may be thought of as an illustration of the intrinsic evolutive advantage of long genomes (as argued in Sec. 11.5). Would evolution be solely driven by natural selection acting on phenotypes, the number of genes rather than the genome length would be the most significant parameter.

A more comprehensive discussion of molecular genetics will be found in Ch. 4. Let us now begin questioning the template-replication paradigm. We saw that a phenotype hosts and shields its genome. Clearly, membranes can to some extent protect DNA against mechanical and chemical damages. But they are unable to shield it from the radiations to which the Earth habitat is subject, of terrestrial (natural radioactivity), solar and cosmic origin. Moreover, the DNA molecule is about 2-nanometer wide, so it is a quantum object in two of its dimensions. It suffers thermal noise and no deterministic description of it can be given. Both the very large size of genomes and the huge time intervals involved (the origin of life dates back to at least 3.5 billion years, and certain genes are shared by species, e.g., flies and humans, which have diverged in the evolution tree several hundred million years ago), make the conservation of the genetic message through the ages appear as almost miraculous. However, the genome is subject to many damages which occur at the much shorter time scale of individual lifetimes and are responsible for ageing or cancer. The genomic message cannot be both vulnerable to errors at a very short timescale and extremely well conserved at the timescale of geology, hence the template-replication paradigm cannot account for the faithful conservation of genomes, which is nevertheless a major biological fact. Computing the DNA capacity in Ch. 7 will dramatically confirm this observation. Faithfulness can only result from completely different mechanisms: information theory tells us that they must rely on error-correcting codes.

CHAPTER 3

An Overview of Information Theory and Error-Correcting Codes

3.1 INTRODUCTION

At variance with molecular genetics, I am afraid that popular books dealing with information theory pale in comparison to that of Dawkins's. Indeed, it is especially difficult to give a fair account of information theory without having recourse to mathematical concepts and notations. For the following rough discussion of information theory, I assume that the reader is reasonably familiar with mathematics at the undergraduate level.

In the literature, I often find concepts alleged as belonging to information theory but which engineers do not recognize as such; then, 'information theory' is used as a label for topics more or less loosely related to it. I must state that in this lecture I'll restrict information theory to the *science of literal communication*, as Shannon expounded it in [72], to its further elaboration and its engineering applications. Other attempts to build information theories were made, some of them rather speculative, but only Shannon's theory has proved to be both rigorous and fully operational (if we except the algorithmic information theory to be briefly discussed in Sec. 5.4 below). Restricting information theory here to Shannon's legacy is not attempting to impose some kind of orthodoxy, but just recognizing that the engineers' experience selected it according to a kind of Darwinian process, as being extraordinarily fruitful for analyzing and designing communication devices, processes, and systems. The restriction to 'literal communication', which is indeed the purpose of communication engineering, entails that information theory ignores semantics: it discards the meaning of messages as irrelevant to their transportation. Although this postulate may seem to severely restrict the relevance of information theory, it turns out that it has been necessary for setting it up as the science of communication engineering and has never hampered its development.

Besides the formalized results of information theory, the engineers' collective experience has provided some principles which remain almost implicit. Making them explicit for a reader foreign to the field is perhaps the most difficult task when trying to give him/her a useful insight about communication engineering. According to Pierce [66], information theory brought to its practitioners wisdom rather than knowledge. I would rather say that it brought wisdom besides knowledge. Communicating wisdom is a difficult and may be a paradoxical task.

I first introduce a scheme named *Shannon's paradigm* in order to make precise the concept of communication. It involves three distinct entities: the source, the channel, and the destination. The source generates a message intended to the destination. This message necessarily passes through the channel, which is generally not faithful because perturbations prevent its output to be identical to its input. I'll then discuss how information theory quantitatively measures information as considered from the sole point of view of literal communication. It then becomes possible to associate information-theoretic quantities with the elements of Shannon's paradigm, namely its *entropy* with the source and its *capacity* with the channel. Going back to this paradigm, I'll then introduce the operations of source—and channel coding as means to transform the message generated by the source so as to make it more easily communicated. Two fundamental theorems show that the entropy and the capacity are the limits of what is possible as regards these coding processes, thus endowing these quantities with an operational significance.

3.2 SHANNON'S PARADIGM

The scheme of a communication shown in Fig. 3.1 is referred to as *Shannon's paradigm*. A *source* generates a *message* intended to a *destination*. A message is a sequence of events, each unambiguously signaling the choice of a particular symbol among a given finite set of symbols, named alphabet. Source and destination are distinct entities, hence spatially separated, but there exists between them a *channel* which, on the one hand, exhibits propagation phenomena such that an input to it results in an output response which can be observed by the destination; and, on the other hand, perturbing phenomena. Due to these perturbations, the input to the channel does not suffice to determine its output with certainty. The destination has no means to perceive the transmitted message but observing the channel output.

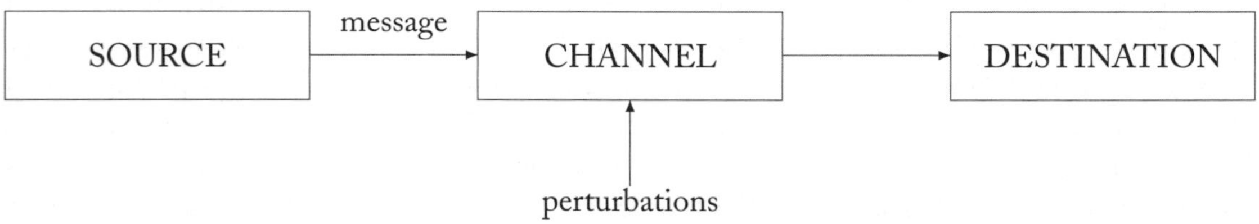

Figure 3.1: Basic scheme of a communication: Shannon paradigm.

Notice that the scheme of Fig. 3.1 is still relevant if the source and destination are separated in time rather than in space. Propagation is then replaced by recording and the channel consists of some medium which saves long-lasting modifications specified by writing signals. Receiving consists of later reading the same medium. This interpretation of the recording and reading process as a channel in the information theoretic sense is important, since it makes information theory relevant to this case as well as to communications through space. It will enable us to deal with the DNA molecule as a channel for communicating through time.

In communication through space a feedback channel (i.e., from the destination to the source of Fig. 3.1) is often available, and many simple and efficient communication processes rely on the use of such a feedback. In communication through time, however, communicating from the future to the past is impossible since it would violate causality. No feedback channel is available, hence no such communication processes can exist. Only transmitting the message as a word of an error-correcting code and recovering it from the received message, a process referred to as 'forward error correction,' can protect the message against the channel perturbations (see Sec. 3.5 and Ch. 6).

3.3 QUANTITATIVE MEASUREMENT OF INFORMATION

For defining its quantitative measurement, information is considered only from the point of view of its literal communication. The basic principle of this measurement is that the more unexpected is a message, the more information it provides. Then the information quantity borne by a message is measured by some increasing function of its improbability.

3.3.1 SINGLE OCCURRENCE OF EVENTS

Let us first consider a single event x which occurs with probability p. It will bring an information quantity equal to $f(1/p)$ where $f(\cdot)$ is a positive and increasing function to be properly chosen. Let us now consider two events which occur independently from each other with probabilities p_1 and p_2, respectively. Due to their assumed independence, the probability that both occur, referred to as their joint probability, equals the product $p_1 p_2$ of their individual probabilities. It is reasonable to state that their joint occurrence provides an information quantity equal to the sum of the information quantities they bear individually, so we should have

$$f(1/p_1 p_2) = f(1/p_1) + f(1/p_2) .$$

This is the main reason why the logarithmic function, which satisfies this equality, is chosen for $f(\cdot)$. We denote it by $\log_\beta(\cdot)$ in order to make explicit that it depends on a positive real number β named the 'logarithmic base,' such that $\beta^{\log_\beta(a)} = a$ for any positive real number a. Changing it merely multiplies the logarithm by a positive factor, hence choosing the logarithmic base defines the unit of information quantity. Shannon [72] proposed to take this base equal to 2 for measuring information, and we shall do so in the following. This choice defines the unit of information as that brought by resolving the alternative between two equally probable events. Shannon called this unit "bit" but this acronym is very often used to mean 'binary digit.' It makes generally no sense to associate an information quantity with a binary digit, most often for lack of a probability being assessed to it; even if it can be given a meaningful probability, the corresponding information quantity equals the binary information unit in the sole case where this probability equals 1/2. To avoid the use of a same word to designate the unit of information itself and an object which does not necessarily bear information and, if it does, bears the unit of information only in a particular case, the International Standards Organization (ISO) proposed the name *shannon*, abbreviated as Sh, for the binary information unit. I'll use this name for the binary information unit in the following, contrary to the current information theory literature which still most often names it 'bit.' It will be most convenient in the sequel to use the word 'bit' in order to exclusively mean 'binary symbol.'

The quantity of information brought by the single occurrence of the event x, with probability p, will thus be measured by

$$h(x) = \log_2(1/p) = -\log_2(p) \tag{3.1}$$

shannons, a positive quantity which cancels only if $p = 1$, i.e., if the considered event occurs with certainty. This is consistent with the above principle that the more unexpected is an event, the more information it provides: no information at all is brought by an event which occurs with certainty.

A single event does not suffice to describe the operation of a channel. Due to the perturbations, the output of a channel depends only probabilistically on its input, so we must consider a pair of nonindependent events: the channel input, say x, and the corresponding output, say y. We wish to measure how much information about x is provided by the occurrence of the event y. The information quantity provided by the event x is given by Eq. (3.1), but the knowledge that y occurred does not suffice to determine with certainty that x occurred. A residual uncertainty about x remains when y occurred, and it can be measured by $-\log_2[\Pr(x|y)]$, to be denoted by $h(x|y)$, where $\Pr(x|y)$ denotes the conditional probability of x given that y occurred. The output resolves the uncertainty about x, $h(x)$ as given by Eq. (3.1), except for the remaining uncertainty $h(x|y)$. The information that the occurrence of y brings about x may thus be measured by the *difference*:

$$i(x; y) = h(x) - h(x|y) .$$

This quantity is referred to as the *mutual information* (quantity) of x and y. The adjective 'mutual' is relevant because this expression is actually symmetric in x and y. Indeed, *Bayes rule* states that

$$\Pr(x, y) = \Pr(x|y)\Pr(y) \tag{3.2}$$

where $\Pr(x, y)$ denotes the joint probability of x and y (i.e., the probability that both occur), which is symmetric in x and y: $\Pr(x, y) = \Pr(y, x)$). It follows that $h(x, y) = h(x|y) + h(y)$, so $i(x; y) = h(x) - h(x|y) = h(x) + h(y) - h(x, y)$, a symmetric expression. Therefore,

$$i(x; y) = h(x) - h(x|y) = h(x) + h(y) - h(x, y) = h(y) - h(y|x) = i(y; x) . \tag{3.3}$$

If x completely determines y, then $i(x; y) = h(x)$ and no information loss results from the channel use, but if x and y are mutually independent, then $i(x; y) = 0$, as expected: the channel is then completely inefficient.

Actually, $i(x; y)$ as just defined is seldom used. The mean of this quantity, $I(X; Y)$, to be defined in the next section by Eq. (3.10), is on the contrary a quantity of paramount importance in information theory. It should be referred to as 'average mutual information' but for brevity's sake the word 'average' is generally dropped, so we'll use 'mutual information' in the sequel in order to name the average quantity $I(X; Y)$.

3.3.2 ENTROPY OF A SOURCE

Repetitive Events, Average Quantities

Events which occur only once are not truly relevant to communication engineering, and the interesting case is that of a source which repeatedly (for instance, but not necessarily, periodically) generates outputs. From now on, a source output is assumed to result from the choice of a particular element among some set of q symbols named alphabet. The finite number q will be referred to as the alphabet size. Instead of the information brought by a single outcome, one considers the average information quantity which is generated by the source operation. It has been named by Shannon the *entropy* of the source, but its relation with the entropy of physics is not straightforward and will not be discussed here. We moreover assume to begin with that the successive outputs are mutually independent. Then the source is referred to as 'memoryless.'

Entropy of a Memoryless Source

Let us first define a finite *random variable* for describing the output of a memoryless source. It is a quantity, say X, which can assume a finite number of different values, or outcomes, say $x_1, x_2, \ldots, x_q$, each occurring with some definite probability: $\Pr(X = i) = p_i$, for $i = 1, 2, \ldots, q$. The probabilities associated with all outcomes sum up to 1 since one of the outcomes occurs with certainty:

$$\sum_{i=1}^{q} p_i = 1 \,. \tag{3.4}$$

It is usual and convenient to represent a random variable by a capital letter and a particular value it assumes by a lowercase one. For instance, $X = x$ or $X = 1$ means that the random variable X is given the particular value x or 1. If the source output is the random variable X, then its entropy $H(X)$ is the average of $h(\cdot)$ as given by Eq. (3.1), namely:

$$H(X) = \sum_{i=1}^{q} p_i \log_2(1/p_i) = -\sum_{i=1}^{q} p_i \log_2(p_i) \,. \tag{3.5}$$

Here, and in the following, we denote average information quantities by capital letters. As a particular case, the entropy of a binary random variable where the symbol probabilities are $\Pr(0) = p$ and $\Pr(1) = 1 - p$ is

$$\mathcal{H}_2(p) = -p \log_2(p) - (1-p)\log_2(1-p) \,, \tag{3.6}$$

which is represented in Fig. 3.2 as a function of p.

Let us give some examples of sources and calculate their entropy. A binary source with equiprobable symbols, say **S1**, i.e., such that $\Pr(0) = \Pr(1) = 1/2$ has an entropy of 1 shannon: $H_{S1} = \mathcal{H}_2(1/2) = -(1/2)\log_2(1/2) - (1/2)\log_2(1/2) = 1$. Unequal symbol probabilities decrease the entropy. For instance, source **S2** for which $\Pr(0) = 0.9$ and $\Pr(1) = 0.1$ has as entropy $H_{S2} = \mathcal{H}_2(0.9) = -0.9\log_2(0.9) - 0.1\log_2(0.1)$ which approximately equals 0.469 Sh. A quaternary source with equiprobable symbols ($\Pr(0) = \Pr(1) = \Pr(2) = \Pr(3) = 1/4$), say **S3**, has an entropy of 2 Sh.

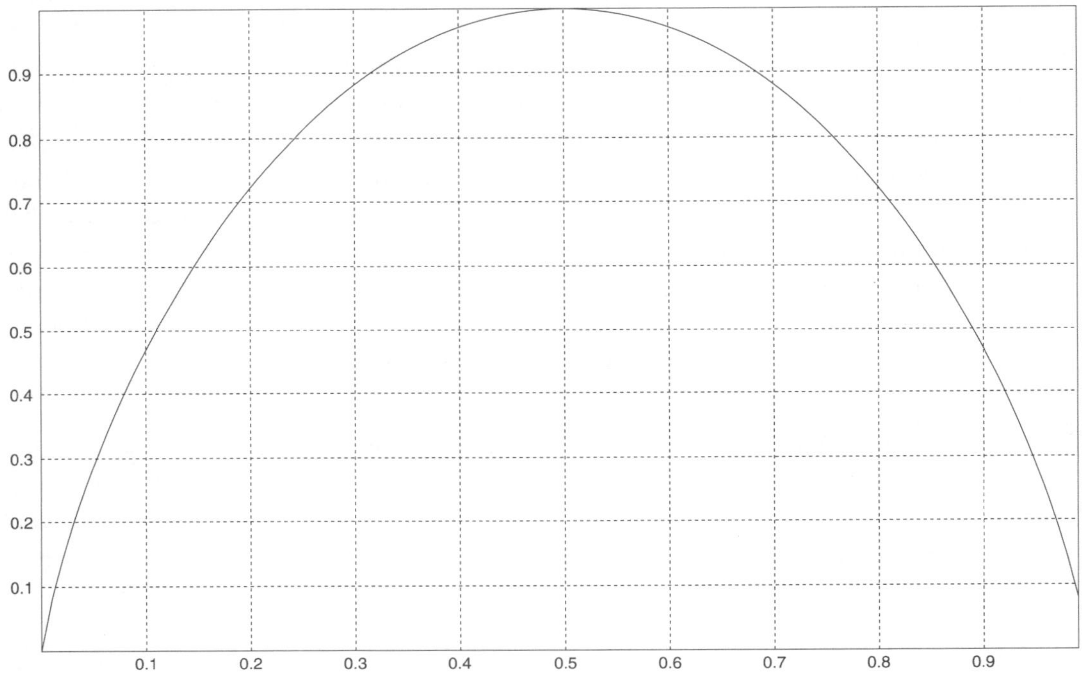

Figure 3.2: Binary entropy function $\mathcal{H}_2(p)$.

Entropy of a Source with Memory

It is necessary to extend the concept of entropy to sources with memory, i.e., where the successive output symbols are mutually dependent. The entropy of such a source can be defined as the limit for s approaching infinity of $H_{\underline{c}}(s)/s$, where $\underline{c}$ denotes a sequence of source output symbols of length s. $H_{\underline{c}}(s)$ is defined as:

$$H_{\underline{c}}(s) = -\sum_{\underline{c}} \Pr(\underline{c}) \log_2[\Pr(\underline{c})] , \tag{3.7}$$

where $\Pr(\underline{c})$ denotes the probability that sequence $\underline{c}$ occurs, and the sum in the above expression is extended to all possible output sequences of length s. The entropy of a source with memory is thus defined as:

$$H(X) = \lim_{s\to\infty} \frac{1}{s} H_{\underline{c}}(s) . \tag{3.8}$$

This limit exists provided the source is stationary, i.e., provided the probabilities which describe its operation remain constant. An especially important application of this extension will be found in the case of error-correcting codes.

The simplest examples of sources with memory are Markovian. For instance, a binary Markovian source has been represented in Fig. 3.3.

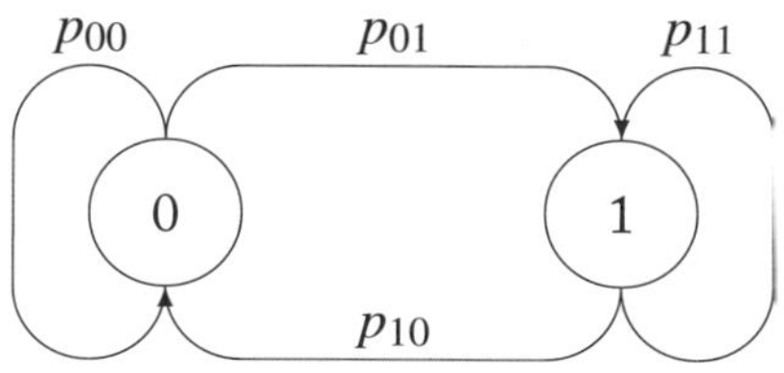

Figure 3.3: A binary Markovian source.

When the source output is x, the probability that the next symbol is y is $\Pr(y|x)$, denoted in the figure and below by p_{xy}, where both x and y are binary symbols 0 or 1; p_{01}, for instance, is the probability that the output 1 follows the output 0. Obviously, $p_{xy} + p_{x\bar{y}} = 1$, where $\bar{y}$ denotes the binary complement to y. The source is thus fully defined by two independent probabilities, e.g., p_{01} and p_{10}. Although the source operation depends to some extent on the initially transmitted symbol, its influence vanishes as time passes and a steady state is reached where the symbols 0 and 1 occur with constant probabilities referred to as steady or stationary, which are easily shown to be

$$\Pr(0) = \frac{p_{10}}{p_{10} + p_{01}}, \quad \Pr(1) = 1 - \Pr(0) = \frac{p_{01}}{p_{10} + p_{01}} . \tag{3.9}$$

Notice that if $p_{10} = p_{01}$, we have $\Pr(0) = \Pr(1) = 1/2$ as for source **S1**, although the probability distribution of the generated sequences generally differs from that of **S1**.

The entropy of the above Markovian source is the average entropy of the two states 0 and 1, i.e.,

$$H = -\Pr(0)[p_{00} \log_2(p_{00}) + p_{01} \log_2(p_{01})] - \Pr(1)[p_{10} \log_2(p_{10}) + p_{11} \log_2(p_{11})] ,$$

or, according to Eq. (3.9):

$$H = -\frac{p_{10}}{p_{10} + p_{01}}[p_{00} \log_2(p_{00}) + p_{01} \log_2(p_{01})] - \frac{p_{01}}{p_{10} + p_{01}}[p_{10} \log_2(p_{10}) + p_{11} \log_2(p_{11})] .$$

In the following, we shall consider as examples the binary Markovian sources where $p_{10} = p_{01} = 0.1$ (say **S4**), and where $p_{10} = 0.9$ and $p_{01} = 0.2$ (**S5**). The entropy of source **S4** is about 0.469 Sh, and that of **S5** is about 0.676 Sh.

The Information Measures Depend on the Ascribed Probabilities

In Sec. 3.3.1 above, we assumed that a probability could be ascribed to a single event, while in Sec. 3.3.2 a probability has been ascribed to each of the outcomes of the source, the output of which is thus a random variable with this probability distribution. Saying that the information measures defined above depend on the ascribed probabilities is merely stating the obvious, but it needs to be emphasized since its consequences are very often forgotten. It should first be recalled that it is not always possible to ascribe a probability to an event. Second, arbitrarily ascribing a probability to an event results in an information measure which is no less arbitrary than this probability itself. In any case, the information measures are relative to the probabilities ascribed to events and by no means absolute.

The probabilistic information measures apply well to situations where highly repetitive events occur with stable frequencies. Then the probabilities can rightfully be estimated from frequency measurements. It is why the entropy of a source as defined in Sec. 3.3.2 can be considered as experimentally accessible if certain conditions are met (stationarity and ergodicity; see Secs. 5.1.1 and 5.1.2). Then meaningful probabilities can be ascribed to the source outputs. It should be emphasized, however, that the information borne by a single event has no meaning unless the source from which it originates is specified, i.e., it solely depends on the set of events to which it belongs and on the probabilities which are ascribed to them. We now illustrate this important point by examples concerning outputs from sources **S1** to **S5** above.

Consider, for instance the occurrence of the symbol 0. If it has been generated by source **S1** or by source **S4**, its probability is 1/2 so it bears an information of 1 Sh. For source **S2**, however, its probability is 0.9, so the corresponding quantity of information is $-\log_2(0.9)$, i.e., approximately 0.152 Sh. From source **S3**, the probability of occurrence of the same symbol 0 is 1/4 so it now bears 2 Sh. The occurrence of 0 from source **S5** occurs with the steady probability of 0, equal to 9/11 according to Eq. (3.9), and thus corresponds to approximately 0.290 Sh.

Now consider the occurrence of the 4-symbol sequence 0101. As an output of source **S1**, this event has as probability 1/16 and the corresponding information quantity is 4 Sh. From source **S2**, however, its probability equals $0.9 \times 0.1 \times 0.9 \times 0.1 = 0.0081$ and the corresponding information quantity is approximately 6.948 Sh. If it is generated by the quaternary equiprobable source **S3**, the same sequence occurs with probability 1/256 hence bears an information quantity of 8 Sh. For the Markovian source **S4**, the probability of generating 0101 is $0.5 \times 0.1 \times 0.1 \times 0.1 = 1/2000$ because the probability of the first 0 is the steady probability of zero, while the conditional probability that it is followed by 1 is 0.1, that this 1 is followed by 0, and then that 0 is followed by 1 are again 0.1. The corresponding information quantity is thus approximately 10.966 Sh. For source **S5**, the probability of this sequence is, similarly, $9/11 \times 0.2 \times 0.9 \times 0.2$ and the corresponding information quantity is approximately 5.085 Sh.

The examples above clearly show that the information borne by a message does not depend on this message itself, but on the set of messages from which this particular message is taken. A source generates such a set of messages, and its entropy, i.e., the information-theoretic significant quantity associated with it, only depends on the probabilities ascribed to the elements of this set. The measure of information is only meaningful for events endowed with known probabilities, a stringent condition which limits the usefulness of Shannon's information theory. Failing to clearly identify the relevant probabilistic set of events often nullifies statements allegedly using information-theoretic measures.

The *algorithmic information theory* is a possible alternative to Shannon's information theory, originated in computer science, to be briefly discussed in Sec. 5.4. It does not base the measure of information on probabilities, which makes it of great conceptual interest. Its practical usefulness in communication problems is however limited, and it is why this lecture relies on Shannon's probabilistic measure of information. It should moreover be realized that the algorithmic information theory results in essentially the same information measures as Shannon's in instances where both theories are relevant, at least for long enough sequences. Rather than competing, the algorithmic information theory and Shannon's appear as two facets of a same science of information.

3.3.3 AVERAGE MUTUAL INFORMATION, CAPACITY OF A CHANNEL

Average Mutual Information

Let us now consider the average of the mutual information $i(x; y)$ between the input x and the output y of a channel, given by Eq. (3.3). We obtain:

$$I(X;Y) = H(X) - H(X|Y) = H(X) + H(Y) - H(X,Y) = H(Y) - H(Y|X) = I(Y;X), \quad (3.10)$$

where the entropy $H(X)$ has already been defined in Eq. (3.5). The joint entropy $H(X, Y)$ is the straightforward extension of the entropy definition to a pair of variables, namely:

$$H(X, Y) = -\sum_{i=1}^{q} \sum_{j=1}^{r} \Pr(x_i, y_j) \log_2[\Pr(x_i, y_j)] ,$$

assuming that Y has as possible outcomes $y_1, y_2, \ldots, y_r$, and the conditional entropy $H(X|Y)$ is similarly the average of $h(x|y)$, namely:

$$H(X|Y) = -\sum_{i=1}^{q} \sum_{j=1}^{r} \Pr(x_i, y_j) \log_2[\Pr(x_i|y_j)] .$$

Notice that the coefficients in the above expression are the *joint* probabilities of X and Y, as associated with a pair of random variables, while the probabilities in argument of the logarithms are *conditional* probabilities. Conditioning decreases the prior uncertainty hence reduces the entropy, so the inequality $H(X|Y) \leq H(X)$ implies that $I(X; Y)$ is positive (zero if and only if X and Y are independent).

The above expressions of $H(X, Y)$ and $H(X|Y)$, as well as the expression Eq. (3.10) of the average mutual information, are valid for a memoryless source and a memoryless channel. They can, in principle, be extended to sources and channels with memory according to Eq. (3.8) and Eq. (3.7). The corresponding expressions can be very complicated, but the important fact is that the relevant quantities are perfectly defined, hence exist in the mathematical meaning of this word under comparatively mild conditions.

Capacity of a Channel

The average mutual information depends on both the source and the channel. In order to obtain a quantity which solely characterizes the channel, one defines its *capacity* as the average mutual information obtained by choosing the source in order to maximize it. It is thus the largest possible average mutual information that the channel can convey. The channel capacity cannot of course exceed the maximum entropy of a source connected to its input, hence cannot exceed $\log_2 q$ shannons when q is the size of the channel input alphabet. The channel perturbations actually diminish the capacity.

Let us give two examples of channels and of their capacity. For simplicity's sake, we restrict ourselves to memoryless channels. In the following figures, a memoryless channel is represented by the set of arrows associated with the transitions from the symbols of the input alphabet to symbols of the output alphabet. Each arrow is labeled with the probability that the transition it represents occurs. The channels are memoryless in the sense that each transition occurs independently from the previous ones. The figure at left depicts the *binary symmetric channel*, that at right the *binary erasure channel*.

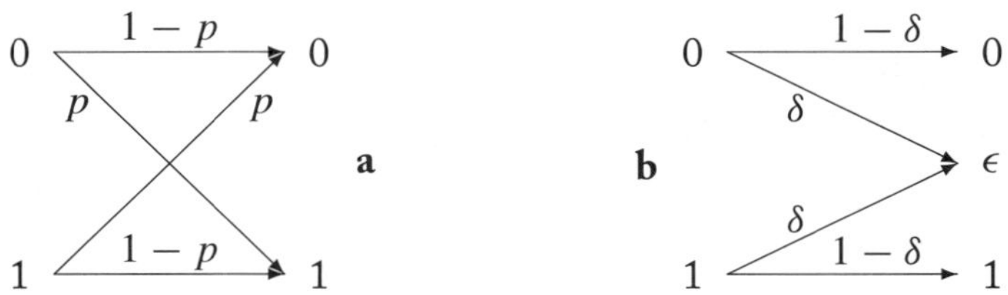

Figure 3.4: Binary-input channels. On the left, binary symmetric channel with error probability p. On the right, binary erasure channel , with erasure probability δ.

The binary symmetric channel is depicted in Fig. 3.4-**a**. Both its input and output alphabets are binary. The word 'symmetric' means here that the probability p of a 0 being transformed into a 1 equals that of a 1 being transformed into 0. We shall refer later to this kind of error as 'substitution.' It turns out that the mutual information of a symmetric channel is maximized when the source is memoryless and equiprobable, i.e., the probability of occurrence of 0 equals that of 1, so both probabilities equal 1/2. We have also, by symmetry, $\Pr(Y = 0) = \Pr(Y = 1) = 1/2$. One thus has $H(X) = H(Y) = 1$ shannon. The joint probabilities are

$$\Pr(X = 0, Y = 0) = \Pr(X = 1, Y = 1) = (1 - p)/2$$

and

$$\Pr(X = 0, Y = 1) = \Pr(X = 1, Y = 0) = p/2 ,$$

hence

$$H(Y|X) = -(1 - p)\log_2(1 - p) - p\log_2(p) = \mathcal{H}_2(p) ,$$

where the function $\mathcal{H}_2(p)$ is the entropy of a binary random variable which assumes one of its values with probability p, as already defined in Eq. (3.6). The capacity of the binary symmetric channel, i.e., $I(X; Y)$ for $\Pr(X = 0) = \Pr(X = 1) = 1/2$, is thus:

$$C_{\text{bsc}} = H(Y) - H(Y|X) = 1 - \mathcal{H}_2(p) . \tag{3.11}$$

One notices that $C_{\text{bsc}} = 0$ for $p = 1/2$. In this case, the channel output has become independent of its input, making the channel useless. The capacity achieves its maximum value of 1 Sh for $p = 0$ and $p = 1$. If the first case is not surprising, the second one is a bit more, but it suffices to swap the labels '0' and '1' of the output symbols (which are in fact arbitrary) to revert to the case where $p = 0$. Due to symmetry, no restriction of generality results from assuming that $p \leq 1/2$.

For the binary erasure channel, which is represented in Fig. 3.4-**b**, the input alphabet is binary but the output alphabet is ternary, its symbols being denoted by 0, 1, and ϵ. Transitions can occur from an input 0 to an output 0 or ϵ, and from an input 1 to an output 1 or ϵ, but never from 0 to 1 or vice versa. The probability of a transition from 0 to ϵ is the same as that from 1 to ϵ and is denoted by δ. The third output symbol ϵ may be interpreted as indicating that the output symbol has not been recognized as belonging to the input alphabet, hence has been 'erased.' This channel is symmetric, too, so calculating its capacity still consists of computing $I(X;Y)$ for $\Pr(X=0) = \Pr(X=1) = 1/2$. The joint probabilities are then:

$$\Pr(X=0, Y=0) = \Pr(X=1, Y=1) = (1-\delta)/2$$

and

$$\Pr(X=0, Y=\epsilon) = \Pr(X=1, Y=\epsilon) = \delta/2 \,,$$

all others being zero. It then results that

$$\Pr(Y=0) = \Pr(Y=1) = (1-\delta)/2$$

and

$$\Pr(Y=\epsilon) = \delta \,,$$

hence

$$H(Y) = -(1-\delta)\log_2[(1-\delta)/2] - \delta\log_2(\delta) = 1 - \delta + \mathcal{H}_2(\delta) \,,$$

and

$$H(Y|X) = -(1-\delta)\log_2(1-\delta) - \delta\log_2(\delta) = \mathcal{H}_2(\delta) \,.$$

The capacity C_{ec} has thus the very simple expression:

$$C_{\text{ec}} = H(Y) - H(Y|X) = 1 - \delta \,. \tag{3.12}$$

It may be interpreted as follows: in the average, a fraction δ of the symbols is lost. Each of the remaining ones bears an information quantity of one shannon, so the average information quantity borne by the channel output is $1 - \delta$ Sh per input binary symbol.

3.4 CODING PROCESSES

3.4.1 VARIANTS OF SHANNON'S PARADIGM

Let us go back to Shannon's paradigm as depicted in Fig. 3.1. We now introduce variants by including in this scheme devices intended to transform the message.

If the source, the channel, and the destination are arbitrary, nothing *a priori* guarantees the mutual compatibility of the source and the channel, on the one hand, and of the channel and the destination, on the other hand. For instance, in radiotelephony, the source and destination are human beings, but the channel is provided by the ability of electromagnetic waves to propagate. Humans, however, have no natural means for transmitting and receiving such waves, except that we perceive light radiations. We thus have to augment the scheme of Fig. 3.1 with devices intended to match the blocks of the initial scheme to each other. One then obtains the scheme of Fig. 3.5-a. It is a mere variant of the scheme of Fig. 3.1, since the set comprising the source and the transmitting device on the one hand, that comprising the receiving device and the destination on the other hand, may be interpreted as a new source-destination pair now matched to the initially given channel (Fig. 3.5-b). One may as well consider the set comprising the transmitting device, the channel, and the receiving device as being a new channel, matched to the original source-destination pair (Fig. 3.5-c); thus, in the previous example we considered as a channel a radiotelephonic link, made of the set of transmitting devices, the propagation medium, and the set of receiving devices.

A more fruitful point of view actually consists of splitting each of the transmission and reception apparatuses into two blocks. One of the blocks of the transmitting device is matched to the source and the other one to the channel input. Similarly, the receiving device is split into a block matched to the channel output and another one matched to the destination. Interestingly, this scheme enables *normalizing* the characteristics of the blocks of Fig. 3.5, redefined as follows: new source before point A in Fig. 3.5-d; new channel between points A and B; new destination beyond point B. The engineering problems then consist of *separately* designing the pairs of matching blocks denoted in the figure by TA1 and RA1 on the one hand, by TA2 and RA2 on the other hand. What this normalization consists of will be precisely stated later.

In general, the borders of the blocks in Fig. 3.5 may be freely redefined for the sake of analysis; cutting any apparatus chain linking a source and a destination in two points such that the origin of the message useful to the destination is in the leftmost block defines a new triplet source-channel-destination, *provided* the *perturbations* only affect devices located in the *central* block, i.e., the channel. To prevent misunderstandings, let us stress that the concepts of information and perturbation are by no means absolute. On the contrary, the message from the source is considered as bearing information only because it can be used by the *destination*. The perturbing events are just those which are harmful with respect to this particular choice of the information message. The usefulness to the destination is the *only* criterion which differentiates the information message from the perturbations, meaning that their distinction is *arbitrary*. They may well exchange their roles. For example, the sun is a source of noise for a receiver intended to the signal of a communication satellite. However, the satellite signal perturbs the observation of a radioastronomer who studies solar electromagnetic radiations, hence it generates perturbations for this particular destination. When we refer to the usefulness or harm that a sequence of symbols has for the destination, we necessarily consider its *purpose*, hence we cannot avoid some kind of teleonomy or subjectivity.

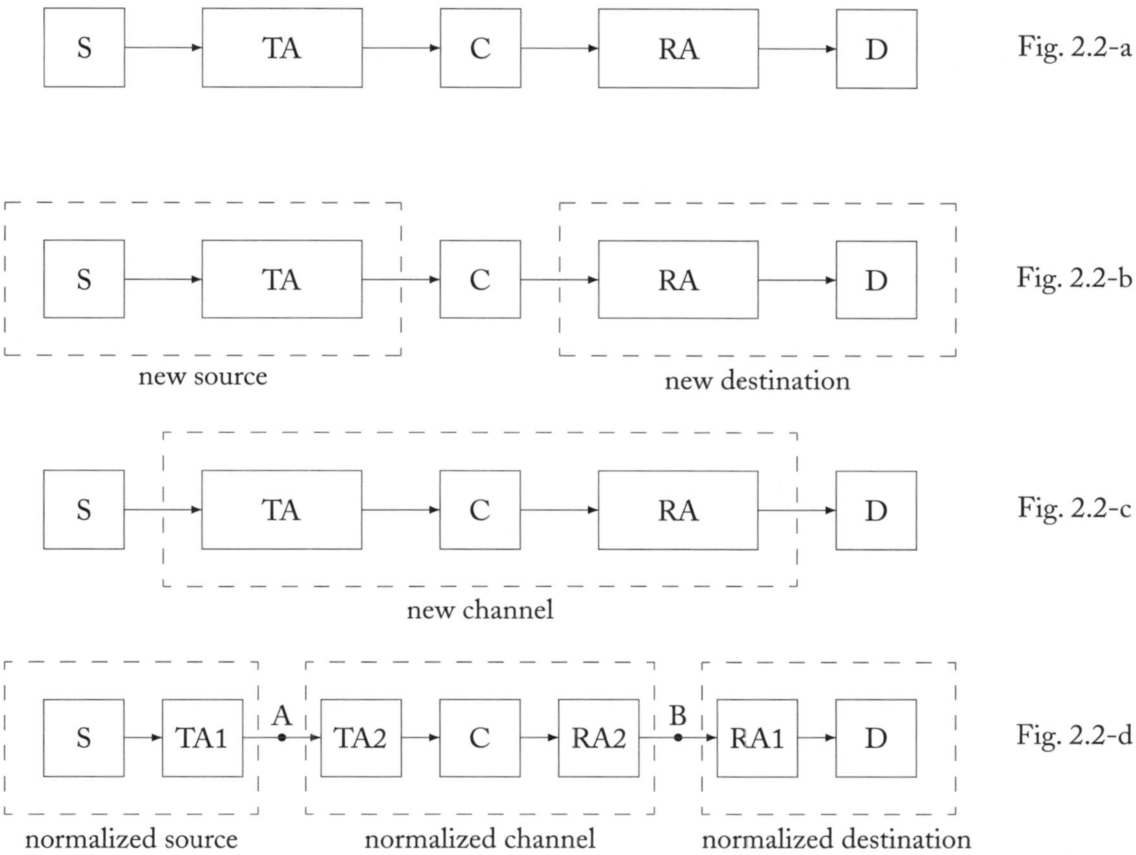

Figure 3.5: Variants of Shannon's paradigm. S stands for "source," C for "channel," and D for "destination." TA means "transmission apparatus" and RA, "reception apparatus."

It is possible to transform any message into a strictly equivalent one which possesses some properties more favourable to its communication than the original ones. Such transformations are referred to as coding processes. I'll briefly discuss the main two ones, namely *source coding* and *channel coding*. The information-theoretic quantities which have been associated with the blocks of Shannon's paradigm, namely the *entropy* of the source and the *capacity* of the channel, then appear as the limits of what is possible when the encoding operations are ideally performed. These two quantities acquire therefore an operational significance, which confirms their adequacy to communication problems and enlightens their meaning.

3.4.2 SOURCE CODING

Source coding aims at *maximum conciseness*, i.e., at substituting an equivalent message as short as possible for the original message according to a reversible transformation, in the sense that the original message should exactly be recovered given the encoded one. Source coding is a very important topic of information theory, but it has probably little relevance to the field of genetics. It seems indeed that DNA is quite 'cheap' since even the shortest genomes are very long, so there is no obvious biological interest in performing source coding. Although it is unlikely that nature implements source coding, it is a concept extremely important for understanding information theory as a whole, and especially the algorithmic information theory to be briefly discussed in Sec. 5.4. It is why the principle of source coding is discussed in Sec. 5.2.3.

3.4.3 CHANNEL CODING

The aim of *channel coding* is completely different: it is intended to faithfully communicate the message despite the channel perturbations. It replaces the original message by an equivalent one from which the original message can be recovered even if it has been altered by channel perturbations. As a very simple example, let us assume that we wish to communicate a binary message within a channel where errors occur: a transmitted 0 is received as 1, or vice versa, with some nonzero probability p_e, for instance $p_\mathrm{e} = 10^{-3}$. We transmit each of its symbols in triplicate and take a decision by majority voting among the received triplets. Then the probability p_d of an erroneous 'decoding decision' is equal to the probability of 2 or 3 errors in the received triplet, namely:

$$p_\mathrm{d} = 3p_\mathrm{e}^2(1 - p_\mathrm{e}) + p_\mathrm{e}^3 = 3p_\mathrm{e}^2 - 2p_\mathrm{e}^3 .$$

With an error probability of $p_\mathrm{e} = 10^{-3}$ in the channel, we obtain that the decoding decision is wrong with the much smaller probability $p_\mathrm{d} = 2.998\ 10^{-6}$. A still smaller error probability would have been obtained by repeating 5, 7, ... times the transmitted binary symbols, the decision rule remaining majority voting.

This example shows both that it is possible to protect a message against perturbations by coding, and the price to pay for this: an increase in the message length referred to as *redundancy*. This is only a rudimentary coding process and similar results could have been obtained at the expense of much less redundancy, by using more sophisticated 'error-correcting codes'. It is nevertheless plainly true that protecting a message against the channel perturbations necessarily implies introducing redundancy.

The aims of source coding and channel coding are thus incompatible. There exists even an antagonism (or complementarity) between them, since source coding actually increases the vulnerability to errors as well as it increases conciseness. In source coding, a single binary digit of the coded message being in error may result in a large fraction of the recovered message to be wrong. This simple remark shows that decreasing redundancy and protecting against errors may not be considered independently of each other.

As regards genetic applications, and contrary to source coding, channel coding is crucial as being the only possible way for ensuring the faithful conservation of genomes.

3.4.4 NORMALIZING THE BLOCKS OF SHANNON'S PARADIGM

It now becomes possible for us to make precise what should be intended by the normalization of the blocks of Fig. 3.1. The message generated by the source is first transformed by *source coding* into a message (ideally) *devoid of redundancy*, i.e., achieving the largest possible entropy, where all successive symbols are independent and where all the alphabet symbols occur with the same probability. The encoding thus performed only matches the source characteristics. Its result is highly vulnerable to perturbations since any of its symbols is essential to the message integrity. *Channel coding* then becomes mandatory so as to make the message generated by the source encoder (ideally) invulnerable to the channel perturbations, which necessarily demands reintroducing redundancy.

We may assume that source coding has been performed ideally. Then, channel coding just protects a redundancy-free message against the channel perturbations. If the message to be encoded is not completely devoid of redundancy, the protection achieved is at least that for a redundancy-free message: exploiting some remaining redundancy can be but beneficial to the protection against channel perturbations.

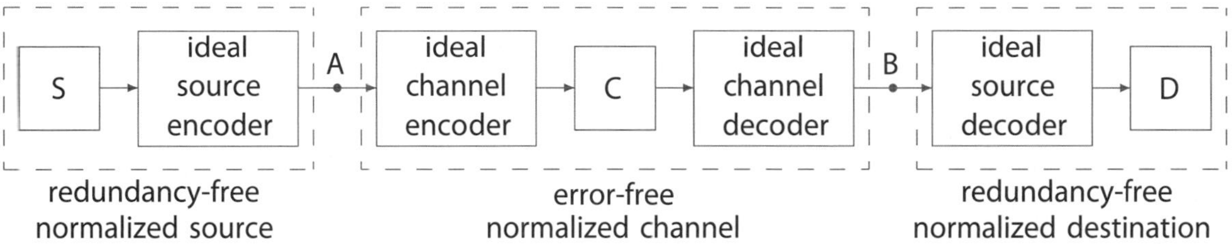

Figure 3.6: Normalization of the blocks 'source,' 'channel,' and 'destination' of Shannon's paradigm. S, C, and D denote the original source, channel, and destination, respectively.

One may then redraw Fig. 3.5-d as in Fig. 3.6, where the normalized source generates a redundancy-free message and where the normalized channel is error free.

The approach which consists of suppressing the redundancy of the original message by source coding, and then of reintroducing redundancy by channel coding, may look self-contradictory but the kind of redundancy of the original source is not necessarily well adapted to the properties of the channel to which it is connected. Rather than globally designing coding systems in order to match a particular source to a particular channel, the normalization as just defined enables dealing with the source and the channel separately, instead of the source-channel pair. This normalization has also an auxiliary advantage: the message alphabet at points A and B of Fig. 3.6 is arbitrary. We may thus assume it to be the simplest possible, i.e., binary, without significantly restricting generality.

3.4.5 FUNDAMENTAL THEOREMS

To summarize, coding operations ideally result in:

— a redundancy-free (or maximum-entropy) message, although the original message is generated by a redundant source, as regards source coding;

— an error-free message after decoding, although the coded message is received through a perturbed channel, as regards channel coding.

These possibilities are stated by the *fundamental theorems* of information theory, under conditions they precisely define. Their proof does not demand that the means for obtaining these results are made explicit. In fact, except for some simple cases, the optimum coding means remain unknown especially as regards channel coding.

The fundamental theorem of *source coding* tells that it is possible to eliminate all redundancy from a stationary source. Coding should be effected on the k-th extension of this source, for k large enough, the stated result being possibly reached only asymptotically. The k-th *extension* of a source S with an alphabet of size q (this source being said q-ary) is the q^k-ary source which results from the original source by grouping its symbols by blocks of k, each of these blocks being interpreted as a symbol of the q^k-ary alphabet. Denoted by S^k, this extension is just another way of representing S, not a different source. When optimized, the resulting message has an *average length* per symbol of the original source which tends to its *entropy* expressed using q for defining the information unit, i.e., using logarithms to the base q. Efficient algorithms which perform source coding are known for long and, for simple enough sources, their performance is close to the theoretical limit. Moreover, they can easily be made *adaptive* so as to deal with sources of slowly varying or even *a priori* unknown parameters.

The fundamental theorem of *channel coding* is asymptotic as the length n of the words associated with blocks of k source symbols tends to infinity. Keeping constant the code rate $R = k/n$ demands that k increases proportionally to n, i.e., that larger and larger blocks of the message that the source generates are encoded. (This statement is not the most general possible, but only concerns the class of so-called 'block codes', and it is moreover assumed here that the channel input and output alphabets are the same.) The necessary condition for the error probability to approach 0 as n increases can be stated as follows: the source *entropy* should be *less* than the channel *capacity* (source and channel are assumed to satisfy proper regularity conditions which are mandatory to guarantee the existence of these quantities). Thus, the presence of channel perturbations does not limit the reliability with which the message is communicated, as measured for instance by how small is the probability that it is not correctly recovered (to be referred to in the sequel as residual error probability), but only the rate at which information can be communicated using this channel. The highly desirable result of an arbitrarily small error probability can however be obtained only when using an appropriate code, for which no practically usable construction process is proposed by information theory. At least from a practical point of view, the turbocodes (invented in 1993) exhibit small residual error probability at rates close to the channel capacity so they can be considered as approaching the theoretical limit as closely as desired. They also provide an experimental evidence that the capacity is a very sharp limit and strikingly confirm the relevance of information theory, and especially its concept of channel capacity, to the communication problems.

3.5 A BRIEF INTRODUCTION TO ERROR-CORRECTING CODES

3.5.1 REDUNDANT CODE, HAMMING DISTANCE, AND HAMMING SPACE

Let us consider the set S_n of n-symbol binary sequences. Their total number is 2^n. A redundant code $C(n, k)$, with $k < n$, is a subset of S_n which contains $M = 2^k$ elements, to be referred to as codewords. We may define a one-to-one correspondence between each of these codewords and each of the possible 2^k binary messages. Transmitting one of the codewords is thus equivalent to transmitting one of these messages, and we assume that only codewords of $C(n, k)$ are actually transmitted. Since $k < n$, the proportion of the words in $C(n, k)$ among the elements of S_n is $2^k/2^n = 1/2^{n-k}$: they are only a minority subset of S_n if $n - k$ is large enough. For instance, we may assume as an order of magnitude that $k = n/2$. In this case, the proportion of sequences of S_n which belong to $C(n, k)$ is $1/1,024$ for $n = 20$, about 8×10^{-31} for n=200 and 9×10^{-302} for $n = 2,000$. For large values of n and if the code rate $R = k/n$ is kept smaller than 1, only a tiny fraction of the n-symbol binary sequences belongs to the code.

I just defined a redundant code as a subset of the set S_n of the n-symbol binary sequences. The word 'code' can be used in at least two other different meanings. As a subset of S_n, $C(n, k)$ is generally defined by constraints specifically satisfied by its words, and I'll occasionally use the word 'code' for referring to this set of constraints. The encoding rule which establishes a correspondence between the 2^k information messages and the codewords is also sometime referred to as a 'code' but I'll avoid doing so.

The restriction to binary sequences is unnecessary, and an alphabet of any size can be used. Also notice that we defined redundancy by the scarcity of the codewords among all the sequences of S_n. This definition of redundancy is much more comprehensive than the usual reference to repeated parts. It simply means that the number of symbols n in a codeword is larger than the number k of message symbols which strictly suffices to specify it. The restriction to memoryless sources and channels was just for simplicity's sake, and the above concepts and results generalize to source and channels with finite memory, at the expense of an increased complexity. We may think of each of the k information symbols as chosen at random with uniform probability $(1/q)$ among the q alphabet symbols independently of the others, a choice which maximizes the information quantity they bear. Then, the total information (expressed in shannons) borne by an n-symbol word is $k \log_2 q$ where q is the alphabet size, so the entropy per symbol of the encoded word is $H = k \log_2 q/n = R \log_2 q$. The condition of the fundamental theorem that the source entropy should be less than the channel capacity becomes $R \log_2 q < C$.

When the encoding rule is such that the k information symbols appear explicitly at k defined positions in the encoded word (e.g., but not necessarily, in the first k positions), the code is said to be in *systematic form*. This is a feature of the encoding rule and not of the code as a set of words, so the often-used term 'systematic code' to designate this case is an abuse of language.

3.5.2 RECEPTION IN THE PRESENCE OF ERRORS

Assume again for simplicity's sake that the code alphabet is binary. Suppose now that the codeword transmitted in order to represent some message suffers errors in the sense that the symbols are transformed into their complements, i.e., 0 into 1 or 1 into 0, with some nonzero probability. If $n - k$ is large enough, the probability that a codeword is transformed into another one is very small due to the extreme scarcity of the codewords within the set S_n. Checking whether a received n-symbol binary sequence belongs or not to the code is thus a very efficient means of checking whether one or more symbol errors affected the transmitted sequence which, by hypothesis, belongs to the code $C(n, k)$. Then the code enables *error detection*. Let us now introduce a metric into S_n, i.e., define a distance between its elements. The *Hamming distance* between two sequences is defined as the number of positions in the sequence where their symbols differ (for instance, the Hamming distance between the two 7-symbol words 1101000 and 0110100 is 4). The set S_n endowed with this metric will be referred to as the n-dimensional Hamming space. Two different sequences of length n are at a distance of at least 1 and at most n. Since the codewords are very sparse among the sequences of S_n for $n - k$ large enough, we may choose them so as to make the smallest distance between any codeword and all the other ones as large as possible. This smallest distance d is referred to as the *minimum Hamming distance* of the code. If e symbols are affected by errors, the received sequence is at a distance e from the transmitted word, while it is at a distance of at least $d - e$ from any other codeword. The set of locations of the symbol errors is referred to an *error pattern* and the number e of these errors is referred to its *weight*. Provided $e < d - e$, or $e < d/2$, the codeword actually transmitted is thus the closest to the received sequence so it can be unambiguously determined despite the occurrence of e errors in the received n-symbol sequence.

The process of decoding can thus be decomposed into two steps: the first and most complex one, the 'codeword recovery,' consists of looking for the word of the code the closest to the received sequence. We may name it *regeneration* since it is intended to restore the original encoded sequence. In the presence of random errors it can fail with some nonzero probability hence result in a codeword different from the transmitted one. The second step, the 'message recovery,' delivers the message associated with the codeword obtained in the first step. The encoding rule being a one-to-one correspondence, this is a trivial task which can be performed by merely reading an entry in a table. Figure 3.7 represents decoding as a two-step process.

Provided the minimum distance of the code is large enough with respect to the codeword length, the probability of an erroneous decoding decision can be made as low as desired by using appropriate codes of increasing length and adequate decoding algorithms. Then, within these limits, error-correcting codes perform the paradoxical function of enabling *reliable communication through unreliable channels*. 'Reliable' means that it is possible to make the residual error probability as low as desired, thus freeing the communication faithfulness from the perturbing events which occur in the channel. In other words, contrary to intuition, the existence of errors in the channel does not hinder communication to be performed with an arbitrarily small error probability provided an efficient and redundant enough error-correcting code is used.

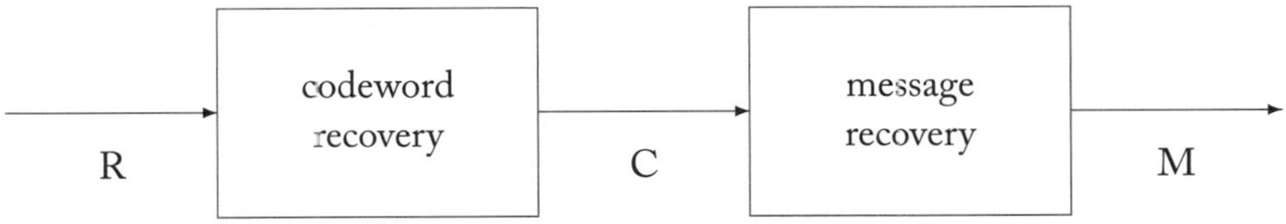

Figure 3.7: Decoding as a two-step process. R stands for 'received sequence,' C for 'codeword,' and M for 'message.'

Notice that any symbol in a codeword contributes to the correction of errors, hence to its own conservation. No parts of genomes, acting as codewords according to our main hypothesis, can be superfluous and the idea of 'junk DNA' is a fallacy.

3.6 A VARIANT OF SHANNON'S PARADIGM INTENDED TO GENETIC APPLICATIONS

In Figs. 3.1 and 3.6, the source is clearly identified as the origin of the message. Sending an information message is intentional when the source is a human being. We now propose a variant of Shannon's paradigm which better fits the absence of intentionality which is typical of the biological context (where, at least, we do not know if there is an intention).

In accordance with a remark above, we first remove from Fig. 3.6 the source encoder and source decoder which are unlikely to exist in genetics. We thus only keep the channel encoder and decoder, simply referred to as 'encoder' and 'decoder.' Then we merge the two blocks 'source' and 'encoder' into a single block (inside a dashed box) labeled 'redundant source' and, similarly, the two blocks 'decoder' and 'destination' into a single one (again inside a dashed box) labeled 'redefined destination.' We obtain the variant of Shannon's paradigm depicted in Fig. 3.8.

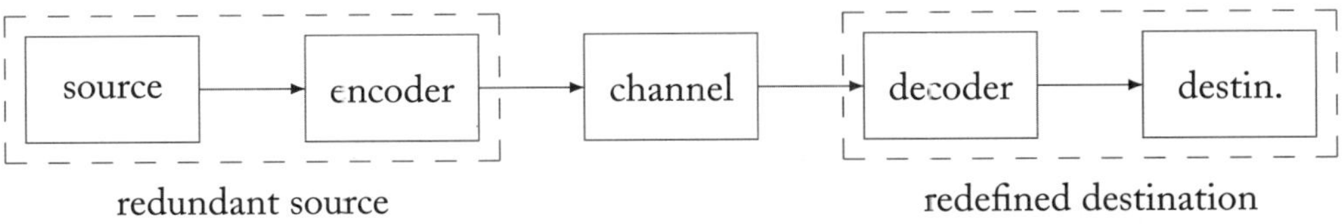

Figure 3.8: A variant of Shannon's paradigm.

The inclusion of the channel encoder in the redundant source endows it with error-correcting capabilities which can be exploited by the redefined destination. The entropy of the redundant source, according to the extension of the entropy concept to sources with memory, is $H = \log_2(q)k/n$ shannons, where q is the alphabet size, k the number of information symbols of the message, and n

the code length. According to the fundamental theorem of channel coding, errorless communication is theoretically possible provided the entropy H is less than the channel capacity C.

Let us now consider the case of a genetic message incurring successive regenerations. The scheme at top of Fig. 3.9 depicts the case where two successive channels are used. Cascading the decoder which follows the first channel and the encoder which precedes the second channel results, according to Fig. 3.7, in the box labeled 'message revovery' of the decoder being followed by the encoder. But the latter just undoes what the former operated, so they cancel each other: both can be omitted. We may replace the decoder followed by the encoder by a single box labeled 'regenerator' where the encoded message no longer explicitly appears, as depicted at bottom of Fig. 3.9.

The concept of *regeneration* is better fitted to the context of genetics than the engineering concept of decoding which refers to an explicit 'information message' to be communicated. Regeneration is a necessary step, the most important indeed, of the error-correction process: identifying the presumed transmitted codeword is a nontrivial task which incurs a risk of failure.

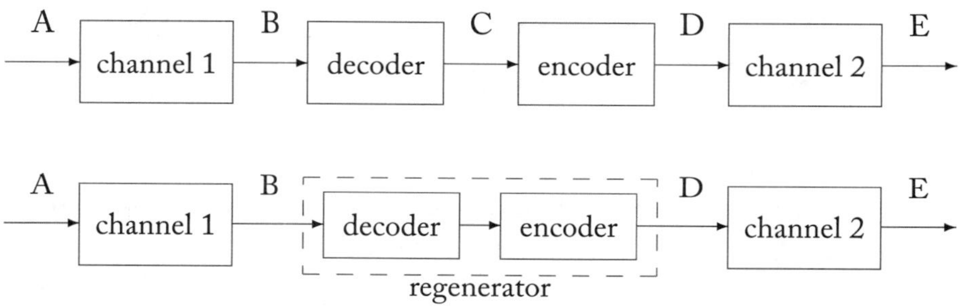

Figure 3.9: The regeneration function. Channels 1 and 2 have to be used successively. The upper picture is relevant to a conventional error-correcting code. The sequences found at the points designated by letters are: in A, an encoded sequence; in B and E, received sequences; in C, an information message; and in D, the sequence which results from the encoding of the decoded information message, hence restoring the initial sequence if no decoding error occurs. In the lower picture, the decoder and encoder have been merged into a single entity labeled 'regenerator' where the information message no longer appears.

We may now describe a chain of successive regenerations performed on a recorded message (e.g., the genetic message of DNA) as depicted in Fig. 3.10. An original redundant source at left delivers a redundantly encoded message which is written on a first support (labeled 'channel 1'), regenerated in the device labeled 'regen. 1,' written again on 'channel 2,' etc. The last step considered is the i-th regeneration, where i is a finite but possibly large number.

The redundant source has been defined in Fig. 3.8 and regenerators in Fig. 3.9. If the number of regeneration steps i is very large, the initial encoded message from the redundant source is likely to have been modified and even maybe 'forgotten,' insofar as regeneration errors likely occurred. As the number of regenerations i increases, the current encoded message thus depends less and less on the original redundant source, and more and more on the regeneration errors which occurred, i.e.,

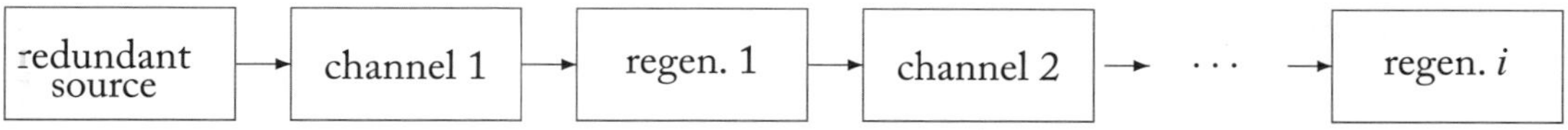

Figure 3.10: A chain of i successive regenerations.

on a succession of contingent events. In the context of genetics, of course, natural selection always operates on the corresponding phenotypes, letting only the fittest ones survive. This illustrates the ambivalence of regeneration errors. On the one hand, they should occur very unfrequently for the sake of genome conservation, which is necessary for maintaining a species and multiplying the number of the individuals which belong to it. On the other hand, they give rise to new phenotypes when they occur, thus exploring at random the field of what is possible. These new phenotypes are targets of natural selection and the rough material which makes evolution possible. (The above statements are more precisely justified by computations in the analysis of the 'toy living world' introduced in Ch. 8.)

Remember that, as we noticed above, the roles of information and perturbations are relative to each other, entirely depending on their usefulness or harm for the destination in Shannon's paradigm. In genetics, the destination consists of the cell machinery which processes the genetic message borne by DNA. Therefore, the regeneration errors *create new genetic messages* in a very objective sense. There is no paradox in this statement since there is no difference of nature between information and perturbations. Because we liken the error-correcting codes to the multiple constraints which make a DNA string admissible as a genetic message (see Ch. 10 below), the DNA strings having suffered regeneration errors are just as plausible to the cell machinery as the error-free ones.

3.7 COMPUTING AN UPPER BOUND OF DNA CAPACITY

The ability of DNA to convey genetic information through time can be measured by its capacity, in the information-theoretic meaning of this word. To compute it, we need to know what kind of events impair the basic information-bearing elements, namely, the nucleic-base pairs; and at which frequency these 'error events' occur.

As regards the kind of error events which affect a nucleic-base pair, only four types need be distinguished: *substitution*: a wrong symbol replaces the correct one; *erasure*: the received symbol is recognized as foreign to the alphabet; *deletion*: a symbol is removed from the sequence; and *insertion*: a spurious symbol is inserted into it. We already gave as examples channels where substitutions and erasures occur: the binary symmetric channel and the binary erasure channel, respectively (Sec. 3.3.3; see Fig. 3.4). Error events of all four types may occur, but it is difficult to estimate in what proportion they do. Instead of precisely computing the DNA capacity, we can instead try to compute an *upper bound* on it. Information theory tells us that the mildest of all kind of errors is erasure[1] (in the above meaning) and we may, very optimistically, assume that only erasures occur.

How frequently should erasures be assumed to occur? Let the probability that an erasure occurs during the infinitesimal time interval dt be denoted by νdt, where ν is referred to as the erasure frequency. We have no reliable estimate of ν, which moreover may not be constant. (We will attempt in Sec. 7.3 to evaluate the frequency of substitutions, which seem to be the most current type of symbol error, using published biological literature and some assumptions.) For simply obtaining an upper bound on the DNA capacity, we may use a lower bound of ν, which results in further overestimating the capacity.

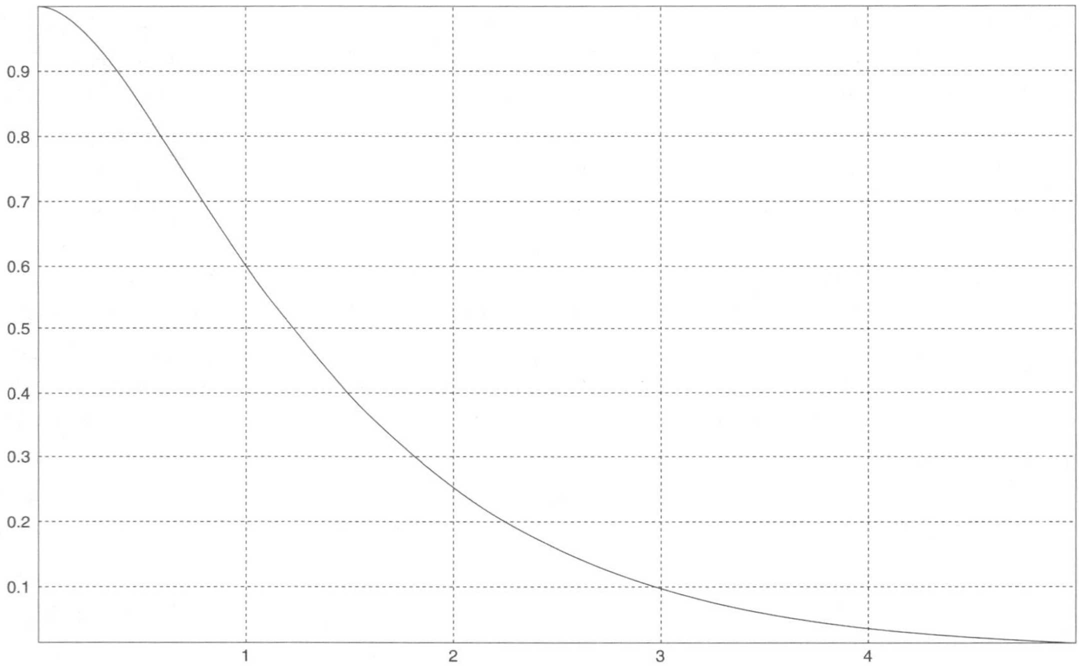

Figure 3.11: Upper bound on the DNA capacity as a function of time. The unit in the horizontal (time) axis is $1/\nu$.

The capacity of double-stranded DNA, assuming that only erasures occur at a constant frequency ν, is computed in Ch. 7, and it is represented as a function of time in Fig. 3.11. The upper bound on the capacity thus obtained is an exponentially decreasing function of the product $\tau = \nu t$, where t denotes time. This product can be interpreted as a measure of time using as unit the recip-

[1]Notice that the two words 'erasure' and 'deletion,' which are almost synonymous in the usual language, are given here very distinct meanings. Contrary to an erasure, the deletion of a symbol changes the sequence length from its initial value, say n, into $n - 1$.

rocal of the frequency, $1/\nu$. The frequency ν is actually unknown, but it cannot be extremely small, since mutations have deleterious effects at the time scale of individual lifetimes (genomes in somatic cells are presumably not strongly protected against errors, at variance with those of the germinal line). The time unit $1/\nu$ thus does not belong to the geological timescale.

We conclude from the fast decrease of the DNA capacity as time passes that mere copying does not suffice to ensure the faithful communication of genetic information. Genomes must be endowed with intrinsic error-correcting properties which enable their regeneration before the DNA capacity has become too low (remember that the capacity is an impassable limit). A genome must thus belong to an *error-correcting code*. An error-correcting code can reliably correct errors provided their cumulated number remains less than a threshold close to half its minimum distance. The conservation of a genome does not only imply that it belongs to an error-correcting code, but that the time interval between successive regenerations is as small as to keep the cumulated number of symbol errors below this threshold. The existence of an intrinsic genomic error-correcting code with a small enough time interval between successive regenerations is thus a mandatory condition of the faithful conservation of genomes, already referred to as our *main hypothesis*. A *subsidiary hypothesis* has been further introduced in order to account for the observed better conservation of very old parts of the genome (e.g., the *HOX* genes), which needs that the genomic code more than compensates the decrease of the DNA capacity. It assumes that the genomic error-correcting code actually consists of a combination of several nested component codes which have originated successively in time at the geological timescale. A new component code then does not only protect some new information against errors, but also provides further protection to the informations already encoded by the yet existing component codes. A layered structure results, where the information is the better conserved, the more central (the older) it is. Some uncoded information may be located in the most peripheral layer.

We already formulated the main and subsidiary hypotheses in Sec. 1.2.1 and briefly examined in the remainder of Sec. 1.2 their consequences on the living world and its evolution. These topics will be further developed in Ch. 11.

3.8 SUMMARY OF THE NEXT CHAPTERS

The remainder of this lecture is organized as follows. Part II is intended to complement the introduction as regards genetics and information theory. It begins with a rough overview of molecular genetics (Ch. 4). We then provide a short discussion of information theory, presented in a manner fitted to its application to genetics (Ch. 5), and then of error-correcting codes (Ch. 6). The third part establishes the need of genomic error-correcting codes and describes their impact on the living world. It begins with Ch. 7 where the information-theoretic capacity of a complementary pair of nucleotides is computed. We find that it vanishes exponentially fast as time passes, showing that DNA is an ephemeral memory at the timescale of geology. The faithful conservation of genetic information thus needs error-correction coding, our main hypothesis. A simple model referred to as 'toy living world' which illustrates this main hypothesis is introduced and quantitatively analyzed

(Ch. 8). It shows that only genomes involving error-correcting codes can be conserved at the geological timescale. We then state the subsidiary hypothesis that the genomic error-correcting code actually involves a multiplicity of component codes combined into a layered structure, referred to as *nested system*, in order to account for the better conservation of the older genomic information (Ch. 9). We introduce the concept of 'soft code', which weakens but widely extends that of error-correcting code, as plausibly describing how nature implements error correction (Ch. 10). From both the main and subsidiary hypotheses we then derive a number of consequences which are shown to properly fit actual features of the living world and of its evolution (Ch. 11). The actual mechanisms of genomic error-correction remain however to be discovered, a task which requires a close collaboration of engineers and biologists as discussed in Ch. 12 together with other open problems. Finally, Ch. 13 presents concluding remarks.

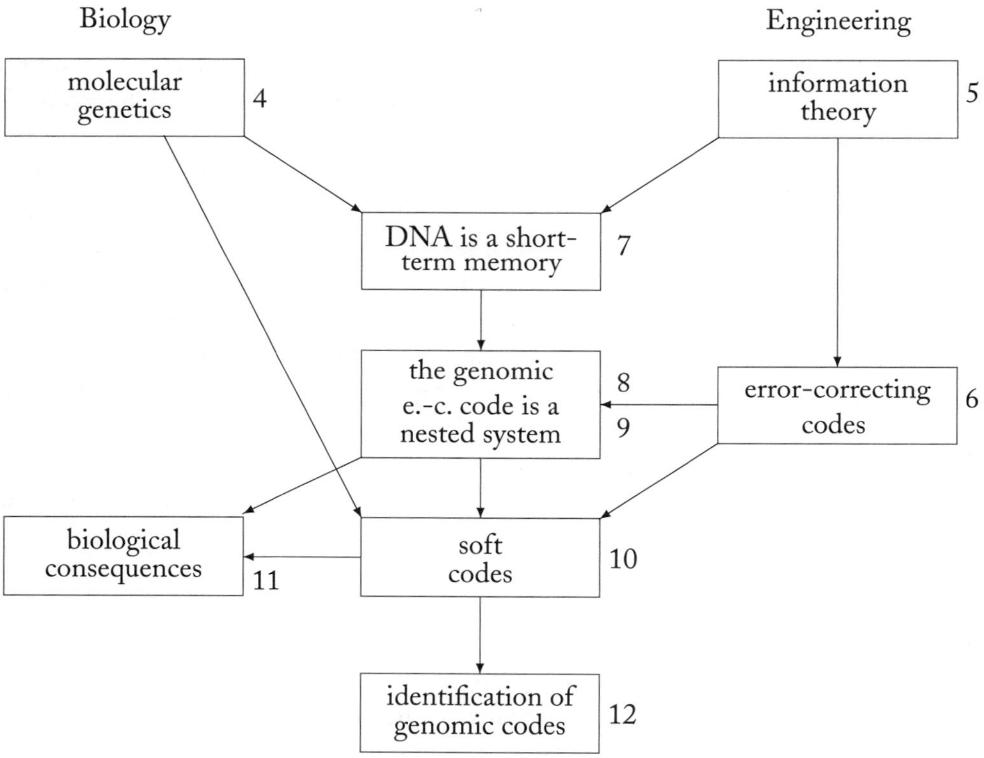

Figure 3.12: Connection between the topics of the lecture. The topic which labels a box needs as prerequisite that of the upstream boxes according to the arrows. The numbers near the boxes indicate the relevant chapter number.

Figure 3.12 schematically indicates the connection between the main topics of the lecture. It is intended to help the reader to find his/her own path in it, depending on his/her background and interest.

Part II

Facts of Genetics and Information Theory

CHAPTER 4

More on Molecular Genetics

As with biology in general, molecular genetics is a complex topic. We shall deal very roughly with it: almost all statements in this survey, besides being simplified, actually have exceptions which are not mentioned for the sake of brevity and readability; similarly, only few references to the texts used for compiling this survey are given besides those of historical significance. We shall mention only the details of chemical structure which are relevant to our purpose.

4.1 MOLECULAR MEMORIES: DNA AND RNA

4.1.1 UNIDIMENSIONAL POLYMERS AS HEREDITARY MEMORIES

Since the works of Avery et al. [2] and Chargaff [26], it is known that the hereditary information is borne by *DeoxyriboNucleic Acid* (DNA) molecules in all living beings (except for certain viruses where ribonucleic acid, RNA, performs this role). DNA has been shown by Franklin and Gosling [38] and Watson and Crick [76] to assume the famous double-helix shape.

To convey information, a molecule must be a unidimensional polymer, i.e., a chain of an arbitrary number of links, each bearing a small molecule from a given finite repertoire, or alphabet. These information-bearing molecules succeed to each other in an arbitrary order; their sequence constitutes the genetic message. Similarly, a written text is a unidimensional sequence of letters, digits, separating spaces, and punctuation marks, i.e., symbols from a given alphabet, in the order that the author of the text chose. Unidimensionality is necessary in both a genetic message and a written text, since an order must exist between its symbols (information-bearing molecules or letters): given any two symbols, one of them should either precede or follow the other one. This is necessary for establishing syntactic rules, which are themselves necessary for the written message to convey a semantic content. No such 'relation of order' exists in two-dimensional space or in spaces having a larger number of dimensions.

4.1.2 STRUCTURE OF DOUBLE-STRAND DNA

The DNA molecule consists of two unidimensional chains (or strands) each made of alternating a phosphate group and deoxyribose (a pentose, i.e., a sugar having 5 carbon atoms) where each sugar molecule is covalently bound to one of 4 nitrogenous bases, referred to as nucleic bases or nucleotides, namely adenine (**A**), thymine (**T**), guanine (**G**), or cytosine (**C**). The nucleic bases in the two strands are such that **A** in one strand always faces **T**, and **G** in one strand always faces **C**, thus appearing in the DNA molecule as *complementary pairs*. The nucleic bases **A** and **G** are purines (abbreviated to **R**), i.e., two-cycle molecules, while **T** and **C** are pyrimidines (**Y**), single-cycle molecules. The bases **A** and **T** bound to facing strands are tied together by two hydrogen bonds, while three hydrogen bonds tie together **G** and **C**. **A–T** and **G–C** are referred to as complementary pairs, each made of a purine and a pyrimidine. It will be convenient to refer to these complementary pairs as weak (**W**) for **A–T** and strong (**S**) for **G–C**. The atoms in each complementary pair are contained in a same plane. The chemical formula of both nucleotide pairs is shown in Fig. 4.1.

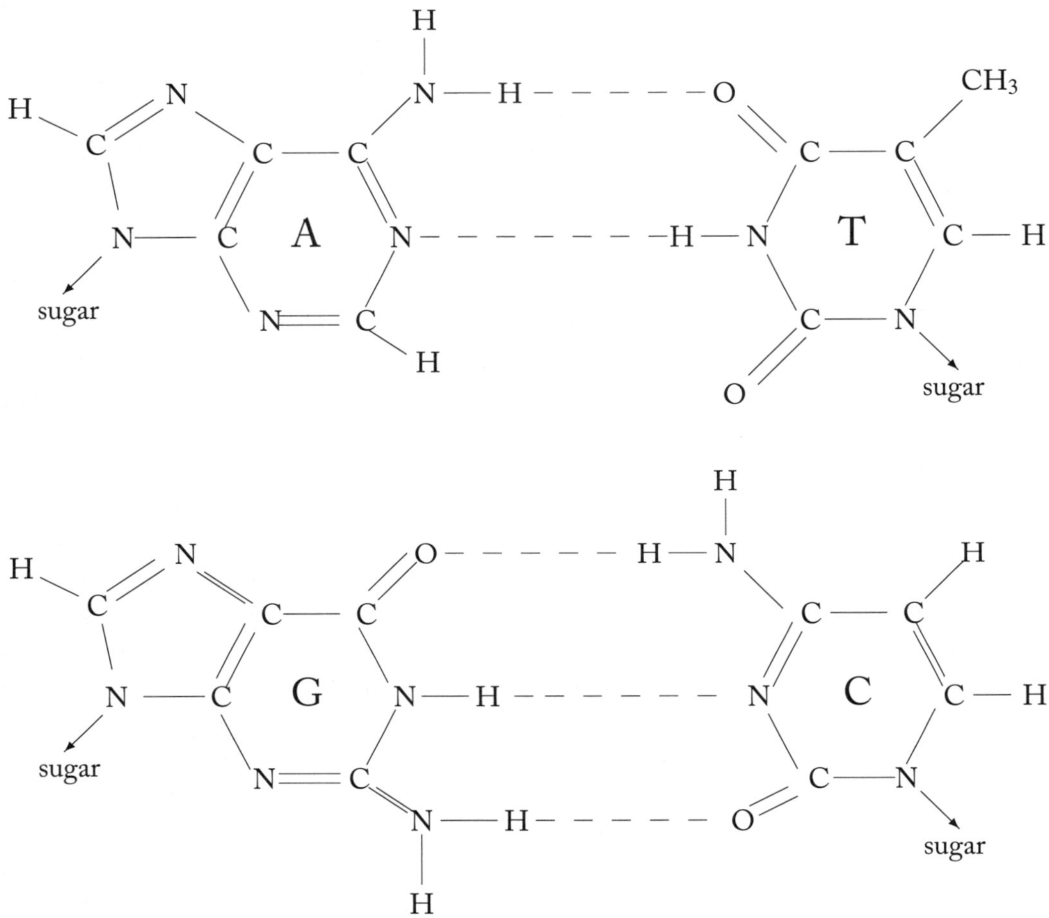

Figure 4.1: Chemical structure of the complementary base pairs **A–T** (top) and **G–C** (bottom) of DNA. The purines (**A** and **G**) are at left and the pyrimidines (**T** and **C**) at right. In RNA, the uracil molecule (**U**) replaces thymine (**T**), which is identical to it except that the methyl group CH_3 at top right is replaced by a hydrogen atom H. Dashed lines represent hydrogen bonds. The mutual distances of the atoms are roughly respected, showing that the two complementary base pairs have almost equal lengths.

The spatial length of the two complementary pairs is almost the same, which results in maintaining a constant distance between the two strands, and the angles of the covalent bonds with the sugar molecules of the backbones are also almost the same. Any complementary base pair can thus take place everywhere in the double-helix structure. The whole DNA molecule appears as a twisted ladder (a right double-helix) where the uprights are the two phosphate-sugar chains (also referred to as 'backbones') and the rungs are the pairs of complementary nucleic bases, which are located in a plane perpendicular to the axis of the double-helix. The width of the ladder is of about 2 nanometers, and the distance between two rungs of about 0.34 nanometers. A full turn of the double-helix corresponds to approximately 10 base pairs. A schematic ladder representation of the double-strand DNA molecule (ignoring the twist) is shown in Fig. 4.2.

The genetic information of a living being is borne by the sequence of nucleotides along one of the two strands of the DNA contained in its cell(s). Just like a written text is read in a direction defined by a convention (e.g., from left to right for texts in the Latin alphabet), the DNA molecule is read in the 5' to 3' direction, referring to the conventional numbering of the carbon atoms of the sugar molecule. Since the atoms of the backbones in complementary pairs are symmetric with respect to the center, the reading direction in one of the two strands of DNA is the inverse of that of the other strand.

4.1.3 RNA AS ANOTHER MOLECULAR MEMORY

The RNA molecule, very similar to DNA, is made of a 'backbone' to which nucleic bases are attached. Instead of deoxyribose, the sugar in RNA is the ribose, another pentose which differs from it by the presence of one more oxygen atom, and the nucleic base thymine (**T**) of DNA is replaced in RNA by uracil (**U**), a pyrimidine which only differs from thymine by a hydrogen atom replacing the methyl group (see Fig. 4.1). As **T**, **U** can be bound to **A** in a complementary pair. Double-strand RNA thus exists, although it is less stable than double-strand DNA. The main genetic memory in certain viruses is made of RNA. Except for these viruses, the main genetic memory of all living beings is made of DNA. RNA is more versatile: it is mainly found as an auxiliary memory in very many instances, especially in the cell replication machinery (see below). RNA has also some catalytic activity. An 'RNA world,' where RNA was acting both as a memory (like modern DNA) and as a catalyst (like modern proteins) is hypothesized to have predated the extant living world. Due to its double competence, RNA could catalyze its own replication.

4.1.4 DNA AS A LONG-LASTING SUPPORT OF INFORMATION

Regardless of its chemical and structural details, we may think of DNA as bearing the genetic message as a sequence of symbols from the quaternary alphabet {**A**, **T**, **G**, **C**}. Then heredity consists of communicating the genomic message through time, from the past to the present. We may thus deal with it as a problem of communication engineering. The DNA molecule is subject to errors due to chemical reactants and radiations, and it is moreover a quantum object in two of its dimensions, hence its deterministic description is impossible. The time scale involved is that of geology (remember that life originated on Earth about 3.5 billion years ago, and maybe even earlier), which makes communicating the genomic message a formidable problem. It turns out that its difficulty has been overlooked by biologists, who moreover failed to understand that information theory is

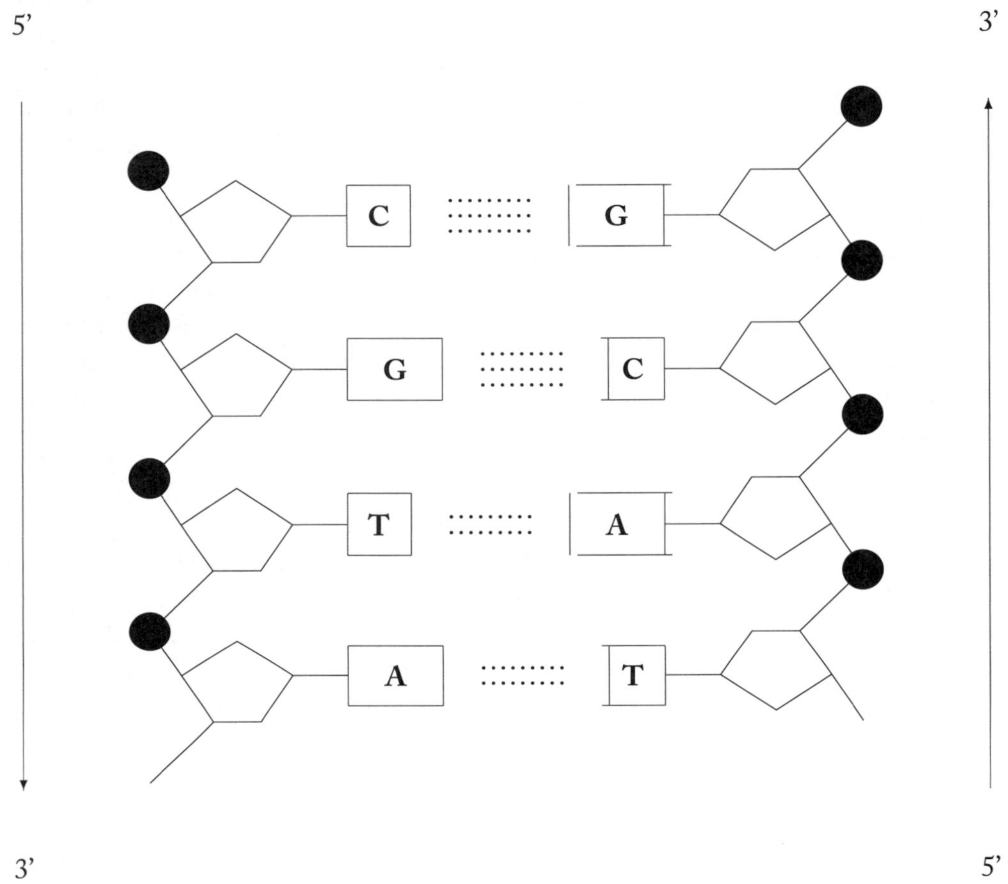

Figure 4.2: Ladder representation of double-strand DNA. The black disks represent phosphate groups and the pentagons, deoxyribose molecules (sugars). The polymers made of alternating phosphate groups and sugars are referred to in the text as 'backbones.' Rectangular boxes represent the nucleotides **A**, **T**, **G** and **C**. The dotted horizontal lines between nucleotides of opposite strands represent hydrogen bonds. The vertical arrows indicate the direction of reading.

relevant to it. Most of them believe that DNA replication according to the template-replication paradigm, i.e., splitting it into its two strands and appending to each of them its complementary sequence, suffices to faithfully communicate the genome through geological times (e.g., [29, 30]). Many DNA repair mechanisms are known to biologists. Their description is too often inaccurate in terms of communication engineering and the parameter values and performance measures given in the biological literature are doubtful. Taken together, these repair mechanisms succeed with an astonishing accuracy. It must be emphasized, however, that they concern means to avoid errors occurring *within the replication process*, a problem entirely different from making the genome resilient to casual errors: clearly, the faithful copy of a wrong original is wrong.

At the *geological time scale*, it is known that some parts of the genome (i.e., the *HOX* genes) are conserved with *great faithfulness*. Errors in DNA nevertheless occur, and moreover have an essential role in biological evolution. They are referred to as *mutations*. They occur in somatic cells at the much shorter time scale of individual life times. The blatant contradiction between the long-term faithful conservation of the genomes and the existence of mutations at a much shorter time scale has been our main motivation for undertaking the research reported in this book. The faithful communication of a genome through time demands that it is made resilient to errors, which can only be obtained by the use of error-correcting codes which enable *regenerating* it, not merely copying it, however faithfully; see Sec. 1.3 above.

4.1.5 ERROR-CORRECTION CODING AS AN IMPLICIT HYPOTHESIS

The existence of error correction mechanisms is indeed implicit in the biological literature. Biologists consider as self-obvious the fact that the mutation rate depends on the species, the locus or the nature of the gene (for instance, the *HOX* genes control the expression of other genes and determine the organization plan; since they are shared by almost all multicellular living beings, they are considered as devoid of any mutation). Biologists also admit that the mutation rate of a gene can be controlled by factors like other genes. The lowest mutation rate reported is found in nonsynonymous mutations (i.e., substitutions of a single nucleotide which change an amino-acid into another; since the genetic 'code' is a many-to-one mapping, different codons may correspond to the same amino-acid), namely 10^{-9} substitutions per year and per nucleotide in the genome of mammals. In contrast, mutations of the same kind occur at a rate of 14×10^{-3} substitutions per year and per nucleotide in a region of *env*, one of the 9 genes of the HIV virus [28]. The effects on the phenotype of a nonsynonymous mutation are highly variable. It sometimes destroys or changes the biological function of the protein, but has often a more limited effect.

Similarly, the repeated use of the rather convincing metaphor of a written text to explain the role of the genome by Dawkins [30] and many others implies the existence of genomic error-correcting codes. Since a language is subject to many constraints at several levels (morphological, grammatical, semantic, etc.), it may rightfully be interpreted as a kind of highly redundant error-correcting system. Interestingly, this several-level system resembles the nested system we assumed according to our subsidiary hypothesis. Moreover, we find the same kind of object (to be referred to as 'soft code' in Ch. 10) when trying to guess what the error-correcting codes we hypothesized look like.

4.2 PLACE AND FUNCTION OF DNA IN THE CELL

4.2.1 CHROMOSOMES AND GENOMES

The double-strand DNA sequences are combined with certain proteins (histones) and more or less closely packed in large aggregates named *chromosomes*. The number of chromosomes depends on the species, e.g., a bacterium has a single circular chromosome but the living beings higher in the evolutionary scale generally have several rod-shaped ones. For instance, the somatic (nonsexual) human cells contain 2×23 chromosomes. The sequence of base pairs in the whole set of chromosomes of a living being is referred to as its *genome* and the total number of base pairs it contains, as the genome length. The genome length varies from about a million base pairs for certain bacteria up to billions in eukaryotic beings. Genomes are thus extremely long (e.g., the total length of double-strand DNA in each somatic human cell, which contains two copies of the genome, is about two meters; much longer genomes moreover exist). Viruses generally have shorter genomes, typically of a few thousands base pairs and, as stated above, some of them are made of RNA. There is clearly a considerable variation in the genome length from a living being to another. The longest genomes are not related to the highest complexity beings: as the most striking example, the unicellular *Amoebia dubia* has the longest known genome, more than 200 times longer than the human one [73, 28, 47]. In *prokaryotes* like bacteria, the genome is not physically separated from the remainder of the cell, whereas in *eukaryotes* (all multicellular organisms and certain unicellular ones like yeasts or amiboeas) it is contained in the *nucleus*, a region separated from the remainder of the cell by a membrane. The eukaryotic cell is much larger and more complicated that the prokaryotic one and probably younger by almost 2 billion years. Being contained inside the cell (for prokaryotes) or the cell nucleus (for eukaryotes), which are both a few micrometers long, DNA is thus very densely packed, although variably along chromosomes: less densely in euchromatin than in heterochromatin. In the nucleus of eukaryotic cells, the double-stranded DNA is packed into nucleosomes, which consist of a histone octamer acting as a spool with almost two turns (about 165-nucleotide long) of double-stranded DNA wrapped around it. (Histones are proteins associated with DNA in chromosomes.) Moreover, nucleosomes themselves are packed together in higher order structures generally referred to as 'fibers.'

4.2.2 PRINCIPLE OF DNA REPLICATION

The whole genome can be faithfully reproduced. The hydrogen bonds which tie together the nuclotides of complementary pairs are much weaker than the covalent bonds which bind the nucleotides to the backbones as well as the atoms of the backbones themselves. Splitting the double-strand DNA into two single strands is thus easy (requires few energy), and the nucleotides of each single-strand DNA can be endowed with their complementary nucleotides so as to restore two double-strand DNA molecules, identical in principle to the initial double-strand one. This is the *template-replication paradigm*. In the transcription process (see below), one of the strands is used as a template to generate a messenger (or premessenger) RNA molecule which is a copy of the other strand, except that uracyl, **U**, replaces thymine, **T**.

4.2.3 GENES INSTRUCT THE SYNTHESIS OF PROTEINS

Besides its own replication, the genome also directs the construction of a *phenotype*. The best understood function in this respect is the synthesis of *proteins* which are the main constituents of living matter and actors of life processes. Whereas replication concerns all the DNA present in the cell in the case of a prokaryote, or its nucleus for a eukaryote, only specific regions of the DNA strands can direct the synthesis of proteins. They are named *genes*. Each of them is precisely located at some *locus* in a chromosome. Non-identical genes which can be found at a same locus in different individuals of the same species are referred to as *alleles*. Proteins and other genes inhibit or enable the synthesis of the protein for which a given gene 'codes' (its *expression*). In multicellular organisms, cells are highly specialized so the on/off switching of the gene expression at a given locus depends on the cell type and history, although each cell contains the whole genome.

The fraction of the genome actually made of genes, hence actually used for synthesizing proteins, is greatly variable, from about 100% in bacteria to 0.02% or less in beings like fritillary or lungfish which have a very long genome. It is of about 1.3% in the human genome [73, 28, 47, 4]. The number of genes varies from about 5,000 in bacteria to about 29,000 in humans (the exact figure is not yet precisely known, all the more since there is no agreement on the gene definition [47, 4]). There is thus much less variation in the number of genes than in the length of genomes of living beings.

The synthesis of a protein results from the gene directing the assembly of a sequence of amino-acids (the polypeptidic chain) to be later properly folded. This process is performed in two successive steps: 'transcription' and 'translation'. Transcription consists of copying the gene into an RNA string referred to as *messenger RNA* (mRNA). Translation then consists of reading the mRNA in an organelle named *ribosome* where triplets of successive nucleotides, referred to as 'codons,' determine amino-acids according to a rule which is referred to as the *genetic 'code'*[1]. Each codon gives rise to a short molecule of 'transfer RNA' which binds itself to the amino-acid which corresponds to this particular codon according to the genetic 'code.' Figure 4.3 is a chart of the genetic 'code.'

4.2.4 AMINO-ACIDS AND POLYPEPTIDIC CHAINS

An amino-acid is made of a carbon atom (referred to as C_α) which is bound to a hydrogen atom, an amino group (NH_2), a carboxy group (COOH), and a specific side chain. The amino-acids which correspond to successive codons are linked together by binding the carboxy group of an amino-acid and the amino group of the other one, with elimination of a water molecule (see Fig. 4.4). The order of the amino-acids is thus determined by that of the codons in the messenger RNA. A 'polypeptidic chain' results. It becomes a protein after its proper folding.

Table 4.1 lists the 20 amino-acids which constitute polypeptidic chains. They differ from each other as regards their side chain. Depending on the properties of these side chains, the amino-acids can be put into four main classes: hydrophobic (abbreviated as h-phobic), hydrophilic (h-philic), acid

[1] In the vocabulary of information theory, it is an encoding rule rather than a code *stricto sensu*. We are led to assume the existence of a *genomic code* in a quite different meaning, so when referring to it we always use quotes.

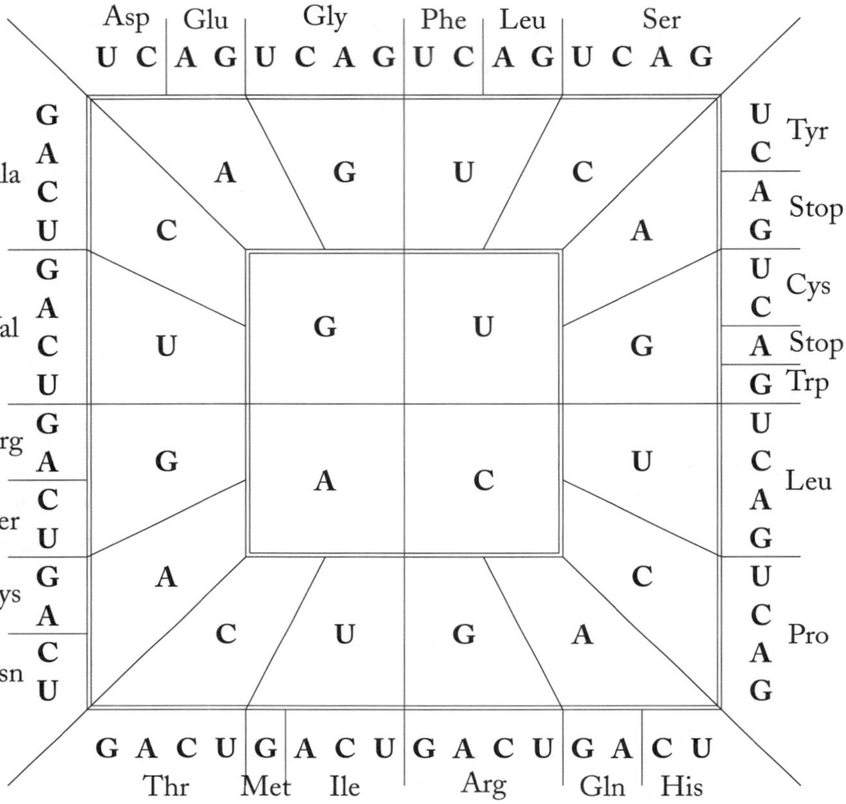

Figure 4.3: The genetic 'code'. This chart indicates how the codons of messenger RNA specify amino-acids or stop the synthesis of a polypeptidic chain. The letters **A**, **C**, **G**, and **U** denote the nucleic bases of RNA. Any codon is a three-nucleic-base word read from center to periphery. Each three-letter acronym at the periphery indicates to which amino-acid a codon corresponds, if it is not a stop command. For instance, **UAC** 'codes' for the amino-acid tyrosine (Tyr), while **UAA** stops the translation process.

(bearing a negative electric charge), and base (bearing a positive charge). Except for glycine which has a single hydrogen atom as side chain, they all are chiral molecules (i.e., they can assume two forms which are symmetric with respect to a plane in the three-dimensional space, like an object and its image in a mirror), but only their L-form occurs in proteins. The reason why the amino-acids listed above were 'chosen' rather than others is unknown, as well as why the L-form has been 'preferred.' Cystein, whose side chain is the group SH, can form disulphur bridges with another cystein molecule close to it in the three-dimensional space due to the protein folding.

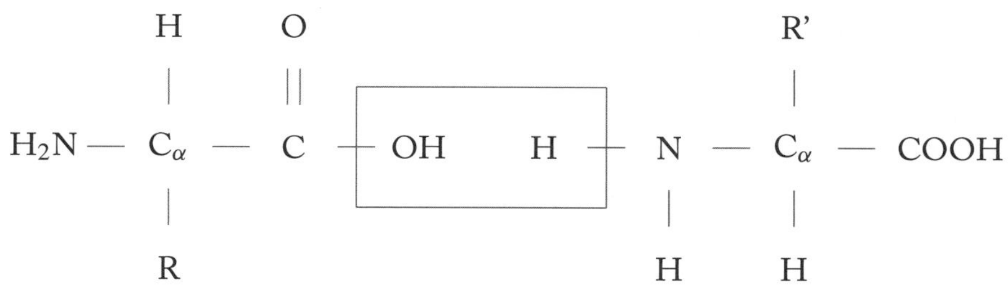

Figure 4.4: Two amino-acids and their linking into a polypeptidic chain. R and R' denote their side chains. Their linking results from eliminating the water molecule shown inside the box.

Table 4.1: Amino-acids which constitute polypeptidic chains

acr.	name	class	#	acr.	name	class	#
Met	methionine	h-phobic	1	Tyr	tyrosine	h-philic	2
Trp	triptophan	h-phobic	1	Ile	isoleucine	h-phobic	3
Asp	aspartic acid	acid	2	Ala	alanine	h-phobic	4
Glu	glutamic acid	acid	2	Gly	glycine	h-phobic	4
Asn	asparagine	h-philic	2	Pro	proline	h-phobic	4
Cys	cysteine	h-phobic	2	Thr	threonine	h-philic	4
Phe	phenylalanine	h-phobic	2	Val	valine	h-phobic	4
Gln	glutamine	h-philic	2	Arg	arginine	base	6
His	histidine	base	2	Leu	leucine	h-phobic	6
Lys	lysine	base	2	Ser	serine	h-philic	6

4.2.5 SYNTHESIS OF A POLYPEPTIDIC CHAIN

In eukaryotic genes, DNA transcription is not directly followed by translation. It results in a 'pre-messenger RNA' which has to be further transformed before it directs the synthesis of a polypeptidic chain. Some parts of it are cutout and the remaining parts are attached together, resulting in the mature messenger RNA. This intermediate step is named 'splicing.' The removed parts (from the premessenger RNA, or the corresponding parts of DNA) are referred to as 'introns,' and the remaining ones as 'exons.' Only exons thus contribute to the synthesis of proteins. A further complication results from the possible rearrangement of the spliced segments of RNA according to several distinct patterns, resulting in possibly different polypeptidic chains, hence in different proteins. The rule formerly believed to be true 'a gene, a protein' is then violated and a single gene then generates several proteins.

The process of transcription is performed by an enzyme which reads one of the DNA strands in a well defined direction (5' to 3'), just like a reading head reads a magnetic tape. Translation similarly proceeds in the same direction on the messenger RNA (the 'reading head' is then the ribosome). Particular sequences referred to as 'promoters' are first recognized which warn that the translation will have to take place farther. Then, it actually starts when the particular codon **AUG**, which 'codes' for the amino-acid methionine, is met. The position of the codon **AUG** also indicates the 'reading frame,' i.e., how the mRNA strand has to be segmented into three-nucleotide codons. The translation continues until one of the three stop codons **CAA**, **CAG**, or **CGA** is found.

4.2.6 PROTEINS

A protein results from the proper folding of a polypeptidic chain, which results itself from the translation of a messenger RNA into a sequence of amino-acids. Typically, a protein is made of a few hundreds of amino-acids, each of them being one of the 20 amino-acids listed in Tab. 4.1.

A protein results from a *polypeptidic chain* by appropriate foldings which give it a unique three-dimensional shape determined by chemical affinities of the side chains of its constituents, especially hydrogen bonds and disulphur bridges, which result in establishing bonds between amino-acids at a distance along the polypeptidic chain. The biological properties of a protein heavily depend on its spatial structure (e.g., anomalous folding of the prion is responsible for the BSE disease). A protein combines several substructures like α-helices and β-sheets, which themselves are combined into higher order structures named 'domains.'

Proteins are the most important constituents of the living matter, having both a structural and catalytic role. When they act as catalysts, they are referred to as *enzymes*. DNA replication, especially, as well as DNA transcription and translation, are catalyzed by proteins. Therefore, proteins catalyze their own synthesis. Most enzymes have a roughly globular shape with an 'active site' where the catalyzed reaction takes place.

4.3 GENOME AND PHENOTYPE

4.3.1 A GENOME INSTRUCTS THE DEVELOPMENT AND MAINTENANCE OF A PHENOTYPE

The sequence of protein syntheses in a reproductive cell like a fertilized egg results, if the environment is favourable enough, in the development of a *phenotype* through very complex processes which are still poorly understood. During these processes the genes are expressed at the appropriate place and time.

The phenotype hosts and protects the genome. Both together constitute a living being. The ability of the genome to reproduce itself implies that an initial DNA chain can generate other ones, identical to it in the absence of replication errors, which will give rise to other phenotypes. An initial genome can thus generate a population of ideally identical living beings (except for replication errors), its children, which themselves generate grandchildren, etc. Any descendant ideally bears the same information as the initial genome although it was born later so we may think of this property as *communication through time*. The hereditary features of living beings are solely determined by the information borne by the genome thus communicated through time. There is evidence that the origin of life has been a single event so the present beings bear (as genomes), and are made according to (as phenotypes), an information which dates back to geological times. Remember that the age of the Earth is about 4.6 10^9 years and the very beginning of life probably occurred about 3.5 10^9 years ago and maybe earlier.

4.3.2 A PHENOTYPE HOSTS THE GENOME FROM WHICH IT ORIGINATES

A phenotype contains the genome from which it originates and provides the environment and machinery necessary to its maintenance and replication. Its membranes shield it against mechanical and chemical aggressions. The genome cannot be conserved outside the phenotype. In prokaryotes like bacteria, the genome is contained inside the cell, which is separated from the outside by a single membrane. In eukaryotes, a second membrane delimitates inside the cell a nucleus which contains the genome, which is thus separated from the outside of the cell by two membranes. Membranes do not perfectly shield genomes against damages of all kinds hence errors may occur. Errors may also result from inaccuracies in the replication process, although many 'proof-reading' mechanisms almost always result in a faithful copy, but even a perfect replication faithfully reproduces the changes suffered by the genome itself hence does not prevent errors of other kind.

According to the *central dogma of molecular biology*, the role of memory is exclusively devoted to DNA. It states that information can never go from proteins to DNA although the polypeptidic chain from which a protein originates, too, is a sequence of elements belonging to a set of a few molecules. Not in contradiction with this 'dogma,' there exists a process of *reverse transcription* such that DNA can be synthesized from RNA. It is believed to have generated a sizeable portion of the genome in many organisms, including the human one [4]. It appears as a tool for lengthening the genome by repeating some parts of it. The central dogma excludes any influence of the phenotype on the genome so it rules out the Lamarckian type of evolution according to which characters acquired by phenotypes could be inherited. Evolution can thus only be Darwinian i.e., genomes suffer variations, resulting in a number of phenotypes which are the target of natural selection. The only remaining genomes are those associated with the surviving phenotypes.

4.4 DNA RECOMBINATION AND CROSSING OVER

In higher beings with sexual reproduction, e.g., in mammals, the somatic cells are diploid, i.e., each one contains two homologous copies of each chromosome, inherited from each of their parents, at variance with their sexual cells (gametes) which are haploid, i.e., contain a single copy of each of the chromosomes. The process of gamete formation, referred to as *meiosis*, thus implies that a diploid cell with two copies of each of its chromosomes gives birth to two gametes having each a single copy of each chromosome. This single copy is actually a mixture of the two parental copies, made of alternating segments from either the paternal or the maternal copy. The choice of these segments seems to be random. The process which assembles these segments into a new chromosome is referred to as *crossing-over*. The alleles of both parental copies of each chromosome are thus combined into a new sequence (remember that alleles are different alternative versions of a same gene). Fertilization will eventually restore a diploid cell, the fertilized egg, with two copies coming from each of the gametes, each of them being itself such a mixture of segments inherited from two parents. Meiosis and fertilization together result in generating new genomes by combining alleles. Each of these genomes may be considered as unique because the number of possible combinations of alleles is immense.

The ability of double-strand DNA to recombine is an important property which is not limited to the crossing-over which takes place at meiosis. If two double-strand DNA strings bear close enough sequences of base pairs, they can similarly exchange segments of genetic material when put together (see Ch. 14 in [37]).

CHAPTER 5

More on Information Theory

This chapter gathers sections intended to complement the overview of information theory presented in Ch. 3. Examples will often be given using the binary alphabet. Remember that we use the word 'bit,' as an abbreviation of 'binary digit,' in order to designate the symbols of the binary alphabet and never as meaning the binary information unit which we name 'shannon' (see Sec. 3.3.1).

5.1 ALPHABET, SOURCES, AND ENTROPY

An information is communicated by causing the occurrence of an event belonging to some predetermined set of M possible events which can be recognized from each other. According to the quantitative measure of information introduced in Sec. 3.3.1, a single occurrence brings an information quantity of at most $\log_2 M$, a finite quantity regardless how large is M. The solution for communicating an unlimited quantity of information consists of causing the occurrence of *successive* events, the number of which is unlimited. If the information-generating event is the choice of a particular symbol among an alphabet of size q, $M = q^n$ possible distinct messages result from n successive such choices, providing the quantity of information $\log_2 M = n \log_2 q$ shannons which is unlimited as proportional to the natural number n. An unlimited information quantity of n shannons is obtained even for the smallest possible alphabet size $q = 2$. Moreover, the necessary condition that the events be recognizable is more easily fulfilled for small values of q. Notice that the symbol choices need be successive, which implies their ordering, hence that they occur within a unidimensional space. It is so for ordinary communication (from a point of space to another one) since in this case the information bearing-events successively occur in time. Using more than one dimension would make this ordering impossible. As a matter of fact, heredity as well as the written human language use unidimensional sequences although they could conceivably use the available three spatial dimensions of physics in order to communicate through time. Moreover, the very existence of grammatical rules and semantics needs ordering of symbols in the case of the human language and, very probably, of DNA-borne heredity.

5.1.1 MEMORYLESS SOURCES, MARKOVIAN SOURCES, AND THEIR ENTROPY

We introduced in Ch. 3 a simple example of Markovian source. We now give a more comprehensive definition of such a source.

Figure 5.1 represents two types of sources: a memoryless source (Fig. 5.1-**a**) and a Markovian source (Fig. 5.1-**b**). A q-ary memoryless source chooses repeatedly (for instance, at each pulse of a periodic clock) a symbol from an alphabet of size q with constant probabilities $p_1, p_2, \ldots, p_q, p_i$

being the probability that i is chosen (without loss of generality, we denote the alphabet symbols by $1, 2, \ldots, q$); we assumed $q = 2$ in the figure and we denoted the probability of choosing 1 by p, so that of choosing 0 is $1 - p$). These probabilities sum up to 1, i.e., $\sum_{i=1}^{q} p_i = 1$, and are assumed not to vary with respect to time, so this source is said to be *stationary*. The information quantity associated with the choice of symbol i is $h(i) = -\log_2(p_i)$ shannons, so the average amount of information associated with the source is the mean (or expectation) of this quantity, namely

$$H = -p_1 \log_2(p_1) - p_2 \log_2(p_2) - \ldots - p_q \log_2(p_q) = -\sum_{i=1}^{q} p_i \log_2(p_i) , \tag{5.1}$$

referred to as the entropy of the source. It is well defined for a stationary source. For historical reasons, Shannon chose the name 'entropy' which was already that of a fundamental thermodynamic entity, but the relation between both is not straightforward and will not be discussed here. The information $h(i) = -\log_2(p_i)$ is a positive quantity since a probability is less than or equal to 1, hence its logarithm is negative. H is thus positive except if one of the outcomes has probability 1; in this case $H = 0$ so, in more rigorous terms, H is actually nonnegative. It is easy to show that the maximum $H_{\max}$ of H is obtained when $p_1 = p_2, = \ldots = p_q = 1/q$ and thus equals $\log_2(q)$ shannons. The difference $r = H_{\max} - H$ between this maximum and the actual source entropy measures the *redundancy* of the source.

A Markovian source has been represented in Fig. 5.1-**b**. It combines two elements: a memoryless source and an m-symbol memory. This memory records the last m symbols generated by the source and can thus assume q^m distinct states. The probabilities of symbol choice of the associated memoryless source now depend on the state s of the memory: symbol i is chosen with probability $p_i(s)$. This scheme can generate much more complex sequences than a memoryless source since they depend on the q^m probability distributions on the alphabet symbols associated with the memory state, instead of a single one. The probabilities $p_i(s)$ are constant for all possible values of i and s so the source is again stationary. The above definition of the entropy needs to be extended. The entropy of the source when it is in state s is expressed according to Eq. (5.1) as:

$$H(s) = -p_1(s) \log_2[p_1(s)] - p_2(s) \log_2[p_2(s)] - \ldots = -\sum_{i=1}^{q} p_i(s) \log_2[p_i(s)]$$

and the overall entropy of the source is defined as the average of this quantity with respect to all possible states, namely:

$$H = \sum_{s=0}^{q^m - 1} \Pr(s) H(s) , \tag{5.2}$$

where $\Pr(s)$ is the steady probability of state s and $H(s)$ has just been defined. Notice that the entropy defined by Eqs. (5.1) or (5.2) is an entropy *per symbol*. The entropy of a sequence of length n, which measures the average information it bears, is thus nH.

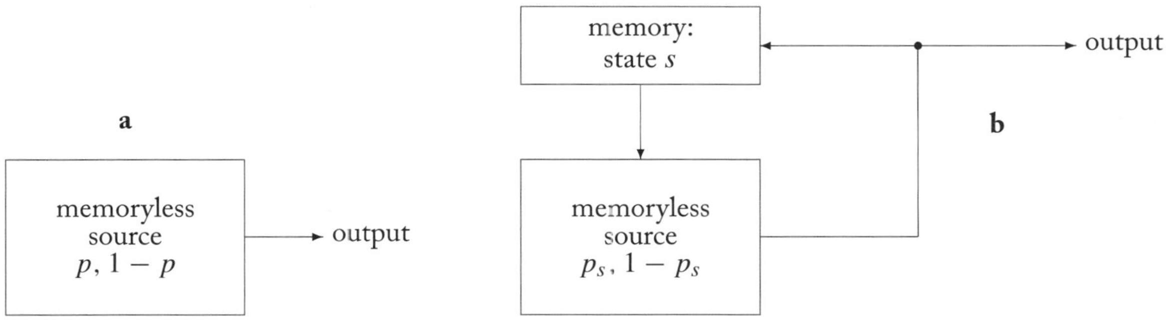

Figure 5.1: Two examples of binary sources. **a**: memoryless source. Its output is 1 with probability p, 0 with probability $1 - p$. **b**: Markovian source. The last m output bits are recorded in the box labeled 'memory,' and the memoryless source below (identical to **a**) outputs 1 with probability p_s (0 with probability $1 - p_s$) where probability p_s depends on the state s, $s = 0, 1, \ldots, 2^m - 1$, of the memory.

A detailed example of Markovian source is given in Appendix 5.A below. Less simple sources can be defined, but the memoryless and Markovian sources will be sufficient for us to understand the main properties of sources in general.

The concept of Markovian source is also useful to illustrate the important notion of *ergodicity*. If all the probabilities $p_i(s)$ are different from 0 or 1, each of the possible states of the memory is sooner or later visited and observing the generated sequence enables in principle to assess the frequency of occurrence of each state, meaning that accurate estimates of the probabilities $p_i(s)$ can be obtained when the observation time is long enough. This may be thought of as a consequence of the (weak) law of large numbers. The source is then said to be *ergodic*: the indefinitely lengthened observation of the source output provides sufficient knowledge of it. Let us assume on the contrary that one of the $p_i(s)$ equals 0 or 1, for instance $p_0(0) = 1$. Then, once the 0-state is reached, the source only transmits zeros. The uninformative all-0 sequence is then generated, and the existence of states other than the 0-state cannot be deduced from observing the generated sequence. In other words, the all-0 state acts as a trap which hinders the source evolution towards other states. Such a source is nonergodic. *A contrario*, the beginning of the operation of an ergodic source, which implies some initial state, is assumed to have occurred in an arbitrarily remote past, making the observed source operation independent of the initial memory content. In general, a Markovian source is ergodic if there is an uninterrupted path between any two of its states, 'uninterrupted' meaning that no transition between the states of the path has zero probability. Although stationarity and ergodicity are often associated, there is no relationship between these two concepts.

For more general sources (i.e., which are neither memoryless nor Markovian), the entropy can be defined as the following limit, for n approaching infinity:

$$H = - \lim_{n \to \infty} \frac{1}{n} \sum \Pr(\underline{s}) \log_2[\Pr(\underline{s})] \tag{5.3}$$

where $\underline{s}$ denotes a sequence of length n and $\Pr(\underline{s})$ its probability; the sum is extended to all possible sequences of length n that the source can generate. This limit exists provided the source is stationary.

5.1.2 A FUNDAMENTAL PROPERTY OF STATIONARY ERGODIC SOURCES

Let us consider the set of all sequences of length n that a stationary ergodic source can generate, n being an arbitrarily large integer. According to the *Shannon-McMillan theorem*, these sequences can be divided into two mutually exclusive groups: that of *typical sequences*, each of which occurring with a probability close in some sense to 2^{-nH}, where H is the entropy as defined by Eq. (5.2); and the sequences which are not typical and thus referred to as *atypical*. The theorem states that the probability that an atypical sequence occurs vanishes as the sequence length n approaches infinity. Only typical sequences thus occur with a nonnegligible probability, which is said for short *almost always*. This result is far from trivial, since the total number of typical sequences is of about 2^{nH}, which shows that they are a minority among all the $q^n = 2^{n \log_2(q)} = 2^{nH_{\max}}$ *a priori* possible sequences for an alphabet of size q. The last equality is true because the maximum possible value of the entropy, for this alphabet, is $H_{\max} = \log_2(q)$. The ratio of the number of typical sequences to that of all possible ones is $2^{n(H-H_{\max})}$, or $1/2^{nr} = 1/(2^n)^r$, if we define the redundancy r as $r = H_{\max} - H$. The redundancy then appears as measuring how few are the typical sequences with respect to all possible ones. The proof of this theorem is rather easy for Markovian sources, but it involves somewhat lengthy calculations, so we omit it here. It can be found in any textbook on information theory, e.g., [27]. Moreover, this theorem holds true for any stationary ergodic source, whether it is Markovian or not.

Let us compare a redundant source as just described with an error-correcting code. We found that the sequences of length n generated by the source can be divided into two mutually exclusive groups: that of typical and atypical sequences. Almost all generated sequences belong to the first group, although they are only a minority among the possible q^n sequences of length n. We also defined an error-correcting code of length n as a strict subset of sequences, say of q^k sequences with $k < n$, among the possible q^n sequences of length n. The two previous sentences have the same meaning except that using the word 'almost' in the first one expresses a reservation which becomes less and less relevant as the sequence length increases. The set of sequences generated by a redundant source is thus clearly similar to an error-correcting code, which may be thought of as an information-theoretic justification of the concept of 'soft code' to be introduced in Ch. 10. The major difference is that the mutual Hamming distances between the sequences generated by a redundant source may not be favorable to error correction as not being designed for this purpose, but codes chosen at random are known to be good error-correcting codes (see Sec. 3.5.1). Then soft codes defined by constraints not intended to error correction are nevertheless likely to be good for this purpose[1]. Moreover, the soft codes of biology result from an evolutive process: they have been subjected to natural selection so it is likely that those which survived are good.

5.2 ABOUT SOURCE CODING

Although it is unlikely that source coding is directly implemented in genomes, it is interesting in order to simply introduce some important information-theoretic concepts. They will especially be useful for understanding some aspects of the algorithmic information theory (see Sec. 5.4).

5.2.1 SOURCE CODING USING A SOURCE EXTENSION

The k-th extension of a q-ary source S, denoted by S^k, is the source which results from grouping the successive symbols of its output messages by blocks of k symbols and dealing with them as symbols of a q^k-element alphabet. Notice that the k-th extension S^k of a source S is an alternative way of describing it, not a different source.

For instance, let a memoryless binary source generate the message

$$000\ 100\ 101\ 101\ 011\ 100\ 110\ \ldots$$

The corresponding sequence generated by its third extension is

$$\underline{0}\ \underline{4}\ \underline{5}\ \underline{5}\ \ \underline{3}\ \underline{4}\ \underline{6} \ldots,$$

the symbols of which now belong to the 8-element alphabet $\{\underline{0}, \underline{1}, \ldots, \underline{7}\}$. (They are denoted by underlined digits in order to distinguish them from the symbols of the original source.) Each of its symbols is represented by the corresponding block in natural binary numeration. Let us assume for instance that a memoryless binary source, say $\mathcal{S}$, generates the symbol '0' with probability p_0 and symbol '1' with probability $p_1 = 1 - p_0$. Then, its third extension $\mathcal{S}^3$ generates symbols $\underline{0}$ to $\underline{7}$ with probabilities $p_0^{k-w} p_1^w$ where w denotes the weight of the corresponding 3-bit block of $\mathcal{S}$, i.e., the number of ones it contains (here and in the sequel we use the word 'bit' as an acronym for 'binary digit'). For instance, if $p_0 = 0.9$ in $\mathcal{S}$, the probabilities associated with the symbols of $\mathcal{S}^3$ are 0.729 for $\underline{0}$, 0.081 for $\underline{1}$, $\underline{2}$ and $\underline{4}$, 0.009 for $\underline{3}$, $\underline{5}$ and $\underline{6}$, and 0.001 for $\underline{7}$.

The k-th extension of a source can be used in order to perform source coding. It basically consists of replacing the blocks of the k-th extension by *words* the shorter, the most probable is the corresponding block. A *variable length code* results. Using binary words, we may for instance encode the third extension $\mathcal{S}^3$ defined above according to the code defined by the following table. Then a single-bit word is associated with the most probable symbol of the 3-rd extension, which in the above example is $\underline{0}$, and words with other symbols the longer, the less probable they are. Using this code results in an average length of 1.598 per triplet of the original source, meaning that the length of the message is reduced by the factor $1.598/3 = 0.53266\ldots$

[1]This sounds paradoxical, but the difficulty with randomly chosen codes is the prohibitive complexity of their decoding. The soft codes can be useful for error correction if the constraints they obey favor their easy decoding.

Table 5.1: A variable length code for source coding of $\mathcal{S}^3$ as defined in the text

symbol	codeword	prob.
$\underline{0}$	0	0.729
$\underline{1}$	100	0.081
$\underline{2}$	101	0.081
$\underline{3}$	11100	0.009
$\underline{4}$	110	0.081
$\underline{5}$	11101	0.009
$\underline{6}$	11110	0.009
$\underline{7}$	11111	0.001

To be useful, a variable length code must enable uniquely separating the codewords, i.e., recognizing the beginning and the end of each of them when they are transmitted successively. A variable length code endowed with this property is said *uniquely decipherable*. Several means to design a code with this property exist, including the trivial use of a symbol exclusively intended to mean the word separation (as the space used in written human language). An especially interesting family of uniquely decipherable codes are the *instantaneous codes*, also referred to as *irreducible*. (The word 'instantaneous' means that a sequence which is a word of the code is recognized as such as soon as it has been received.) These codes obey the constraint that no codeword begins with a sequence which is another codeword, referred to as the 'prefix condition' for no codeword may be a prefix to another one. One easily checks that this condition is actually fulfilled in the code given as an example in Tab. 5.1. This code was actually designed using the Huffman algorithm, which results in an instantaneous code having the minimum possible average length [46].

5.2.2 KRAFT-MCMILLAN INEQUALITY

Let an instantaneous code consist of N codewords of respective lengths $\ell_1, \ell_2, \ldots, \ell_N$ (in nondecreasing order) and q denote the alphabet size. This code then obeys the Kraft inequality

$$\sum_{i=1}^{N} q^{-\ell_i} \leq 1 . \qquad (5.4)$$

Conversely, if N integers $\ell_1, \ell_2, \ldots, \ell_N$ obey (5.4), an instantaneous code exists which has these integers as word lengths.

This inequality can be given a very simple proof. It relies on a graphical representation of all the q-ary words of length ℓ_N by means of a *tree*. Such a tree is depicted in Fig. 5.2. We assumed for drawing it $q = 2$ and $\ell_N = 4$; q branches diverge from a point referred to as the root of the tree. The other extremities of these branches are referred to as the 1-st level nodes. Then q branches diverge

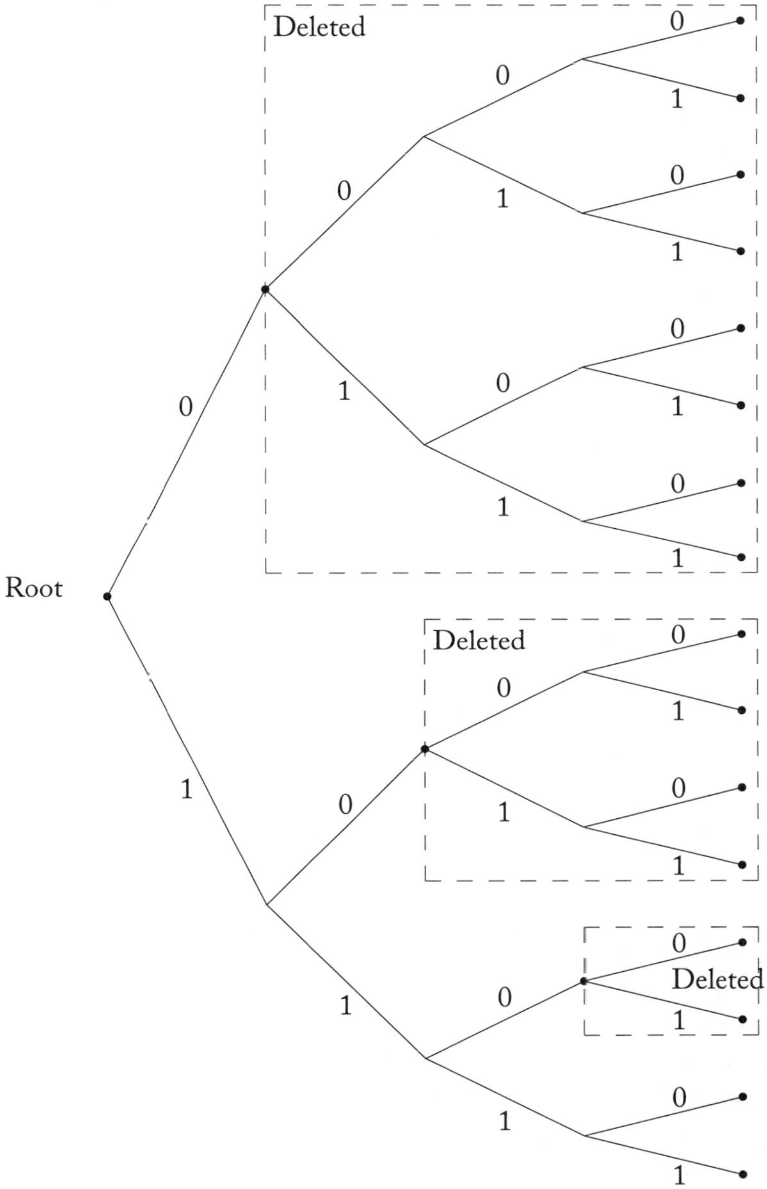

Figure 5.2: Binary tree of length 4. An ascending branch represents the bit 0 and a descending one the bit 1. Deletion of the partial trees as indicated in the figure results in the tree representing the five-word irreducible code {0,10,110,1110,1111}.

again from each of the 1-st level nodes and their other extremities are referred to as the 2-nd level nodes. This process is repeated until the ℓ_N-th node level is reached. Clearly there are q^{ℓ_i} nodes of level ℓ_i. Each branch in the tree is labeled with one of the alphabet symbols. The q^{ℓ_N} nodes of level ℓ_N are referred to as 'terminal nodes.' Each path along the branches which connect the root to a terminal node is labeled with one of the q^{ℓ_N} sequences of q-ary symbols of length ℓ_N, with a one-to-one correspondence between the terminal nodes and these sequences. Assume now that a first word of length $\ell_1 < \ell_N$ is chosen as belonging to an instantaneous code. It corresponds to it a unique path from the root to some node of level ℓ_1. Then the prefix condition forbids that sequences beyond this node be included in other codewords, meaning that the partial tree having this node as root and terminal nodes common with those of the original tree must be deleted in order to account for the choice of the first word of the instantaneous code. This entails that the terminal nodes of the partial tree, hence $q^{\ell_N-\ell_1}$ terminal nodes of the original tree, are deleted. Choosing another word of length ℓ_2 similarly entails that $q^{\ell_N-\ell_2}$ of the remaining terminal nodes are further deleted, and this process is repeated. The number of deleted terminal nodes of the original tree cannot exceed their total number q^{ℓ_N}, hence:

$$\sum_{i=1}^{N} q^{\ell_N-\ell_i} \leq q^{\ell_N} ,$$

and (5.4) results from dividing both sides by q^{ℓ_N}.

McMillan has shown that the Kraft inequality (5.4) is actually much more general, being satisfied by *any* uniquely decipherable code regardless of the means which endow it with this property.

5.2.3 FUNDAMENTAL THEOREM OF SOURCE CODING

The fundamental theorem of source coding can be stated as follows. The messages generated by a stationary source, with or without memory, of entropy H shannons per symbol, can be transformed according to a uniquely decipherable source encoding process into shorter (compressed) ones from which the original messages can be exactly recovered. The ratio $\lambda = \overline{\ell}/m$ of the average length after coding $\overline{\ell}$ to that before coding, m, i.e., the average number of symbols of the compressed message per symbol of the original one, is at least $\lambda = H/\log_2 q = H_q$, where q denotes the alphabet size (assuming to be the same in both the original and the compressed messages) and H_q is the entropy of the source expressed using logarithms to the base q. By encoding extensions of the source of increasing order k, $\overline{\ell}/m$ can approach as closely as desired its lower bound H_q.

First notice that the compression factor λ cannot be less than H_q. Indeed, the largest possible entropy of a q-ary source is $\log_2 q$ shannons, so the source which results from the source encoding of the original source has at most this entropy. Then the average information quantity borne by a message after source coding is at most $\overline{\ell}\log_2 q$ shannons. Each message generated by the original source bears in the average an information quantity equal to mH shannons. An encoding process does not create, and can at best avoid losing, information. The most efficient source coding thus results in an average length $\overline{\ell}$ such that:

$$\overline{\ell} \log_2 q \geq mH \; . \tag{5.5}$$

Dividing both sides by $m \log_2 q$ results in

$$\lambda \geq H_q \; ,$$

as stated.

This lower bound can be approached as closely as desired by a source coding process operating on a source extension of arbitrarily increasing order k. In order to prove this statement, we consider the source coding of the k-th extension of the original source. A word of a variable-length code is associated with each symbol of this extension. Let $\ell_i^{(k)}$ denote the length of the codeword which encodes the i-th symbol of the extension, $1 \leq i \leq q^k$, $\overline{\ell}^{(k)}$ denote its average, and $p_i^{(k)}$ denote its probability. Since $\ell_i^{(k)}$ represents a single symbol of the extension, we have $m = 1$ in (5.5). Provisionally assuming that $\overline{\ell}^{(k)}$ equals its lower bound results in

$$\overline{\ell}^{(k)} = H_q^{(k)} \; ,$$

where $H_q^{(k)}$ denotes the entropy of the k-th extension of the original source. This equality implies

$$\sum_{i=1}^{q^k} p_i^{(k)} \ell_i^{(k)} = - \sum_{i=1}^{q^k} p_i^{(k)} \log_q p_i^{(k)} \; .$$

The condition expressed by this equality is fulfilled only if

$$\ell_i^{(k)} = - \log_q p_i^{(k)}$$

for all possible values of i, but this equality does not in general result in ℓ_i being an integer as the length of a codeword should be. However, we may look for a set $\{\ell_i\}$ of q^k integers which for the purpose of source coding approximates the set of q^k positive real numbers $\{- \log_q p_i^{(k)}\}$.

It is always possible to find an integer $\ell_i^{(k)}$ such that

$$\ell_i^{(k)} \leq - \log_q p_i^{(k)} < \ell_i^{(k)} + 1$$

for each value of i ($1 \leq i \leq q^k$). Multiplying the terms of this double inequality by $p_i^{(k)}$ and summing for i from 1 to q^k results in

$$\sum_{i=1}^{q^k} p_i^{(k)} \ell_i^{(k)} \leq - \sum_{i=1}^{q^k} p_i^{(k)} \log_q p_i^{(k)} < \sum_{i=1}^{q^k} p_i^{(k)} \ell_i^{(k)} + \sum_{i=1}^{q^k} p_i^{(k)}$$

which can be rewritten

$$\overline{\ell}^{(k)} \leq H_q^{(k)} < \overline{\ell}^{(k)} + 1 ,$$

where $\overline{\ell}^{(k)}$ is the average length of the codewords; the rightmost '1' results from $\{p_i^{(k)}\}$ being a probability distribution, hence $\sum_{i=1}^{q^k} p_i^{(k)} = 1$. Dividing by k results in

$$\lambda^{(k)} \leq H_q^{(k)}/k < \lambda^{(k)} + 1/k ,$$

where $\lambda^{(k)}$ is defined as $\lambda^{(k)} = \overline{\ell}^{(k)}/k$. If the original source is memoryless, $H_q^{(k)} = kH/\log_2 q$, where H is the entropy of the original source expressed in shannons, and the double inequality above becomes

$$\lambda^{(k)} \leq H_q < \lambda^{(k)} + 1/k ,$$

so $\lambda^{(k)}$ approaches H_q as k approaches infinity. If the original source is not memoryless, remember that we defined the entropy of such a source as the limit of $H_q^{(k)}/k$ as k approaches infinity (see Eq. (5.3) above), so the same conclusion holds.

The recourse to an extension of the source of high-order results in an exponential increase of the alphabet size, hence enables a finer and finer matching of the integer lengths of the codewords to the probabilities of its symbols. Most is gained, however, in the first extension orders. For instance, in the above example of Tab. 5.1, the compression factor $\lambda^{(3)} = 1.598/3 = 0.53266\ldots$ obtained with the 3-rd extension of the source is fairly close to the entropy of the original source in shannons, 0.469..., i.e., the theoretical limit.

There exist many source coding algorithms—infinitely many, in fact, since most of them depend on numerical parameters. Some of them are adaptive, i.e., do not demand that the probabilities which characterize the source operation are exactly known. These probabilities are estimated from measured frequencies, the measurement of which is an integral part of the algorithm operation. These probabilities may even be *a priori* completely unknown. Similarly, the assumption of stationarity need not be strictly satisfied, and significant compression factors can be obtained with slowly varying sources, or sources varying by steps. Given an arbitrary sequence there is no general means to know what algorithm, if any, can the most efficiently compress it.

5.3 ABOUT CHANNEL CODING

5.3.1 FUNDAMENTAL THEOREM OF CHANNEL CODING

The proof of the fundamental theorem of channel coding is rather difficult, which makes giving a convincing sketch of its proof without lengthy and somewhat technical mathematical details rather challenging.

That errorless communication cannot involve a source entropy larger than the channel capacity is rather intuitive. The mere possibility of communicating through a channel an information quantity at most equal to its capacity C does not solve the problem of reliably communicating through this channel the message generated by a source of entropy smaller than or equal to this capacity. The expression of mutual information:

$$I(X; Y) = H(X) - H(X|Y) \tag{5.6}$$

appears as the difference between two terms: the information quantity $H(X)$ present at the channel input minus the uncertainty as regards X which remains when Y is given, measured by $H(X|Y)$. Clearly, the actual communication of a message implies that this term be zero or negligible when $H(X)$ measures the information from the source which should be received by the destination. In other words, the messages delivered to the destination should satisfy a criterion of recovery quality such that $H(X|Y)$ be zero or negligible. But $H(X|Y)$ solely depends on the channel once the probability distribution of X has been given. If the channel is affected by errors, $H(X|Y)$ has some nonzero value which cannot in general be neglected. Then for faithfully communicating the messages it generates, the source may not be directly connected to the channel input. Its output should be transformed according to some coding process and the inverse transformation should be performed on the channel output so as to recover the message. Intermediate devices, namely an encoder and a decoder, should thus be interposed between the source and the channel input on the one hand, between the channel output and the destination on the other hand.

Figure 5.3: The channel is preceded by an encoder receiving the input variable U and followed by a decoder delivering the output variable V. The channel input and output are X and Y, respectively.

Figure 5.3, again an avatar of Shannon's paradigm, depicts the situation. The output of a memoryless source is connected to the input of an encoder, the output of which is connected to the input of a memoryless channel. The channel output is connected to a decoder, the output of which is the input to the destination. We assume that the same q-ary alphabet is used everywhere in the figure. The channel has as input and output the q-ary variables X and Y, respectively. The mean mutual information $I(X; Y)$ is expressed by Eq. (5.6) with $H(X|Y) > 0$, depending on the channel. Let U and V denote the random variables at the input of the encoder and at the output of the decoder, respectively. Then the relation homologous to Eq. (5.6) is true:

$$I(U; V) = H(U) - H(U|V) .$$

The inequality

$$I(U; V) \leq I(X; Y)$$

is true because the encoder and the decoder do not create, but at best avoid to lose, information. The equality thus obtains for a well-designed encoding system, so

$$H(X) - H(X|Y) = H(U) - H(U|V) .$$

The criterion of recovering quality demands $H(U|V) < \varepsilon$, where ε is a given positive constant smaller than $H(X|Y)$, which results in:

$$H(X) - H(U) = H(X|Y) - H(U|V) > 0 .$$

The entropy $H(U)$ must thus be less than $H(X)$, which means that the encoding is necessarily redundant.

This argument needs further comments. The random variable X can actually be given two different meanings. As the channel input, it should be considered as a q-ary random variable since the channel deals separately with each of its input symbols. However, the source and the encoder taken together can be considered as a new source. Contrary to the original source which is assumed memoryless, the new source has memory since the encoder acts on sequences of input symbols, not separately on each of them. In the above expressions, X should be understood as the *memoryless restriction* of the encoder output[2]. If on the contrary X is interpreted as the output of the source with memory made of the original source and the encoder, say X_{seq}, then by the same token as above we have

$$H(U) \geq H(X_{\text{seq}}) ,$$

with equality for a well-designed encoder. For short, the channel input must thus be interpreted as made of independent symbols, hence represented as the q-ary random variable X, although as the output of the channel (hence at the decoder input) it consists of a sequence X_{seq} having at most the same entropy as the random variable U at the encoder input.

The channel capacity C is the largest possible value of $I(X; Y)$. If we assume an errorless communication, $U = V$ implies that $H(U|V) = 0$. The equality of $I(X; Y)$ and $I(U; V)$ entails that

$$C \geq H(X) - H(X|Y) = H(U) ,$$

showing that errorless communication cannot be performed if the source entropy $H(U)$ exceeds the channel capacity C.

5.3.2 CODING FOR THE BINARY SYMMETRIC CHANNEL

It is more difficult to prove the positive statement of the theorem. Shannon's proof of the channel coding theorem relies on the extraordinary idea of *random coding* [72]. Since no one knew (and no one still knows) how to design the best possible code, and *a fortiori* no one could compute the probability of erroneous recovery associated with it, Shannon considered, instead of a single code, a broad set of randomly chosen codes. Then he computed the average error probability for this ensemble of

[2]By 'memoryless restriction,' we mean that the mutual dependence between the sequence symbols is ignored.

codes and he showed that, provided the condition of the theorem is met (i.e., the source entropy is smaller than the channel capacity), then this average error probability can be made arbitrarily small by indefinitely increasing the codeword length. The ensemble of random codes contains at least a code as good as the average, which shows that 'errorless' communication is possible. *Errorless* should be understood in an asymptotic sense, i.e., the probability of error can be made as small as desired by increasing the codeword length. The original proof given by Shannon was criticized by pure mathematicians who pointed out its lack of formal rigour and expressed doubts about the validity of the result itself. American mathematicians were especially sceptical, and the first mathematically sound proofs of the theorem came years later from the Soviet Union [50]: Shannon's conclusions were fully confirmed but the conditions of validity of the theorem were more precisely stated.

Let us first present an argument which will give us some insight, although it is not entirely satisfactory. As a simple case, we consider the binary symmetric channel already mentioned in Sec. 3.3.3. We consider sequences of a given length n and we first assume that the number of errors is somehow kept constant: exactly t bits, chosen at random among the n bits of a sequence, are always wrong. This is a very unrealistic assumption, but it will be useful for our purpose and we shall see later that it is approximately true when n and t are very large. The bit error probability of the channel is thus $p_{su} = t/n$, a constant, where the subscript 'su' stands for 'substitution' since in case of an error a wrong bit is substituted for the correct one. Thanks to this simplifying assumption, the total number of different erroneous words possibly received when a given codeword is sent is exactly the number of different possible choices of t objects among n, or combinations, generally denoted by $\binom{n}{t}$. Each of the possible transmitted n-bit sequences, i.e., each codeword, can thus be transformed by the channel errors into one among $\binom{n}{t}$ other n-bit sequences. No more than $M = 2^n/\binom{n}{t}$ distinguishable codewords can thus exist. But

$$\binom{n}{t} = \frac{n!}{t!(n-t)!}$$

and it is shown in Appendix 5.B that Stirling's formula results in the approximation:

$$\binom{n}{t} \approx 2^{n\mathcal{H}_2(t/n)} ,$$

valid for n and t large enough.

The maximum number of distinguishable codewords, $M = 2^n/\binom{n}{t}$, has thus the approximate expression:

$$M \approx 2^{n[1-\mathcal{H}_2(t/n)]} .$$

The entropy H of a source connected to the channel input can achieve errorless communication if it is less than $(1/n)\log_2(M)$, thus approximately if:

$$H < 1 - \mathcal{H}_2(t/n) .$$

The right-hand side equals the channel capacity of the binary symmetric channel of error probability p_{su} given by Eq. (3.11), so the equality

$$H < 1 - \mathcal{H}_2(p_{su}) \tag{5.7}$$

holds true asymptotically for n approaching infinity.

The assumption that exactly t bits are in error in any received word becomes itself asymptotically true as t and n approach infinity. Indeed, when the errors occur at random with probability p_{su}, the error frequency tends to a constant p_{su} when n approaches infinity as a mere consequence of the (weak) law of large numbers.

Although the calculations based on Stirling's formula are not especially intuitive, one can give a geometric interpretation of the above argument. The space of binary sequences of length n contains 2^n elements, or points, and we may define a distance (the Hamming distance) between two of them as the number of positions in the sequence where they differ. Assuming that exactly t errors occur means that there is a distance of t between the transmitted sequence and the received one. Let us define the 'volume' of some subset of this space as the number of sequences, or points, it contains. The volume of the whole space is 2^n. We may think of the set of $\binom{n}{t}$ possible received sequences when some sequence $\underline{c}$ is transmitted as an 'error sphere' of radius t centered in $\underline{c}$. The volume of this sphere is $\binom{n}{t}$. If we use M codewords, the total volume of the error spheres, $\binom{n}{t}M$, cannot exceed the volume of the whole space, namely, 2^n. A geometric interpretation of inequality (5.7) results.

This argument actually shows that it is impossible to achieve errorfree communication if the source entropy is larger than the channel capacity (a conclusion we already arrived at by other means in Sec. 5.3.1). But can we show that it can be achieved if the entropy is smaller than the capacity, however, close to it may be? Intuitively, the error spheres should be as equally distributed in the whole space as possible. No explicit rule is in general available to this end. However, it turns out that choosing the codewords at random results, asymptotically as n approaches infinity, in the most regular possible configuration. By 'choosing at random,' we mean that each bit of a sequence is drawn with probability 1/2 independently of the choice of other bits, and that each n-symbol sequence is thus chosen independently of the other sequences. The obtained regularity may look surprising since any random choice is completely unexpected hence highly irregular. Do not forget, however, that when very many random choices occur with constant probabilities, a statistical regularity appears. Probability theory states precise laws which concern random choices successively effected very many times. The 'laws of large numbers,' roughly speaking, tell that the observed frequencies tend to the probabilities when the number of considered occurrences increases, and moreover tell how closely and how fast they do.

As another simple example, the binary erasure channel considered above has the very simple capacity [Eq. (3.12)] which is quite different from that of the binary symmetric channel, [Eq. (3.11)]. Let us try to understand why. First of all, we may notice that, if we choose arbitrary binary symbols at the n_{er} places where the original ones have been erased, only $n_{er}/2$ erroneous bits result in the average. This shows that, in some sense, an erasure is equivalent to half an error. The capacity of the erasure channel with erasure probability $p_{er} = 2p_{su}$ cannot thus be less than that of the binary symmetric channel of error probability p_{su}. It is even significantly larger since the location of the unidentified symbols is an information in itself which is destroyed when arbitrary binary symbols replace the signs ϵ which pinpoint the locations where the erasures occurred. Destroying information contained in the channel output necessarily reduces its capacity. Notice that the 'equivalence' of two erasures with a single error only holds in the binary case. If the alphabet size is q, the number of erroneous symbols which result in the average from arbitrary choices becomes $(q-1)n_{\text{er}}/q$ so an erasure is then 'equivalent' to $(q-1)/q$ errors.

The simplest interpretation of this case is that the number of dimensions of the space of sequences, i.e., the length n of these sequences, is reduced in the average by the factor p_{er}. The largest information quantity that a sequence of length n can bear, $n \log_2 q$ shannons, is thus reduced by the occurring erasures to $p_{\text{er}} n \log_2 q$.

5.3.3 GENERAL CASE: FEINSTEIN'S LEMMA

Up until now, we just gave the sketch of a proof of the fundamental theorem of channel coding for the particularly simple example of the binary symmetric channel. For any source and channel, provided they satisfy the regularity conditions of being stationary and ergodic, the fundamental theorem of channel coding states that the inequality

$$H < C \tag{5.8}$$

is the necessary and sufficient condition of asymptotically errorfree communication, where H denotes the source entropy and C the channel capacity.

The simple case of the binary symmetric channel is merely a particular instance of a general result named *Feinstein's lemma*. Assuming a stationary and ergodic source and a discrete stationary channel with finite memory, it states that a one-to-one correspondence between the typical sequences generated by the source (see Sec. 5.1.2) and asymptotically disjoint sets of points in the space of the output sequences can be established provided the source entropy is less than the channel capacity, when the sequence length approaches infinity. The 'error spheres' in the case of the binary symmetric channel are mere examples of such asymptotically disjoint sets of points in the channel output space. The proof of Feinstein's lemma is unfortunately rather difficult and implies lengthy mathematical developments. The statement of the fundamental theorem of channel coding then results from the lemma in a straightforward manner: observing the channel output enables uniquely identifying the channel input sequence if it belongs to the set of typical sequences. If it does not (an event of vanishingly small probability according to the Shannon-McMillan theorem), a particular distinguishable sequence can be transmitted in order to warn the destination that an atypical sequence occurred.

5.4 A SHORT INTRODUCTION TO THE ALGORITHMIC INFORMATION THEORY

As a convenient notation, sequences will be denoted in this section by an underlined letter to distinguish them from single symbols or numbers.

5.4.1 PRINCIPLE OF THE ALGORITHMIC INFORMATION THEORY

Up until now, we defined all the quantities of information theory, and especially the entropy of a source, in terms of probabilities. A very different point of view consists of measuring the information borne by a sequence by its *algorithmic complexity*, defined as the smallest length of a program which can describe this sequence. This program is written in a given alphabet, assumed here binary, and instructs a given machine to generate the sequence.

This idea was introduced by the great mathematician Kolmogorov [51], to whom we owe the axiomatization of probability theory, but also, independently, by Gregory Chaitin who was then 15 years old. Chaitin recently wrote a very readable book which popularizes the algorithmic information theory and discusses its philosophical impact [25].

Contrary to intuition, the algorithmic complexity of a sequence is almost independent of the language and of the machine used in order to describe it. Then, it becomes possible to associate with a single sequence a measure of information which does not rely on probabilities. One may think of the notion of algorithmic complexity as *prior* to that of probability and maybe able to found its theory. Let us first give a few examples of sequences and consider how they can be described.

a) The sequence

10101010101010101010101010101010

is made of the periodic repetition of '10' 16 times. Clearly, a program describing this sequence may be as short as: "Print 16 times '10' ." An equivalent way of describing this sequence is "Print the first 32 digits of the development of 2/3 in binary numeration." Would the sequence consist of repeating '10' infinitely many times, then these descriptions become: "Print repeatedly '10' " or "Print the digits of the binary development of 2/3."

b) The sequence:

01101010000010011111001100110011

is not periodic and looks random. However, it consists of the first 32 binary digits of the development of the irrational number $\sqrt{2}-1$, so the program "Print the first 32 digits of the binary development of $\sqrt{2}-1$" suffices for describing it. Indefinitely continued, this sequence is described by the shorter program "Print the digits of the binary development of $\sqrt{2}-1$."

c) The sequence:

01001010101111000011101100101100

was obtained by drawing the bits '0' and '1' at random with the same probability 1/2, independently of each others. It is likely that no program for its description exists which is much shorter than

"Print '01001010101111000011101100101100' "

and, furthermore, one could not describe a continuation of this sequence without lengthening the program, for instance by simply writing the following outcomes.

The first two sequences, (a) and b), have thus a description almost independent of their length, at variance with sequence (c) for which one may not expect to find a description much shorter than itself.

d) The sequence

0110101111111101101111100101110

also resulted from drawing '0's and '1's at random, but the probability of a '1' was 3/4 instead of 1/2. One may give a description of it shorter than its length n by counting the number of '1's, say k, and by indicating the rank ρ of this sequence in the list of all binary sequences of length n comprising k '1's, arranged in lexicographic order. For example, with $n = 5$ and $k = 3$, this list would be:

00111, 01011, 01101, 01110, 10011, 10101, 10110, 11001, 11010, 11100,

which comprises $\binom{5}{3} = 10$ entries. Describing the original sequence thus demands[3] a number of bits equal to $\log_2 n + \log_2 \binom{n}{k}$. For n and k large enough, $\binom{n}{k}$ is close to $2^{n\mathcal{H}_2(k/n)}$, where $\mathcal{H}_2(x) = -x \log_2 x - (1-x)\log_2(1-x)$ is the binary entropy function defined above by Eq. (3.6), as shown by applying Stirling formula (see Appendix 5.B). In this case, describing the sequence does not demand that the program length be equal to that of the sequence itself, n, but this length approximately equals n times the binary entropy $\mathcal{H}_2(k/n)$, which is the smaller, its argument is more different from 1/2. One could as well use a source encoding process, for instance dealing with an extension of the sequence by the Huffman algorithm, which would result in a random equiprobable sequence, similar to sequence c), of length close to $n\mathcal{H}_2(k/n)$, too (see Sec. 5.2).

We said above that the algorithmic complexity is almost independent of the machine and of the language used in order to describe a sequence. More precisely, the machine should be universal, i.e., able to simulate the operation of any computer. Any modern general purpose computer belongs to this category. Using a given universal machine and a given language as references makes the length of a program for describing a sequence a well defined quantity. The Turing machine[4] is the prototype of such a *universal machine*. This machine can read, erase or write symbols on strips of paper. It executes a *program* written in a certain alphabet assumed here binary. This program $\underline{p} = p_1\ p_2\ p_3\ \ldots$ is written on a strip of paper (the input strip) which is always read in the same direction. At a given step of executing the program, the machine reads a bit on this strip, changes its internal state according to its own instructions, reads, or erases and writes, a bit on a second strip of paper, the working strip, which it may shift one position to the left or the right, and possibly writes on a third strip: the output strip. It finally moves one position forward on the strip which contains the program in order to read its following bit. The program is read always in the same direction, with no possible return, which results in the programs constituting an *irreducible* set in the sense that none of them is the prefix of another one (see Sec. 5.2). Indeed, when the execution of one of

[3]The number of bits needed to represent k and ρ should be understood as the smallest integer larger than or equal to the written logarithms. Representing both k and ρ by a fixed number of bits is necessary to unambiguously separate them; the figures given result from $k \le n$ and $\rho \le \binom{n}{k}$.

[4]This machine is useful in abstract reasonings but its practical use would be inconvenient and slow.

the programs ends, the machine stops and thus does not read the strip beyond the last symbol of the program just executed. As an important consequence, all the possible programs to the Turing machine have lengths which obey the Kraft-McMillan inequality (5.4).

Then the algorithmic complexity $K(\underline{x})$ of a sequence $\underline{x}$ is used as a measure of the information quantity it bears. More precisely, the average value of $K(\underline{x})/l(\underline{x})$, where $l(\underline{x})$ denotes the length of the sequence $\underline{x}$, taken in the limit as $l(\underline{x})$ approaches infinity, is an algorithmic measure of information homologous to the entropy of a source in conventional information theory but it is associated with a single sequence, not with a source, i.e., a set of sequences.

5.4.2 ALGORITHMIC COMPLEXITY AND ITS RELATION TO RANDOMNESS AND ENTROPY

Let $\underline{x}$ denote a finite binary sequence of length $l(\underline{x})$. Let $\mathcal{U}$ be the universal machine used as reference, e.g., the Turing machine. The sequence generated by the machine when it executes some program $\underline{p}$ is denoted by $\mathcal{U}(\underline{p})$; we assume it is a finite binary sequence.

The algorithmic complexity (also referred to as Kolmogorov complexity) of the sequence $\underline{x}$ with respect to the universal machine $\mathcal{U}$ is defined as:

$$K_{\mathcal{U}}(\underline{x}) = \min_{\underline{p}:\, \mathcal{U}(\underline{p})=\underline{x}} l(\underline{p}) \,. \tag{5.9}$$

In other words, it is the shortest length of a program such that the machine prints the sequence $\underline{x}$ and stops. The conditional complexity $K_{\mathcal{U}}(\underline{x}|l(\underline{x}))$ is defined the same way, but it is further assumed that the length $l(\underline{x})$ is available to the machine $\mathcal{U}$ prior to processing.

We have seen in Sec. 5.4.1 that very long sequences could be printed by a machine executing a short program such as "Print the first n bits of the development of $\sqrt{2}-1$." Translated into a binary language, this program is much shorter than n if n is large, and its length weakly depends on n. On the contrary, describing a random sequence demands a program approximately as long as the sequence itself. Here are a few important results as regards measuring the algorithmic complexity of a sequence by a program length.

If $\mathcal{U}$ is a universal machine and if $\mathcal{A}$ is another machine, then

$$K_{\mathcal{U}}(\underline{x}) \leq K_{\mathcal{A}}(\underline{x}) + c_{\mathcal{A}} \,, \tag{5.10}$$

where $c_{\mathcal{A}}$ is a constant independent of $\underline{x}$. This constant is the length of the program which teaches the universal computer $\mathcal{U}$ how to simulate $\mathcal{A}$. If $\mathcal{A}$ is also a universal machine, then $|K_{\mathcal{U}}(\underline{x}) - K_{\mathcal{A}}(\underline{x})| < c$, where c is a constant independent of $\underline{x}$. This property establishes that the complexity is *universal*, in the sense that the length difference between the programs describing $\underline{x}$ with machines $\mathcal{U}$ and $\mathcal{A}$ is negligible if $\underline{x}$ is large enough. The subscript which tells what machine has been used in order to define the complexity may thus be dropped, as it will be in the sequel.

The conditional complexity of $\underline{x}$ is smaller than its length, except for a constant:

$$K(\underline{x}|l(\underline{x})) \leq l(\underline{x}) + c \,. \tag{5.11}$$

This result is obvious since the program may contain the sequence $\underline{x}$ itself. If the length of $\underline{x}$ is not known, hence for nonconditional complexity, one has the rough upper bound:

$$K(\underline{x}) \leq l(\underline{x}) + 2\log_2 l(\underline{x}) + c\ . \tag{5.12}$$

In order to prove it, it suffices to return to the previous case by feeding the machine with the length $l(\underline{x})$, besides $\underline{p}$. One needs $\log_2 l(\underline{x})$ bits for expressing this length, but the number of bits which is necessary for this purpose is itself unknown. If one repeats each bit twice consecutively [which accounts for the factor 2 in (5.12)], it becomes possible to unambiguously mark the end of the binary representation of this number with the doublet '01.' More subtle means can be used for still shortening the expression of $l(\underline{x})$, resulting in a slightly smaller bound.

Another important result is that the low complexity sequences are few, in the sense that:

$$\#\{(\underline{x} \in (0, 1)^* : K(\underline{x}) < k)\} < 2^k \tag{5.13}$$

where $\#\{\cdot\}$ denotes the number of elements of a set and where $(0, 1)^*$ represents the set of possible binary sequences of any length (i.e., λ, 0, 1, 00, 01, 10, 11, 000, ..., where λ denotes the empty sequence). In other words: the total number of sequences of algorithmic complexity less than k is less than 2^k. The proof of this result is obtained by merely counting the programs of length k. Their number is less than 2^k, and each of them can produce at most a sequence $\underline{x}$.

Still another result on algorithmic complexity is the bound:

$$K(\underline{x}|n) \leq n\mathcal{H}_2\left(\frac{1}{n}\sum_{i=1}^{n} x_i\right) + 2\log_2 n + c\ , \tag{5.14}$$

where $\underline{x} = x_1x_2\ldots x_n$ is a binary sequence of length n. $\mathcal{H}_2(\cdot)$ is the binary entropy function defined by Eq. (3.6). The symbols of $\underline{x}$ are dealt with as integers in (5.14), so $\sum_{i=1}^{n} x_i$ involves the ordinary addition and merely equals the number of '1's in $\underline{x}$ (its weight). This bound may be proved using a construction similar to the example of compressing the binary sequence drawn with a probability of bit '1' different from 1/2, given in Sec. 5.2.

The algorithmic complexity of a random sequence (per symbol) is close to its entropy. In order to prove this statement, one first notices that the set of programs have lengths which satisfy the Kraft-McMillan inequality, i.e.,

$$\sum_{\underline{p}:\,\mathcal{U}(\underline{p})\text{ stops}} 2^{-l(\underline{p})} \leq 1\ . \tag{5.15}$$

Let $\underline{X} = X_1X_2\ldots X_n$ be a sequence of length n of random variables, each drawn at random independently of the others with a same probability distribution $f(x)$, where x is an element of some finite alphabet of size q. Sequences of length n have as probability distribution $\prod_{i=1}^{n} f(x_i)$, which is denoted by $f(x^n)$. Then there exists a constant c such that

$$H(X) \leq \frac{1}{n}\sum_{x^n} f(x^n)K(\underline{X}|n) \leq H(X) + \frac{q\log_2 n}{n} + \frac{c}{n} \tag{5.16}$$

for any n, where $\sum_{x^n}$ means that the sum is effected for all sequences of length n. $H(X)$ denotes the entropy associated with the distribution $f(x)$. The expectation of the complexity $K(\underline{X}|n)$, namely $\sum_{x^n} f(x^n)K(\underline{X}|n)$, tends to the entropy $H(X)$ as n tends to infinity since then the two rightmost terms vanish.

Difficulties nevertheless exist so one may not liken without precautions Shannon's entropy and the algorithmic complexity per symbol. For example, the familiar formulas

$$H(X, Y) = H(Y, X)$$

and

$$I(X; Y) = H(X) - H(X|Y)$$

have an equivalent for algorithmic complexity except for an additional term which is of the same order of magnitude as the logarithm of their left-hand sides [51].

Let $\underline{X} = X_1X_2 \ldots X_n$ be a sequence of binary independent random variables of probability 1/2. One then has:

$$\Pr[K(\underline{X}|n) < n - k] < 2^{-k} \,, \tag{5.17}$$

for any $k < n$. This result means that most sequences have a complexity which is close to their length.

A sequence $\underline{x} = x_1x_2 \ldots x_n$ is said to be *algorithmically random* if

$$K(\underline{x}|n) \geq n \,. \tag{5.18}$$

Counting the sequences shows that there exists at least one such sequence for each value of n.

A sequence $\underline{x} = x_1x_2 \ldots x_n$ is said to be *incompressible* if

$$\lim_{n\to\infty} \frac{K(\underline{x}|n)}{n} = 1 \,. \tag{5.19}$$

The law of large numbers implies for incompressible sequences:

$$\frac{1}{n}\sum_{i=1}^{n} x_i \to \frac{1}{2} \,, \tag{5.20}$$

meaning that the proportion of the '0's and the '1's in such a sequence is almost the same.

For a random binary sequence $\underline{X}$ of length n with unequal probabilities of '0' and '1,' $\Pr(X = 1) = \theta \neq 1/2, 0 < \theta < 1$, one obtains that

$$\Pr[\frac{1}{n}K(\underline{X}|n) - \mathcal{H}_2(\theta) \geq \varepsilon] \to 0, \; n \to \infty \,. \tag{5.21}$$

5.4.3 SEQUENCES GENERATED BY RANDOM PROGRAMS

Let us now consider a universal machine fed by a *random* program. By random program, we mean that each of its bits is chosen at random with probability 1/2, independently of the other bits. There exists a nonzero probability that such a randomly chosen program be significant for the machine and thus eventually results in the machine printing some output sequence.

But how does the execution of a program stop? A theorem of theoretical computer science (quite similar to Gödel's theorem) states that no algorithm can predict whether the execution of a program will stop after printing a sequence, or will continue to run indefinitely without providing an output. This theorem implies that the algorithmic complexity of a sequence $\underline{x}$ is *uncomputable*. The only means to find the shortest possible program describing a sequence $\underline{x}$ would be to try all possible programs, but the execution of certain of them would never stop.

From a point of view closer to that of conventional information theory, the uncomputability of a sequence complexity of is not so surprising. We actually noticed in Sec. 5.2 that infinitely many possible source coding algorithms exist. It is thus impossible, when the description $\underline{p}$ of a sequence $\underline{x}$ has been obtained using some source coding algorithm, to assert that a shorter one cannot be obtained using another algorithm, for lack of being able to try all algorithms. One can only assert that $K(\underline{x}) \leq l(\underline{p})$.

Chaitin introduces a number he denotes by Ω [25], equal to the probability that the execution of a program stops, namely:

$$\Omega = \sum_{\underline{p}:\ \mathcal{U}(\underline{p})\text{ stops}} 2^{-l(\underline{p})} . \tag{5.22}$$

Since the lengths $l(\underline{p})$ of possible programs satisfy the Kraft-McMillan inequality (5.4), one actually defines by this means a number between 0 and 1. There exists no effective means for calculating Ω, just like the complexity of a sequence cannot be computed.

The probability of a random binary program $\underline{p}$ of length $l(\underline{p})$ is $\Pr(\underline{p}) = 2^{-l(\underline{p})}$, assuming its bits are chosen independently of each other with probability 1/2. A short program thus more probably occurs than a long one. If a short program produces a long sequence, this sequence cannot be algorithmically random since it has a simple description. One is led to define the *universal probability* of a sequence $\underline{x}$ as:

$$P_{\mathcal{U}}(\underline{x}) = \sum_{\underline{p}:\ \mathcal{U}(\underline{p})=\underline{x}} 2^{-l(\underline{p})} = \Pr[\mathcal{U}(\underline{p}) = \underline{x}] . \tag{5.23}$$

In this case, too, this equality defines a probability since the lengths $l(\underline{p})$ of the possible programs satisfy the Kraft-McMillan inequality. It is the probability that a program resulting from the choice of random successive independent bits with the same probability 1/2 results in the universal machine $\mathcal{U}$ printing the sequence $\underline{x}$. This probability is not exactly independent of the machine, but only weakly depends on it.

To conclude this section, the algorithmic information theory has a great theoretical interest. The uncomputability of the algorithmic complexity makes it, however, poorly fitted to practical applications and somewhat frustrating. Moreover, the algorithmic information theory essentially appears as a theory of sequences becoming *infinitely long*. The equivalence between machines, which justifies the universality of the algorithmic complexity, is established by Eq. (5.10) only apart from adding a constant, and this constant may be of the same order of magnitude as the length of practically interesting sequences.

There is apparently a wide difference between information theory in the sense of Shannon, which is based on probabilities, and the algorithmic information theory which apparently involves a single sequence. It may thus look surprising that the algorithmic complexity and the entropy of a random sequence are close to each other, as stated in Sec. 5.4.2. Apparently, the algorithmic complexity is based on a single sequence and the entropy on a set of sequences endowed with probabilities. However, estimating probabilities involves frequency measurements, which for long sequences can only concern a very tiny fraction of all possible sequences of that length. On the other hand, the algorithmic complexity of a sequence implicitly refers to a collection of sequences, i.e., the subset of all sequences which share the same complexity.

5.5 INFORMATION AND ITS RELATIONSHIP TO SEMANTICS

Shannon and the pioneers of information theory had an empirical approach since they did not attempt *defining* information, nor explicating its connection with semantics. They just proposed means for its quantitative measurement. Communication engineering being concerned only with *literal communication*, it was possible to discard semantics as irrelevant. In the very first page of the paper which founds information theory [72], Shannon wrote:

> The fundamental problem of communication is that of reproducing at one point either exactly or approximately a message selected at another point. Frequently the messages have *meaning*; that is they refer to or are correlated according to some system with certain physical or conceptual entities. These semantic aspects of communication are irrelevant to the engineering problem. The significant aspect is that the actual message is one *selected from a set* of possible messages. The system must be designed to operate for each possible selection, not just the one which will actually be chosen since this is unknown at the time of design. (Shannon's italics.)

In short, just like a messenger has not to know about the message he/her carries, a communication system should ignore meaning and operate regardless of the message. Replacing in the above quotation the word 'point' which refers to a location in space with the word 'instant' moreover extends its relevance to communication in time, hence to heredity.

We try here rather naively to define information, hopefully shedding some light on the relationship between information and semantics. Contrary to the remainder of this chapter, the remarks below do not express any consensus among information theorists.

Let us consider a digital message, i.e., a sequence of symbols from some finite-size alphabet. Such a message is a mathematical abstraction which needs a physical support for having any interaction with the real world, and especially for being communicated. The physical support of a message can take a variety of forms, either material (e.g., ink deposits on a sheet of paper according to conventional patterns, i.e., letters, holes punched in a card, local states of magnetization of some ferromagnetic medium, shallow tiny holes in a compact disk (CD), etc.), or consisting of the variation of some physical quantity as a function of time (e.g., air pressure, or electrical current, or electromagnetic field, etc.). The material supports are actually used for recording, i.e., communication through time consisting of writing a message to be read later, while the variation in time of a physical quantity can be propagated as a wave hence enables communication through space. Regardless of its support, each of the alphabet symbols needs only be unambiguously distinguishable from the other ones.

A same message can be supported by different physical media which, moreover, can be converted from one to another. For instance, the message recorded on a computer memory or a CD can be converted into an acoustic wave (i.e., a sound), or emitted as an electromagnetic wave. Similarly, the alphabet size can be changed: a musical record in a computer memory uses the binary alphabet, but a large-size alphabet is needed for converting it into an audible acoustical signal. The message itself can moreover be changed so as to improve its characteristics as regards some desired property. For instance, it may be compressed in order to reduce the size of the memory needed for its recording (this is source coding as briefly considered in Sec. 5.2), or on the contrary, expanded by a redundant encoding making it resist transmission errors. This is channel coding, defined in Sec. 5.3 and a major topic of this lecture, to be developed in the next chapter.

A message can thus exist in a variety of equivalent forms, depending on its possible encoding, alphabet size, and physical support. We refer to the underlying entity which is common to all these forms as *an information*. We may thus define an information as the equivalence class of all the messages which can be converted into each other by changing their encoding, alphabet size, or physical support. Each of these messages will be said to *bear* this information. The equivalence class associated with an information clearly contains infinitely many elements. Any of these messages is a representative of this class. An information is thus an abstract object which manifests itself in the physical world by any of its representatives, or *realizations*.

As an example, the sequence of Latin letters and spaces:

Information theory discards semantics (1)

and the binary sequence
1001001 1101110 1100110 1101111 1110010 1101101 1100001 1110100 1101001 1101111
1101110 0100000 1110100 1101000 1100101 1101111 1110010 1111001 0100000 1100100
1101001 1110011 1100011 1100001 1110010 1100100 1110011 0100000 1110011 1100101
1101101 1100001 1101110 1110100 1101001 1100011 1110011
share the same *information*, since the latter just resulted from transforming sequence (1) using the ASCII (American Standard Code for Information Interchange) 'code' which is currently used in computer memories. Each Latin letter or space is replaced by a 7-bit word according to a one-to-one correspondence.

The binary sequence
10010011 11011101 11001100 11011110 11100100 11011011 11000011 11101000 11010010
11011110 11011101 01000001 11101000 11010001 11001010 11011110 11100100 11110011

01000001 11001001 11010010 11100111 11000110 11000011 11100100 11001001 11100111
01000001 11100111 11001010 11011011 11000011 11011101 11101000 11010010 11000110
11100111

also bears the same information as sequence (1) since an 8-th bit has been appended to each of the 7-bit words of the first binary sequence, equal to the sum modulo 2 of its bits thus making the total number of '1's even. This may be thought of as a rudimentary means of error control: if an error affects a symbol in a 8-bit word, the number of '1's becomes odd so counting the '1's in each word enables detecting a single-symbol error. Of course, the first binary sequence could be transformed by sophisticated error-correcting codes into an equivalent one made resilient to errors (up to a limit) and bearing again the same information.

As a counterexample, the sentence:

La théorie de l'information exclut la sémantique (2)

is a French translation of the English sentence (1). Although it looks close to it, articles and a preposition have been appended to comply with the French grammar and the word order is different, so sentence (2) does not bear the same information as (1). However, both sentences share the same *meaning*.

The multiplicity of messages which bear the same information leads to ask the following question: given some alphabet with an arbitrary physical support, is there a minimal-length realization of a given information? The simplest alphabet, i.e., the binary one, is a natural choice. The problem becomes whether a minimal-length binary message exists within the equivalence class associated with the given information. Shannon's information theory asserts this existence when the given representative is a message generated by a stationary probabilistic source. The algorithmic information theory extends this statement to any message that a universal computer can generate. We'll refer to the minimal binary realization of an information as its *information message*.

The fundamental theorem of source coding of Shannon's information theory states that any n-symbol message generated by this source can be transformed by source coding into a binary message of average length at least $\overline{\ell} = nH$ bits where H denotes the source entropy per symbol, expressed using binary information units, i.e., shannons. This minimal-length realization of an information generated by a probabilistic source is its information message. It results from optimal source coding, which entails that its bits are probabilistically independent and equiprobable. In the algorithmic information theory, the length of the information message is used for defining the algorithmic complexity associated with the given information. The first case is much more restrictive than the second one, which may be considered as general. However, the source stationarity enables effectively estimating the probabilities of the sequences it generates, hence its entropy, by making frequency measurements. In the second case, on the contrary, the existence of a minimal-length realization is mathematically proven, but the algorithmic complexity is actually an uncomputable quantity which thus can generally not be evaluated.

The set of all binary messages of length ℓ can be represented by a tree like that of Fig. 5.2 with each of its branches labeled with a bit according to some convention. Let us interpret the i-th bit of the information message as the answer to a dichotomic question (i.e., answerable by yes-or-no), '0' meaning for instance 'yes' and '1' meaning 'no.' Any information message of length ℓ may thus be interpreted as one of the 2^ℓ paths of a binary tree of length ℓ taken from the root to the leaves, the choice of a branch at any fork being made at random with probability 1/2. A path in this tree, i.e., an information message of length ℓ, can be interpreted as *an integer* i, $0 \leq i \leq 2^\ell - 1$. The questions associated with the successive bits of the information message may for instance be those needed for identifying the species to which some given living being belongs. Provided the set of species is ordered according to a binary hierarchical taxonomy, using $\ell = nH$ properly chosen successive dichotomic questions enables distinguishing from each other $2^{nH} = (2^H)^n$ species. The entropy H of the source then measures the ability of the messages it generates to discriminate among objects of some outer world, hence to bear some kind of semantics. The corresponding information quantity is simply the number of binary digits which are needed to represent it.

Establishing a correspondence between each bit of the information message and a dichotomic question makes it eventually identify or represent an object which belongs to some outer world, like living beings as in the above example of taxonomy, making possible to distinguish between them. Then information in the above meaning is given a semantic content according to an external convention which we may refer to as its *meaning*.

This is a rather crude kind of semantics, apparently restricted to representing material objects which can be ordered in a tree-like fashion. However, the possible semantic content can be widely extended if we notice that:

— Describing an outer reality by a binary message is not limited to answering dichotomic questions. Data of various other kind can also be represented. For instance, grouping k bits of the information message into a block may be used to specify that the value of some parameter is one of 2^k predetermined levels. If k is large enough, this amounts to approximately specify the value of a continuously varying parameter. This is for instance currently used in telephony to represent instantaneous values of the speech signal (referred to as samples) by a sequence of bits, a process referred to as 'pulse code modulation' (PCM). For frequent enough samples (8 kilohertz, i.e., a sample every 125 microseconds) and k as low as 8 (hence $2^8 = 256$ levels), a sufficient quality of speech transmission is achieved.

— Moreover, a relation between material objects can be represented by the same means as the objects themselves, then opening the semantic field to abstract as well as material objects.

Information theory, either Shannon's or algorithmic, uses the length of the information message as a quantitative measure of information. Since an information message of length ℓ enables distinguishing 2^ℓ different objects, ℓ is a logarithmic measure of the discriminating ability of the information message, regardless of the distinguished objects, a matter of semantics. We may thus understand a quantity of information as the number of semantic instances it can represent, or as the number of dimensions of some space which represents semantic objects. It should be kept in mind that the information quantity is by no means an exhaustive description of an information, just like the mass is only one of the many possible attributes of a material object (shape, temperature, electric charge, color, texture, internal organization, etc.).

5.6 APPENDICES

APPENDIX 5.A. DETAILED ANALYSIS OF A MARKOVIAN SOURCE

We now consider the example of a binary Markovian source, as depicted in Fig. 5.1-**b**, having memory $m = 3$ hence $s = 2^3 = 8$ states, for a more detailed analysis. Let us assume that the initial content of the memory is 000. Let x (0 or 1) denote the next output symbol. It changes the memory content into $00x$. The next symbol, say y, shifts again the content which becomes $0xy$, and one more symbol, say z, results in the content xyz from which all the initial bits were removed. It is convenient to define the state of the memory as the integer $s = 4x + 2y + z$, which assumes values from 0 to 7 and uniquely labels the 8 possible contents of the memory. The probability of transmitting 1 in state s is denoted by p_s, that of transmitting 0 in the same state is $q_s = 1 - p_s$. The following table plots the transition probabilities between the states.

Table 5.2: Each entry in this table is the probability of the transition from the corresponding state in the leftmost column into that of the top row

s	0	1	2	3	4	5	6	7
0	q_0	p_0						
1			q_1	p_1				
2					q_2	p_2		
3							q_3	p_3
4	q_4	p_4						
5			q_5	p_5				
6					q_6	p_6		
7							q_7	p_7

The steady probability of state s is denoted by x_s. Thus, we have $\sum_s x_s = 1$, where the sum is extended to all values of s. The transitions between the states imply that:

$$x_0 = q_0x_0 + q_4x_4,\ x_1 = p_0x_0 + p_4x_4\ ,$$

$$x_2 = q_1x_1 + q_5x_5,\ x_3 = p_1x_1 + p_5x_5\ ,$$

$$x_4 = q_2x_2 + q_6x_6,\ x_5 = p_2x_2 + p_6x_6\ ,$$

$$x_6 = q_3x_3 + q_7x_7,\ x_7 = p_3x_3 + p_7x_7\ .$$

This is a homogeneous system of linear equations, the solution of which is defined only up to a common factor. The probabilities x_s which solve this system can thus all be expressed in terms of one of them, say x_1, and the common factor is eventually determined by solving $\sum_s x_s = 1$. We define the coefficients $\{a_s, s = 0, 1, \ldots, 7\}$ as $a_s = x_s/x_1$. Then, $a_0 = q_4/p_0$, $a_1 = 1$, $a_4 = 1$, $a_3 = a_6$, $a_7 = a_3 p_3/q_7$. The three remaining unknowns a_2, a_3 and a_5 satisfy the three equations:

$$a_2 + a_3 = 1 + a_5 \ ,$$

$$a_2 = q_1 + a_5 q_5$$

and

$$a_5 = a_2 p_2 + a_3 p_6 \ .$$

This system has as solutions:

$$a_5 = \frac{p_1 p_6 + q_1 p_2}{1 - p_2 q_5 - p_5 p_6} \ ,$$

$$a_2 = q_1 + q_5 \frac{p_1 p_6 + q_1 p_2}{1 - p_2 q_5 - p_5 p_6} = q_1 + q_5 a_5 \ ,$$

and

$$a_3 = p_1 + p_5 \frac{p_1 p_6 + q_1 p_2}{1 - p_2 q_5 - p_5 p_6} = p_1 + p_5 a_5 \ .$$

The equality $\sum_s x_s = 1$ results in

$$x_1 = 1/(a_0 + 2 + a_2 + 2a_3 + a_5 + a_7) \ .$$

All the other probabilities x_s follow from $x_s = a_s x_1$. If we let $p_s = 1/2$ for any s all the coefficients $a_0, a_1, \ldots, a_7$ become 1, so that all the probabilities x_s become 1/8, as expected. Then the source is actually a memoryless source which transmits 0 or 1 with probability 1/2.

Calculator simulation. For a numerical application, we assumed that the probability of transmitting symbol '1' when the memory is in state s is $p_s = [1 + 2w(s)]/8$, which only depends on the weight $w(s)$ of the state, i.e., on the number of '1's in its binary representation, and is an increasing function of it. This choice was intended to favor the clustering of either '0's or '1's in long segments. The obtained sequences are very different from that resulting from the independent choice of '0' or '1' with probability 1/2, which evidences the memory effect. Computing the state probabilities in this case results in $x_0 = x_7 = 5/16$, $x_1 = x_2 = x_3 = x_4 = x_5 = x_6 = 1/16$. One easily checks that the steady probabilities of '0' and '1' in the generated sequence are both 1/2. The operation of this Markovian source has been simulated using a calculator. Starting from the initial state $s = 0$, the following sample sequence has been obtained:

00000000000101000000000000000000010000000100100110111110000000000001000000001111
1100000000000000010000000000000001000000000100000000000001100000001000000000000101001
01011111111010001010111010011111111111111010000100100011100110110111101111100
1111011011111111101011111111110111111111110000010000000000000110111111000010000
001110110000000000000000000000111100010011111111111111111 ...

APPENDIX 5.B. APPROXIMATIONS BASED ON THE STIRLING FORMULA

Factorial n, denoted by $n!$, is the product $1 \times 2 \times \ldots \times n$ of the first n positive integers. It is the number of possible different permutations of n objects. An approximate expression of $n!$ valid when n is large is provided by Stirling's formula:

$$n! \approx (n/\mathrm{e})^n \sqrt{2\pi n}(1 + \varepsilon_n) \tag{5.24}$$

where ε_n approaches $1/12n$ as n approaches infinity, $\mathrm{e} = 1 + \sum_{i=1}^{\infty} 1/i!$ is the base to the 'natural' logarithms, denoted by $\ln(\cdot)$, i.e., $\ln(\mathrm{e}) = 1$, and π is the ratio of the perimeter of a circle to its diameter. This approximation is asymptotically tight for n large, meaning that it is comparatively the closer to the exact value, the larger n.

The number of combinations of t objects among n, denoted by $\binom{n}{t}$, is equal to

$$\binom{n}{t} = \frac{n!}{t!(n-t)!} \, .$$

If we assume that both n and t are large, an approximate, but asymptotically tight, expression of $\binom{n}{t}$ results from Stirling formula, namely:

$$\binom{n}{t} \approx \frac{1}{\sqrt{2\pi t(n-t)}} \left(\frac{n}{t}\right)^t \left(\frac{n}{n-t}\right)^{n-t}$$

which can be rewritten as

$$\binom{n}{t} \approx 2^{n\mathcal{H}_2(t/n) - (1/2)\log_2(2\pi t(n-t))} \, ,$$

where $\mathcal{H}_2(\cdot)$ has been defined by Eq. (3.6). If n is large enough it approaches

$$\binom{n}{t} \approx 2^{n\mathcal{H}_2(t/n)} \tag{5.25}$$

since $\log_2(x)/x$ approaches 0 as x approaches infinity.

CHAPTER 6

An Outline of Error-Correcting Codes

6.1 INTRODUCTION

This chapter provides an introduction to error-correcting codes, considered as tools for making *reliable communication through unreliable channels*. A lot of works were devoted to error-correcting codes but only a few of them are actually accounted for below. Even when dealing with well-known codes, the approach for presenting them is rather unconventional: much of the vocabulary and notation used in this text departs from the standard ones, which have their origin in the history of the discipline rather than in a logical development. Our starting point will be an analysis of mere repetition, which is a rather crude error-correcting means. Then more elaborated error-correcting codes will be introduced. Interestingly, this approach will enable us to meet and understand practically important families of error-correcting codes, including the most efficient ones. We need first to describe the basic communication problem and the kind of channels which will be considered.

6.2 COMMUNICATING A MESSAGE THROUGH A CHANNEL

6.2.1 DEFINING A MESSAGE

A *message* consists of an arbitrary sequence of elements, referred to as *symbols*, taken from some alphabet, i.e., some finite collection of signs. For instance, the previous sentence is made of letters from the Latin alphabet, plus the interval which enables separating the words. If we ignore the punctuation marks and typographic differences (some letters are in italic or capitalized), we may thus think of the alphabet as containing 27 symbols (we shall say that its size is 27). Similarly, the genomic message is made of a sequence of nucleic bases which belong to the 4-symbol alphabet {**A**, **T**, **G**, **C**}. The only requirement on the alphabet is that it be made of unambiguously distinguishable symbols, so the smallest possible size of an alphabet is 2. We shall assume in the following that we consider messages from the binary alphabet and we shall represent its symbols by the digits 0 and 1, often referred to as *bits*, an acronym for 'binary digits'[1]. If a given message uses an alphabet of size larger than 2, it can be converted into the binary one. For instance, if we represent the nucleic bases according to the correspondence rule **A** $\rightarrow$ 00, **G** $\rightarrow$ 01, **T** $\rightarrow$ 10 and **C** $\rightarrow$ 11, we can replace, e.g., the 6-symbol message from the nucleic-base alphabet **AATCGG** by the 12-bit binary message 000010110101, where each block of two successive bits represents a nucleic base according

[1] Remember that we use 'bit' with the meaning of binary symbol, never to name the binary information unit which is referred to as 'shannon.'

to the following (arbitrary) rule: the first bit represents the chemical structure of the nucleotide, 0 meaning 'purine' (**R**) and 1 'pyrimidine' (**Y**); the second bit indicates to what complementary pair the nucleotide belongs, 0 meaning **A–T** (or **W**) and 1 **C–G** (or **S**). The conversion is especially easy here since 4 is a power of 2, but conversion from an alphabet of any size into the binary alphabet is possible by slightly less simple means. If, for instance, the alphabet is of size 3, say $\{0, 1, 2\}$, the correspondence rule may be $0 \to 00$, $1 \to 01$ and $2 \to 1$. Then, a symbol 0 or 1 of the ternary alphabet is represented by a block of two bits which begins with the bit 0, while the ternary symbol 2 is represented by the single bit 1 (this rule obeys the prefix condition stated above in Sec. 5.2 hence enables unambiguously separating the binary representations of ternary symbols). Assuming the messages to be binary thus results in little loss of generality. We shall do so in the sequel for the sake of simplicity.

When a k-symbol binary message is represented by an n-symbol codeword, the average information quantity which is borne by a transmitted symbol is $R = k/n$, which is referred to as the code rate. Thus, R equals the entropy of the 'redundant source' made of the channel encoder fed by the original source (see Fig. 3.8 in Sec. 3.6). The fundamental theorem of channel coding thus implies that errorless communication is possible if and only if $R = k/n < C$, where C is the channel capacity in shannons. If the message symbols belong to the q-ary alphabet, then the above inequality becomes $(k/n)\log_2(q) < C$ since each of the k information symbol now bears an information quantity of $\log_2(q)$ shannons.

6.2.2 DESCRIBING A CHANNEL

A *channel* is any medium where symbols can be propagated or recorded. For instance, if the symbols 0 and 1 of the binary alphabet are represented by the polarity of a voltage, say $+v$ corresponding to 0 and $-v$ to 1, measuring the polarity at the extremity of a pair of conducting wires will let know what binary symbol was represented by the voltage at the other extremity. Changing the polarity at regularly spaced intervals according to the sequence of bits of some message, at the transmitting end, will enable to reconstruct the transmitted message at the receiving end. We are interested in the case where the channel is *unreliable*, i.e., where observing the voltage at the receiving end does not ensure with certainty that the transmitted bit is identified, due to the addition to the proper voltage of a spurious one, referred to as *noise*. Then the channel is characterized by its *signal-to-noise ratio* which is the ratio of the signal power to that of the noise. For a given constant signal-to-noise ratio, the decisions taken about the binary symbols at the receiving end incur a constant probability of error which is the smaller, the larger the signal-to-noise ratio. We then obtain the simple model to be introduced now.

The simplest model of a binary channel is referred to as the *binary symmetric channel*. A channel can be described by the probabilities of the transitions between its input and output symbols, as depicted in Fig. 6.1 (already drawn as Fig. 3.4-**a** in Ch. 3). This channel is *memoryless*, i.e., deals separately with the successive input symbols. Its input symbol X and its output symbol Y are both binary. If $X = 0$, there is a probability p that $Y = 1$, and if $X = 1$ that $Y = 0$ with the *same*[2] probability p, referred to as the channel error probability.

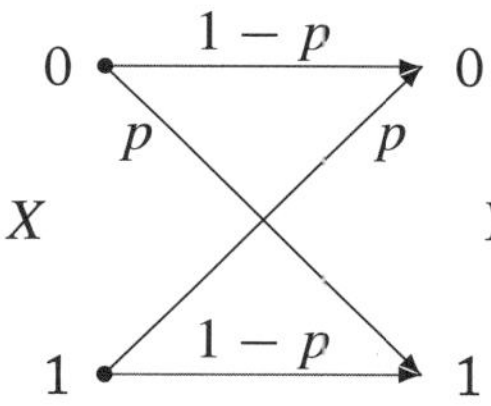

Figure 6.1: Binary symmetric channel with error probability p. All the transition probabilities from an input symbol to an output symbol are plotted on the figure.

Another description can be given of the same channel. Let E denote a binary random variable which assumes the value 1 with probability $p < 1/2$ and hence the value 0 with probability $1 - p$. The assumption that $p < 1/2$ does not incur any loss of generality, since exchanging the arbitrary signs 0 and 1 suffices to replace the initial error probability $p > 1/2$ with $1 - p < 1/2$; the case $p = 1/2$ may be discarded as equivalent to the absence of any channel, since then the output does not depend on the input. Then the channel operation can be described as the addition modulo 2 of E to the input binary random variable X, namely:

$$Y = X \oplus E \tag{6.1}$$

where $\oplus$ denotes the addition modulo 2: $0 \oplus 0 = 0, 0 \oplus 1 = 1 \oplus 0 = 1$ as in ordinary addition, but $1 \oplus 1 = 0$. (The addition modulo 2 may be interpreted as representing the parity of the ordinary addition, the binary variables '0' and '1' meaning 'even' and 'odd,' respectively.) Then, adding 1 modulo 2 results in changing a bit to its binary complement as does an error in the binary symmetric channel.

When endowed with the addition modulo 2 and the ordinary multiplication (such that $0 \times x = 0$ and $1 \times x = x$, where x is any binary symbol), the binary set $\{0, 1\}$ acquires the mathematical structure of a *field*: any of its elements has an inverse for the addition, there exists a neutral element for the addition (0), any nonzero element has a multiplicative inverse. Here, 1 is the single nonzero element and is its own inverse, so it is the neutral element for multiplication. The binary set $\{0, 1\}$ is the smallest set which can be endowed with the field structure.

The error probability p depends on the noise sample which affects the transmitted voltage which, for instance, has some known probability distribution (typically, a Gaussian one). At the receiving end the noise sample is not known but each received voltage is measured and used to assess the reliability of the corresponding symbol. Then the output of the channel is a real quantity, say a, which for Gaussianly distributed noise is related to the transmitted binary symbol x and to the

[2] It is why the channel is referred to as 'symmetric.'

instantaneous error probability p (up to a constant factor) according to:

$$a = (-1)^x \ln \frac{1-p}{p}, \tag{6.2}$$

which implies

$$p(a) = \frac{1}{\exp[(-1)^x a] + 1} = \frac{1}{\exp(|a|) + 1}. \tag{6.3}$$

By 'instantaneous error probability,' we mean the probability that an error affects the received symbol when the measured channel output is the real number a. Its dependence on a has been explicated by the notation $p(a)$. Assuming $p(a)$ to be less than 1/2, $\ln \frac{1-p(a)}{p(a)}$ is positive so the sign of a is positive for $x = 0$ and negative for $x = 1$. The sign of a thus bears a binary decision. The magnitude $|a| = \ln \frac{1-p(a)}{p(a)}$ measures the reliability of this decision since it is a decreasing function of the error probability, with $|a| = 0$ for $p = 1/2$ (the decision is indeed not reliable at all in this case) and $|a|$ approaching infinity if the error probability approaches 0 (then the decision is fully reliable). It will be convenient to refer to a quantity like (6.2) as the *real value*[3] of the corresponding received binary symbol and its use is referred to as a 'soft decision.'

If the measured received voltage is not used in the subsequent process, but only the binary decision based on its sign, $p(a)$ must be replaced by its average $\bar{p}$ and the channel is described by the scheme of Fig. 6.1 with constant $\bar{p}$. The performance is significantly degraded but the process is much simplified. This case is referred to as 'hard decision.'

Reliably communicating through an unreliable channel means the ability to communicate a message (for instance, a sequence of binary symbols) with an arbitrarily small probability of error through a noisy channel (for instance, a binary symmetric channel as just described, having a symbol error probability $\bar{p} > 0$, or the channel with binary input and real output, namely, the real value a of each received symbol).

6.3 REPETITION AS A MEANS OF ERROR CORRECTION

6.3.1 ERROR PATTERNS ON REPEATED SYMBOLS AND THEIR PROBABILITY

To begin with, let us assume that we wish to reliably communicate a single binary symbol x through a binary symmetric channel of error probability p, as depicted in Fig. 6.1. We may repeat this symbol, say, 3 times. Then, we may consider that we send a 'word' of the form xxx, where x belongs to the binary alphabet $\{0, 1\}$. At the receiving end, 8 possible words may correspond to the transmitted word xxx, depending on the error pattern $\underline{e}$ which occurred due to the channel noise. The components of $\underline{e}$ are independent binary variables due to the assumption that the channel is memoryless. The possible error patterns are listed together with their probability in the following table, where $\bar{x}$ denotes the binary complement of x, $\bar{x} = x \oplus 1$.

[3] In earlier works, and especially in [7], we named this quantity 'algebraic value' for translating the French 'valeur relative' used in [6], just meaning that it is endowed with a sign; it has however no relationship with algebraic codes and we hope that the word 'real' helps avoiding misunderstanding.

received word	error pattern	probability
xxx	000	$(1-p)^3$
$\bar{x}xx$	100	$p(1-p)^2$
$x\bar{x}x$	010	$p(1-p)^2$
$xx\bar{x}$	001	$p(1-p)^2$
$\bar{x}\bar{x}x$	110	$p^2(1-p)$
$\bar{x}x\bar{x}$	101	$p^2(1-p)$
$x\bar{x}\bar{x}$	011	$p^2(1-p)$
$\bar{x}\bar{x}\bar{x}$	111	p^3

One immediately notices that the probabilities do not depend on the transmitted symbol x, but only on the *weight* of the error pattern $\underline{e}$, defined as the number of symbols '1' it contains. An obvious generalization to n-fold repetition is that the probability of *a given* error pattern $\underline{e}$ of weight $w(\underline{e})$ equals

$$P_{\underline{e}} = p^{w(\underline{e})}(1-p)^{n-w(\underline{e})} . \tag{6.4}$$

Assuming the channel error probability p to be less than 1/2, the probability $P_{\underline{e}}$ of an error pattern of weight $w(\underline{e})$ is a decreasing function of $w(\underline{e})$ for any codeword length n. This conclusion holds, of course, independently of the transmitted symbol.

We also notice that the probability of occurrence of *any* error pattern of weight w equals

$$P_w = \binom{n}{w} p^w (1-p)^{n-w} \tag{6.5}$$

since there are $\binom{n}{w}$ distinct possible error patterns of weight w, each of which of probability given by Eq. (6.4). This is a Bernoulli distribution with expected value (or mean) np and variance (i.e., the expected value of the square of the difference between its actual value and its mean) $np(1-p)$.

6.3.2 DECISION ON A REPEATED SYMBOL BY MAJORITY VOTING

Let us go back to the case of a triply repeated bit x, and assume that the real values of the received bits are not available. Their common probability of error is denoted by p. If the received word consists of nonidentical bits, for instance 001, it is *sure* that a nonzero error pattern occurred since the transmitted word consisted of a same repeated symbol. On the other hand, if we receive a word made of a repeated bit, say 000, we *do not know* whether 000 was actually transmitted without error or if the actually transmitted word was 111 but the error pattern 111 occurred, transforming 111 into 000. We can say only that 000 is more likely to have actually been transmitted, since the probability of the error pattern 000, $(1-p)^3$, is larger than the probability p^3 of the error pattern 111, because $p < 1/2$. Therefore, the best assumption we can make is that 000 was transmitted. Similarly, the probability of the error pattern 001, $p(1-p)^2$, is larger than that of 110 which is $p^2(1-p)$, so the best assumption when 001 is received is that the error pattern 001 actually occurred so that the transmitted word was 000. In both cases, the best decision $\hat{x}$ as regards the transmitted bit is $\hat{x} = 0$.

We may thus formulate the optimum decision rule: the best assumption $\hat{x}$ as regards the transmitted bit is obtained by *majority voting*. For a symbol repeated n times (instead of 3 times in the example), majority voting clearly defines a decision procedure if n is odd. For n even, a tie occurs if $n/2$ bits equal 1 (and the remaining $n/2$ equal 0). It is optimally solved by choosing the bit $\hat{x}$ at random (0 or 1 with probability 1/2). This optimum decision rule has as *average* probability of error the probability that an error pattern occurs with weight larger than $n/2$, namely:

$$P_{n\text{rep}} = \sum_{w=(n+1)/2}^{n} \binom{n}{w} p^w (1-p)^{n-w}, \; n \text{ odd}, \tag{6.6}$$

$$P_{n\text{rep}} = \sum_{w=n/2+1}^{n} \binom{n}{w} p^w (1-p)^{n-w} + \frac{1}{2}\binom{n}{n/2} p^{n/2}(1-p)^{n/2}, \; n \text{ even}$$

as a consequence of Eq. (6.5).

We notice that no improvement of the error probability results from repetition if $n = 2$ since then $P_{2\text{rep}} = p(1-p) + p^2 = p$. If $n > 2$, however, repetition at the transmitting end and majority voting at the receiving one always improve the error probability since $P_{n\text{rep}} < p$, and moreover $P_{n\text{rep}}$ is a decreasing function of n which approaches 0 when n approaches infinity. The lack of improvement of the decision error probability in the case of $n = 2$ does not mean that twofold repetition is always useless: when the received word is 01 or 10, we know that no reliable decision can be taken on the transmitted bit so we may think of this bit as being erased. This is better than taking a wrong decision and can be exploited, e.g., when twofold repetition is combined with some other coding process.

Using a decision rule which results in a symbol belonging to the alphabet of the transmitting end, like majority voting, is referred to as a *hard decision*. We know that it is the best possible decision, but no assessment of the probability that it is correct or wrong is available. For instance, in the case of n-fold repetition with n even, a binary decision is taken with a probability of 1/2 to be wrong if a tie occurs, but the bit which results from this decision cannot be distinguished from those which result from decisions taken with a large margin, hence having a small probability of error.

6.3.3 SOFT DECISION ON A REPEATED SYMBOL

A different situation arises with the use of *soft decisions*. A soft decision consists (ideally) of assessing the probability of a symbol by taking into account measurements on the corresponding received physical quantity as well as the constraints this symbol possibly obeys. At the single-bit level, measuring the channel output voltage implies a soft decision since we know that it is related to the probability of an error according to Eqs. (6.2) and (6.3). When we consider n-fold repetition of a bit, the fact that the repeated bits are identical is a constraint which can be taken into account. One easily shows that the optimum decision rule in this case is:

$$\hat{a} = \sum_{i=1}^{n} a_i \, , \tag{6.7}$$

where a_i is the real value (referred to as *a priori*) of the i-th individual received symbol and where $\hat{a}$ is the *a posteriori* real value of the decision on the repeated symbol, optimally taking account of the fact that n identical symbols were transmitted. The extreme simplicity of this decision rule is the main motivation for using the real value formalism. Interestingly, replacing the *a priori* real value a by the *a posteriori* one $\hat{a}$ in Eqs. (6.2) and (6.3) still expresses the relationship of the *a posteriori* real value $\hat{a}$ of the decision on the repeated symbol with the error probability of this decision. Therefore, the output of a decoder has the same form as its input and the same relation to the error probability, which enables cascading several decoders, or iterated decoding. By comparison, taking a hard decision implies the loss of the reliability information borne by the magnitude of the real value. Multiple successive hard decisions thus incur considerable information loss, which practically hinders the use of schemes combining several codes. It turns out, however, that these schemes are very efficient provided successive soft decisions are implied in their decoding.

6.4 ENCODING A FULL MESSAGE

6.4.1 INTRODUCTION

The trouble with n-fold repetition is that reliably communicating a *single* symbol needs that *many* symbols are sent. The *rate* R, defined as the ratio of the number of symbols actually communicated to the number of sent symbols is here $R = 1/n$. This scheme provides a vanishingly small probability of error when n approaches infinity, but at the expense of a vanishingly small rate. In any realistic situation, the rate should be as large as possible. Is it possible to encode a message of length k into a codeword of length n so as to reduce the error probability of the message? Moreover, is it possible to let k and n together approach infinity so as to keep a nonzero value of the rate $R = k/n$, while at the same time the probability of a decoding error vanishes?

The researches on error-correcting codes were intended to answer both questions and they succeeded in finding adequate solutions. We shall first introduce a simple example where the mathematical formalism already discussed can be employed, and we shall then generalize.

6.4.2 A SIMPLE EXAMPLE

Given a codeword length n and a message length k, with $n > k$, we assume that we associate a codeword of length n to any message of length k, resulting in a set of words named (n, k) *block code*. We first consider small values of k and n. Let us assume for instance that we wish to communicate a message of $k = 3$ bits (referred to as *information bits*) using a word of length $n = 6$ bits. Let $u_1u_2u_3$ denote this message. Let us assume that we append to it $n - k = 3$ redundancy bits (also referred to as 'check bits') r_4, r_5 and r_6, so we obtain the codeword $u_1u_2u_3r_4r_5r_6$. The redundancy bits are computed in terms of the information bits according to the following rule:

$$\begin{aligned} r_4 &= u_1 \oplus u_2 \,, \\ r_5 &= u_2 \oplus u_3 \,, \\ r_6 &= u_3 \oplus u_1 \,. \end{aligned}$$

The addition modulo 2 denoted by $\oplus$ has already been defined in Sec. 6.2.2. Its properties enable writing these equalities in the form of the three *parity checks*:

$$\begin{aligned} u_1 \oplus u_2 \oplus r_4 &= 0, \quad (c_{12}) \\ u_2 \oplus u_3 \oplus r_5 &= 0, \quad (c_{23}) \\ u_3 \oplus u_1 \oplus r_6 &= 0, \quad (c_{31}) \end{aligned} \tag{6.8}$$

which express that the number of bits '1' in each of the left-hand sides is zero or even. The check on the information bits u_i and u_j is denoted by c_{ij}, $i, j = 1, 2$ or 3. With respect to the case of mere repetition dealt with in Sec. 6.3, we can say that each redundancy bit repeats an information bit in combination with another one. This enables a higher rate, and is almost equivalent to the mere repetition if the probability that two bits are simultaneously in error is low enough, i.e., if the error probability p is small enough.

Notice that the above encoding process immediately generalizes to an arbitrary number k of information bits by associating a redundancy bit to each of the $\binom{k}{2} = k(k-1)/2$ pairs of information bits. A code of length $n = k(k+1)/2$ and minimum distance k results, to be referred to as a 'pair-checking code'. Such a code is very redundant for large k. We'll denote it by $\mathcal{PC}(k)$, so the code taken as example is $\mathcal{PC}(3)$.

A matrix formalism is convenient for describing this code. Let us represent the message by the row matrix (vector) $\underline{u} = [u_1\ u_2\ u_3]$ and the encoded word by the row matrix $\underline{c} = [u_1\ u_2\ u_3\ r_4\ r_5\ r_6]$. Then the encoding rules can be expressed by the matrix relation (we assume that the entries of all matrices are dealt with as elements of the binary field, i.e., their addition is effected modulo 2):

$$\underline{c} = \underline{u}\mathrm{G}\,, \tag{6.9}$$

where G is the 'generator matrix'

$$\mathrm{G} = \begin{bmatrix} 1 & 0 & 0 & 1 & 0 & 1 \\ 0 & 1 & 0 & 1 & 1 & 0 \\ 0 & 0 & 1 & 0 & 1 & 1 \end{bmatrix}. \tag{6.10}$$

Similarly, the set of parity checks can be summarized by the matrix relation:

$$\underline{c}\mathrm{H}^{\mathrm{t}} = \underline{0}\,, \tag{6.11}$$

where the superscript t denotes transposition, $\underline{c}$ denotes a codeword and $\underline{0}$ denotes the all-0 $(n-k)$-vector (row matrix), and where H is the 'parity-check matrix'

$$\mathrm{H} = \begin{bmatrix} 1 & 1 & 0 & 1 & 0 & 0 \\ 0 & 1 & 1 & 0 & 1 & 0 \\ 1 & 0 & 1 & 0 & 0 & 1 \end{bmatrix}. \tag{6.12}$$

Then, the matrix equality (6.9) tells how to generate a codeword in terms of the message to be transmitted, while the matrix equality (6.11) describes the constraints that any codeword $\underline{c}$

specifically obeys. The set $\mathcal{PC}(3)$ of words which correspond to all possible messages according to Eq. (6.9), which is also the set of words which obey constraints Eq. (6.11), constitutes what is referred to as a *linear code*. The term 'linear code' means that all codewords are obtained by linear combinations of the rows of the generator matrix, i.e., if $\underline{g}_1, \underline{g}_2, \ldots, \underline{g}_k$ are rows of G, then $\lambda_1\underline{g}_1 \oplus \lambda_2\underline{g}_2 \oplus \cdots \oplus \lambda_k\underline{g}_k$ is a codeword, where $\lambda_1, \lambda_2, \ldots, \lambda_k$ are elements of the binary field and the componentwise addition modulo 2 is denoted by $\oplus$. This implies that the all-0 word, which results from choosing $\lambda_1 = \lambda_2 = \cdots = \lambda_k = 0$, belongs to any linear code.

The table below lists all the codewords of the code $\mathcal{PC}(3)$ chosen as example. They are generated according to Eq. (6.9) with (6.10) as generator matrix and they obey the parity-check relation (6.11) with H given by (6.12).

message	codeword
000	000000
100	100101
010	010110
001	001011
110	110011
101	101110
011	011101
111	111000

The code has been designed so that the first three bits of any codeword are identical to the corresponding message. A code such that the encoded message appears in a given set of positions in its codewords is referred to as being in *systematic form*. Any linear code can be put in systematic form, and assuming it to be so does not incur any loss of generality as regards its performance.

Let us define the *Hamming distance* between two codewords as the number of bit positions where these words differ. The smallest Hamming distance between any two of the listed codewords, referred to as the *minimum distance* of this code, is 3. We do not need to check this property for any pair of codewords. It suffices to notice that the weight of the codewords (as defined in Sec. 6.2.2) is at least 3: if $\underline{c}_1$ and $\underline{c}_2$ belong to a linear code, then due to the linearity $\underline{c}_1 \oplus \underline{c}_2$ also belongs to this code, where the operation denoted by $\oplus$ has been defined above. The weight of $\underline{c}_1 \oplus \underline{c}_2$ equals the Hamming distance between $\underline{c}_1$ and $\underline{c}_2$ due to its very definition, but it is also the distance of the codeword $\underline{c}_1 \oplus \underline{c}_2$ to the codeword $\underline{0}$. Therefore, the set of Hamming distances between the words of a linear code reduces to the set of their weights.

Since the all-zero word $\underline{0}$ belongs to any linear code, no loss of generality results from assuming that the message to be transmitted is the sequence $\underline{0}$. The only 'ones' found in the output sequence then result from channel errors.

6.4.3 DECODING THE CODE TAKEN AS EXAMPLE USING THE SYNDROME

We first briefly discuss a decoding process of the code $\mathcal{PC}(3)$ taken as an example which is typical of the hard decoding of binary linear block codes. In such a hard-decision decoder the input as well as the output are symbols of the binary alphabet. We'll consider later 'soft-decision' decoders involving real values, which have significantly better performance at the expense of a more difficult implementation.

Let $\underline{c}'$ denote the word received when the word $\underline{c}$ is transmitted. We may write $\underline{c}' = \underline{c} \oplus \underline{e}$, where $\underline{e}$ is the error pattern. Let us compute the matrix product

$$\underline{s} = \underline{c}'\mathrm{H}^{\mathrm{t}} = \underline{e}\mathrm{H}^{\mathrm{t}} , \tag{6.13}$$

where the second equality is a consequence of Eq. (6.11). The $(n - k)$-vector $\underline{s}$ is referred to as the *syndrome*. If all its components are 0 ($\underline{s} = \underline{0}$), then $\underline{c}'$ belongs to $\mathcal{PC}(3)$ so the most likely assumption is that no errors occurred. Now, if an error pattern of weight 1 occurred, e.g., $\underline{e} = [1\,0\,0\,0\,0\,0]$, we have $\underline{s} = [1\ 0\ 1]$ i.e., the transpose of the first column of matrix (6.12). Similarly, had the error occurred on the i-th bit of $\underline{c}'$, the syndrome would be the transpose of the i-th column of matrix (6.12). Since all the columns of matrix H are distinct, the syndrome thus designates the column of matrix (6.12) at the position in $\underline{c}'$ where the error occurred, provided the error pattern is of weight 1. Since the alphabet is binary, a located error can immediately be corrected by replacing the corresponding bit by its complement. Of course, if an error pattern of weight larger than 1 occurs, the decoding rule above results in a wrong decision.

Computing the syndrome is the first step of the algebraic decoding of any linear code. In the above example, the relationship of the syndrome and the most likely error pattern is straightforward, because $\mathcal{PC}(3)$ was designed in order to obtain it. In more complicated codes, determining the most likely error pattern in terms of the syndrome is much more difficult. For short enough codes, a precomputed table can be used to associate the possible 2^{n-k} syndromes with the corresponding minimum-weight error patterns.

6.4.4 REPLICATION DECODING OF THE CODE TAKEN AS EXAMPLE

Let us now look at another means for decoding $\mathcal{PC}(3)$. We notice that the situation is not very different from the three-fold repetition of a single bit, except that the code rate is now 1/2 instead of 1/3. Indeed, we can solve the parity-check relations (6.8) in terms of u_1 when they contain it. Then, (c_{12}) results in $u_1 = u_2 \oplus r_4$ and (c_{31}) in $u_1 = u_3 \oplus r_6$. This means that both $u_2 \oplus r_4$ and $u_3 \oplus r_6$ repeat u_1 so it is possible to apply the above decision rules discussed in Sec. 6.3.3 to the 3 available received *replicas* of u_1: u_1', $u_2' \oplus r_4'$ and $u_3' \oplus r_6'$, where the prime denotes the received symbols. We shall refer to the first replica as the 'trivial' one (it is the information symbol to be decoded itself) and the other two as 'compound replicas,' made of a combination of other symbols. The hard decision rule consists of majority voting among the above three replicas, and its result is the same as that of the decoding rule described in Sec. 6.4.3. We now consider soft decision decoding. The 'soft decision' rule consists of applying Eq. (6.7) to the corresponding *a priori* real values, namely a_1 which is associated with u_1, and the *a priori* real values associated with $u_2' \oplus r_4'$ and $u_3' \oplus r_6'$. One easily shows that the real value a associated with the sum modulo 2 of two received binary symbols of real values a_i and a_j is such that

$$t(a) = t(a_i)t(a_j) \tag{6.14}$$

where $t(a)$ is defined as

$$t(a) = \tanh(a/2) = \frac{\exp(a) - 1}{\exp(a) + 1} .$$

Then this definition and Eq. (6.14) result in

$$a = \ln \frac{1 + t(a_i)t(a_j)}{1 - t(a_i)t(a_j)} = \ln \frac{\exp(a_i + a_j) + 1}{\exp(a_i) + \exp(a_j)} . \tag{6.15}$$

The real value associated with the sum modulo 2 of $\ell > 2$ binary variables is such that

$$t(a) = \prod_{i=1}^{\ell} [t(a_i)] , \tag{6.16}$$

hence its explicit expression is

$$a = \ln \frac{1 + \prod_{i=1}^{\ell} [t(a_i)]}{1 - \prod_{i=1}^{\ell} [t(a_i)]} . \tag{6.17}$$

Let $a_1, a_2, \ldots$ and a_6 denote the *a priori* real values of $u'_1, u'_2, u'_3, r'_4, r'_5$ and r'_6, respectively. Using Eq. (6.15), we can write the soft decision rule for optimally decoding u_1 as

$$\hat{a}_1 = a_1 + \ln \frac{1 + t(a_2)t(a_4)}{1 - t(a_2)t(a_4)} + \ln \frac{1 + t(a_3)t(a_6)}{1 - t(a_3)t(a_6)} , \tag{6.18}$$

where $\hat{a}_1$ is the *a posteriori* real value associated with u_1. Similarly, the decoding rules of u_2 and u_3 are

$$\hat{a}_2 = a_2 + \ln \frac{1 + t(a_3)t(a_5)}{1 - t(a_3)t(a_5)} + \ln \frac{1 + t(a_1)t(a_4)}{1 - t(a_1)t(a_4)} , \tag{6.19}$$

and

$$\hat{a}_3 = a_3 + \ln \frac{1 + t(a_2)t(a_5)}{1 - t(a_2)t(a_5)} + \ln \frac{1 + t(a_1)t(a_6)}{1 - t(a_1)t(a_6)} . \tag{6.20}$$

Figure 6.2 represents the encoder which generates the code taken as example in Sec. 6.4.2 if we interpret the square boxes as memories containing a single bit (1-bit memories) and the circles as modulo 2 adders. The 3 leftmost boxes are initially loaded with the 3 information bits u_1, u_2, and u_3, while the 3 boxes labeled r_4, r_5, and r_6 are initially left empty. The sums modulo 2 of the contents of the 1-bit memories which are connected to each modulo 2 adder are computed and load the 3 boxes labeled r_4, r_5, and r_6. Then the six 1-bit memories contain a codeword.

Interestingly, Fig. 6.2 also represents a bit-by-bit decoder for the same code $\mathcal{PC}(3)$. We consider for instance decoding the information bit u_1. Let the 1-bit memories of the upper row be initially loaded with the received bits $u'_1, \ldots, r'_6$. We now assume that c_{12} and c_{31} compute

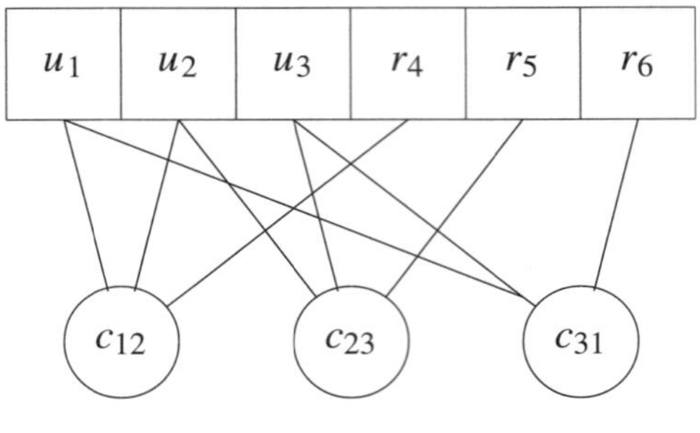

Figure 6.2: Implementation of the code defined by the parity-check equations [(6.8)].

the sum modulo 2 of the contents of the 1-bit memories connected to them, u_1 excepted. These sums modulo 2 are the two compound received replicas of u_1, so majority voting on u_1', $u_2' \oplus r_4'$ and $u_3' \oplus r_6'$, where u_1' is present in the first 1-bit memory while the other two are computed in the checks c_{12} and c_{31} and are applied to this 1-bit memory through the indicated connections. The decoded bit results from majority voting among u_1', $u_2' \oplus r_4'$ and $u_3' \oplus r_6'$, and is denoted by $\hat{u}_1 = \text{maj}[u_1', (u_2' \oplus r_4'), (u_3' \oplus r_6')]$.

We have just described hard decision decoding of the code $\mathcal{PC}(3)$. The same figure also describes its soft decision decoding, provided we interpret the square boxes as containing real numbers and the circles as processors which compute the function defined by Eq. (6.15) in terms of the real quantities present at their inputs. The square boxes are initially loaded with the *a priori* real values a_1, $a_2, \ldots, a_6$ associated with bits u_1, u_2, u_3, r_4, r_5, and r_6, respectively, and the circular boxes c_{12} and c_{31} compute the real values associated with the 2 compound replicas of u_1. The real quantities present in the first square box and those coming from the circular boxes c_{12} and c_{31} are added, resulting in the soft decision (6.18) about u_1. The soft decision on the other two information bits are obtained similarly. A decoder like the one just described which both accepts and delivers soft decisions, i.e., real numbers the sign of which bears a binary decision and the magnitude of which measures the reliability of this very decision is referred to as a *soft-input soft-output*, or SISO, decoder.

The decoding procedures just described for the code $\mathcal{PC}(3)$ taken as example directly generalize to any pair-checking code $\mathcal{PC}(k)$ as defined in Sec. 6.4.2.

6.4.5 DESIGNING EASILY DECODABLE CODES: LOW-DENSITY PARITY CHECK CODES

Of the two operations of encoding and decoding, the latter is by far the most difficult. It is why it is expedient to design a code so as to make its decoding as easy as possible. To this end, let us now consider the generalization to much longer messages and codewords of the scheme of Fig. 6.2, as depicted in Fig. 6.3. The length of the row of 1-bit memories, i.e., the codelength n, is now much larger, say, 1000 or more. The row of $(n-k)$ parity checks is of proportionate size, say, 500 or so. Now, connect *at random* the 1-bit memories and the parity checks, only obeying the following conditions:

— each parity check is connected to at least 2 of the k 1-bit memories which contain the information bits and to 1 of the $(n - k)$ ones which contain the check bits;

— the 1-bit memories connected to each parity check are few: they are at least 3 but do not need to be much more numerous (their number needs not be the same for all parity checks);

— each 1-bit memory is connected to at least one parity check.

The codes thus defined are referred to as *low density parity check* (LDPC) codes. They have been introduced and studied by Gallager [39]. Their presentation here is inspired by Berrou [23].

Just like the scheme of Fig. 6.2, that of Fig. 6.3 represents a decoder as well as an encoder. The hard decision decoder using majority voting has only a limited interest. Now replace the 1-bit memories with 1-real-number memories containing the real values of the corresponding received bits, and interpret the parity checks as processors acting on real numbers to compute real values of sums modulo 2 according to Eqs. (6.15) or (6.17). The real value thus computed in a particular parity-check processor is added to that already contained in the 1-real-number memory to which it is connected. We thus obtain a SISO decoder as defined in Sec. 6.4.4. Then a major fact is that decoding can be *iterated*. If the code rate k/n is low enough with respect to the channel capacity, the decoding process described above results in an average improvement, the magnitudes being changed and the sign of certain real values (hence the corresponding binary decisions) being possibly inverted. Once these improved results have been obtained, the decoding process may be repeated in terms of the new real values and result in a further average improvement. The process can be iterated many times, until a stable enough result has been obtained. No theorem ensures that this process converges towards a well defined limit, but it actually does in practice. The performance thus obtained is close to that of turbocodes (see Sec. 6.7), i.e., small decoding error probabilities are obtained at rates close to the channel capacity.

Strictly speaking, this decoder is not optimal since decoding equations like Eqs. (6.18) to (6.20) are valid only provided the replicas of each information bit are disjoint, but no provision was made in order to ensure they are so. However, the scarcity of the bits combined in each parity check entails that only a small fraction of the replicas of information bits are not disjoint, so dealing with them as disjoint is an acceptable approximation which entails little impairment with respect to the optimal decoding.

Interestingly, although the low-density parity check codes are intended to make bit-by-bit decoding easy, their design also results in a distance distribution close to that which is obtained by random coding, as shown in [14].

Connecting in the scheme of Fig. 6.3 some of the information bits to many checks and other information bits to few checks, or even to none, results in an unequal-protection error-correcting code as the nested system to be discussed in Ch. 9. Moreover, it can be extended to provide an increased protection to already present information bits by including more information and check bits, increasing the number of parity checks and connecting the already present information bits to

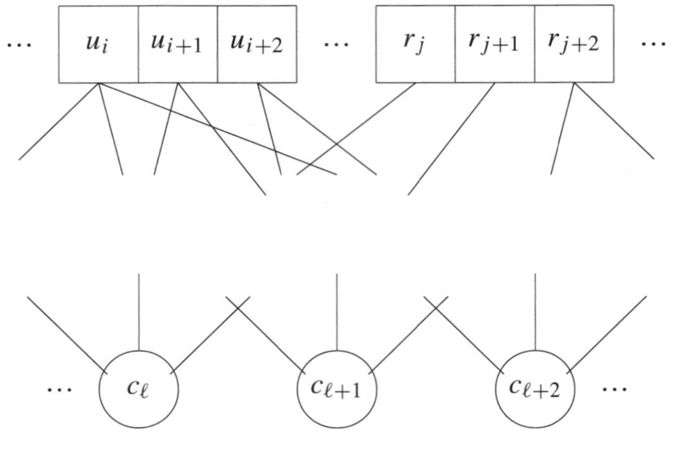

Figure 6.3: Scheme of a low-density parity check code.

the newly introduced parity checks, while keeping the already existing connections. This process is similar to that described in Sec. 9.1 to account for the setup of a nested system.

6.4.6 SOFT DECODING OF OTHER BLOCK CODES

The soft decoding rule based on replicas takes the simple form of Eqs. (6.18), (6.19), and (6.20) only if the replicas obtained by combining bits are disjoint in the sense that no bit belongs to more than one combination, or if they can be considered as approximately so. It turns out, unfortunately, that the codes such that disjoint replicas of the information symbols can be written are few, so this simple rule is far from being the general case. For nondisjoint replicas, a similar decoding rule can be written or implemented as an algorithm, but it is not as simple as in the above example. However, it can be always written as

$$\hat{a}_i = a_i + f_i(a_1, a_2, \ldots, a_{i-1}, a_{i+1}, \ldots, a_n) \tag{6.21}$$

where $\hat{a}_i$ denotes the *a posteriori* real value associated with the information bit u_i, $a_1, a_2, \ldots, a_n$ are the *a priori* real values associated with the n received bits, and $f_i(\cdots)$ is a function of *a priori* real values other than a_i. We may refer to it as the *extrinsic* real value of u_i, i.e., contributed by all the other bits of the codeword. The explicit expression of this function in the general case is

$$f_i(a_1, a_2, \ldots, a_{i-1}, a_{i+1}, \ldots, a_n) = \ln \frac{\sum_{\underline{c} \in C_{i0}} \exp(-\sum_j c_j a_j)}{\sum_{\underline{c}' \in C_{i1}} \exp(-\sum_j c'_j a_j)} - a_i \ , \tag{6.22}$$

where C_{i0} (resp. C_{i1}) denotes the set of words of the given code C having 0 (resp. 1) as i-th symbol, $\underline{c}$ denotes a codeword, c_j its j-th bits, and a_j denotes the corresponding *a priori* real value. This formula is rather complicated but fairly simple algorithms enable computing the function, at least for simple enough codes (we shall briefly describe some of these algorithms in Sec. 6.6.3). Moreover, approximations of $f_i(\cdots)$ can be computed with little performance degradation. The simplest one consists of keeping only the largest terms in the numerator and the denominator of the argument of the logarithm in Eq. (6.22), which results in:

$$f_i(a_1, a_2, \ldots, a_{i-1}, a_{i+1}, \ldots, a_n) \approx \min_{\underline{c} \in C_{i1}} \sum_j c_j a_j - \min_{\underline{c}' \in C_{i0}} \sum_j c'_j a_j - a_i \,. \tag{6.23}$$

From the definition of C_{i0} and C_{i1}, it is clear that $\hat{a}_i$ takes the form of the sum of a_i and a term $f_i(a_1, a_2, \ldots, a_{i-1}, a_{i+1}, \ldots, a_n)$ which does not depend on a_i, since $\exp(-a_i)$ is a factor of the denominator of the fraction in the logarithm argument and its numerator does not depend on a_i. Therefore, Eqs. (6.22) and (6.23) may look trivial. Writing the decision rule (6.21) as the sum of a_i and a term which does not depend on it is, however, meaningful due to the existence of algorithms which directly compute $f_i(a_1, a_2, \ldots, a_{i-1}, a_{i+1}, \ldots, a_n)$, exactly or approximately, in terms of real values other than a_i.

The search for block codes resulted in the invention of a large number of code families. The simplest one is that of the Hamming codes, which can correct a single bit in error. The code taken as example in Sec. 6.4.2 is actually a slightly modified Hamming code. Much more powerful and complex code families were designed using the algebra of finite fields, also referred to as Galois fields, collectively known as 'algebraic codes'. All of them were designed according to the *minimum distance criterion* for which a code is the better, the larger its minimum distance, and their decoding generally used the hard decision procedure of computing the syndrome and determining the most likely error pattern in terms of it. Although soft decoding using Eq. (6.21) is always possible for these codes (a means for doing so consists of applying the soft-output Viterbi algorithm or the BCJR algorithm to the trellis of a block code, see Sec. 6.6.3), computing the function $f_i(\cdots)$ given by Eq. (6.22) [even its approximate expression (6.23)] is quite complex for long codes and, moreover, soft decoding was seldom contemplated by the researchers on algebraic codes as involving probabilistic methods quite foreign to the algebra of finite fields, their discipline. We shall not discuss algebraic codes although they generated a plentiful literature, but consider instead the case of convolutional codes which has been fruitful in the design of practically efficient codes. Before doing so, we consider the more general framework provided by information theory.

6.5 ERROR-CORRECTING CODES IN THE FRAMEWORK OF INFORMATION THEORY

6.5.1 AN OUTLOOK ON THE FUNDAMENTAL THEOREM OF CHANNEL CODING

As yet, we have just introduced repetition as an error correction means and shown on a simple example that indirect repetition, i.e., involving combination of symbols by means of addition modulo 2, enables using less redundancy bits for a same number of replicas of each information bit. However, the sum modulo 2 of two or more unreliable bits is still less reliable than each of them, which means that the performance is degraded when the channel has a high probability of error (a low signal-to-noise ratio). We have seen also two very different decoding methods: hard decoding, where the received symbols are dealt with as symbols of the code alphabet, and soft decoding, where the *a priori* probabilities of the symbols are taken into account and where decoding moreover consists of reassessing the probability of the decoded symbols. More emphasis will be laid on the latter which is mandatory when several codes are combined (as in turbocodes to be considered later), in order to prevent a prohibitive performance degradation. The implementation of soft decoding is more complex than that of hard decoding, but the semi-conductor technology made such progresses since the invention of the transistor in 1948 that it is by now inexpensive and reliable.

Examples are necessarily simple, but efficient codes imply large codewords: a vanishingly small probability of decoding error can be obtained only asymptotically as the length of the codewords tends to infinity. Unlike the case of a symbol encoded by mere repetition, a good code should maintain a strictly positive rate $R = k/n$ when n approaches infinity, where k is the number of information bits, so k also should approach infinity. That an errorless communication is asymptotically possible at any nonzero rate is actually a major result of information theory, referred to as the *fundamental theorem of channel coding* and already discussed in Sec. 5.3.1. It predated the attempts of building practical error-correcting codes (hence of finite length). Actually, this theorem acted as a strong incentive to the research of practical codes. The challenge was moreover that the proof of the theorem assumed *random coding*, the decoding of which is far too complex to be used.

Although the proof of the fundamental theorem of channel coding in its more general form involves rather difficult mathematics, we believe that it can be easily understood at the light of the *law of large numbers* in its weakest form, indeed a very fundamental concept. This law simply tells that the experimental outcomes have frequencies (defined as the ratio of the number of occurrences of given outcomes to the total number of trials) which approach with high probability the probabilities of the corresponding events when the number of trials approaches infinity. For instance, tossing a fair coin 10,000 times may result in the all-head (or all-tail) sequence, but with a probability[4] of only $2^{-10,000} \approx 10^{-3,010}$: a miraculous event! In strong contrast, the probability that the frequency of the heads or tails in the sequence thus obtained is close to 1/2 (say, in the range 0.47–0.53), is very large (see Eq. (6.25) in Sec. 6.5.3). Similarly, obtaining an error pattern of weight close to np when n symbols are transmitted over a binary symmetric channel of error probability p has a large probability, provided the number of errors expected in the average, np, is large enough. So, although of course we do not know *where* the errors occur, we know that their *total number* is probably close to np provided this number is large enough.

[4]This probability is actually that of *any given* sequence of that length. The difference is that the all-head sequence is unique, while there are $\binom{10,000}{5,000} \approx 7.98 \times 10^{3,007}$ sequences having exactly 5,000 heads and 5,000 tails.

6.5.2 A GEOMETRICAL INTERPRETATION

Let us consider the finite space of n-bit words, which has been referred to as the Hamming space of dimension n and denoted by S_n in Sec. 3.5.1. Its contains 2^n elements referred to as its points. In geometrical terms, knowing with high probability the total number of errors means that the received word is represented by a point which is probably close to the surface of an n-dimensional hypersphere of S_n centered on the transmitted word, the radius of which is the expected number of errors np. If the minimum distance d of the code is larger than twice this number, a point at the surface of this hypersphere is closer to the actually transmitted word than to any other codeword, hence unambiguously identifies it. Being with high probability close to this surface, the point which represents the received word is with high probability closer to the transmitted word than to any other codeword. The optimum decoding rule can thus be simply stated: *choose the codeword the closest to the received word.* The probability that this decoding rule results in an error is the smaller, the larger is n, and it vanishes as n approaches infinity.

Again in geometrical terms, the design of the best code can be thought of as intended to *spread* $M < 2^n$ points belonging to S_n so as to make them as far apart to each other as possible for the Hamming distance. For a given value of the error probability p (assuming the channel to be binary symmetric), there is clearly a limit to the number M of points which can be put in S_n while keeping the distance between these points larger than $d = 2np$. Let $M_{\max}$ denote this number. The quantity

$$C = \lim_{n\to\infty} \frac{\log_2(M_{\max})}{n}$$

is the largest possible information rate which can be communicated through this channel. It is referred to as the *capacity* of the channel, expressed in shannons per bit, a quantity less than 1 since $M < 2^n$ is a necessary condition for the code to be redundant. It has been shown in Sec. 3.3.3 that the capacity of a binary symmetric channel of error probability p is given by:

$$C_{\text{bsc}} = 1 + p\log_2(p) + (1-p)\log_2(1-p) = 1 - \mathcal{H}_2(p) \text{ Sh/bit} .$$

The channel with binary input and real output, in the case of additive Gaussian noise, has a capacity significantly larger for the same value of the average error probability. It is expressed in terms of the signal-to-noise ratio S/N by a more complicated formula which we do not give here.

6.5.3 DESIGNING GOOD ERROR-CORRECTING CODES

Random Coding

Random coding, i.e., the design of a code by choosing at random M points into the n-dimensional Hamming space, is a means for regularly spreading points which is asymptotically optimum as n approaches infinity. Random coding ensures that all the n dimensions of the space S_n are dealt with on an equal footing, in the average, and it results in achieving an average rate equal to the channel capacity. In the case of a code made of M n-bit words, it consists of drawing at random each bit of a word independently of the other ones, with probability 1/2 of being 1 or 0, the M words which make up the code being so drawn independently of each other. Then, the probability of a particular codeword $\underline{c}$ is $P_{\underline{c}} = 2^{-n}$, and the probability of obtaining any codeword of weight w is:

$$P_w = \binom{n}{w} 2^{-n} . \tag{6.24}$$

The expected value, or mean, of this weight is $n/2$, and its variance, i.e., the expected value of the square of the difference between its actual value and its mean, is $n/4$. For very large n, a good approximation of this distribution is the Gaussian probability density: if we approximate w/n by the continuous random variable X, the probability that X belongs to the interval $(x, x + \mathrm{d}x)$ is $p_X(x)\mathrm{d}x$, where

$$p_X(x) = \sqrt{\frac{2n}{\pi}} \exp[-2n(x - 1/2)^2] . \tag{6.25}$$

This probability density function has a maximum at $x = 1/2$, hence for $w = n/2$, and assumes symmetrically decreasing values as x differs from 1/2. It is concentrated around $x = 1/2$, and the width of the region where it has nonneglible values decreases as $1/\sqrt{n}$, hence vanishes as n approaches infinity.

Let us also mention that the fundamental theorem of channel coding has been proved by Gallager, assuming a randomly chosen error-correcting code [40].

Largest Minimum Distance of Error-Correcting Codes

No explicit means for designing a code with the largest possible minimum distance d is known. However, asymptotically for n approaching infinity, we know from random coding arguments that the largest possible minimum distance of a code is at least equal to the Gilbert-Varshamov bound d_{GV}, defined in the binary case by the implicit equation:

$$1 - k/n = \mathcal{H}_2(d_{\mathrm{GV}}/n) , \tag{6.26}$$

where the binary entropy function $\mathcal{H}_2(p)$ has been defined in Eq. (3.6). For instance, if we assume as above that $k = n/2$, d_{GV} is very close to $0.11 \times n$. Notice that the left-hand side of Eq. (6.26) measures the redundancy of the code and that $\mathcal{H}_2(\cdot)$ is an increasing function when its argument is smaller than 1/2, so the larger the redundancy, the larger d_{GV}. For k/n approaching 0, d_{GV} tends to $n/2$.

For a nonbinary code, i.e., with $q > 2$, the Gilbert-Varshamov bound becomes:

$$1 - k/n = \mathcal{H}_q(d_{\mathrm{GV}}/n) + (d_{\mathrm{GV}}/n)\log_q(q - 1) , \tag{6.27}$$

with

$$\mathcal{H}_q\,(x) = -x\log_q(x) - (1 - x)\log_q(1 - x) = \frac{\mathcal{H}_2(x)}{\log_2(q)} .$$

Notice the presence in Eq. (6.27) of the additive term $(d_{\mathrm{GV}}/n)\log_q(q - 1)$ which vanishes for $q = 2$. For very redundant codes k/n vanishes. Then d_{GV}/n approaches $(q - 1)/q$, a result easily deduced from Eq. (6.27).

Designing Effective Codes
Decoding a code obtained by random coding, however, is practically impossible, since decoding a single received word would imply to compare it with *all* possible codewords. Moreover, long codewords are necessary for good performance, which also means that the number of possible codewords, 2^{Rn}, hence the number of needed comparisons, is extremely large for large n (remember that $0 < R < 1$). It is why researches on error-correcting codes had to look for deterministic design rules, which enable their decoding with reasonably low complexity.

No general method is known for designing a code having $M_{\max}$ codewords, for an arbitrary value of n and a given error probability p. Only a few coding schemes are known, or more often conjectured, to be optimal for specific values of M and n, for some simple channels. In the absence of a general design rule, codes have been sought which are good for the minimum distance criterion, and the relevance of this simple criterion has not been questioned until the late 1980s. However, this criterion does not take into account the number of codewords at the minimum distance (or 'multiplicity') although a large value of the multiplicity results in performance impairment. The best codes known as yet, the turbocodes to be considered in more detail in Sec. 6.7, were not designed to comply with the minimum distance criterion. Their minimum distance is indeed often fairly small (at least with respect to the known bounds on the largest minimum distance) but their multiplicity is very small. These properties result in an 'error floor,' i.e., a decrease of the decoding error rate in terms of the signal-to-noise ratio much less steep when the signal-to-noise ratio is large than when it is small. Since the aim of an error-correcting code is to improve communication when the channel is poor, we may say that turbocodes are good where they are useful but bad where they are not.

We have seen that long words are needed for obtaining a small probability of decoding error. The design of block codes, their encoding and, above all, their decoding become complicated when an n-symbol codeword represents a message of k information symbols where both k and n are large. There exists fortunately a family of linear codes which enable low complexity encoding, where only a small number of information symbols, say k_0, generate n_0 output symbols at a time. Repeating this encoding process K times results in an arbitrarily long sequence of $n = Kn_0$ symbols as a function of $k = Kk_0$ information symbols: the codes thus obtained are referred to as *convolutional codes*. Besides the ease of their encoding, certain of these codes have very good weight properties, and their decoding is comparatively easy. Moreover, combining such codes according to efficient concatenation schemes and using adequate means for their decoding result in the most powerful error-correcting devices known as yet, the 'turbocodes', dealt with below in Sec. 6.7. The practical performance of turbocodes being close to the theoretical limit stated by information theory, i.e., the channel capacity, no significant improvement can be expected from other code families.

6.6 CONVOLUTIONAL CODES

6.6.1 CONVOLUTIONAL ENCODING

The *convolutional codes* are made of arbitrarily long codewords which can be generated by a simple encoding device. The simplest way for defining these codes consists of describing the encoder which generates them. They have no predetermined length, and their encoder consists of a device which receives the sequence of information bits and outputs the corresponding encoded sequence. For simplicity's sake, we assume that $k_0 = 1$, i.e., that the information bits enter the encoder one by one. Then there is no need to segment the input data stream into blocks, and the encoder is directly fed by the information bits. The heart of a convolutional encoder is a *shift register*, i.e., a set of a (generally small) number m of binary memories, or cells, which are arranged in a row and connected to each other so that when a new bit enters the leftmost cell, the content of the memory of any cell is transferred into the cell immediately at its right. The content of the rightmost cell is lost unless it is explicitly connected to some other device. We shall represent a shift register according to Fig. 6.4. We assume that the binary symbols are represented by some physical quantity (e.g., a voltage, the polarity of which represents the binary symbol) which is maintained constant during some time interval T referred to as 'bit duration,' after which the same physical quantity takes on the value which represents the next bit. For this reason, we interpret a memory cell as a delay element, the output of which is delayed with respect to its input by the bit duration T, and a shift register as made of a number of such delay elements.

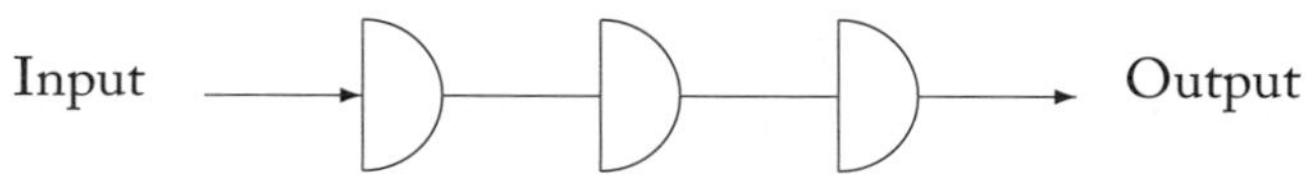

Figure 6.4: Binary shift register of size $m = 3$. It consists of 3 binary memories, or cells. Each cell consists of a delay element of one-bit duration, represented by a D-shaped box. When a new bit enters the leftmost cell, the content of the memory of any inner cell is transferred into the cell immediately at its right; that of the rightmost cell is output.

A convolutional encoder is made of such a shift register which directly receives the information sequence. The contents of some of its memory cells are added modulo 2 to compute output bits on the one hand, and possibly 'recursion bits' which are fedback to the register input via a modulo 2 adder on the other hand. In the two examples to be presented, each encoder delivers two output bits for each input information bit. These two bits are generated in parallel, but they are most often transmitted in alternance (serially), each during a time interval $T/2$. The first example of a convolutional encoder is nonsystematic (the input bit is not output) and nonrecursive (there is no feedback) and has a memory $m = 2$ (Fig. 6.5). Despite its simplicity, it generates sequences which are at a Hamming distance of 5 to each other, hence they constitute a code which enables correcting all double errors. This convolutional code is a standard example in textbooks.

The encoder depicted in Fig. 6.6 is systematic since the input bit is directly output, and it also generates and outputs another bit, referred to as 'check bit.' The shift register and its connections will be referred to as a 'check bit generator' or 'rate-1 encoder.' That of Fig. 6.6 has been represented

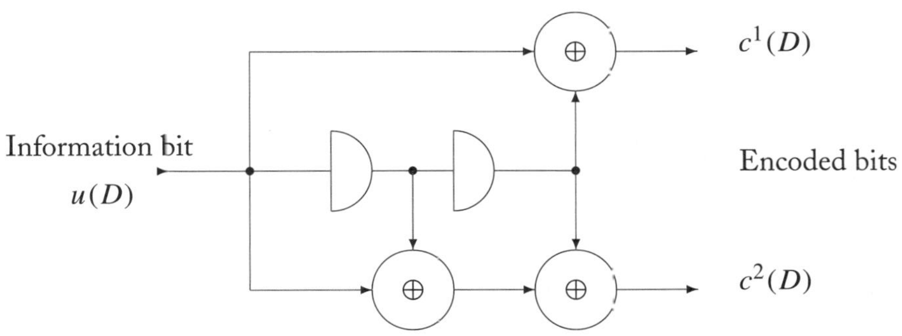

Figure 6.5: Example of a convolutional encoder of rate 1/2 with memory $m = 2$. For each entering information bit $u(D)$, two encoded bits $c^1(D)$ and $c^2(D)$ are output (the dependence on the dummy variable D will be explicated later). Each point may assume one of two states, denoted by 0 and 1. The elements drawn as half-circles are one-bit delay operators (which together make a register), and $\oplus$ means addition modulo 2.

inside a dashed box. It uses a feedback and is thus referred to as 'recursive.' The feedback does not increase the minimum distance of a code, so it has no advantage in terms of the minimum distance criterion. However, it is very important in shaping the weight distribution of the code, since this feedback entails that the Hamming distance between most of the sequences it generates linearly increases with time. A code generated by such a recursive encoder is typical of a component code of a turbocode (see next section).

A very convenient formalism, referred to as the 'D-transform,' can be used to deal with shift registers and the sequences they generate instead of the vector notation used as yet for block codes. It uses polynomials and formal series of an indeterminate D (which stands for 'delay,' meaning a delay of one bit duration with respect to some origin of time; D is also referred to as 'delay operator'). We use a polynomial $N(D)$ in the indeterminate D with coefficients belonging to the binary field to represent how the output is computed as the sum modulo 2 of bits contained in the register. The coefficients of $N(D)$ are 1 or 0, depending on the output of the memory cell corresponding to the degree of the indeterminate being connected or not to the modulo 2 adder which computes the encoder output. For instance, $N(D) = 1 + D^2 + D^3$ for the encoder depicted in Fig. 6.6. Similarly, a different polynomial $Q(D)$ specifies the recursion bit, where the coefficients are 1 or 0, depending on the output of the corresponding memory cell being used or not to compute the recursion bit fedback to the register input ($Q(D) = 1 + D + D^3$ for the encoder depicted in Fig. 6.6). We shall moreover assume that the degree of both polynomials $N(D)$ and $Q(D)$ equals the register memory m and that both polynomials are irreducible, i.e., they cannot be expressed as a product of polynomials. Then it can be shown that the sequence generated by the encoder made of the shift register with its feedback determined by $Q(D)$ and its output computed according to $N(D)$, provided the initial state, i.e., the content of all the memory cells of the register, is 0 at the origin of time, is represented

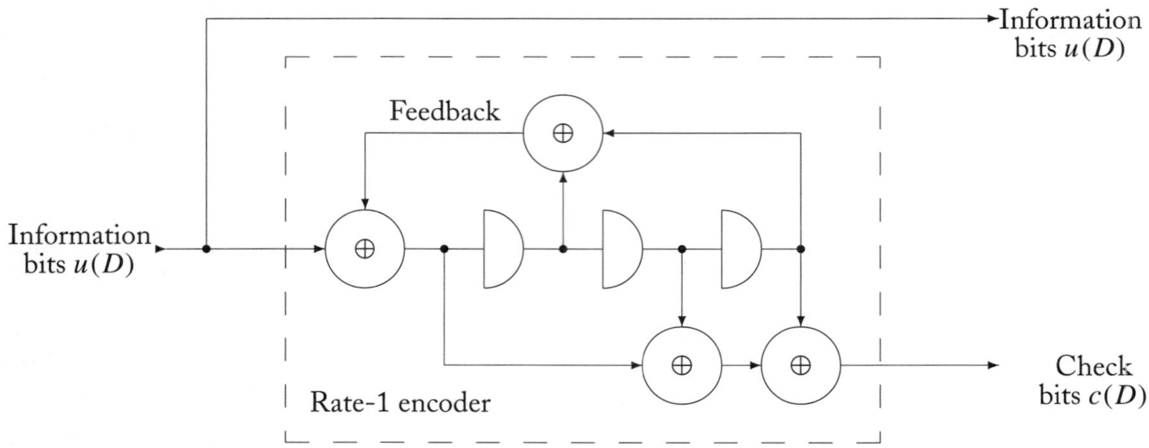

Figure 6.6: Example of a recursive systematic encoder of rate 1/2 with memory $m = 3$. Each entering information bit is output together with a check bit computed in the dashed box labeled 'rate-1 encoder.' This box contains the register and its connections. Each point may assume one of two states, denoted by 0 and 1. The elements drawn as half-circles are delay operators (of an information bit duration); they form together a register, and $\oplus$ denotes addition modulo 2. This encoder is referred to as 'recursive' because it involves a feedback.

as:

$$c(D) = u(D)\frac{N(D)}{Q(D)}, \tag{6.28}$$

where $u(D)$ represents the input (information sequence) and $c(D)$ is the output sequence. For instance, if $u(D) = 1$, i.e., if the input sequence consists of a single 1 at the origin of time, the encoded sequence is represented as

$$c(D) = \frac{N(D)}{Q(D)},$$

which, since $Q(D)$ does not divide $N(D)$, can be expanded in a formal series with infinitely many terms, which thus represents an infinitely long sequence. For instance, we have in the above example:

$$c(D) = \frac{1 + D^2 + D^3}{1 + D + D^3}.$$

The reciprocal of the polynomial $Q(D)$ is

$$\frac{1}{1 + D + D^3} = 1 + D + D^2 + D^4 + D^7 + D^8 + D^9 + D^{11} + D^{14} + \cdots$$

where the right-hand side is periodic (the period here is the largest possible, $7 = 2^3 - 1$; the largest possible period for an m-cell shift register is $2^m - 1$), so

$$\begin{aligned} c(D) &= \frac{1 + D^2 + D^3}{1 + D + D^3} \\ &= 1 + D + D^4 + D^5 + D^6 + D^8 + D^{11} + D^{12} + D^{13} + D^{15} + D^{18} + D^{19} + D^{20} + \cdots \end{aligned}$$

which is also periodic with period 7 except for the first two terms.

The requirement that the initial state of the shift register be zero implies that means for controlling the initial state are employed. For instance, a termination made of m properly chosen bits may be appended to an N-bit information message so as to make the register content return to the zero state. Another solution to the register initialization problem consists of choosing a nonzero initial state such that the same state is reached after an N-bit message has been fed to the register, resulting in a 'tail-biting' convolutional code. The initial state needs then to be pre-computed in terms of the message. This is an elegant solution to the problem of initializing both component encoders of a turbocode (see next section).

Going back to the sequences generated by a convolutional encoder, we first notice a very important property: when an encoder uses feedback, i.e., is recursive, a finite-weight input sequence can generate an infinite-weight output sequence. This occurs for any input polynomial $u(D)$ which is not a multiple of the denominator $Q(D)$, i.e., for any input $u(D)$ such that $u(D)N(D)/Q(D)$ does not reduce to a polynomial, for instance if the input sequence is made of a single bit 1. On the other hand, in an encoder without feedback, any finite-weight input sequence would generate a finite-weight output sequence.

The multiples of the polynomial in denominator $Q(D)$, of degree m, constitute a fraction of 2^{-m} among all polynomials of any degree greater than or equal to m. This means that only a fraction 2^{-m} of all possible inputs to the encoder are encoded into sequences represented by polynomials and hence of finite weight, while all the others are represented by a formal series hence by an infinite-weight codeword. One could think of using a register of very long length m so as to ensure that almost all generated encoded sequences have infinite weight, resulting in an almost errorless communication. Unfortunately, the complexity of decoding varies as 2^m (at least if optimum decoding is sought), which severely limits the possible values of m. Excellent weight properties can however be obtained if several short encoders are combined according to the *turbocode* scheme, which moreover can be almost optimally decoded with fairly low complexity. Performance close to the theoretical limit can indeed be obtained by this means with only two combined registers. Before we start describing the turbocodes, let us consider the crucial question of decoding convolutional codes.

6.6.2 SYSTEMATIC CONVOLUTIONAL CODES AND THEIR DECODING

Given the output sequence $c(D)$ given by Eq. (6.28), how can we decode it, i.e., recover the input sequence $u(D)$? Inverting this formula seems to be an answer. Doing so, we formally obtain:

$$u(D) = c(D)\frac{Q(D)}{N(D)} . \tag{6.29}$$

However, this relation fails to provide adequate means to decode $c(D)$. Remember that encoding is intended to protect the information message against errors. Applying Eq. (6.29) works only if no error occurred. What is actually received in the presence of errors is $c(D) \oplus e(D)$, where $e(D)$ is the D-transform representing the error pattern. Trying to use Eq. (6.29) with $c(D) \oplus e(D)$ instead of $c(D)$ results in $u(D) \oplus e(D)Q(D)/N(D)$. Except for the no-error case $e(D) = 0$ and a fraction 2^{-m} of all nonzero error patterns, $e(D)Q(D)/N(D)$ is a sequence of infinite weight. Instead of correcting an error, using Eq. (6.29) severely increases the number of errors, since a simple error pattern would generally result in the 'decoded' sequence having a larger weight error pattern than $e(D)$, often an infinite-weight one, for instance in the simplest case of a channel error pattern of weight 1. Most channel-error patterns would have such a catastrophic effect on the recovered sequence.

The set of check sequences $c(D)$ generated by a recursive encoder (i.e., involving feedback) does not actually constitute an error-correcting code. Remember that we have already stated that redundancy is a necessary condition of error correction. Since one bit of $c(D)$ is output every time an information bit (represented by a single term in $u(D)$) enters the register, there is no redundancy if we transmit only $c(D)$. We must actually transmit more mutually dependent bits, for instance both $u(D)$ and $c(D)$. Then, since two bits are transmitted every time an information bit enters the register, we have a redundant code of rate 1/2. Other rates can be obtained if information bits enter the register k_0 at a time and result in n_0 output bits at a time, with $n_0 > k_0$, so as to generate a redundant code of rate $R = k_0/n_0$. If the output contains the information bits at some specified places, the encoder is said to be 'systematic,' and it can be shown that no loss of generality results from considering only codes in systematic form. The output bits other than the information bits are referred to as 'check' or 'redundancy bits.' We may think of a systematic convolutional code, say of rate 1/2, as a kind of twofold repetition of the information bit where one of the repeated bits is transmitted while the other one is combined with others due to the encoder operation. The information bits and the check bits are most often transmitted according to a time-invariant pattern (e.g., in alternation for a 1/2-rate encoder), but what is important is that bits of both kinds are actually transmitted. Assuming from now on the code to be systematic, we shall denote any of its codewords by $[u(D), c(D)]$. At the decoding end, the systematic and check bits need to be distinguished, i.e., their location in the assumed time-invariant pattern should be known. This pattern is the exact equivalent of the 'reading frame' for properly identifying the three-nucleotide codons in the DNA molecule.

Then, the decoding devices or algorithms should not merely implement an inversion rule like (6.29), but must (ideally) determine the sequence which belongs to the code, (i.e., which can be generated by the encoder as described), say $[\hat{u}(D), \hat{c}(D)]$, the closest for the Hamming metric to the received (hence possibly erroneous) sequence, say $[u'(D), c'(D)]$ where $u'(D) = u(D) \oplus e_u(D)$ and $c'(D) = c(D) \oplus e_c(D)$ denote the sequence of received information and check bits, respectively, $e_u(D)$ and $e_c(D)$ denoting the error patterns which affect the information and check sequences, respectively. We may think of the process which leads to determine $[\hat{u}(D), \hat{c}(D)]$ as a means for *regenerating* the transmitted codeword $[u(D), c(D)]$. If it is successful, and it is almost always so if the code design is well fitted to the statistical characteristics of the channel errors, we may say that the effect of these errors is cancelled. In other words, thanks to its association with the check sequence, the information sequence has been made *resilient* to the channel errors.

We now discuss how convolutional codes are decoded. The main tool for describing decoding algorithms is the trellis diagram considered in the next section. A trellis diagram can also be drawn for block codes, so the same algorithms still work for block codes. We shall not describe the algorithms in detail, but only indicate their general principle.

6.6.3 THE TRELLIS DIAGRAM AND ITS USE FOR DECODING

Trellis of a Convolutional Code

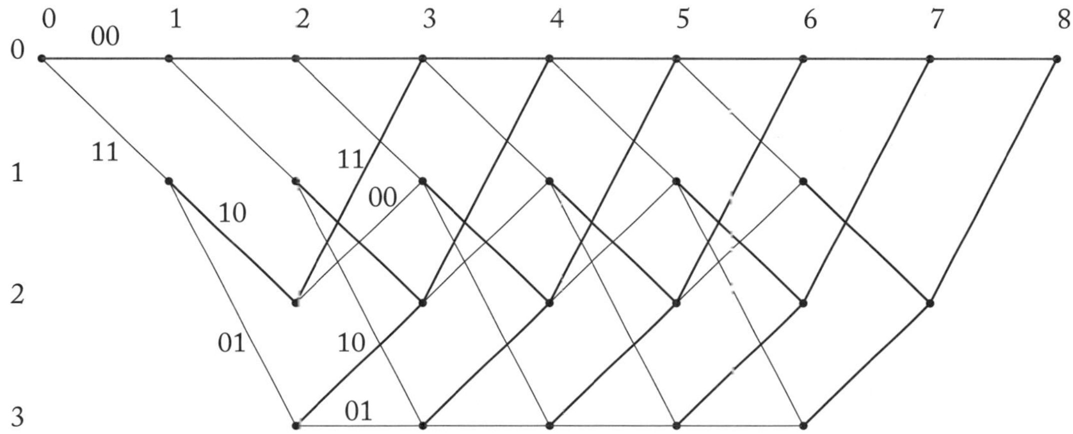

Figure 6.7: Trellis diagram corresponding to the encoder of Fig. 6.5. The states are the binary representations of the numbers from 0 to 3, plotted in terms of the step number. The branches drawn as thick lines correspond to information bits '0,' the thin ones to '1.' The labels of the branches are the bits output when the corresponding transition occurs. They have only been written for the first branches but periodically repeat themselves. Beyond the node level 6, we have drawn only the shortest paths going back to the zero state.

The trellis diagram graphically depicts the constraints that sequences generated by an encoder satisfy. Let us first consider the simple convolutional encoder represented in Fig. 6.5. How this encoder works can be graphically represented according to the trellis diagram of Fig. 6.7. The points in this figure represent the possible states of the encoder register at each step, i.e., every time a new information bit enters the encoder, and the lines between the points, referred to as 'branches,' represent the possible transitions from a state to another one. We assumed that both the initial and final states are 0. Apart for the first and last m steps, the trellis is made of a repetitive pattern.

Trellis of a Block Code

A trellis can be similarly drawn for binary linear block codes. Let us consider, again, the binary linear block code $\mathcal{PC}(3)$ taken as an example in Sec. 6.4.2. Its parity-check matrix is given by (6.12) and is rewritten here for convenience:

$$\mathrm{H} = \begin{bmatrix} 1 & 1 & 0 & 1 & 0 & 0 \\ 0 & 1 & 1 & 0 & 1 & 0 \\ 1 & 0 & 1 & 0 & 0 & 1 \end{bmatrix}.$$

One notices that all the columns of this matrix are binary representations of the numbers from 1 to 6. We define the state for a node of level i, which will play the same role as in the trellis of a convolutional code, as the number which has as binary representation the partial result of computing the product $\underline{c}\mathrm{H}^t$ for the first i symbols of the codeword $\underline{c}$. The state is plotted in terms of the node level in Fig. 6.8. If the first bit of $\underline{c}$ is '0,' the state at node level 1 is 0 (represented by vector [000]). If the first bit of $\underline{c}$ is '1,' one finds that the state at node level 1 is represented by the first row of H^t, namely [101], so the state is 5. If the second component is also '1,' the state obtained at node level 2 is represented by $[101] \oplus [110] = [011]$, hence is 3, etc. Only the paths which end at the state given by Eq. (6.11), i.e., 0, have been drawn as corresponding to possible codewords. The trellis thus obtained is represented in Fig. 6.8.

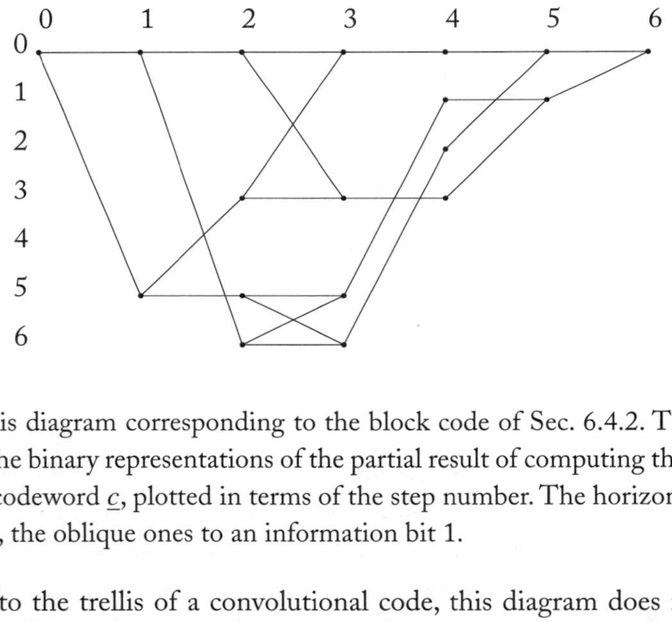

Figure 6.8: Trellis diagram corresponding to the block code of Sec. 6.4.2. The states are the numbers from 0 to 6, i.e., the binary representations of the partial result of computing the product $\underline{c}\mathrm{H}^t$ for the first i symbols of the codeword $\underline{c}$, plotted in terms of the step number. The horizontal lines correspond to an information bit 0, the oblique ones to an information bit 1.

Contrary to the trellis of a convolutional code, this diagram does not periodically repeat a same pattern. However, as expressing the coding constraints, it can be used as a support for decoding algorithms. These can be understood as means for computing the function $f_i(\cdots)$ in Eq. (6.21), using either the exact expression (6.22) or its approximation (6.23).

Decoding ultimately involves a decision which determines the most likely transmitted sequence among all the codewords. In the implementation of decoding or regeneration, however, it is more practical to proceed symbol by symbol, i.e., to decide what is the most likely symbol at each location in the sequence. Then the choice of an element among q^k possible sequences is replaced by n successive choices of a symbol among the q-ary alphabet (assuming an (n, k) block code). Of course, each decision about a symbol is taken in terms of other symbols. An interesting possibility, which is especially exploited in turbocodes (see Sec. 6.7), consists of using previous provisional decisions to improve later ones, according to an iterated process (see Sec. 6.7.2).

A Decoding Algorithm Based on a Trellis: The Viterbi Algorithm

Let us first consider how the trellis diagram can be used for word-by-word hard decoding, i.e., determining the transmitted word the closest for the Hamming distance to the received one. Any transmitted word is represented as a path in the trellis, which depends on the input sequence of information bits. Examining a trellis shows that the encoder states being in finite number, there are as many paths which diverge from each node than paths which converge in it, namely two for a binary code (except for the beginning and end of the trellis). When two paths converge at a same node, their distance can be computed and one of them (ignoring a possible tie) is definitively better than the other. We may thus, at each node, keep only the best converging path, which is referred to as the *survivor*. The Viterbi algorithm consists of keeping only the survivor at each node of the trellis to which paths converge, an event which occurs for all nodes at each step after it has begun. After a sufficiently large number of steps, a single path survives and it corresponds to the best codeword which is thus decoded [74, 33].

Instead of using the Hamming metric, the distance can be computed between the paths so as to take account of the *a priori* probabilities of the received bits, resulting in 'soft-input decoding.' As regards the possibility of soft-output decoding, one can first notice that the distance between the paths which converge to a node measures the reliability of the decision on the corresponding information bit. Moreover, one can deal with the previous decisions as conditioned on the new ones, so the assessment of their reliability can be updated as the algorithms proceeds, resulting in symbol-by-symbol soft-input, soft-output decoding [9, 44].

Another Decoding Algorithm Based on a Trellis: The BCJR Algorithm

The Bahl, Cocke, Jelinek, and Raviv (BCJR) algorithm also uses the trellis diagram [3]. It consists of considering the reliability of a decision at each decoding step as conditioned on both the previous and future decisions (once a sequence is registered the future is available as well as the past). Then, the reliability of each symbol is evaluated starting from the knowledge of the initial state, based then on the previous states but still depending on the future ones. At the last step, the final state is known so the algorithm exactly determines the probability of the last symbol and can proceed in the reverse direction, providing the lacking information on the reliability of the future state at each decoding step. For this reason, this algorithm is also referred to as the 'forward-backward algorithm.'

6.7 TURBOCODES

6.7.1 DESCRIPTION AND PROPERTIES

The basic scheme for generating a turbocode involves the combination of two[5] convolutional systematic recursive encoders with an interleaver. Systematic recursive convolutional encoders have been defined in the previous section. An *interleaver* of length N is a device having as input a sequence of length N, and an output which contains the same symbols as the input, hence also of length N, but in a different order. If its input sequence is $u_1, u_2, \dots, u_N$, where u_i is an element of some alphabet, for instance the binary one, its output is $u_{j_1}, u_{j_2}, \dots, u_{j_N}$, meaning that the j_i-th element of the input sequence became the i-th in the output sequence. An interleaver Π is completely defined by a set of N different indices $\{j_i, 1 \leq j_i \leq N, i = 1, 2, \dots, N\}$. For example, for $N = 7$, the interleaver defined by the set of indices $\{3, 6, 1, 2, 7, 5, 4\}$ transforms the word 0101100 into 0001011 and the word 1101011 into 0111110. The encoders currently used have a small memory (typically, $m = 3$) but, unlike the above example, the interleaver length N is long (say, a few thousands of bits), so there is an extremely large number of possible interleavers, as many, i.e., $N!$, as the permutations of N objects. When operating on some sequence $u(D)$ (represented by a polynomial of degree $N - 1$ in the indeterminate or delay operator D), the interleaver Π transforms it into $\Pi[u(D)]$. The device which recovers $u(D)$ in terms of $\Pi[u(D)]$, often referred to as the de-interleaver, is another interleaver of length N, to be denoted by Π^{-1}, such that $\Pi^{-1}[\Pi[u(D)]] = u(D)$.

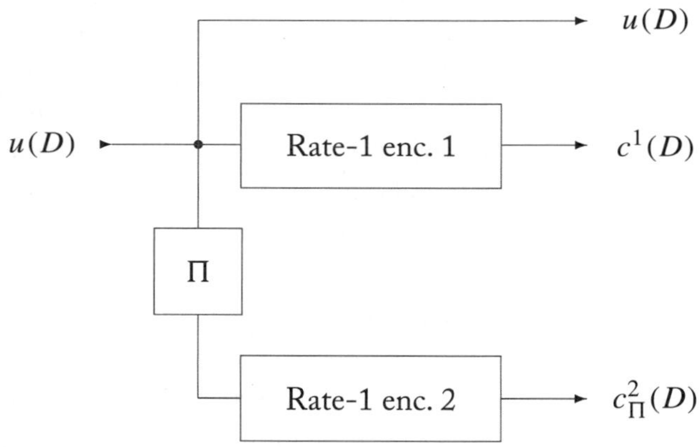

Figure 6.9: Encoder of a turbocode of rate-1/3. The blocks labeled 'Rate-1 enc.' designate rate-1 encoders as represented by the dashed box in Fig. 6.6. The block labeled Π represents the interleaver which implements permutation Π.

Then the encoder of a turbocode is organized according to Fig. 6.9: the input sequence of length N, represented by a polynomial in D of degree $N - 1$, say $u(D)$, is output since the encoder

[5] Or more, but then some specific difficulties are met; two-component codes suffice for obtaining results close to the optimum.

is systematic. At the same time, each of two recursive rate-1 encoders like that of Fig. 6.6 computes check bits. These encoders may be identical to each other, or not. It will be convenient for us to distinguish them by the use of a superscript 1 or 2. One encoder is fed directly by the input sequence and the check sequence it generates is denoted by $c^1(D)$, while the other one is fed by the output of the interleaver, the input of which is the information sequence $u(D)$. The check sequence it generates is denoted by $c^2_\Pi(D)$. This basic scheme has inherently a rate of 1/3, but several tricks can be used so as to obtain higher rates (the most usual of them, referred to as 'puncturing,' consists of suppressing some of the generated bits according to a given periodic pattern). Considering the encoding scheme of a rate-1/3 turbocode will however suffice to understand the principle of turbocodes so we shall not consider other rates. We shall denote the output of this rate-1/3 encoder by $[u(D), c^1(D), c^2_\Pi(D)]$. It is a sequence of length $3N$, the weight of which is the sum of the weights of $u(D)$, $c^1(D)$ and $c^2_\Pi(D)$. This is a 'parallel' representation of the encoder output. In practice the encoder would generally involve a conversion into a 'serial' representation, consisting of successively transmitting the three binary outputs at three times the frequency of information bits.

An information sequence $u(D)$ made of a single bit '1' followed by '0's, i.e., of weight 1, results in both $c^1(D)$ and $c^2_\Pi(D)$ having a large weight (it would be infinite without the necessary truncation of the sequences to the interleaver length N, as we already noticed in Sec. 6.6.1). Then the overall output has large weight. If now we consider an information sequence of weight 2, we obtain $c^1(D)$ having a finite weight only if $u(D)$ is a multiple of the denominator $Q^1(D)$, which occurs if the two bits of the information sequence are located an integer number of periods apart in the sequence, i.e.,

$$u(D) = D^\alpha(1 + D^{\beta P}) , \tag{6.30}$$

where α and β are integers, and P denotes the period of the formal series $1/Q^1(D)$. If $Q^1(D)$ is a 'primitive' polynomial[6], the content of the register successively represents in binary numeration all possible nonzero integers from 1 to $2^m - 1$ (in an order which depends on $Q^1(D)$) and we get a sequence of the longest possible period that can be generated by a register of length m, namely $P = 2^m - 1$. Then $Q^1(D)$ is a factor in the decomposition of $1 + D^{2^m-1}$ as a product of irreducible polynomials. It is so in the example of Fig. 6.6, where one easily checks that $1 + D^7 = (1 + D + D^3)(1 + D^2 + D^3)(1 + D)$ (do not forget that the coefficients are computed modulo 2), where both $1 + D + D^3$ and $1 + D^2 + D^3$ are primitive. Then the weight of the check sequence $c^1(D)$ corresponding to the information sequence (6.30) is β times the weight of the periodic part of the generated sequence plus a small constant, hence it is the larger, the larger β. Similarly, $c^2_\Pi(D)$ of small weight is obtained if, and only if, $\Pi[u(D)]$ has the form (6.30) as above. It is possible to choose the permutation performed by the interleaver so that $u(D)$ and $\Pi[u(D)]$ do not simultaneously assume this form for small values of β. Then the weight of the generated sequence $[u(D), c^1(D), c^2_\Pi(D)]$ is large. For input sequences of larger weight, it may be very difficult to strictly avoid the simultaneous occurrence of $c^1(D)$ and $c^2_\Pi(D)$ having low weight, but the probability of

[6] A polynomial of degree m is said to be primitive if taking the successive powers of one of its roots generates all the 2^m elements of a finite field.

such an event can be made very small. Although the design of a particular interleaver so as to obtain a specified weight performance is very hard, and although even the exact weight distribution which results from the use of a particular interleaver is difficult to compute, almost any interleaver will result in very few codewords of small weight and thus the overall performance will be good. Indeed, thanks to the interleaver operation and the fact that recursive convolutional encoders generate sequences of arbitrarily large weight, the set of sequences generated by a turbocode encoder resembles a set of sequences generated at random, so this encoding scheme appears as an approximate means for implementing random coding.

It turns out that a code generated at random is good with high probability. Indeed, random coding is the code construction that Shannon used in order to prove the fundamental theorem of channel coding (see Sec. 3.4.5) but, as we already mentioned, its decoding is practically impossible. The interest of the turbocode scheme is that its almost optimum decoding is comparatively easy, as we shall now show, while its weight distribution is close to that of random coding, hence intrinsically good.

6.7.2 SYMBOL-BY-SYMBOL SISO DECODING OF TURBOCODES

Let us first redefine the aim of decoding. Instead of trying to determine the codeword which has been transmitted with the highest probability, given the received word, let us extend the decoding role to determine the probability of each symbol of the code[7] (or at least of each of the bits in the information sequence) to be 0 or 1. Moreover, we may know the *a priori* probability of each of the symbols of the received word, as measured by the demodulator output (see Sec. 6.2.2). This kind of decoding is generally referred to as 'soft-input soft-output decoding' (SISO). We have already met SISO decoding in the very simple case of mere repetition, and in the simple example of Sec. 6.4.2. The general soft decision rule (6.21) was moreover of this kind. This redefinition of the decoding aim is an essential step for dealing with turbo decoding. It can be thought of as consisting of reassessing the probabilities of all the received bits, or at least of the information bits, given the *a priori* probabilities of all the received bits, so as to take account of the constraints of the code. The resulting probabilities will be referred to as *a posteriori* probabilities. We shall use the formalism of real values, as defined in Eq. (6.2), as we already did when dealing with mere repetition. Then, we may think of the decoding process as consisting of computing the *a posteriori* real value of each symbol (or of each information symbol) in terms of the *a priori* real values of all the received symbols. By *a priori* real value, we now mean the real value known at some present stage of the decoding process, prior to further processing, while *a posteriori* real value means that it resulted from taking into account some constraints yet ignored, or somehow improving the way constraints are accounted for.

Let us also notice that the *a posteriori* real value of any symbol, say $\hat{a}$, can be written as the sum of two terms, referred to as 'intrinsic' and 'extrinsic,' respectively. The first one, a_{in}, is the *a priori* real value of the symbol itself (which was referred to as the trivial replica in Sec. 6.4.4). The second

[7]The word made of the most probable bits is not necessarily identical to the codeword with the largest probability of having been transmitted. If the channel is not too bad, however, both almost always coincide.

one, a_{ex}, originates from the 'compound' replica obtained by expressing the same symbol in terms of others according to the constraints newly taken into account. The two replicas are independent as written in terms of disjoint sets of symbols. According to Eq. (6.21) but with different notation, the corresponding *a posteriori* real value is thus the sum of the intrinsic and extrinsic real values:

$$\hat{a} = a_{in} + a_{ex} \ . \tag{6.31}$$

We shall consider in the sequel elementary SISO decoders which implement the decoding rule (6.31) and are organized as in Fig. 6.10.

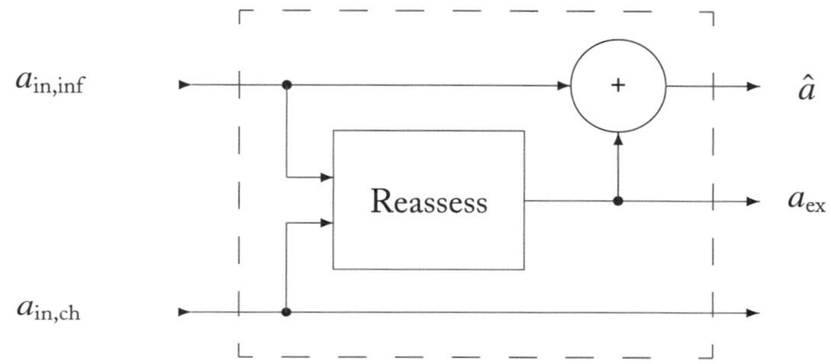

Figure 6.10: Elementary SISO decoder. Its two inputs are the *a priori* real values of the information bit and of the check bit, denoted by $a_{in,inf}$ and $a_{in,ch}$, respectively. Its three outputs are the *a posteriori* real value of the information bit $\hat{a}$, its extrinsic real value a_{ex}, and the *a priori* real value of the check bit which passes through the device without change. All these quantities are real and + denotes ordinary addition. The box labeled 'Reassess' implements the function $f_i(\cdots)$ in Eq. (6.21) so as to compute the extrinsic real value a_{ex}.

Let us now assume that we have an optimum (or nearly optimum) decoding rule for each of the two systematic codes (to be referred to as 'component codes') consisting of the two check sequences generated by the rate-1 encoders of the turbo encoder associated with the information sequence which, apart from the order of the symbols, is common to both. According to Eq. (6.31), we may write the *a posteriori* real value of any symbol which results from decoding the first code as:

$$\hat{a}^1 = a_{in} + a^1_{ex} \ , \tag{6.32}$$

where a^1_{ex} only takes account of the constraints of the first code. The same information symbol is present in the input sequence to the second encoder, at a location determined by the interleaver operation. We may thus consider the *a posteriori* real value from the first decoder (fed by the check sequence generated by the first rate-1 encoder) as the *a priori* real value of this symbol for the decoder which operates on the second component code, and write the decoding rule of this symbol with respect to the second code:

$$\hat{a}^2 = \hat{a}^1 + a^2_{ex} = a_{in} + a^1_{ex} + a^2_{ex} \ , \tag{6.33}$$

where the extrinsic real value a_{ex}^2 is computed in terms of *a priori* real values consisting of the *a posteriori* real values delivered by the first decoder. The second equality results from Eq. (6.32).

A very important fact is now that the decoding process may be *iterated*. Indeed, once all the symbols have been decoded by the second decoder, we may think of the *a posteriori* real values from the second decoder as being improved estimates which can be used as *a priori* real values in the first one. However, the term a_{ex}^1 in the *a posteriori* real value as expressed by Eq. (6.33) has already been calculated when the first decoder was used, so it should be *subtracted* from the real values fed to it. Provided the extrinsic real value originating from a decoder is systematically subtracted[8] from the *a posteriori* real value from the other one used as input to this decoder, as shown in Fig. 6.10, the decoding process can in principle involve an arbitrary large number of iteration steps. It almost always converges, although the conditions for this to occur are very hard to express. In practice, the number of iteration steps has to be limited. Moreover, criteria for stopping the iteration as soon as a good enough result has been obtained can be used. The complex task of decoding the turbocode as a whole has been reduced to a succession of alternate decodings of the component codes, each of which is easily performed.

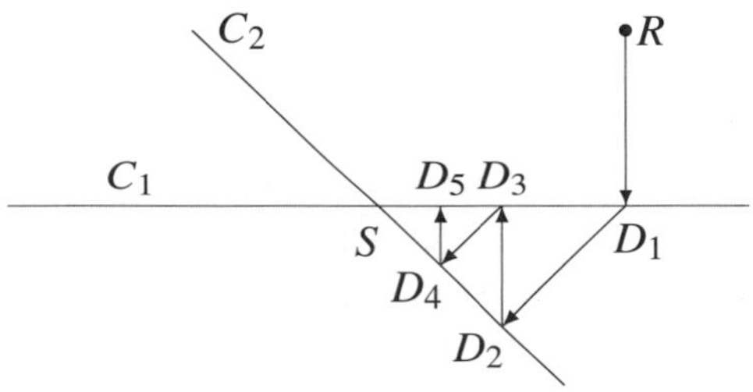

Figure 6.11: Geometric interpretation of the iterated decoding of two combined codes of a turbocode. Any point of the space represents the probability distribution of word components, initially conditioned on the received signals and then resulting from soft-output decodings. The point representing the optimally decoded codeword is denoted by S, the one which represents the received word by R. The loci of the points satisfying the constraints of each of the combined codes C_1 and C_2 are C_1 and C_2, respectively. By hypothesis, they both pass by S. The result of a first decoding which only takes account of C_1 is represented by D_1, that of the second decoding, which only takes account of C_2, by D_2, etc. The sequence of points $D_1,\ D_2,\ D_3,\ \ldots$ tends to S as the number of iteration steps increases.

The iteration process can be given a very simple picture in order to illustrate why it improves decoding (Fig. 6.11). We assume that we can define a space where a probability distribution is represented by a point and where the closeness of two distribution can be interpreted as the distance between the corresponding points (this can be done using the information-theoretic concept of 'cross-entropy,' also referred to as 'Kullback-Leibler divergence,' of two distributions). Then the

[8]Failing to do so would increase the magnitude of the *a posteriori* real value without improving its reliability; remember that the magnitude of a real value is intended to measure its reliability.

coding constraints can be represented by subspaces, say lines, and a probability distribution which satisfies the constraints of both codes is represented by the point where the corresponding lines intersect. The first decoding performed may be thought of as determining the point of the line which represents the constraint of the first code the closest to the one which represents the received point. To take account of the constraint due to the second encoding, we must start from this point and determine the point on the second line the closest to it. Since the second line only represents the constraint due to the second code, we must again determine the point on the first line the closest to the lastly obtained one, etc. After a sufficient number of iteration steps, a point close enough to the intersection of the two lines is reached and almost optimum decoding of the two combined codes has been performed. At variance with the figure, the two loci associated with the codes have more than a single intersecting point. There are indeed as many intersecting points as codewords, so the convergence indicated here is only a local property.

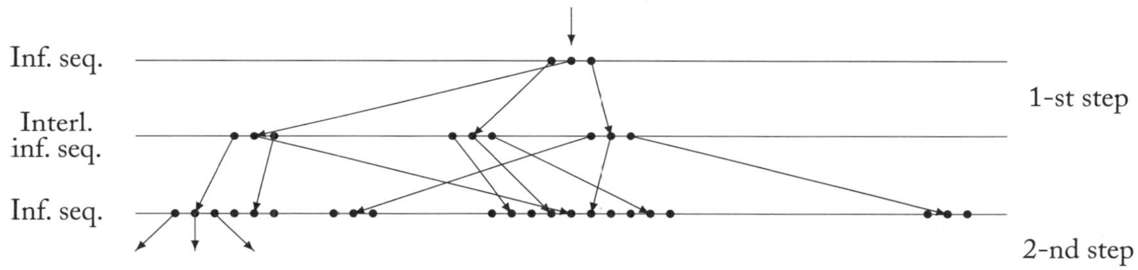

Figure 6.12: Diffusion of the dependence in the iterated decoding process. The lines labeled 'Inf. seq.' and 'Interl. inf. seq.' refer to the original information sequence and the one which results from its interleaving, respectively, at the 1-st iteration step and the 1-st half of the second one, from top to bottom. The points locate particular bits in the sequence. See the text for comments.

We can also illustrate how the iterated decoding proceeds in the information sequence with the help of Fig. 6.12. Remember that we interpreted the encoding as a kind of indirect repetition, i.e., where an information bit is repeated as combined modulo 2 with other information and check bits. Let us assume that this combination only concerns the two neighbors of each information bit (the combination also involves check bits which are not represented in the figure). The top horizontal line represents the information sequence at the beginning of the first iteration step. We consider the particular bit indicated by the vertical arrow at the top of the figure. As assumed, the encoding by the first component code made it dependent on its two neighbors. This means that, at the first half of the first iteration step (i.e., decoding in terms of the 1-st component code), the real value of the bit initially considered as well as those of its neighbors have been recomputed in terms of their own *a priori* real values and that of the corresponding checks bits. The *a posteriori* real values thus obtained are used as *a priori* real values for the second half of the first iteration step, which consists of decoding the interleaved information sequence in terms of the second component code. The arrows indicate the location of the interleaved bits. Due to the encoding by the second code, the neighbors of each of these bits in the interleaved information sequence have been made dependent and their *a posteriori* real values have been recomputed according to the second component code. At the first half of the second iteration step, the information bits are again ordered according to the original sequence, thanks to the de-interleaver operation. The bits initially considered have recovered their original place, but the neighbors of the corresponding bits in the interleaved sequence are located somewhere, generally far apart in the de-interleaved sequence. Iteration consists of repeating this process. Clearly, there is a diffusion of the dependence relationships through the entire information sequence, which eventually makes the *a posteriori* real value of an information bit depend on increasingly many other ones in an increasingly more complex fashion. We have a kind of 'diffusion of the dependence' as the decoding iteration proceeds. Remember that, in the average, the decoding process results in an increase of the magnitude of the *a posteriori* real values, hence of the reliability of the decoding decisions. Even if the improvement is small at the first iteration steps, a large number of steps eventually results in an almost sure decision if the code rate does not exceed the channel capacity. We may think of the decision involving each single bit as cumulating more and more information from the remainder of the sequence of received bits as the iteration of decoding proceeds.

As regards the implementation of the iterative decoding process, we may schematically think of cascading several decoding modules, each performing an iteration step. One of these modules is represented in Fig. 6.13. Its inputs can be connected to the outputs of the previous one, and its outputs to the inputs of the next one. Besides SISO decoders for each of the component codes, this elementary decoder involves an interleaver and a de-interleaver so as to ensure that both SISO decoders operate on the same information bit. The *a posteriori* real values of successive bits are correlated due to the dependency created by encoding and, in particular, the errors of each SISO decoder appear in bursts. The interleavers used in the decoding device spread out the *a posteriori* real values (and their possible errors occurring in a burst) before the sequence is fed to the other one, thus ensuring that the successive input *a priori* real values are uncorrelated, a condition for the validity of SISO decoding. This is a second important role of the interleaver, besides that of shaping the overall weight distribution of the turbocode which we already noticed.

6.7.3 VARIANTS AND COMMENTS

In the device represented in Fig. 6.13, we assumed that the *a priori* real values of the check bits pass through the elementary SISO decoders without change. As represented in Fig. 6.10, these decoders only update the *a posteriori* real values of the information bits. However, the formulas which express the *a posteriori* real value of any received bit in terms of the *a priori* real values of all the received bits, Eqs. (6.21) and (6.22), enable computing the *a posteriori* real values of the check bits, too. The choice of not doing so has the advantage of keeping samples of the channel output throughout the iterated decoding process, to the benefit of its stability, but schemes where both the information and check real values are updated can be contemplated.

The iterated decoding of turbocodes may actually be implemented quite differently from the scheme of Fig. 6.13 since the same device may be used several times at a speed greater than the input

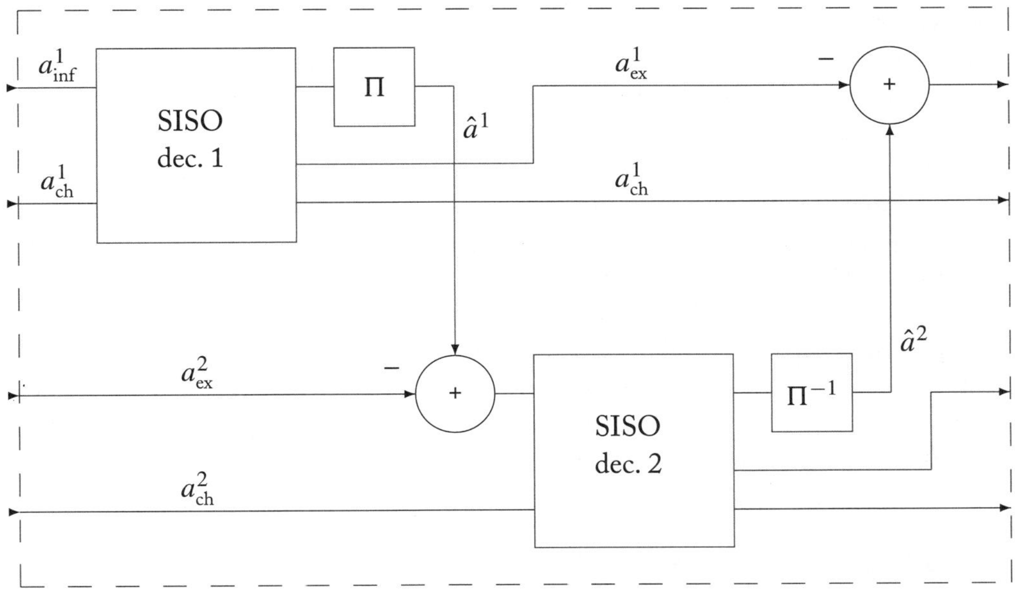

Figure 6.13: Decoding module (dashed box) for the turbocode of rate 1/3 generated by the encoder of Fig. 6.9. Iteration results from cascading a number of such modules. Contrary to Fig. 6.9, all the quantities considered are real: real values a^i_{inf} and a^i_{ch} of the information and check bits, respectively, and extrinsic real values as defined in the text, a^i_{ex}, with $i \in \{1, 2\}$. The boxes labeled 'SISO dec.' are identical to the elementary SISO decoder of Fig. 6.10. The extrinsic real value input is 0 in the first module, since then no extrinsic real value has yet been generated.

bit-rate. Then the iteration can be implemented using a single elementary decoder connected in a feedback loop.

Figure 6.14: A fully serial schematic representation of a turbo encoder of rate $1/n$. The box named n-replicator designates a device which successively delivers n times its input symbol. The interleaver scrambles the sequence it receives and the rate-1 encoder creates dependency between a number of successive symbols. Among the 3 devices, only the n-replicator generates redundancy.

We already stated that the information bits and the check bits, which appear as separate outputs in the encoder of Fig. 6.9, are generally transmitted 'serially,' i.e., in alternation according to a regular time pattern. We may redraw Fig. 6.9 in a fully serial fashion, as in Fig. 6.14, where three blocks are connected to each other: an n-replicator, an interleaver and a rate-1 encoder. Of course, each of these blocks results from a serial rearrangement of the elementary components (memory cells, connections, etc.) of the corresponding blocks of Fig. 6.9. Interestingly, each of the blocks of Fig. 6.14 performs one of the three functions which can be expected from a good encoder since they provide *redundancy, randomness*, and *mutual dependence*, respectively. We may thus think of Fig. 6.14 as describing a kind of paradigmatic encoder. Moreover, it turns out that some improvements can be obtained from variants of this scheme, especially if the number of copies generated by the n-replicator is made irregular, some symbols being repeated more than others while keeping the overall rate constant. The ease of decoding demands, however, that the decomposition of the overall code into short component codes be conserved, so as to enable splitting the whole task of decoding into alternate iterated decodings of simple codes.

6.8 HISTORICAL OUTLOOK

We now give a historical outlook on error-correcting codes, restricted to the descent of turbocodes. It is very sketchy and many other references could have been cited. Of course, the papers by Shannon are the starting point of the researches on error-correcting codes [72]. Pioneering works concerning block codes were due to Hamming and Golay in the late fourties [45, 42]. They originated in the algebra of finite fields and developed as 'algebraic coding', which is well accounted for in books like [65, 55], but has not been dealt with here. Convolutional codes were invented by Elias in 1955 [31]. The most important means for decoding convolutional codes has been the Viterbi algorithm [74, 33]. The trellis of a block code, enabling its SISO decoding, was introduced by Battail [8] and Wolf [78]. Battail [9] and Hagenauer and Hoeher [44] described a soft-output version of the Viterbi algorithm. Another SISO-type decoding algorithm was published earlier by Bahl et al. [3] and is now widely used. Replication decoding, by Battail et al. [6, 7], can be considered as a SISO-type version of Massey's threshold decoding [57]. Questioning the minimum distance criterion for the design of codes and suggesting that a good code should mimic random coding appeared in [10, 11], an idea which was later developed in [12]. The original paper on turbocodes by Berrou et al. dates back to 1993 [21]; turbocodes were given later a more comprehensive discussion in [22]. Also, see [43] for an excellent paper popularizing turbocodes. It was belatedly recognized that Gallager's 'low-density parity-check codes' [39] alluded to above in Sec. 6.4.5 are related to turbocodes as regards their decoding process and their performance. Since 1993, turbocodes were probably the topic which prompted the most important amount of work since the very beginning of researches on error-correcting codes: thousands of relevant bibliographic references about them are now available.

6.9 CONCLUSION

This chapter provided an introduction to error-correcting codes which hopefully enables understanding how they work and why they are efficient. It has been restricted to a few important topics hence is very far from being exhaustive. It kept the mathematical apparatus to a minimum and relied on the reader's intuition to deal with the practically most successful codes. Also, the strong connection of error-correcting codes with basic concepts of information theory has been emphasized. Much of this text relies on rather unconventional concepts, vocabulary and notation elaborated by the author for many years. To some extent, they paved the way to the invention of turbocodes, which shows that their value is not merely subjective.

Part III

Necessity of Genomic Error Correcting Codes and its Consequences

CHAPTER 7

DNA is an Ephemeral Memory

We now present computations of DNA capacity. Remember that information theory enables measuring the ability of any channel or memory element to convey information through space or time by its capacity, a quantity defined in Sec. 3.3.3 above. To compute it, we need to know what kind of events impair the basic information-bearing elements, namely, the nucleic-base pairs; and at which frequency these 'error events' occur. As regards the type of error to be considered, we already mentioned in Sec. 3.7 that it will be restricted to substitution and erasure, namely the cases where a wrong symbol replaces the correct one and where the actual symbol cannot be identified as belonging to the alphabet, respectively. In both cases, the length of the sequence remains unchanged, at variance with the cases of deletion or insertion where a symbol is removed from the sequence or inserted in it, resulting in its length decreasing or increasing by one. These latter events are not so uncommon in DNA, but it seems that their occurrence is less frequent than that of substitutions and erasures. Moreover, these cases are significantly more difficult to deal with so substitutions or erasures have been more thoroughly studied by information and channel coding theorists. To begin with, we compute the probability of symbol erasure or substitution as a function of time, assuming the errors are either substitutions or erasures, moreover occurring at a given constant frequency. We defer to Sec. 7.3 a closer examination of how the actual frequency of these error events can be estimated.

7.1 COMPUTING THE PROBABILITY OF SYMBOL ERASURE OR SUBSTITUTION AS A FUNCTION OF TIME

7.1.1 SYMBOL ERASURE PROBABILITY

We assume here that only errors of a single type occur, either substitution or erasure. We begin with erasure, the simplest case, assumed to occur with a constant frequency ν_{er}, meaning that a symbol is erased during the infinitesimal time interval $\mathrm{d}t$ with probability $\nu_{\text{er}}\mathrm{d}t$. We first compute the probability $p_{\text{er}}(t)$ that a symbol has been erased at time $t \geq 0$. A symbol can be erased only if it has not been already erased, so we can express the probability $p_{\text{er}}(t + \mathrm{d}t)$ as

$$p_{\text{er}}(t + \mathrm{d}t) = p_{\text{er}}(t) + \nu_{\text{er}}\mathrm{d}t[1 - p_{\text{er}}(t)]$$

which is equivalent to the differential equation

$$p'_{\text{er}}(t) = \nu_{\text{er}}[1 - p_{\text{er}}(t)]$$

where $p'_{\text{er}}(t)$ denotes the derivative of $p_{\text{er}}(t)$ with respect to time. Its solution satisfying the initial condition $p_{\text{er}}(0) = 0$ is

$$p_{\text{er}}(t) = 1 - \exp(-\nu_{\text{er}}t) \ . \tag{7.1}$$

The slope of the graph of $p_{\mathrm{er}}(t)$ at the origin, $p'_{\mathrm{er}}(0)$, equals ν_{er} and $p_{\mathrm{er}}(t)$ tends to the horizontal asymptote $p_{\mathrm{er}}(\infty) = 1$. Thus, after a long enough time the given symbol is eventually erased. Notice that this result does not depend on the alphabet size q.

7.1.2 SYMBOL SUBSTITUTION PROBABILITY

Consider a symbol from an alphabet of size q. In the case of DNA, $q = 4$ for the alphabet made of the nucleotides **A**, **T**, **G**, and **C**, or $q = 2$ if we only distinguish as symbols, for instance, their chemical structure purine, **R**, or pyrimidine, **Y**. We assume that this symbol incurs a substitution during the infinitesimal time interval $\mathrm{d}t$ with probability $\nu_{\mathrm{su}}\mathrm{d}t$, where ν_{su} is constant. We moreover assume that the symbol which is substituted for the initial one is randomly chosen with probability $1/(q-1)$ among the other $q-1$ symbols of the alphabet (this is the worst probability distribution in case of error, according to information theory).

We now compute the probability of error of a given symbol as a function of time. Let $p_{\mathrm{su}}(t)$ denote the probability that this symbol differs from the initial (correct) one at time $t \geq 0$. The given symbol is identical to the initial one with probability $1 - p_{\mathrm{su}}(t)$, and in this case the probability of error increases during the interval $(t, t + \mathrm{d}t)$ by an amount of $\nu_{\mathrm{su}}\mathrm{d}t$. But if the given symbol is already in error, an event of probability $p_{\mathrm{su}}(t)$, the probability of error *decreases* by an amount of $\nu_{\mathrm{su}}\mathrm{d}t/(q-1)$ since the newly occurring error can recover by chance the initial symbol. We can thus express the probability $p_{\mathrm{su}}(t + \mathrm{d}t)$ as

$$p_{\mathrm{su}}(t+\mathrm{d}t) = p_{\mathrm{su}}(t) + \nu_{\mathrm{su}}\mathrm{d}t[1 - p_{\mathrm{su}}(t)] - \nu_{\mathrm{su}}\mathrm{d}t\frac{p_{\mathrm{su}}(t)}{q-1} = p_{\mathrm{su}}(t) + \nu_{\mathrm{su}}\mathrm{d}t\left[1 - \frac{q}{q-1}p_{\mathrm{su}}(t)\right],$$

which is equivalent to the differential equation

$$p'_{\mathrm{su}}(t) = \nu_{\mathrm{su}}\left[1 - \frac{q}{q-1}p_{\mathrm{su}}(t)\right], \tag{7.2}$$

where $p'_{\mathrm{su}}(t)$ denotes the derivative of $p_{\mathrm{su}}(t)$ with respect to time. Its solution satisfying the initial condition $p_{\mathrm{su}}(0) = 0$ is

$$p_{\mathrm{su}}(t) = \frac{q-1}{q}\left[1 - \exp\left(-\frac{q}{q-1}\nu_{\mathrm{su}}t\right)\right]. \tag{7.3}$$

The slope of the graph of $p_{\mathrm{su}}(t)$ at the origin, $p'_{\mathrm{su}}(0)$, equals ν_{su} and $p_{\mathrm{su}}(t)$ tends to the horizontal asymptote $p_{\mathrm{su}}(\infty) = (q-1)/q$. This asymptotic behavior for t approaching infinity means that after a long enough time the given symbol no longer depends on the initial one and becomes random with uniform probability over the alphabet.

7.2 CAPACITY COMPUTATIONS

7.2.1 CAPACITY COMPUTATIONS, SINGLE-STRAND DNA

Assuming the errors to be all of the same type (either erasures or substitutions) it is a simple matter to compute the capacity of DNA in terms of the error probabilities (7.1) and (7.3) just obtained. The channels are symmetric in both cases, so their capacity equals the mutual information $I(X;Y) = H(Y) - H(Y|X)$ for equiprobable input symbols.

The capacity of the erasure channel has already been computed in Sec. 3.3.3, namely, p_{er} denoting the erasure probability:

$$C_{q,\mathrm{er}} = [1 - p_{\mathrm{er}}] \log_2 q \tag{7.4}$$

shannons per symbol hence, letting $p_{\mathrm{er}} = p_{\mathrm{er}}(t)$ according to Eq. (7.1):

$$C_{q,\mathrm{er}} = \exp(-\nu_{\mathrm{er}} t) \log_2 q \ , \tag{7.5}$$

where the subscript indicates the alphabet size q and the nature of the assumed error, 'er' standing for 'erasure.'

Similarly, if we assume in the case of q-ary symbols that all errors consist of substitutions of the type specified above, occurring with probability p_{su}, we obtain after straightforward computations the capacity:

$$C_{q,\mathrm{su}} = \log_2 q - \mathcal{H}_2(p_{\mathrm{su}}) - p_{\mathrm{su}} \log_2(q-1) \tag{7.6}$$

shannons per symbol, where the binary entropy function has been defined as $\mathcal{H}_2(p) = -p\log_2(p) - (1-p)\log_2(1-p)$. The subscript is similarly intended to indicate the size q of the alphabet and 'su' stands for 'substitution.' Letting in the above expression $p_{\mathrm{su}} = p_{\mathrm{su}}(t)$ as given by Eq. (7.3) results in

$$\begin{aligned} C_{q,\mathrm{su}}(t) &= \frac{q-1}{q}\left[1 - \exp\left(-\frac{q}{q-1}\nu_{\mathrm{su}} t\right)\right] \log_2\left[1 - \exp\left(-\frac{q}{q-1}\nu_{\mathrm{su}} t\right)\right] \\ &+ \frac{1}{q}\left[1 + (q-1)\exp\left(-\frac{q}{q-1}\nu_{\mathrm{su}} t\right)\right] \log_2\left[1 + (q-1)\exp\left(-\frac{q}{q-1}\nu_{\mathrm{su}} t\right)\right] , \end{aligned} \tag{7.7}$$

which expresses the capacity of the genomic channel as a function of time in the presence of a constant substitution frequency ν_{su}.

Notice that the error probabilities $p_{\mathrm{er}}(t)$ and $p_{\mathrm{su}}(t)$ in Eqs. (7.1) and (7.3), hence the capacities $C_{q,\mathrm{er}}(t)$ and $C_{q,\mathrm{su}}(t)$ in Eqs. (7.5) and (7.7), depend on time through the product $\tau = \nu t$ (ν stands here for either ν_{er} or ν_{su}), a dimensionless quantity, which can be interpreted as a measure of time using $1/\nu$ as unit, to be referred to as *time constant*. The formulas (7.5) and (7.7) account for the degradation of the channel capacity due to the accumulated errors. The capacities decrease from $\log_2 q$ for $\tau = 0$, with a slope equal to $-\nu_{\mathrm{er}}$ for erasures and $-\infty$ for substitutions, down to 0, *exponentially* for τ approaching infinity.

For the purpose of illustration, let us assume that the relevant alphabet is binary (say, {**R**,**Y**}, i.e., only distinguishes the purine or pyrimidine chemical structure of a nucleotide). The capacity given by Eq. (7.7) for $q = 2$ is:

$$\begin{aligned} C_{2,\mathrm{su}}(t) &= \frac{1}{2}\{[1 - \exp(-2\nu_{\mathrm{su}} t)] \log_2[1 - \exp(-2\nu_{\mathrm{su}} t)] \\ &+ [1 + \exp(-2\nu_{\mathrm{su}} t)] \log_2[1 + \exp(-2\nu_{\mathrm{su}} t)]\} . \end{aligned} \tag{7.8}$$

Similar expressions would be obtained with other alphabet sizes.

7.2.2 CAPACITY COMPUTATIONS, DOUBLE-STRAND DNA

The capacities just computed are only relevant to single-strand DNA. However, the availability of pairs of nucleotides as elements of the alphabet enhances the capacity.

In the presence of erasures, a receiver dealing with pairs of nucleotides as elements of the alphabet will deal with a nucleic-base pair as erased only if its nucleic bases are both erased. Assuming the erasures of each base in the pair to be mutually independent events results in the pair erasure probability being the square of the erasure probability of a single base, so $p_{\text{er}}^2(t)$ should be substituted for $p_{\text{er}}(t)$ in Eq. (7.4). The capacity thus obtained is:

$$C_{q,\text{er,ds}} = \exp(-\nu_{\text{er}}t)[2 - \exp(-\nu_{\text{er}}t)] \log_2 q \ , \tag{7.9}$$

where the appended subscript 'ds' stands for 'double strand.' It has been plotted in terms of $\tau = \nu_{\text{su}}t$ in Fig. 3.11.

In the case of substitution, an error can affect a single base or both. A substitution which affects a single base necessarily destroys the pair complementarity, so it must be interpreted as an erasure of the pair. Substitutions in both nucleic bases result in either a substitution or an erasure of the pair: a substitution when they result in a complementary pair, an erasure when they do not.

The case $q = 4$ is rather complicated and is not developed here. Assuming again for the purpose of illustration that the alphabet is binary, e.g., {**R**,**Y**}, a straightforward computation of the capacity in the presence of substitutions, taking account of the availability of pairs of nucleotides, results in:

$$C_{2,\text{su,ds}} = 1 + \mathcal{H}_2[2p_{\text{su}}(1 - p_{\text{su}})] - 2\mathcal{H}_2(p_{\text{su}}) \ .$$

Replacing p_{su} with its expression in terms of $\nu_{\text{su}}t$ according to Eq. (7.3) results in:

$$\begin{aligned} C_{2,\text{su,ds}}(t) = \ & \frac{1}{2}\{[1 - \exp(-2\nu_{\text{su}}t)]^2 \log_2[1 - \exp(-2\nu_{\text{su}}t)] \\ & + [1 + \exp(-2\nu_{\text{su}}t)]^2 \log_2[1 + \exp(-2\nu_{\text{su}}t)] \\ & - [1 + \exp(-4\nu_{\text{su}}t)] \log_2[1 + \exp(-4\nu_{\text{su}}t)]\} \ . \end{aligned} \tag{7.10}$$

7.3 ESTIMATING THE ERROR FREQUENCY BEFORE CORRECTION

Applying the computed capacity formulas to actual genomes would of course imply that the kind of errors and their frequency are known. Unfortunately, this is not the case. Although many works were intended to estimate the mutation rate in genomes, none of them can directly be used for our purpose. Indeed, since the existence of genomic error-correcting codes is not recognized by current biology, the error frequencies were measured only on the basis of observed mutation rates hence, according to our point of view, *after* error correction. What we need is the error frequency *before* correction.

We now examine how the frequency of errors can be estimated. Due to the relevance of mutations to pathologies, the human genome has been especially studied in this respect: by Haldane, as early as 1935 [64], and recently by Kondrashov [52] and many others. They all agree on a figure of about 2×10^{-8} error per nucleotide per generation in the average, the errors mainly consisting of substitutions. Interestingly, its reciprocal is 50 million generations, or about 1.5 billion years. This is the proper order of magnitude although this figure is somewhat too small: remember that the origin of life dates back to at least 3.5 billion years.

Although it concerns an error rate after correction, we may try to use this figure in order to estimate the error frequency before correction, but doing so needs some further assumptions. We were already led to the subsidiary hypothesis that the genomic code assumes a nested structure implying a very unequal protection of the nucleotides against errors, depending on their location. We were moreover led to conjecture, for making the genomic code compatible with sexual reproduction, that the outer layer of the nested system is left uncoded (see Sec. 9.4). Let u denote the proportion of nucleotides which belong to this layer and are thus left uncoded. If the overall mutation rate per nucleotide per generation is denoted by μ, we have $\mu = u\nu + (1 - u)\varepsilon$ where ν denotes as above the error frequency per nucleotide (expressed using a generation as time unit) and ε denotes the average error rate after regeneration which is contributed to the error rate by the inner, encoded layers of the nested system. As an approximation we may reasonably assume that ε can be ignored, which results in $\mu = u\nu$. The polymorphism of the human genome concerns about one thousandth of the nucleotides. Then, letting $\mu = 2 \times 10^{-8}$ and $u = 10^{-3}$ results in a frequency of $\nu = 2 \times 10^{-5}$ error per nucleotide per generation, a fairly plausible estimate in the absence of direct experimental data. That this estimate relies on data from the human genome is unlikely to restrict its generality, since the type and frequency of errors are presumably intrinsic properties of DNA in its terrestrial environment, regardless of the organism which hosts it. The time constant $1/\nu$ is then about 50,000 generations or about 1.5 million years. It is very large at the scale of a human life hence can well give the illusion that genomes are permanent objects, although it is short at the geological timescale.

The assumption of a constant frequency of errors is not necessary if we only look for an upper bound on the channel capacity. It suffices to use the smallest value it can assume to obtain an upper bound on the actual capacity. Similarly, as regards the type of errors it is possible to merely compute an upper bound on the DNA capacity, based on hypotheses which systematically minimize the impact of errors. The mildest kind of error is erasure, and an upper bound on the DNA capacity would result from assuming that all errors are of this type. Then, the true capacity of DNA would be less than that which concerns erasures and takes the double-strand structure into account, as given by Eq. (7.9) and illustrated in Fig. 3.11. This bound decreases exponentially down to 0 as time increases, and the actual DNA capacity is less.

7.4 PARADOXICALLY, A PERMANENT MEMORY IS EPHEMERAL

Regardless of the uncertainty about the type and frequency of the errors, the above computations show that the DNA capacity vanishes exponentially fast whatever assumptions are made. DNA is indeed an ephemeral memory, hence unable to conserve the hereditary information during geological times without the help of intrinsic error-correcting means used frequently enough to efficiently regenerate the genome.

This conclusion is by no means specific to DNA. Like any 'permanent' memory, DNA is paradoxically ephemeral because it faithfully keeps the incurring errors as well as its initial content. Errors accumulate up to eventually replacing the initial memory content. The above calculations show moreover that its information-theoretic capacity vanishes exponentially fast. The need for regenerating the memory content from time to time is actually shared by any 'permanent' memory used in the presence of errors. Stated in more adamant words, 'no memory is permanent.' The engineers who design computer memories know that. They have recourse to dynamic processes generally involving error-correcting codes to increase the actual permanence of the recorded data up to the needed value. There are moreover concerns as regards conservation of recorded data beyond a few decades. Compare with the few billion years of geological times!

CHAPTER 8

A Toy Living World for Illustrating the Main Hypothesis

We now introduce a very simple model, referred to as 'toy living world,' where n-symbol sequences referred to as 'genomes' are subjected to random errors. The 'genomes' can be either arbitrary sequences or words of an error-correcting code. They are periodically replicated, after regeneration when an error-correcting code enables it. The model is as simple as to enable easy calculations of the lifetime of a 'genome' and the population size of the corresponding 'species.' If no genomic error-correcting code is used, no discrete species can exist beyond a certain genome length: a property of the true living world as fundamental as the existence of discrete species is lost, which shows the inadequacy of the uncoded model. It turns out on the contrary that the 'toy living world' shares basic features with the actual one in the case where a genomic error-correcting code is used, especially as regards the existence of discrete species. The toy living world just illustrates the main hypothesis. The following chapter will deal with the subsidiary hypothesis that the genomic code takes the form of a 'nested system.'

8.1 A SIMPLE MODEL

According to our *main hypothesis*, genomes act as error-correcting codes, and we have seen above (Ch. 7) that the vanishing capacity of DNA makes it absolutely necessary. We shall first illustrate this main hypothesis with a model as simple as to enable easy calculations, but hopefully realistic enough to mimic some basic features of the actual living world.

Assuming regeneration and replication to be jointly performed, an insight in the transmission of genetic information can be gained by analyzing a 'toy living world' in which 'genomes' consist of words of some binary error-correcting code of length n (we throughout put 'genomes' between quotes to remind their fictitious character). We refer to the code as to the 'genomic code.' The 'genomes' are periodically regenerated and replicated in the presence of symbol substitutions occurring with a constant probability p, giving rise to 'genome' populations. All identical 'genomes' are said to belong to a same 'species' and their number at a given instant is referred to as the 'population' of this 'species.' For comparison, we also consider the case where no intrinsic genomic code exists hence where a 'genome' is an arbitrary n-bit sequence. Since it cannot be regenerated, only its replication is performed. In order to use a single word in both cases, we refer to regeneration followed with replication as 'duplication' in the case where an error-correcting code is used, the same word being used as a synonymous of 'replication' in the uncoded case. We denote by Δt the time interval between successive duplications. For a constant symbol error frequency ν_{su}, the probability of symbol error is $p = p_{su}(\Delta t)$, where $p_{su}(t)$ is given by Eq. (7.3).

In a first step, we do not consider any limit to the population growth, hence no natural selection. When a genomic error-correcting code is assumed to exist, a regeneration error results in a 'genome' which differs from the original one in at least d symbols, the minimum Hamming distance of the genomic code. The occurrence of regeneration errors thus results in discrete 'species' successively appearing. Stemming from a single ancestral 'genome,' regeneration (decoding) errors generate an increasing number of new 'genomes' according to a branching process. In the uncoded case, on the contrary, 'genomes' are comparatively short-lived random sequences which do not constitute discrete 'species' beyond some critical length.

We first develop an analysis of this 'toy living world.' In a second step we introduce in it natural selection. A third step could consist of taking the subsidiary hypothesis (that the genomic error-correcting code takes the form of a nested system) into account. Doing so would introduce a hierarchical taxonomy and result in a model of the living world much more similar to the real one but much less simple. We shall deal with the subsidiary hypothesis more roughly, only qualitatively relying on the analysis of the simple toy living world based on the main hypothesis, in the next chapter.

8.2 COMPUTING STATISTICAL QUANTITIES RELEVANT TO THE TOY LIVING WORLD

We now analyze the toy living world exclusively based on the main hypothesis, assuming 'genomes' of a given length n, each symbol of which incurs an error independently of the others with a probability p which depends only on the error frequency ν and the time interval Δt since the last duplication. The 'genomes' either are not encoded or belong to some error-correcting code of minimum Hamming distance d. In the latter case, the error-correcting code enables regenerating the 'genomes.' Remember that 'duplication' is intended to mean either regeneration followed by replication, or replication alone in the uncoded case. It provides two copies of the possibly regenerated 'genome.'

We say that a 'genome' incurs a mutation when it differs from the original one. Let P denote the probability of this event. In the absence of error correction, this probability is the complement to 1 of the probability that the 'genome' is received without error, hence $P = P_{unc} = 1 - (1 - p)^n$. Then P_{unc} is fairly large for reasonable values of p and n. Moreover, it is an increasing function of the 'genome' length n. When an error-correcting code is used, the probability of a mutation, i.e., of the regeneration process failing to recover the original 'genome,' $P = P_{cod}$, can on the contrary be made *arbitrarily small* through the choice of a long and efficient enough code. Moreover, P_{cod} can be made the smaller, the larger the codeword length n. The precise value of P_{cod} depends on the genomic code so no simple expression of it can be given. However, a plausible approximation of it is $Kp^{d/2}$ where K is a constant which depends on the code. Insofar as the error-correcting code is efficient, P_{cod} is much smaller than P_{unc}.

The following calculations are valid whether an error-correcting code is used or not which helps understanding the specific features brought by encoding. P standing below for P_{cod} or P_{unc}, depending on the case, the following results hold:

- the probability that a 'genome' is conserved (i.e., remains identical to itself) after i successive duplications is $(1 - P)^i$;

- the probability that a 'genome' fails to be regenerated after exactly i successive successful regenerations or errorless duplications is $P(1 - P)^i$. This is also the probability that a 'genome' lasts exactly $i\,\Delta t$, or that its 'lifetime' assumes this value, where Δt is the time interval between duplications;

- the average lifetime $L(P)$ of a 'genome,' expressed using Δt as time unit, to be referred to in the sequel as its *permanence*, is the expectation of its lifetime i affected with its probability $P(1 - P)^i$, namely
$$L(P) = \sum_{i=1}^{\infty} i P(1 - P)^i = (1 - P)/P \; . \tag{8.1}$$
When an adequate error-correcting code is employed, and if Δt is short enough, $P = P_{cod}$ is very small so $1/P_{cod}$ is a simple approximation to the 'genome' permanence. This permanence increases without limit if we use a long and efficient enough error-correcting code such that P_{cod} approaches 0. This is possible provided the condition stated by the fundamental theorem of channel coding that the source entropy is less than the channel capacity is satisfied (see Sec. 5.3.1). Then the average lifetime of a 'genome' can assume values of the order of magnitude of geological times. If no error-correcting code is used, $P = P_{unc} = 1 - (1 - p)^n$ so the permanence $L(P_{unc})$ is approximated by $1/[1 - (1 - p)^n]$ or, for p small enough (i.e., if $p \ll 1/n$), by $1/np$. The permanence of a 'genome' is then the shorter, the longer it is, at variance with the encoded case;

- the total number of 'genomes' which remain identical to the initial one after i duplications, hence the population of the corresponding species at that time, is in the average $[2(1 - P)]^i$ since we assume that each duplication implies a replication which results in two identical 'genomes.' In the absence of any limiting factor, this number exponentially increases only if $P < 1/2$. If no error-correcting code is used this inequality is satisfied only for short enough 'genomes,' such that $P_{unc} < 1/2$, which implies the inequality $(1 - p)^n > 1/2$. Taking logarithms to the base 2 results in $n < -1/\log_2(1 - p)$. An approximation of $-\log_2(1 - p)$ for p small enough is $p/\ln(2)$, so the existence of stable species if no encoding is performed is only possible for 'genome' lengths n satisfying the inequality $n < \ln(2)/p \approx 0.69/p$. For larger values of n, the 'genomes' obtained would be diverse and constantly changing, without any definite species. If on the contrary an error-correcting code is used, P_{cod} can be made much smaller than 1/2 and the persistence of a species is ensured by the exponential increase of the population of individuals which share the same 'genome.'

Before discussing some other properties of the toy living world and making it more realistic by taking into account natural selection, let us notice that the above calculations unambiguously show that the absence of coding is not compatible with the existence of discrete species with long genomes, at variance with the actual living world. The permanence $L(P_{\text{unc}})$ of a species in the absence of coding has been shown above to decrease proportionally to the inverse $1/n$ of the genome length. Moreover, the population of a species grows (hence its persistency is secured) only provided the following condition holds:

$$n < \ln(2)/p \,. \qquad (8.2)$$

8.3 THE INITIAL MEMORY CONTENT IS PROGRESSIVELY FORGOTTEN

It is easy to show that any n-symbol word can result from errors accumulated during a large enough number of duplications, up to the point where the initial word is actually 'forgotten.' By 'any n-symbol word' we mean an arbitrary binary sequence of length n in the absence of coding, or any word of the genomic code when an error-correcting code is used.

Let us first consider the uncoded case. We assume that a 'genome' is replicated g times, and that each time it is replicated a given symbol has some probability p to incur an error, i.e., to be changed into its binary complement. We assume that such an error is truly random, i.e., that p differs from both 0 and 1: $0 < p < 1$. Moreover, we assume that the successive errors occur independently from each other. We wish to compute the probability P_0 that, after g replications, a given symbol is the same as the original one. This occurs if and only if the number of errors it incurs within the g successive generations is zero or even. The probability of i errors within g generations equals $\Pr(i) = \binom{g}{i} p^i (1-p)^{g-i}$, so P_0 is the sum of these probabilities for i zero or even. Notice that $\Pr(i)$ can be obtained as the i-th term (starting with $i = 0$) of the development of $(x + y)^g$, letting $x = 1 - p$ and $y = p$. (Incidentally, we check that $\sum_{i=0}^{g} \Pr(i) = 1$, as expected.) The probability P_0 is the sum of the terms in the development of $(x + y)^g$ where the power of y is zero or even. Now consider $(x - y)^g$. In its development, the terms where the power of y is zero or even are the same as in the development of $(x + y)^g$, while those where the power of y is odd have the opposite sign. Then $(x + y)^g + (x - y)^g$ gives twice P_0. Letting $x = 1 - p$ and $y = p$ results in:

$$P_0 = \frac{1}{2}[1 + (1 - 2p)^g] \,.$$

Since $0 < p < 1$ entails $|1 - 2p| < 1$, $(1 - 2p)^g$ tends to 0 as g increases. P_0 thus becomes the closer to 1/2, the larger g. After a large enough number g of replications, each binary symbol is thus random with a probability of about 1/2, independently of the others. It no longer gives any clue on the initial value of the symbol, which we may think of as *forgotten*. The resulting set of 'genomes' tends to a large number of purely random sequences. In sharp contrast, a species of the toy living world is a set of identical 'genomes.'

The case where an error-correcting code is used is less simple. The probability P_{cod} of a regeneration error, for an efficient enough code and a short enough time interval between successive regenerations, is much smaller than the probability P_{unc} of an error without such a code. P_{cod} depends in a complex way on the parameters of the code. The regeneration errors are much less frequent than in the absence of an error-correcting code, but a regeneration error entails that at least d symbols among the n symbols of the codeword are in error (d is the minimum Hamming distance of the code). If a large number of regeneration errors occur (which implies a very long time interval, say at the geological scale), a similar phenomenon occurs, i.e., an average of one symbol out of two becomes wrong and the 'genome' becomes again asymptotically unrelated to the original one. Let $e(i)$ denote the number of erroneous symbols after i regeneration errors. If one more regeneration error occurs, this number becomes

$$e(i+1) = e(i)(1 - d/n) + (d/n)[n - e(i)] = e(i)(1 - 2d/n) + d\ ,$$

assuming as an approximation that a regeneration error entails exactly d erroneous symbols. This expression results from noticing that, when an error affects a binary symbol, it becomes wrong if it is correct but is corrected if it was already wrong. It relies moreover on the simplifying assumptions that exactly d erroneous symbols result from a regeneration error, and that the successive error patterns which result from regeneration errors are mutually independent. Both are not exactly true but mere approximations. According to the above expression, $e(i)$ approaches $n/2$ regardless of the minimum distance d of the code as i tends to infinity (incidentally, this result is valid for $d = 1$, hence in the uncoded case, too). The use of a genomic error-correcting code makes however a big difference with the uncoded case since the result of a regeneration error is a codeword. Although the 'genome' eventually obtained results from a series of random choices, it is not itself a purely random sequence of symbols since it obeys the specific constraints of the code to which it belongs. Although each regeneration error results in a random jump to a genome at a Hamming distance at least d apart when an error-correcting code is used, the overall speed of evolution is then much slower than in the uncoded case since regeneration errors are very seldom: it proceeds with larger steps which however occur very much less often.

Notice that in the toy living world discrete 'species' originate in regeneration errors just because the 'genome' is a word of an error-correcting code, apart from any selection. The average lifetime $L(P_{cod})$ varies proportionally to the inverse of the regeneration error probability P_{cod}. Hence, the less likely is a mutation, the larger the permanence of the mutated 'genome.' The genomic message which results from a mutation is made just as resilient to casual errors as the original one since both are codewords of the same genomic error-correcting code.

For the purpose of illustration, we'll consider in Sec. 8.5 a toy living world using a very simple code and show some results of its simulation.

8.4 INTRODUCING NATURAL SELECTION IN THE TOY LIVING WORLD

Of course, the toy living world analyzed above is oversimplified. It could maybe model the evolution of a bacterial population in very favorable conditions and in the absence of any 'horizontal transfer' of genetic material. The individuals are assumed to be eternal and no kind of natural selection is considered. Even for potentially eternal individuals, the population increase in the real world would become rapidly slower than exponential as a mere result of the finiteness of available resources, according to the usual scheme of Darwinian selection. We now assume that, analogously to the actual living world, some specific phenotype hosts each 'genome' and is subject to environmental factors more or less favorable to its survival so as to hopefully make the toy living world a bit more similar to the actual one. This case can be dealt with by introducing a probability of death, P', besides the probability P of a mutation. 'Death' means here that the 'genome' is not conserved for another reason than the occurrence of a mutation due to channel errors, for instance its recombination or the destruction of the phenotype which hosts it.

The above computation of the probability that the 'genome' is not conserved is easily extended. It suffices to notice that the survival of the 'genome' as regards its death (in the above meaning) and the absence of a mutation affecting it are mutually independent events, so the probability of the 'genome' conservation is the product of the probabilities of these two events. These probabilities are $1 - P'$ and $1 - P$, respectively, so the 'genome' is conserved with probability $(1 - P)(1 - P') = 1 - P - P' + PP'$. The probability that it is not conserved is thus $P + P' - PP'$. All the results computed in Sec. 8.2 remain valid in the case where natural selection occurs provided the probability P is replaced by $P + P' - PP'$. For instance, the permanence (average lifetime) of the 'genome' becomes

$$L(P, P') = \frac{1 - P - P' + PP'}{P + P' - PP'}, \tag{8.3}$$

closely approximated by $1/(P + P' - PP')$ if P and P' are small enough. One easily checks that $L(P) = L(P, 0) > L(P, P')$ for any $P' > 0$, as expected.

The difficulty of properly evaluating P' in the real world, which obviously depends on very many factors, limits however the practical usefulness of this result. The death probability P' then mainly depends on how well the phenotype corresponding to the considered 'genome' fits its environment. In the absence of error correction, it is only the phenotype fitness which would ultimately cause some species to survive rather than another one. Since mainstream biology ignores error-correcting codes, it is the sole mechanism it contemplates for explaining the discreteness of species, but there is no reason why natural selection should carve well separated and very sparse 'genomes' out of the mass of possible nucleotide sequences. If, on the contrary, a genomic error-correcting code is used, a new 'genome' results from a regeneration error. The nascent species originating in it becomes the target of natural selection, but its discreteness is an intrinsic preexisting property of the genomic code, not the sole consequence of natural selection.

Our main hypothesis thus suffices to account for the existence and evolution of 'species,' even in the absence of any natural selection. In sharp contrast, natural selection is thought of according to the neo-Darwinian orthodoxy as the sole reason why discrete species exist, as discussed, e.g., by Mark Ridley [69]. Neo-Darwinians assert that natural selection results in sharply defined species rather than poorly differentiated living things. This statement is, however, not supported by plausible arguments. There are cases where natural selection operates only within the limits of some hereditary invariant. For instance, the limbs of many vertebrates have 5-digit extremities. This fact cannot result from an adaptation but from the permanence of a hereditary information, since the functions achieved widely differ from a species to another: think of the hand and the foot of a man, the foot of a lizard or a mole, the fin of a dolphin or a whale, the wing of a bat, etc., fitted to as diverse functions as prehension, walking, running, digging, swimming, flying, etc. This adaptation results in considerable variations in the form, size and mutual connection of bones, tendons, muscles, skin ...*within* the 5-digit pattern, which appears as a hereditary invariant.

8.5 EXAMPLE OF A TOY LIVING WORLD USING A VERY SIMPLE CODE

We present here a very simple example of a 'toy living world.' Since the regeneration performance depends on the total number of erroneous symbols in a word, we may assume a very short 'genome' and a very high error rate. We ran simulations assuming a binary 'genome' of length $n = 7$ and a symbol error rate of $p = 0.1$. (In a realistic situation, a given error rate would be obtained by properly adjusting the time interval Δt between successive regenerations, for a given error frequency.) We used either no error-correcting code, or the very simple (7,4) Hamming code which can correct all single errors (see Ch. 6). The probability of a regeneration error is then $P_{\mathrm{cod}} = 1 - (1-p)^7 - 7p(1-p)^6$ which approximately equals 0.16 for $p = 0.1$. We assume that no other events than replications-regenerations occur. Especially, we assume that no natural selection limits the number of replicated 'genomes.' Despite the extreme simplicity of this model and the lack of natural selection, i.e., the factor which is believed the most important for shaping the living world, what we obtain mimics it rather well. Moreover, introducing natural selection (Sec. 8.4) and taking account of the subsidiary hypothesis (see Ch. 9 below) would refine the model and make it closer to the real living world, although of course still oversimplified.

A drawback of the choice of the very short (7,4) Hamming code is that the average number of erroneous bits in a received word is small, hence subject to large statistical fluctuations. The stability of species in the toy living world is thus much less than that of a model more realistically involving much longer genomes. Apart this important exception, we hope that analyzing the toy living world sheds some light on the properties which may be expected from the main hypothesis that genomic error-correcting codes exist.

The Hamming (7,4) code is the first invented nontrivial code (in the late 1940s). It uses the binary alphabet ($q = 2$). The length of its words is $n = 7$. The constraints which tie together its bits

$c_1, c_2, \ldots, c_7$ are

$$c_1 \oplus c_3 \oplus c_4 \oplus c_5 = 0$$

$$c_1 \oplus c_2 \oplus c_3 \oplus c_6 = 0$$

and

$$c_2 \oplus c_3 \oplus c_4 \oplus c_7 = 0$$

where $\oplus$ denotes addition modulo 2. It comprises only $2^4 = 16$ words which are listed in Tab. 8.1 below and numbered from 0 to 15:

Table 8.1: The 16 words of the (7,4) Hamming code

0:	0 0 0 0 0 0 0	8:	0 0 0 1 1 0 1
1:	1 0 0 0 1 1 0	9:	1 0 0 1 0 1 1
2:	0 1 0 0 0 1 1	10:	0 1 0 1 1 1 0
3:	1 1 0 0 1 0 1	11:	1 1 0 1 0 0 0
4:	0 0 1 0 1 1 1	12:	0 0 1 1 0 1 0
5:	1 0 1 0 0 0 1	13:	1 0 1 1 1 0 0
6:	0 1 1 0 1 0 0	14:	0 1 1 1 0 0 1
7:	1 1 1 0 0 1 0	15:	1 1 1 1 1 1 1

Notice that the number made of the first four bits of each word is the binary representation of the word number (with the less significant bits on the left). This is referred to as the 'systematic form' of the code (see Sec. 6.4.2 in Ch. 6).

This code has been designed according to a precise algebraic structure, but to understand how it works, it suffices to notice that its minimum distance is $d = 3$, which means that it can definitely correct any substitution of a single binary symbol for its binary complement, or the erasure of any 2 binary symbols (see Ch. 6).

As an illustration of a hereditary process where a 'genome' is successively replicated several times starting from an ancestral one, and of the introduction of an error-correcting code in this process, we consider a 'toy living world' whose 'genome' is binary and of very short length, namely 7. We assume that the error rate is very high, namely $p = 0.1$, with the errors drawn at random independently of each other. Then the probability that an error pattern of weight e occurs in a word of length 7 is $P(e) = \binom{7}{e} p^e (1-p)^{7-e}$. The numerical values of $P(e)$ are given in the following table.

Table 8.2: Probability P of an error pattern of weight e

e	0	1	2	3	4	5	6	7
P	0.4782969	0.3720087	0.1240029	0.0229635	0.0025515	0.0001701	0.0000063	0.0000001

Simulating this process on a calculator, we obtained the following sequence of 63 error patterns of length 7 (an erroneous symbol is indicated by 1):

0000000 0000000 0000000 0101000 0000000 0000000 0000001 0100000
0000101 0000001 0000000 0010000 0000011 0000000 0100000 0010000
0000000 0000000 0000010 0000001 0000000 0000000 0000000 0000000
0000000 0000000 0000010 0000010 0000000 0010000 0000000 0001000
0000000 0100000 0001000 0000000 0000000 0000010 0001111 0000010
1000000 0100000 0000000 0000100 1100000 0000000 1000000 0000100
0000000 0000010 0000000 0000100 0000000 0000000 1000000 0000000
0000010 0000000 0001001 0000001 0001000 0000000 0000001.

Assuming the ancestral 'genome' to be 1111111, we considered the evolution which results from successive replications of this 'genome' in the presence of mutations due to the above error patterns, either without error-correction or when the (7,4) Hamming code is employed (1111111 is a word of this code). This code has three as minimum distance. Therefore, it corrects all error patterns with a single erroneous symbol. However, all error patterns involving two or more erroneous symbols result after decoding in an error pattern with at least three erroneous symbols. The 'genealogical trees' obtained in both cases are represented in the following two figures.

A sample of the simulated evolution of the population of 'genomes' when no error-correcting code is used is given in Fig. 8.1. The population of 'genomes' contains more and more different individuals as the number of successive replications increases. They are short-lived genomes and no distinct species can be observed. The initial 'genome' rapidly ceases to be a majority in this population which, after a few replications, looks like a set of random words. When the (7,4) Hamming error-correcting code is used, the initial 'genome' remains present during a larger number of regenerations-replications. Regeneration errors give rise to 'genomes' differing from the initial one by at least 3 symbols, which may be interpreted as other 'species.' Moreover, they exhibit the same permanence as the initial 'genome.'

8.6 EVOLUTION IN THE TOY LIVING WORLD; PHYLETIC GRAPHS

Notice that Figs. 8.1 and 8.2 represent genealogical trees, the elements of which are individuals. We consider as a 'species' the set of identical 'genomes' (the model does not take into account the slight individual differences which exist in the real world). One can deduce *phyletic graphs* from the genealogical trees by merging into a single branch all branches which correspond to a same 'genome.' Doing so for the genealogical tree of the uncoded case, represented in Fig. 8.1, results in a rather messy graph. The phyletic graph in the encoded case, deduced from the genealogical tree of Fig. 8.2, is more interesting. It is drawn in Fig. 8.3. New species originate in regeneration errors, represented as dots in the figure. One notices that some branches converge, so this graph is not a tree. The code of the example contains however very few words so there is a high probability that distinct regeneration errors result in the same 'genome,' making such convergences likely events.

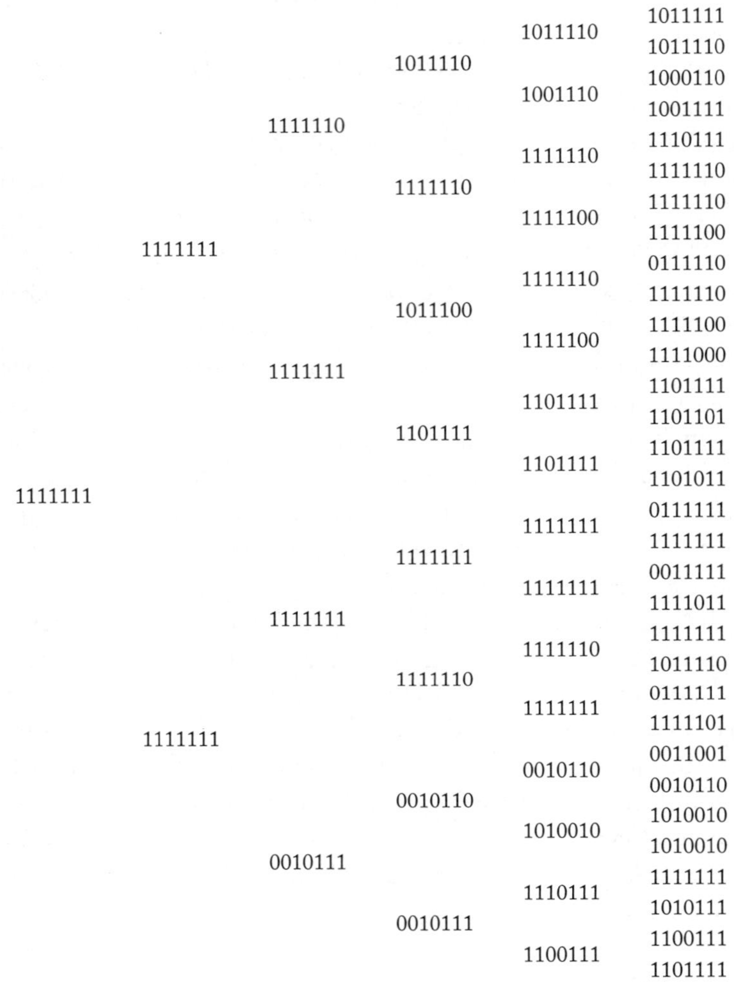

Figure 8.1: Genealogical tree of the toy living world, without error correction.

This probability is much lower for codes having more numerous words close to each other in terms of the Hamming metric, except maybe for the very redundant codes in the innermost layers of the nested system (see Ch. 9). This case excepted, it may be expected that branch convergences are highly unlikely so all phyletic graphs met in practice assume the shape of a tree: branch convergences are highly improbable, although not strictly impossible.

The graph in Fig. 8.3 illustrates the relationship between error-correction coding and evolution for the toy living world: new species originate in regeneration errors. If we ignore the branch convergences (represented by arrows) which occur due to the choice of a very short and simple code

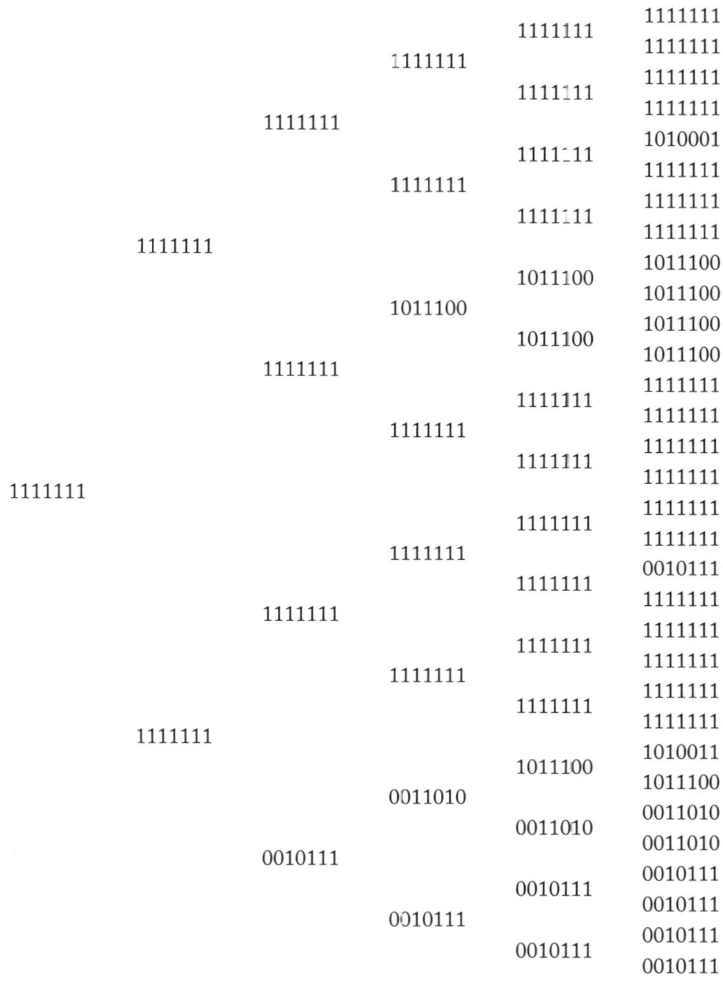

Figure 8.2: Genealogical tree of the toy living world using the (7,4) Hamming code.

but would be very unfrequent for a more realistic code choice, this graph depicts the evolution as a radiative process starting from the ancestral 'genome.' Due to a regeneration error, a chance event which moreover occurs at a random instant, the 'genome' of a new species is chosen among the codewords. Once a new species has been created, it potentially lasts indefinitely (just as the ancestral one) since its extinction due to mutations would imply the occurrence of simultaneous regeneration errors in all the individuals of the species. Its probability is thus P_{cod}^{j}, where j denotes the number of individuals in the species and P_{cod} is the probability of a regeneration error. Since replication is assumed to follow each regeneration, the smallest possible value of j is 2. Moreover, j grows expo-

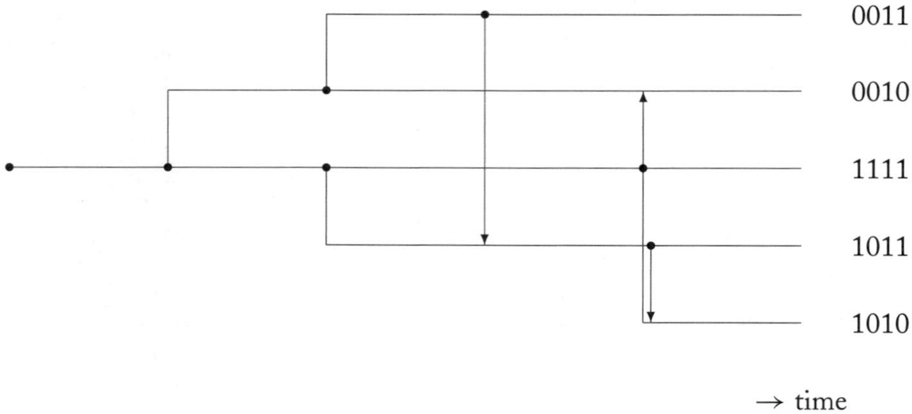

Figure 8.3: Phyletic graph associated with the genealogical tree of Fig. 8.2. Each 'species' is labeled with the four leftmost bits of its 'genome,' i.e., its information bits.

nentially as time passes, making the probability of extinction of a species due to mutations, P^j_{cod}, a miraculous event. Indeed, the extinction of a species can only occur when natural selection operates on the associated phenotype; then, it depends on the probability P' of Sec. 8.4 and results from a differential lack of fitness.

CHAPTER 9

Subsidiary Hypothesis, Nested System

This short chapter is entirely devoted to the concept of nested system basic to the subsidiary hypothesis. Its consequences on the shape of the living world are very important.

9.1 DESCRIPTION OF A NESTED SYSTEM

The main hypothesis that a genomic code exists should be supplemented by the *subsidiary hypothesis* that it actually combines several codes in a *nested system*. Indeed, we saw in Ch. 7 that the capacity of DNA is a rapidly vanishing function of time. Would an error-correcting code equally protect all symbols, then an information would be the less faithfully conserved, the older it is. It is the contrary which is observed in the living world: some of the most faithfully conserved parts of the genome, especially the *HOX* genes, are among the oldest. The subsidiary hypothesis is intended to account for this fact. It assumes that the genomic error-correcting code actually consists of a combination of codes which were successively laid down during the geological ages. Every time a new coding layer has been appended, the information inside it has acquired a much smaller probability of regeneration error. Francis Crick's 'frozen accident' finds here a substantial content.

Prior to examine natural nested systems, we give a detailed example of how an engineer can design a nested system. We assume that we use binary block codes in systematic form[1] as its components. The j-th code is denoted by $\mathcal{C}_j(n_j, k_j)$, where n_j is its length and k_j its 'dimension,' i.e., the length of the message it encodes. We moreover assume that the message M_j it encodes explicitly appears as the leftmost k_j bits of the word. We graphically depict a word of this code as

$$\lfloor\ M_j \ \| \ R_j(M_j) \ \rfloor$$

where $R_j(M_j)$ denotes the set of $n_j - k_j$ redundancy bits which the encoder computes in terms of the k_j bits of the message M_j; the double vertical bar separates the k_j information bits of this code from its $n_j - k_j$ redundancy bits. Using this representation, the following figure describes a four-layer nested system where the fourth layer is left uncoded.

A word belonging to the four-layer nested system is represented at the bottom of Fig. 9.1. Its construction involved four steps. In the first one (top), the i_1-bit information message I_1 is encoded according to a first code $\mathcal{C}_1(n_1, k_1)$ with $k_1 = i_1$, i.e., the set of $n_1 - k_1$ redundancy bits denoted by $R_1(I_1)$ is computed in terms of I_1 and appended to it. In step 2, the i_2-bit information message I_2

[1]Remember that in a code in systematic form the information message explicitly appears at given locations in the codeword.

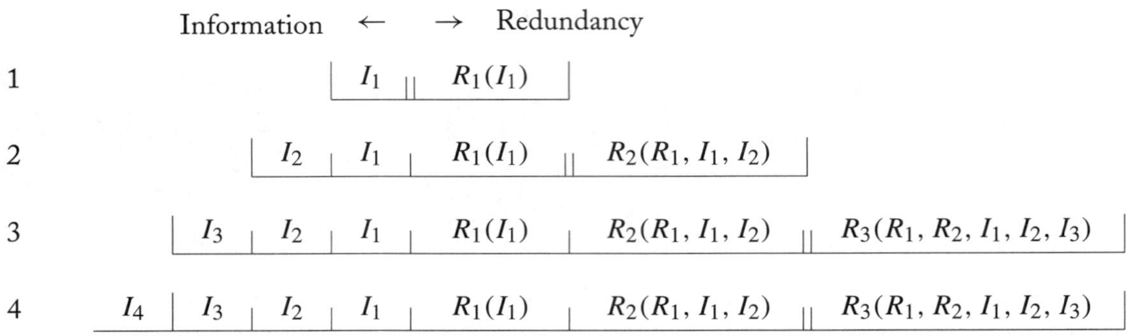

Figure 9.1: A four-layer nested system. The mentions 'information' and 'redundancy' at the top refer to the overall redundancy of the nested system.

is appended to the codeword resulting from the encoding performed in step 1. The $(i_2 + n_1)$-bit message thus obtained is encoded by the code $C_2(n_2, k_2)$ with $k_2 = i_2 + n_1$, i.e., the set of $n_2 - k_2$ redundancy bits denoted by $R_2(R_1, I_1, I_2)$ is computed and appended. Similarly, in step 3 the i_3-bit information message is appended to the result of the encoding performed in step 2, resulting in an $(i_3 + n_2)$-bit message which is again encoded by the code $C_3(n_3, k_3)$ with $k_3 = i_3 + n_2$, i.e., the set of $n_3 - k_3$ redundancy bits denoted by $R_3(R_1, R_2, I_1, I_2, I_3)$ is computed and appended. The last step (step (4) simply consists of appending the information message I_4 to the word resulting from the previous encoding since we assume that the fourth layer remains uncoded.

A simpler and more intuitive illustration uses the fortress metaphor: a code is represented as a wall which protects what is inside it against outside attackers. Several concentric walls have been successively built to enclose information, so the content of the oldest, most central, wall is much better protected than the more recent and peripheral information (see Fig. 9.2). Notice that a very efficient protection of the most central information does not demand very efficient individual codes: the multiplicity of enclosing walls is much safer than each of them separately.

None of the assumptions which we made to illustrate the nested system in Fig. 9.1 is mandatory. The alphabet needs not be binary. Its size can, moreover, be different from a coding layer to another one: a newly performed encoding may well deal with some extension of the information message, i.e., consider blocks of m successive q-ary symbols as symbols of an alphabet of size q^m. The codes do not need to be in systematic form and, if they are, the location of the information bits is arbitrary. The assumption that the successive codes are of the block-code type may also be relaxed (convolutional codes are indeed likely candidates as being in a sense more 'natural' than block codes). In the case of convolutional codes the length of the codewords is no longer precisely defined, which moreover shows that even the condition that the dimension of each code is larger than the codeword length of the previous coding layer has not to be strictly satisfied. It suffices that the encoding performed at the j-th step brings some further protection to the message which has

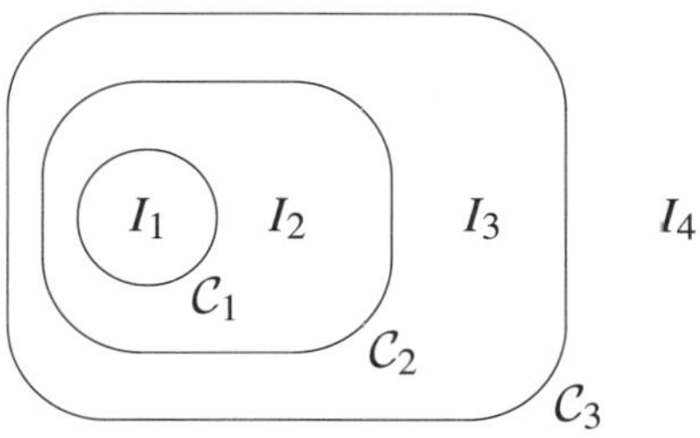

Figure 9.2: The same nested system as Fig. 9.1 represented according to the fortress metaphor. A code C_j, $j = 1, 2, 3$, is represented as a closed wall which protects what is inside it. I_1, I_2, I_3, and I_4 are successive information messages. I_1 is protected by 3 codes, I_2 by 2 codes, I_3 by a single code, and I_4 is left uncoded.

already been encoded at the previous steps. Moreover, the codes actually used by nature do not result from an engineering design but likely belong to the broader class of 'soft codes', to be introduced in Ch. 10.

9.2 RATE AND LENGTH OF COMPONENT CODES

In order to get an insight on how the rate and length of the component codes vary when their number is large, we may assume that the length of the information message appended at each step of the construction of the nested system is constant, say i. We denote by $\rho_j = k_j/n_j$ the rate of the j-th component code. Its reciprocal $\lambda_j = 1/\rho_j = n_j/k_j$ will be referred to as the expansion factor of this code. The cumulated expansion factor for the first j steps is defined as

$$\Lambda_j = n_j/ji$$

since then the cumulated information message has length ji. Clearly, the cumulated expansion factor at step j obeys the recursion equality

$$j\Lambda_j = \lambda_j[(j-1)\Lambda_{j-1} + 1] .$$

We may first assume, for instance, that $\lambda_j = \lambda$, a constant. Then we find that

$$\Lambda_j = (\lambda + \lambda^2 + \ldots + \lambda^j)/j ,$$

or, in more compact form:

$$\Lambda_j = \frac{\lambda(\lambda^j - 1)}{j(\lambda - 1)} ,$$

which shows that the length $ji\Lambda_j$ of the j-th component code then grows exponentially in terms of the step number j.

It is, however, possible to choose the expansion factors $\{\lambda_j\}$ so as to limit the growth of the overall length. For instance, $\Lambda_j = \Lambda$, a constant, is obtained by choosing

$$\lambda_j = \frac{j\Lambda}{(j-1)\Lambda + 1}$$

which results in the rate of the j-th component code being

$$\rho_j = 1 - \frac{\Lambda - 1}{j\Lambda} .$$

Then the overall length of the system at step j is $ji\Lambda$. If $\Lambda = 2$, for instance, the rate of the j-th component code is $\rho_j = (2j-1)/2j$.

9.3 DISTANCES IN THE NESTED SYSTEM

Notice that the walls of the fortress of Fig. 9.2 are purely metaphoric. The presence of a wall just means that the elements enclosed in it are codewords separated by some minimum Hamming distance. Figure 9.3 attempts to visualize this distance distribution in the case of a two-layer nested system. Representing the distribution of points in a high-dimensional space is not easy and can only result in a very rough picture. First, the codewords are represented as points in the plane, according to some kind of projection, although they actually belong to an n-dimensional space. Second, the Hamming distance in the n-dimensional space is represented as the Euclidean distance in the plane of the figure. The restriction to a two-layer nested system aims at keeping the figure simple enough but a more realistic picture involving many nested codes would be extremely complicated.

We represented in Fig. 9.3 the first (older) code $\mathcal{C}_1$ as a 'constellation' of points (big dots of the figure). At the second encoding layer, using the second code $\mathcal{C}_2$ results in putting clusters of points centered on all the points of the initial constellation. It has been found convenient for drawing Fig. 9.3 to take each of these clusters exactly similar to the 'constellation' of points representing the codewords of $\mathcal{C}_1$. Pursuing this process would result in a fractal structure. Although the actual structure of the nested system is probably much less simple, it is interesting to notice that the known correlation properties of the DNA sequences are compatible with a fractal structure.

The experimental analysis of DNA sequences has shown indeed they exhibit long-range dependence. Their power spectral density has been found to behave as $1/f^{\beta}$, asymptotically for small f, where f denotes the frequency and β is a constant which depends on the species: roughly speaking, β is the smaller, the higher the species is in the scale of evolution; it is very close to 1 for bacteria and significantly less for animals and plants [75]. In another study of DNA sequences, the quaternary alphabet of nucleic bases {**A**, **T**, **G**, **C**} was first restricted the to the binary one {**R**,**Y**} by retaining only their chemical structure, purine or pyrimidine. An appropriate wavelet transform was used to cancel the trend and its first derivative. Then the autocorrelation function of the binary string thus obtained has been shown to decrease according to a power law [1], which implies long-range dependence (at variance with, e.g., Markovian processes which exhibit an exponential decrease).

Figure 9.3: Sketch of a two-component nested system in the space of sequences. The big dots represent the codewords of the first (older) component code C_1, and all the dots represent words of the nested system, i.e., combining the first and the second component codes. The oldest information is borne by any of the points in the cluster which surrounds each of the big dots.

The $1/f^{\beta}$ behavior of the spectrum and the long-range dependence exhibited by the autocorrelation function of the DNA sequence are of course compatible with each other. Moreover, they are both compatible with a fractal structure, meaning that the DNA sequence may in some sense be self similar.

9.4 SOME CONSEQUENCES OF THE SUBSIDIARY HYPOTHESIS

We conjecture that individual genomic differences within a species are restricted to the peripheral layer of the nested system, assumed to be left uncoded as in Fig. 9.1. (Assuming that the outer layer of the nested system is left uncoded enabled in Sec. 7.3 estimating the nucleotide error rate in terms of measurements of the mutation rate.) Genomic encoding becomes compatible with sexual reproduction since then the genetic shuffling that crossing-over operates only affects the peripheral (actually uncoded) layer. Sex then promotes fast genetic changes in a population, an evolutive advantage against fast evolving pathogens or parasites, but confined in the peripheral layer of the nested system while the information of inner layers remains unaffected by symbol errors due to the genome error-correcting ability, except for very unfrequent regeneration errors.

The nested structure implies a very unequal protection of the nucleotides against errors, depending on their location. This makes the genome appear as very inhomogeneous as regards the susceptibility of its nucleotides to errors. The very unequal distribution of error rates depending on the location of the nucleotides implied by the subsidiary hypothesis matches the observed reality (see, e.g., [64, 52]). Interestingly, extremely well-conserved sequences of nucleotides are not only found in genes properly said (e.g., the *HOX* genes) but also in portions of the DNA which do not direct the synthesis of proteins (referred to as 'noncoding DNA' in the biological literature) [56]. Assuming that genomic error-correcting codes exist, we have no reason to restrict our attention to 'coding DNA.' On the contrary, any symbol in a codeword both contributes to, and benefits from, the error-correcting ability of the code. Needless to say, the idea of 'junk DNA' is entirely foreign to our hypotheses.

Analyzing the toy living world based on the use of a single code assumed to equally protect its symbols, as we did in Ch. 8, is a simple matter. A similar analysis of a model where the code is a nested system as just described would on the contrary be very complicated. Moreover, it would need several parameters like the number of component codes, their sizes and minimum distances which *a priori* depend on the species and are difficult to estimate. Even a toy living world limited to a two-component nested system would be difficult to analyze. We prefer to qualitatively describe the features of a world where the genomic code consists of a nested system.

Let us consider separately the component codes of the nested system. To begin with, consider a long and very redundant code as found in the deeper layers (the words 'deep' and 'peripheral' refer to the fortress metaphor). We may assume that it is good as a result of natural selection, 'good' meaning that its minimum distance is large, say equal to the Gilbert-Varshamov bound (see Eqs. (6.26) and (6.27) in Sec. 6.5.3). Being highly redundant, this code has comparatively few words separated from each other by large Hamming distances. In sharp contrast, we were already led to assume that the peripheral layer of the nested system is left uncoded. Then errors frequently occur but result in a genome very close to the original one. Between the deepest and the peripheral layers, the codes have less many codewords, the deeper the layer. At the same time, the codewords are the more separated in terms of the Hamming metrics, the code to which they belong lies in a deeper layer. The biological consequences of this nested structure will be more fully developed in Ch. 11.

CHAPTER 10

Soft Codes

10.1 INTRODUCING CODES DEFINED BY A SET OF CONSTRAINTS

In Ch. 6, we defined an error-correcting code $\mathcal{C}(n, k)$ as a subset among the set S_n of n-symbol sequences over some finite alphabet of size q. The number of symbols k of the information message that each codeword represents is referred to as the number of dimensions of the code or, more briefly, as its dimension. The code being necessarily redundant implies the strict inequality $k < n$. There are basically two ways for defining such a code. One way is to give the list of its words; random coding, as used by Shannon and his followers in order to prove the fundamental theorem of channel coding, is a typical example of it. The trouble with a mere list of codewords is that the number of its elements exponentially grows in terms of the codeword length n: assuming a constant information rate $R = k/n$, $0 < R < 1$, the number of codewords is $q^k = q^{Rn} = (q^R)^n$, which makes the decoding process of such a code of prohibitive complexity when n is large enough for the code to be useful. Each received word should indeed be compared with *all* the words of the code! Another way of defining an error-correcting code consists of stating some specific set of intrinsic constraints which its codewords should satisfy. All the sequences of length n which satisfy this set of constraints, and only these sequences, belong to the code. The codewords can, in principle, be listed once this set of constraints is given, but the set of constraints is a much more parsimonious description of the code than the full list of its words.

The recourse to intrinsic constraints for defining an error-correcting code is an absolute necessity for making it tractable. Choosing a set of constraints without introducing a bias detrimental to the code performance is however a tough problem. It has even been thought of as unsolvable. A long-lived popular belief among communication engineers was expressed by the 'folk theorem': 'All codes are good, except those we can think of.' Its fallacy was only revealed with the invention of turbocodes. Presenting these codes in 1993, Berrou et al. reported unprecedented performance [21]. They showed that error-correcting codes based on a set of deterministic intrinsic constraints, but which mimic random coding as regards the distribution of distances between the codewords, exhibit a performance close to the theoretical limit, i.e., the channel capacity. Designing good codes is thus practically possible, although the contrary has long been believed true.

Another important advantage of defining an error-correcting code in terms of a set of constraints (as opposed to a list) is that these constraints are expressed as relations between the symbols of the codewords. Then the decoding procedure (i.e., aiming at recovering the original codeword from its received, possibly erroneous, version) does not need to deal with n-symbol sequences. Instead, decisions can be separately taken about each symbol and, moreover, provisional decisions on symbols can be used in the process of making new decisions, according to an iterative procedure. These possibilities are fully exploited in turbocodes, the decoding of which involves a kind of feedback made possible by symbol-by-symbol decoding (see Sec. 6.7.2).

10.2 GENOMIC ERROR-CORRECTING CODES AS 'SOFT CODES'

10.2.1 DEFINING SOFT CODES

Defining an error-correcting code in terms of a set of intrinsic constraints is an unavoidable necessity in engineering. The genomic error-correcting codes assumed to exist according to our main hypothesis do not escape it. However, this does not mean that natural codes must closely resemble those designed by engineers. The constraints considered by engineers are defined by deterministic *mathematical equalities* because they are thus perfectly defined and easily implemented using physical devices[1]. The implementation means of nature and humans widely differ, making the ease of engineering implementation irrelevant to nature. We thus propose to extend the kind of constraints which define an error-correcting code to inequalities, forbidding rules and constraints expressed in probabilistic terms, as well as physical-chemical constraints, besides mathematical equalities. Steric constraints due to the position and dimensions of molecules in the three-dimensional physical space are especially important. Still other constraints are linguistic. Genomes are assumed to direct the construction and maintenance of living things. This cannot be done without a language, which implies a number of constraints of morphological, syntactic and semantic character. According to this point of view, any constraint to which genomes or parts of genomes are subjected can be exploited to the benefit of error correction, regardless of its nature. We thus call for a wide extension of the concept of error-correcting code, to be referred to as *soft code*.

Even though we think that natural genomic error-correcting codes are not likely to closely resemble those designed by human engineers, the structure of eukaryotic genes with exons (used in the synthesis of a protein) and introns (having no known function) strikingly reminds that of block codes in systematic form which are currently used in engineering, where redundancy symbols only employed for correction purpose are appended to the information symbols. Are exons made of information symbols and introns of redundancy symbols? This question is examined in Sec. 10.5 below.

10.2.2 IDENTIFYING THE ALPHABETS

In engineering problems, the alphabet is often *a priori* given and endowed with some chosen precise mathematical structure. This is not the case for the hypothesized genomic error-correcting codes

[1] An additional advantage is the availability of preexisting mathematical tools. As a striking example, Evariste Galois's theory of finite fields, long believed to exclusively belong to pure mathematics, found important engineering applications for designing error-correcting codes 120 years or so after being formulated. The vast family of algebraic codes rooted in Galois's work has not been dealt with in Ch. 6, but merely alluded to in Sec. 6.8.

where the alphabets themselves and their possible mathematical structure have to be determined. We put here 'alphabets' to the plural since we consider nested soft codes, and we already noticed that the component codes of a nested system may use different alphabets.

At first sight, however, it may seem obvious that the DNA alphabet, as consisting of the set of nucleotides {**A**, **T**, **G**, **C**}, should be considered as quaternary. But with what mathematical structure is it endowed? Liebovitch et al. [53] assumed it to be the ring of integers modulo 4. This choice is arbitrary and the structure considered in the coding literature for properly defining a linear code is that of a Galois field. It is only when the alphabet size q is a prime that the addition rule modulo q and that of the Galois field are identical, which is not the case for $q = 4$.

Even a mathematical structure more appropriate in engineering terms is unlikely to be relevant as being more or less arbitrarily chosen. Since soft codes can be defined by physical and chemical constraints, alphabets having a physical-chemical significance are more likely to be relevant. In this respect, we may consider that any nucleotide simultaneously belongs to two independent binary alphabets. First, the alphabet {**R**,**Y**} whose symbols are the possible chemical structures of the nucleic bases, namely purine (2-cycle molecule, **A** or **G**), denoted by **R**, or pyrimidine (single-cycle molecule, **T** or **C**), denoted by **Y**. Second, the alphabet which we denote by {**W**,**S**} where **W**, standing for 'weak,' represents the complementary pair of nucleotides **A**–**T** which are tied together by two hydrogen bonds, and **S**, standing for 'strong,' the other complementary pair, namely **G**–**C**, where the nucleotides are tied together by three hydrogen bonds. The alphabet {**R**,**Y**} corresponds to nucleic bases of different chemical structure and physical size, while the second one, {**W**,**S**}, indicates how strongly a nucleic base is tied with the complementary one. Then, Forsdyke interpreted a sequence of quaternary symbols as simultaneously bearing two independent binary codes, one over the alphabet {**R**,**Y**} and the other one over {**W**,**S**} [35]. According to the second Chargaff parity rule, the first code is balanced, i.e., the two symbols **R** and **Y** have the same frequency, like almost all codes designed by human engineers. On the contrary, the code over the alphabet {**W**,**S**} is not balanced since the frequency of its symbols (referred to as the **G**+**C** content) varies from a species to another one and, for long and inhomogeneous genomes like the human one, from a region to another one inside the genome. It could be interpreted as a kind of density modulation which maybe is read at several scales. The different number of hydrogen bonds of the two symbols of the alphabet {**W**,**S**} implies that this density modulation results in a variation of the bonding energy between the complementary strands of DNA.

Other constraints are directly expressed in terms of larger alphabets. For instance, constraints induced on DNA by the structural properties of the proteins for which it 'codes' involve triplets of nucleic bases, i.e., the codons of the genetic 'code'. Then the alphabet size is $4^3 = 64$. Genes themselves may even be considered as the symbols of an alphabet [49, 61].

10.2.3 POTENTIAL GENOMIC SOFT CODES

We now list some potential genomic soft codes. Potential soft codes can first be associated with structural constraints of DNA. The very dense packing of DNA implies that any part of the DNA string is spatially close to many distant other parts of the same string (where 'distant' refers here to distance along the DNA string) [62]. The spatial neighborhood of these parts is due to chemical affinities[2], hence constraints, and many such constraints take place everywhere in the DNA string. The distances between the interactive parts along the DNA string may then play the role of interleavers so the whole genomic code would look like a kind of turbocode. Not only chemical affinities of the DNA string with distant parts of itself but also with histones, as constituents of nucleosomes, need be considered in the case of eukaryotic cells. The constraints due to the packing in nucleosomes concerns the 165 or so base pairs which are wrapped around histone octamers acting as spools, and the higher structures result in constraints presumably weaker but covering a wider range.

In the portions of the genome which specify proteins, i.e., in genes in a restricted sense, the sequence of codons (triplets of nucleotides) is furthermore constrained as are the proteins themselves: the structural constraints of proteins induce constraints on the sequence of codons which correspond to the amino-acids according to the 'genetic code'[3]. Physiologically active proteins are not fully characterized by the sequence of amino-acids (the polypeptidic chain) that the sequence of codons specifies. They owe their functional properties to their folding according to a unique pattern, which implies many chemical bonds between amino-acids which are separated along the polypeptidic chain but close together in the three-dimensional space when the protein is properly folded. Proteins are made of a number of three-dimensional substructures, especially α-helices and β-sheets, which are themselves included into higher-order structures named 'domains,' which impose strong constraints of steric and chemical character on the succession of amino-acids. The sequence of amino-acids is thus subjected to many constraints. The sequence of codons which specify the amino-acids is thus itself subjected to related constraints. Due to the universal role of DNA for specifying proteins, such constraints must be present in any living being.

Soft Codes from Linguistic Constraints

We mentioned above that soft codes may be induced by linguistic constraints, too. We already noticed that the message which is needed for unambiguously identifying a biological species and even an individual inside it is very much shorter than the actual genomes, even the shortest ones like those of viruses (see Sec. 2.4). Genomes are thus highly redundant, a necessary condition for them to possess significant error-correcting properties. From another point of view, this redundancy has rather obvious reasons: the genome does not merely identify a living being. Modern biology interprets it as a *blueprint* for its construction. The genome of any living being needs actually contain the *recipe* for its development and its maintenance. Besides those parts of the genome which direct the synthesis of proteins, i.e., the genes in a restricted sense, and the associated regulatory sequences which switch on or off their expression (i.e., make the gene direct or not the synthesis of the protein it specifies), the genome must somehow *describe* the succession of operations which results in the development and the maintenance of a living thing. This demands some kind of *language*,

[2]'Chemical affinities' should be intended in a broad sense, including the forces which determine molecular shapes in an aqueous medium, e.g., the interaction of hydrophobic radicals with water, besides the usual chemical bounds.

[3]We use quotes here in order to remind that it is not truly a code in the information-theoretic sense, but rather a mapping in the mathematical vocabulary.

and a language involves many morphological and syntactic constraints which may be interpreted as generating redundant soft codes having error-correcting capabilities. Moreover, the linguistic constraints appear at several different levels, so a written text assumes the structure of nested soft codes which we were led to hypothesize for the genetic message. In order to illustrate the error-correcting ability of languages, notice that we, humans, can correctly perceive the spoken language in extremely noisy acoustic surroundings like vehicles or crowded places. By 'correctly perceive,' we do not mean to grasp the meaning, which concerns the semantic level, but simply to identify without error the uttered sequence of phonemes. It turns out that the individual phonemes are identified with a large error probability, but the linguistic constraints together with the high processing power of the human brain eventually result in errorless communication even in the presence of noise. We may assert that the daily experience of a conversation experimentally proves the ability of the human language, as a highly redundant soft code, to behave like good error-correcting codes designed by engineers. Biologists consistently use the metaphor of a written text to explain the role of the genetic message, at least in popular science books like [30] and many others. This metaphor is quite convincing, but the consequences it can have on the genome conservation are overlooked: indeed, it turns out that a language involves numerous and diverse constraints which may be interpreted as soft codes having error-correcting capabilities. Of course, it remains to understand how these error-correcting capabilities are exploited. Indeed, recent researches use tools of formal linguistics in order to describe the genomes and proteins [56, 71] but ignore the error correction problem. Moreover, the concept of dependence is actually shared by information and coding theory on the one hand, and formal linguistics applied to genomes and proteins on the other hand. As another interesting analogy, human natural languages have undergone evolution, and methods of the same kind as used for studying evolution in biology have been applied to that of languages.

The connection outlined in Sec. 5.5 between semantics and information implies that a longer genome is not only useful to decrease the error probability, but also provides room for 'more semantics' and thus enables specifying more complex beings. An important and useful tenet of information theory is the separation between information and semantics. However, the hypothesized error-correction mechanisms based on linguistic constraints heavily rely on the genome being a blueprint for the construction and maintenance of a living being, so one could consider the error-correction ability of the genetic message as, at least partially, a by-product of its semantics. But this is only a facet of the question. One can argue equally well that the error-correction property is the main feature of this message, since without it no transmission of hereditary characters would be possible and life could not have developed. Then, the construction and maintenance of living things would be a mere projection in the physical-chemical world of the abstract properties of the genetic message which enable error correction. This is a hen-and-egg problem, as often met in biology.

10.3 BIOLOGICAL SOFT CODES FORM NESTED SYSTEMS

Defining genomic codes by a set of constraints leads to a structure close to that described in Ch. 9.1 as a 'nested system,' although understandably much more complicated. Let us take as an example a eukaryotic gene. Consider that part of an exon which specifies the amino-acid sequence of an α-helix in some protein. Certain amino-acids are preferred or forbidden in α-helices. According to [24], Ala, Glu, Leu, and Met will thus be more often met than average in the considered sequence, while Pro, Gly, Tyr, and Ser will be almost absent. As a consequence, the codons **GC*** for Ala, **GAA** and **GAG** for Glu, **UUA**, **UUG** and **CU*** for Leu, and **AUG** for Met, will be more often met than average in the corresponding sequence of the messenger RNA, while the codons **CC*** for Pro, **GG*** for Gly, **UAU** and **UAC** for Tyr, and **UC***, **AGU**, and **AGC** for Ser, will be almost absent (* stands for any nucleotide **A**, **U**, **G**, or **C**; see Ch. 4). The DNA composition will thus greatly differ from the average in the part of the gene which corresponds to an α-helix. Moreover, there are probably stronger constraints which forbid some successions of amino-acids as not compatible with the α-helix structure. Similar constraints are related to other protein structures like β-sheets. A protein, moreover, involves at a larger scale several α-helices and β-sheets organized in domains, according to a scheme which may be considered as a 'metacode' the symbols of which are words of the previously identified soft codes.

But these soft codes only pertain to the exons which actually direct the protein synthesis. According to a rather plausible conjecture (see Sec. 10.5) the nucleotides which belong to an exon are the information symbols of some block code in systematic form, the redundancy symbols of which are located in the corresponding introns. The previously identified soft codes are embedded into this block code which thus constitutes another coding layer of the nested system.

Only a fraction of the DNA base-pairs actually belongs to proper genes. This fraction is greatly variable from a species to another, and is quite small in, e.g., higher animals. We suggested above that the remainder, besides regulatory sequences, is encoded in linguistic codes necessary to describe the phenotype, which moreover assume themselves a nested structure.

This example shows the multiplicity of potential soft codes and their organization into a nested system or even into several intricate nested systems. The number and complexity of these soft codes is due in part to the choice of a eukaryotic gene as example. Both the block code made of exons and introns and the soft code due to the constraints of packing DNA into nucleosomes are not relevant to a prokaryotic gene. However, the close packing of DNA, although different, would similarly generate specific constraints. Those induced by the protein constraints would remain. The constraints of linguistic character are presumably still present, although less important since the specified phenotype is much simpler.

The number and variety of constraints indeed suggest that many potential genomic error-correcting mechanisms actually exist, which moreover are organized as nested soft codes. It is interesting to compare the resulting nested system with Barbieri's organic codes [5]. The codes we consider protect the DNA molecule against radiations and quantum indeterminism which no phenotypic shielding can ensure, hence are needed to ensure the conservation of hereditary information. Barbieri's concept of organic codes, on the other hand, does not refer to the necessity of error correction but results from a deep reflection on biological facts. He considers as an organic code the correspondence which exists between unidimensional[4] strings of completely different molecules (as a famous example, think of the relationship between triplets of nucleotides and the 20 amino-acids which make up proteins, referred to as the 'genetic code'). This correspondence does not result from any physical or chemical law, but should be considered as a pure convention or artifact, just like conventional rules are found in linguistic or engineering. (We conjectured that such rules are maintained thanks to 'semantic feedback loops' [17].) According to our point of view, the specific constraints obeyed by each of the strings which are tied together by a correspondence rule induce specific constraints in the DNA string, hence act as soft codes with error-correcting ability. Barbieri's organic codes actually assume the structure of nested codes. Especially significant in this respect is Fig. 8.2 in [5], p. 233, to be compared with Fig. 9.2 above which uses the fortress metaphor to illustrate the concept of nested codes. Moreover, both Barbieri's organic codes and the soft codes which make up our nested system successively appeared in geological times and correspond to major transitions in life evolution. This rather unexpected convergence provides a mutual confirmation of both points of view, which appear as complementary. This may also be thought of as an illustration of 'tinkering' as a typical method of nature, where some biological objects are used to perform functions completely different from those they initially performed. However, since many biological problems assume the chicken-and-egg form, a word like 'multivalence' could be more appropriate than 'tinkering' (although less picturesque) in the absence of a clear knowledge of the chronology.

Another point must be emphasized. The nested system provides a very unequal protection of different symbols depending on their place in the DNA string. Some base pairs are related to many others according to many constraints, and thus are strongly protected against casual errors. In our fortress metaphor, they belong to central coding layers. Others are subjected to few constraints and only related to few other base pairs and belong to outer layers. Some of the base pairs may even escape any constraint and appear as uncoded, belonging to the peripheral layer. Genomes, especially those of higher-living beings, thus appear as very inhomogeneous. This consequence of our hypotheses meets the experimental reality.

10.4 FURTHER COMMENTS ABOUT GENOMIC SOFT CODES

Soft codes do not exactly fit the properties of error-correcting codes which were described in Ch. 6. Since their definition involves probabilistic constraints and constraints expressed as inequalities, the mutual distances between their words become random, and especially the minimum distance d which accounts to a large extent for the performance of a code. On the other hand, when discussing in Ch. 11 the consequences of our hypotheses we'll assume that the properties of genomic error-correcting codes are those of conventional codes. This may be thought of as a simplifying assumption. One may moreover argue that, if the soft codes combined into the nested scheme are numerous enough, and

[4]Unidimensionality is a common feature of messages in engineering and in biology. It appears as necessary for unambiguous semantic communication as well as in engineering.

if moreover their words are long enough, then the law of large number results in a small-variance minimum distance which can rightfully be approximated by a deterministic quantity.

Let us also notice that the soft code concept implies that the biological constraints are also those which enable error correction, at variance with the uncoded case but also with that of hypothetic codes obeying purely mathematical constraints. This may mean that the genomes which are admissible as words of a genomic error-correcting code also specify viable phenotypes. If this is true, decoding (regeneration) errors produce viable, possibly hopeful, monsters. This remark makes rather plausible the explanation of the Cambrian explosion to be suggested in Sec. 11.3.2.

10.5 IS A EUKARYOTIC GENE INVOLVING EXONS AND INTRONS A CODEWORD IN SYSTEMATIC FORM?

The interesting idea that introns are made of redundancy symbols associated with the message borne by the exons was formulated by Forsdyke as early as 1981 [34]. The literature generally states that introns are more variable than exons. A counter-example was however provided in 1995, again by Forsdyke, who experimentally found that the exons are more variable than introns in genes which 'code' for snake venoms [36].

It turns out that both the generally observed greater variability of introns and Forsdyke's counter-example can be explained by the assumption that the system of exons and introns actually acts as a systematic error-correcting code where exons constitute the information message (which directs the synthesis of a protein) and introns are made of the associated redundancy symbols[5]. Interpreted as a decoding error, a mutation occurs with large probability in favour of a codeword at a Hamming distance from the original word equal to the minimum distance of the code or slightly larger. If the exons 'code' for a protein of physiological importance, which is by far the most usual case, it may be expected that only mutations with a few errors within the exons, hence having no or little incidence on the protein, will survive natural selection. Few symbol errors being located in the exons, most of them will affect the introns since the total number of symbol errors after decoding is at least equal to the minimum distance of the code.

The situation is completely different in the case of genes which 'code' for snake venoms. The typical preys of snakes are rodents. Snakes and rodents are involved in an 'arms race': some rodents incur mutations which provide an immunity to snake venom, the population of rodents with such mutations increases as they escape their main predators, and the snakes are threatened with starvation unless mutations in their own genes make their venom able to kill mutated rodents [36]. The genes which 'code' for snake venoms are thus under high evolutive pressure: natural selection favours mutated genes producing proteins as different as possible from the original one. In terms of the Hamming distance, much of the difference should thus be located in the exons. The total number of decoding errors in exons and introns being roughly constant for a given code (equal to the minimum distance or slightly larger), introns are much less variable.

[5] An observer unaware of their role for error correction would easily conclude that the redundancy symbols have no function.

CHAPTER 11

Biological Reality Conforms to the Main and Subsidiary Hypotheses

The simple model of Ch. 8, solely based on the main hypothesis, enabled calculating the lifetime, or *permanence*, of 'genomes' consisting of symbol sequences. Taking the subsidiary hypothesis into account as well as the results of Ch. 8 led in Ch. 9 to a qualitative description of a less simple, but more realistic, model of the living world. Introducing the concept of 'soft code' in Ch. 10 brought some biological relevance to that of genomic error-correcting code, and to that of 'nested system' central to the subsidiary hypothesis. (Remember that we used the word 'hypotheses' for lack of a more appropriate, stronger word, to designate inescapable conditions dictated by information theory.)

For translating properties of error-correcting codes into biological terms, we assume that the channel-encoded part of the genome (i.e., apart from the peripheral layer of the nested system, assumed to be left uncoded and to contain all individual differences, which constitutes only a small fraction of the genome) can be used in order to identify the species. Then, we may think of the information symbols of a genomic code as associated with the independent alternatives or parameters needed to specify a phenotype. The more numerous are the information symbols, i.e., the more numerous distinct dimensions has the genomic code, the more numerous semantic features a genome can instruct (see Sec. 5.5). The redundancy of genomes implies that the number of information symbols is only a small fraction of their total length. For a given redundancy rate, the phenotypic features that a genome determines are thus the more numerous, the larger its length.

We now show that the living world actually exhibits features which are biological consequences of these hypotheses. These are indeed the most basic and conspicuous properties of life, many of them being left essentially unexplained by conventional biology which rather deals with them as unquestioned postulates. Our hypotheses thus provide a broad understanding of the living world as a whole, which can be thought of as a global experimental check of their relevance. Moreover, they answer debated questions, some of them of very general character.

Table 11.1 lists in the left column a number of properties of error-correcting codes while the right column contains the features of the living world which result from each of these properties. We'll examine in the following sections each line of this table.

Table 11.1: Consequences of the main and subsidiary hypotheses: properties of error-correcting codes entail main features of the living world

Error-correcting codes	Biology
Redundancy is necessary	Genomes are very redundant
Distance between codewords	Existence of discrete species
Nested codes	Hierarchical taxonomy
Limited correcting ability of a code	Necessity of successive regenerations
Regeneration errors result in distant codewords	Evolution proceeds by jumps
Nested codes imply a hierarchy of distances	Gradualism in periphery, saltationism in deep coding layers
The longer a code, the more efficient	Evolutive advantage of long genomes
Regeneration errors are chance events	Evolution is contingent
Close words do not correspond to close information messages	Close genomes do not correspond to close phenotypes (overlooked?)

11.1 GENOMES ARE VERY REDUNDANT

The main hypothesis implies that genomes are redundant, in the sense that the sequences which belong to an error-correcting code, hence obey its constraints, are less than all the possible sequences of same length written using the same alphabet. We have already shown in Sec. 2.4 that genomes are extremely redundant. Indeed, the existing genomes are only a tiny fraction of all possible nucleotide sequences. Nobody knows the number of species which now live on Earth, because the identified ones are an unknown fraction of all extant species. An order of magnitude could be 10^7. Let us assume that the extant species are a fraction of only 1/100 of all those which once existed (due to the extinction of many). We obtain 10^9 as an order of magnitude for the number of past and present species. Although this rough estimate may look large (and should even this number be underestimated by a few orders of magnitude), it appears as extremely small with respect to the total number of possible genomes in the absence of redundancy, namely 4^n. Typical values of the genome length n range from thousands to billions and more base pairs, hence 4^n is a huge number. Genomes thus exhibit a very high redundancy.

11.2 LIVING BEINGS BELONG TO DISCRETE SPECIES

11.2.1 A GENOMIC ERROR-CORRECTING CODE IMPLIES DISCRETE SPECIES

The description of an error-correcting code we gave in Sec. 6.5 and the identification of a species with a genome (except for its small uncoded fraction) entails that the main hypothesis implies that *species are sparse*. It was already apparent in Ch. 8 that species discreteness is an essential feature of a genomic error-correcting code. We saw in Sec. 8.4 that the effects of natural selection could formally be taken into account in the toy living world which uses an error-correcting code, and we criticized the claim made by neo-Darwinians that the discreteness of species is the sole consequence of natural selection acting on phenotypes. We propose instead that this discreteness is *intrinsic* to the genomic error-correction system, hence pre-exists natural selection. Taxonomy then just reflects the discreteness of species, which itself follows from the necessary separation of codewords in the Hamming space of genomic error-correcting codes.

11.2.2 SPECIES CAN BE ORDERED ACCORDING TO A HIERARCHICAL TAXONOMY

Implying a nested error-correction system, the subsidiary hypothesis greatly complicates the picture, making it at the same time much more realistic. Instead of just predicting that species differ from each other, it entails a hierarchical taxonomy where species are the more different, they belong to deeper layers of the nested system. The subsidiary hypothesis then explains the existence of a hierarchical taxonomy, a major biological fact, which, moreover, necessarily coincides with the phyletic tree. Instead of the rather simple picture of a phyletic graph as introduced in Sec. 8.6, we find branchings at different hierarchical levels. The species become the more different, the deeper their phyletic trees diverge.

Each symbol of the genome belongs to some layer of the nested system. Considering these layers separately, we observe that going from the periphery to the center also means passing from a component code having very many codewords with a fairly high probability of gradual transitions between them to a component code having comparatively few but very different codewords with an extremely low probability of being transformed into each other.

11.2.3 TAXONOMY AND PHYLOGENY

We assume that the different coding layers appeared successively during geological times, meaning that the nested structure of ancestral forms of life has a number of layers less than that of more recent ones. As a consequence, the individuals of more recent species have both more diversified genomes and a longer time interval between regenerations (a larger lifetime for higher animals) than earlier species. The more numerous are the layers in the 'nested system' scheme, the larger the minimum distance in the innermost layers. The genome being made more resilient to errors, its conservation can be ensured with a larger time interval between regenerations (which has the advantage of more flexibility as regards environmental changes). Hence, we may expect that the longevity of individuals is larger in species where the genome involves many code layers. Similarly, in the same case, more differences between the individuals can exist since they are assumed to differ from each other only in the peripheral layer of the nested system, which is left uncoded. The specific features (i.e., common to all individuals of the species) are encoded into the most central layers where they are adequately protected. These remarks seem to roughly match reality if we think of the number of code layers as an index of how high a species is in the scale of evolution: compare, for instance, the complexity, lifetime, and individual diversity of *Caenorhabditis elegans* with those of vertebrates.

The appearance of the successive coding layers may well, at least for certain of them, coincide with the 'major transitions in evolution' [60] or with the onset of Barbieri's 'organic codes' [5]. The nested system indeed closely resembles Barbieri's organic codes, which, however, were not initially thought of as error-correcting: compare Fig. 9.2 with Fig. 8.3 of [5]. The many organic codes Barbieri convincingly identified may be considered as providing an experimental evidence of the existence of many codes organized according to the nested scheme. Each of them involves specific constraints so, according to the concept of soft code introduced in Ch. 10, they are potential error-correcting codes. The existence of many organic codes also suggests that the sketchy list of possible soft codes in the nested system given in Ch. 9 should be enriched with those associated with other organic codes in Barbieri's meaning.

The onset of the nested structure can be understood since DNA can be replicated. If its copy is appended to the initial genome instead of being separated from it, then an increase in the genome length results and the longer genome can evolve so as to increase its error-correcting ability. Using old means to create new ones, as assumed here, is a typical feature of nature's approach often referred to as *tinkering*. Moreover, many other mechanisms like inverse transcription or horizontal genomic transfer are already known as means for increasing the genome length.

11.3 NECESSITY OF SUCCESSIVE REGENERATIONS

11.3.1 CORRECTING ABILITY OF GENOMIC CODES AND TIME INTERVAL BETWEEN REGENERATIONS AS TARGETS OF NATURAL SELECTION

In the absence of genomic error-correcting code, it has been shown in Ch. 8 that the lifetime of a genome is the less, the larger is its length. Besides the decreasing permanence of individual genomes, it has been shown that no population of identical genomes, i.e., no species, can exist beyond a comparatively short genome length, of about $\ln(2)/p_{su}$ according to Eq. (8.2), where p_{su} is the probability that an erroneous nucleotide is substituted for the correct one. For plausible estimates of p_{su}, this limit is much shorter than the actual genome length of many species, some of which being among the most enduring ones. Using for p_{su} the estimate computed in Sec. 7.3 results in an upper limit of 3.5×10^7 base pairs. Many longer genomes exist in the actual living world, and there is no upper limit to the genome length of stable species: some of the oldest species have very long genomes. For instance, the lungfish genome contains about 1.4×10^{11} base pairs, hence is 4,000 times larger than the above estimate of the limit. The absence of error-correction coding would forbid long genomes. The actual existence of species with long genomes confirms that genomes need error correction.

If, on the contrary, a genomic error-correcting code exists, the genome permanence, as computed in Sec. 8.2, varies approximately as $1/P_{\text{cod}}$, where P_{cod} is the probability of a regeneration error. The genome permanence thus increases without limit if a long and efficient enough error-correcting code is used, provided the condition stated by the fundamental theorem of channel coding that the source entropy is less than the channel capacity is satisfied (see Sec. 5.3.1). Then P_{cod} approaches 0 and the genome permanence has no upper limit.

The probability of a regeneration error, P_{cod}, depends itself on two factors: the intrinsic error-correcting ability of the code, on the one hand, and the time interval between regenerations Δt, on the other hand. The symbol error probability after a time interval Δt is related to the substitution frequency ν_{su}, assumed to be constant, according to Eq. (7.3). For quaternary symbols it reads:

$$p_{\text{su}}(\Delta\ (t)) = \frac{3}{4}\left[1 - \exp\left(-\frac{4}{3}\nu_{\text{su}}\Delta t\right)\right] .$$

Then p_{su} is an increasing function of Δt. The regeneration error probability, P_{cod}, is itself an increasing function of $p_{\text{su}}(\Delta t)$ which depends on the error-correcting code, and especially on its minimum distance d. If $p_{\text{su}}(\Delta t)$ is smaller than some threshold close to $d/2n$, P_{cod} is very small; however, there is a steep increase of P_{cod} when $p_{\text{su}}(\Delta t)$ approaches this threshold (see Sec. 6.5). The time interval Δt is thus a critical parameter which controls the probability of successful regeneration. Too large its value makes the genomic error-correcting code inefficient. As directly controlling the genome permanence, the genomic error-correcting code and the time interval Δt are the first and main targets of natural selection, independently of the associated phenotype.

11.3.2 NATURE MUST PROCEED WITH SUCCESSIVE REGENERATIONS

Not only genomic error-correcting codes must exist, but a genome must be regenerated before the number of cumulated errors exceeds the error-correcting ability of the code. To keep the number of erroneous symbols small enough, the time interval Δt between regenerations should be short enough, or else regeneration errors result in the genome quickly tending to a sequence of nucleotides widely different from the initial codeword and almost uncorrelated with it (see Sec. 8.3). That nature proceeds with successive generations is a trivial biological fact. As a direct consequence of the hypothesis that a genomic error-correcting code exists, we find that genome conservation actually demands that successive **re**generations be performed, a much stronger statement. We moreover assume that the regeneration process concerns the genome as a whole so the several component codes of the nested system are not regenerated separately: the time interval Δt between regenerations is shared by all these component codes.

Since conservation of a genome depends on its error-correcting ability on the one hand, and on the time interval between regenerations on the other hand, the proper matching of these presumably independent factors can only result from natural selection. Let us briefly look at consequences of their possible mismatch. If the time interval Δt between regenerations is too large, the regeneration errors are rather frequent, giving rise to many comparatively short-lived species. The steep increase of the regeneration error probability when the nucleotide error rate approaches the threshold makes an even moderate increase of Δt result in a large increase in mutation probability. Maybe the Cambrian explosion could be explained by too long an interval between regenerations soon corrected by natural selection ('soon' should be understood at the geological timescale, of course). If, on the contrary, Δt is too small, regeneration errors very seldom occur. Then the species genome is very faithfully conserved but does not but very unfrequently originate in other ones. This species lacks flexibility to cope with environmental changes and thus its phylum risks extinction.

11.3.3 JOINT IMPLEMENTATION OF REPLICATION AND REGENERATION

We already noticed that regeneration and replication are conceptually different functions. Replication does not provide in itself any protection against the channel errors. It is its assumed error-correcting ability which enables regenerating the genome. However, a single individual is fragile and the survival of a species implies its replication besides regeneration. Both replication and regeneration of the genome must occur in living beings, as in the toy living world considered in Ch. 8. We now examine the possible modes of their joint implementation. Both can be performed simultaneously but there is no logical necessity for this. Clearly, once both functions of replication and regeneration began to be performed, the process of natural selection was engaged. What we observe is the joint product of regeneration, replication, and selection. Replication alone results in an exponentially increasing number of individuals, and the availability of finite resources necessarily results in selection.

The single cell of a unicellular being must be able to perform both replication and regeneration of its genome. The simplest way would thus be to perform genome regeneration systematically followed by replication at (roughly) periodic time intervals of Δt. It is, however, possible that the two events of regeneration and replication occur separately and are initiated by different means. For instance, replication may depend on an internal clock while regeneration-replication may be triggered by an external event.

In multicellular beings, each cell may behave as a unicellular being, hence be able to perform both regeneration and replication. This seems to be the case of, e.g., the Hydra. Again, replication may be performed apart from regeneration and be initiated by different means. There is, however, a new possibility, that of differentiated cells:

— somatic cells performing only replication in mitosis;

— germinal cells resulting from meiosis where regeneration is performed and followed with replication.

Due to the cost of regeneration (a much more complex operation than replication), differentiation into (comparatively) few germinal cells and many somatic cells has an evolutive advantage.

One may indeed plausibly assume that the genome regeneration occurs during meiosis and/or fertilization. Then some inner clock roughly determines the interval between them. But other mechanisms may be contemplated, especially triggered regeneration. For instance, the finding in *Arabidopsis thaliana* of 'non-Mendelian inheritance' [54] could be explained by assuming that, in this species and probably in other plants, the regeneration process is sporadically triggered by some external factor. Geneticists also suspect that some living beings (which are again plants) can control to some extent the mutation rate in response to external factors [48]. Triggering regeneration only under the control of external factors on the one hand, and favoring mutations to cope with the environmental situation, on the other hand, imply both that the mutation rate, i.e., the regeneration error rate, can be controlled, hence the existence of error-correcting means which can in certain circumstances be inhibited so as to increase the mutation rate. Notice however that increasing the mutation rate can also be obtained by inhibiting repair mechanisms which proofread the replicated genomes.

In the case of multicellular beings with differentiated germinal cells, a completely different way of performing regeneration also becomes possible: producing a large number of gametes using a simplified, hence approximate, regeneration algorithm and screening these gametes as regards their genome [70]. Screening must involve some criterion to distinguish the genomes which do not obey the coding constraints and should thus be discarded. Such a criterion implies testing whether a given nucleotide sequence belongs or not to the genomic code, which then works as *error detecting*. The complexity of error detection is generally much less than that of proper regeneration, i.e., determining what codeword is the closest to the received word. However, the full process involves growing a large population of germinal cells, hence a large number of replications. The high complexity of regenerating a single genome is replaced by many replications and many checks of faithfulness of replicated genomes to the genomic code.

The above remarks also suggest a possible connection with epigenetics. The regeneration process is complex, costs much energy, and is necessarily implemented by the phenotype. The crucial step of the regeneration process when the genomic message is rewritten can be perturbed by external conditions which affect the phenotype. The genome regeneration then depends to some extent on the environment, according to a weak kind of Lamarckian inheritance. This simple explanation avoids the recourse to hypothetical entities: DNA remains the one and only agent of heredity, but the dynamic mechanisms which ensure its conservation can be perturbed by external factors. From the experimental point of view, the study made by Marcus Pembrey et al. concerning inhabitants of the Swedish village of Överkalix reports evidence of transgenerational effects precisely related to well dated famine events and is especially interesting to us [63]. It turns out that the famine events had transgenerational consequences on the observed subjects only when they occurred during the meiosis from which the gametes of their grandparents originated. This is consistent with our hypothesis of a regeneration process which takes place at meiosis and can be affected by external factors. Maybe some phenomena referred to as epigenetic and explained by cytosine methylation or histone modification could find an alternative explanation in this remark.

11.4 SALTATIONISM IN EVOLUTION

11.4.1 REGENERATION ERRORS RESULT IN EVOLUTIVE JUMPS

The wide separation of genomes in terms of their Hamming distance which results from the main hypothesis, and the fact that regeneration errors, although rare events, occur with nonzero probability, entail that regeneration errors originate in new species, as illustrated by the model dealt with in Ch. 8. Insofar as it results from regeneration errors, biological evolution thus proceeds by jumps, i.e., is *saltationist*, although other mechanisms of speciation do exist. The subsidiary hypothesis however leads to complicate the picture as it makes the amplitude of a jump which results from a regeneration error strongly depend on the layer of the nested system where it takes place.

11.4.2 SALTATIONISM DEPENDS ON THE LAYER DEPTH IN THE NESTED SYSTEM

The subsidiary hypothesis that the genomic code is made of nested codes sheds light on its relationship with evolution. Let us consider separately the component codes of the nested system. To begin with, consider a long and very redundant code as found in the deeper layers (the words 'deep' and 'peripheral' refer to the fortress metaphor). Being highly redundant, this code has comparatively few words separated from each other by large Hamming distances, and we may assume that it is good as a result of natural selection. Then the probability of a regeneration error is very low, but its occurrence results in a very different genome, hence the corresponding phenotype is necessarily very different from that of the initial genome. In this layer, thus, strong saltationism occurs with very unfrequent transitions due to regeneration errors. In sharp contrast, we were already led to assume that the peripheral layer of the nested system is left uncoded. Then errors frequently occur but most often result in a genome very close to the original one. Close genomes will result in close phenotypes (apart from an important caveat stated in Sec. 11.7.2).

The peripheral layer of the nested system being left uncoded and the other layers being the more efficiently encoded, the deeper they are, we may think of the successive layers as exhibiting a transition between gradualism (in the peripheral layer) and saltationism, affecting more and more important features as deeper layers are considered. Remember that the protection against errors that the codes provide entirely relies on the distance between codewords. Their minimum distance is only $d = 1$ in the uncoded peripheral layer and increases from the periphery to the most central layers. The larger the distance between two codewords, the more they differ but the more improbable is a regeneration error transforming one of them into the other (see Fig. 9.3). In the nested system, thus, errors are frequent but result in small differences when they affect a symbol which belongs to the peripheral layer, while at the same time they become much less frequent and have more important consequences in deeper layers. Then microevolution which enables adaptation to small environmental changes depends on frequent errors in the peripheral layer, which is also the place where sexual reproduction introduces genetic diversity, while species differentiation and more important taxonomic divisions, i.e., macroevolution, result from very unfrequent but high-weight error patterns affecting deeper layers.

11.5 EVOLUTIVE ADVANTAGE OF LONG GENOMES; TREND OF EVOLUTION TOWARDS COMPLEXITY

11.5.1 EVOLUTIVE ADVANTAGE OF LONG GENOMES

The main hypothesis suffices to explain the poorly understood *trend of evolution towards increasing complexity*. What we mean is that new species or phyla more complex than the species or phyla already existing at the time of their appearance have an evolutive advantage. When dealing with the fundamental theorem of channel coding, we already noticed that the probability of a decoding (or regeneration) failure can be made the lower, the longer the words of the error-correcting code. This statement is paradoxical since increasing the word length also increases the average number of symbol errors in a received word. Appending a new layer in the nested system and provided the code is properly chosen, this theoretically proven property is fully confirmed by the engineering experience.

The genomic error-correcting code is thus presumably the more efficient, i.e., provides a lower probability of regeneration error P_{cod}, the longer it is. But we know from Ch. 8 that the permanence of a species is approximately proportional to its inverse, $1/P_{\text{cod}}$. Increasing the codeword length then provides an immediate evolutive advantage. The nested system postulated according to the subsidiary hypothesis, as described in Ch. 9, is precisely a means to both increase the codeword length and decrease the regeneration error probability (except in the peripheral layer).

11.5.2 INCREASING COMPLEXITY RESULTS FROM LENGTHENING GENOMES

Besides an immediate evolutive advantage, increasing the codeword length results in more numerous information symbols for a given redundancy rate. We already noticed in Sec. 5.5 that it gives room to more independent alternatives or parameters needed to specify phenotypic features. Even if it does not necessarily keep the redundancy rate constant, the onset of the nested system as described in Ch. 9 results in an increase in the number of information symbols every time a new coding layer is appended. More and more complex phenotypes can thus be specified as new coding layers are appended, and natural selection acting on these phenotypes lets the fittest ones survive.

11.6 EVOLUTION IS CONTINGENT

We saw in Ch. 8 that new genomes originate in regeneration errors although it is not the sole possible mechanism which creates new species. Regeneration errors are chance events which moreover occur at random instants, meaning that, to a large extent, biological evolution is contingent. We noticed moreover in Sec. 8.3 that the initial memory content of the genome is progressively forgotten as the number of regeneration errors increases.

Beware of not confusing a codeword chosen by such a chance event with a sequence of random symbols. As belonging to the genomic error-correcting code, the genome is actually subjected to many constraints which tie together its symbols (see Ch. 10). The wrong recovery of a codeword does not mean that each of its symbols is chosen at random. A regeneration error consists of choosing a wrong codeword *as a whole*. Errors on symbols which belong to a same codeword, at two specified locations, can by no means be considered as independent events and the probability of their simultaneous occurrence is larger than the product of the probabilities of these events taken separately. This suffices to refute arguments from the 'intelligent design' upholders on the improbability of errors simultaneously affecting two different nucleotides. Ironically, when biologists refute the 'intelligent design' argument, they implicitly rely on our main hypothesis, they generally ignore.

11.7 RELATIONSHIP BETWEEN GENOMES AND PHENOTYPES

11.7.1 GENOME AS SPECIFYING A PHENOTYPE

We may think of the information symbols of a genomic code as associated with the independent alternatives or parameters needed to specify a phenotype. The more numerous are the information symbols, i.e., the more numerous distinct dimensions has the genomic code, the more semantic features a genome can instruct (see Sec. 5.5). Remember that the redundancy of genomes implies that the information symbols constitute only a small fraction of the their total length. For a given redundancy rate, the phenotypic features that a genome determines are thus the more numerous, the larger its length.

In the nested system described in Ch. 9, the outer layers have a large dimension hence can represent many such phenotypic features. The inner layers in the nested system have however a dimension the smaller, the deeper they are. The most central ones can thus only represent few alternatives or parameters which then necessarily correspond to very basic phenotypic features, while the peripheral layers on the contrary correspond to more numerous but less basic features, in accordance with the taxonomic-phylogenic hierarchy.

11.7.2 NEIGHBORHOOD IN GENOMIC AND PHENOTYPIC SPACES

It is often considered as self evident that small differences between genomes result in small differences in the corresponding phenotypes. Deep mathematical reasons, however, prevent this to always be true. The space of genomes has an extremely large number of dimensions, equal to the genome length. The number of dimensions of the space of phenotypes, although large, is smaller since an error-correcting code, as assumed by the main hypothesis, is necessarily redundant. Then close neighbors in the space of genomes cannot always correspond to close neighbors in the space of phenotypes. Let us illustrate this property by an example where the numbers are likened to phenotypes and their decimal representations to genomes. (The base of the numeration system actually does not matter.) The space of numbers has a single dimension, in sharp contrast with the space of decimal representations which has as many dimensions as digits. Consider for instance the four 10-digit numbers $a = 1,999,999,999$, $b = 2,000,000,000$, $c = 2,000,000,001$, and $f = 9,000,000,000$. The first three are consecutive numbers, hence close to each other. However, while the Hamming distance between the decimal representations of b and c is 1 (the smallest possible distance between the representations of two different numbers), the distance between the decimal representations of a and b is 10, i.e., the largest possible between ten-digit numbers, although a and b are consecutive. The Hamming distance between the decimal representations of b and f also equals 1, although the numbers b and f widely differ. In order to map the numbers into their decimal representations, different weights need be affected to the digits so a difference in a single digit makes a big difference in the corresponding number if its weight is large.

11.7.3 ON GENOME COMPARISONS EXPRESSED AS PERCENTAGES

The previous section is a warning as concerns comparison of phenotypes which correspond to close genomes. Moreover, and independently of the above remark, the closeness of genomes is often expressed as a percentage of their length. For instance, the genome of the chimpanzee is often reported as differing from the human one by only 2%. But the human genome is more than 3 billion base pair long. The absolute difference, more than 60 million base pairs, is still huge and may correspond to a very large amount of information, hence of phenotypic difference.

CHAPTER 12

Identification of Genomic Error-Correcting Codes, and other Open Problems

12.1 NECESSITY OF IDENTIFYING GENOMIC CODES

12.1.1 AN UNUSUAL APPROACH

Up until now, we asserted that (1) a genomic error-correcting code exists, (2) assumes the form of a nested system, (3) the components of which are soft codes. Statements (1) and (2), referred to above as our main and subsidiary hypotheses, express necessary conditions. Their necessity is of mathematical character, hence absolute: the relevance of information theory to literal communication is no more questionable than that of arithmetic or Euclidean geometry in practical situations. We saw in Ch. 11 that many actual basic features of the living world can be interpreted as mere consequences of these statements.

Statement (3) is not that necessary. It is just a plausible guess about what genomic error-correcting codes could look like. We saw in Ch. 10 that many potential soft codes *a priori* arise from biological constraints, and moreover that they assume the form of a nested system as described in Ch. 9. The following question, however, remains unanswered: *How are such constraints actually used in order to perform genome regeneration?* The answer would need experimentally identifying the biological regeneration mechanisms based on constraints.

The usual process of biological research starts with observations and experiments upon which theoretical constructions will possibly be based. Our approach in this work is inverse. Its starting point consists of a main and a subsidiary 'hypotheses' devoid of direct experimental bases but which are needed for conciliating two major but conflicting biological facts: the conservation of genomes and their modification by mutations. Information theory leaving no other alternative, genomic error-correcting codes *must* exist and the better conservation of very old genetic information *needs* that they take the form of a nested system.

If one wonders why such basic facts of life escaped as yet the investigations of biologists and remained unidentified, the answer is simple: they were not found because nobody searched for them. Those who could discover the solution ignored the problem. It will moreover become apparent that the experimental identification of genomic error-correcting codes is an extremely broad and difficult task. These codes must be deliberately searched for in the light of information theory.

12.1.2 A NECESSARY COLLABORATION OF ENGINEERS AND BIOLOGISTS

The experimental identification of the biological error-correction mechanisms, and especially of the means of genome regeneration, thus needs a close collaboration of biologists and information theorists. They should not only work together but actually learn from each others. There are many obstacles in this way; the cultural difference between mathematically educated people and experimental biologists is not the least one. Maybe the most blatant difficulty is the lack of an adequate popularization of information theory.

The very method to be used for identifying genomic error-correcting means has, moreover, to be invented. Even if the identification process is eventually successful, it will undoubtedly be long and uneasy. The progress will probably result from a slow mixing of the disciplines of information theory and genetics, and sudden breakthroughs are unlikely.

12.2 IDENTIFYING ERROR-CORRECTION MEANS

12.2.1 IDENTIFYING AN ERROR-CORRECTING CODE

Given a sequence of symbols, say, of nucleotides, consider the problem of determining whether this sequence is a word of some error-correcting code and, if so, of identifying the code to which it belongs. As stated, this problem has no solution. We defined an error-correcting code in Sec. 3.5 and Ch. 6 as a *collection* of words. For a hypothetical code defined by a list of words, as resulting from random coding, the knowledge of a particular codeword does not provide any clue on the other ones. The situation is a bit more favourable for a code defined by a set of constraints, but a single sequence can be characterized by an extremely large number of constraints. If it belongs to a code, the constraints specific to this code cannot be distinguished from many other possible ones. Identifying a code thus necessarily implies the knowledge of several of its codewords. In the case of a linear (n, k) code (as defined in Sec. 6.4.2) at least k words of the code must be available; k words suffice provided they are linearly independent.

The problem resembles that of source coding of a given sequence without any knowledge of the source from which it originates. It is also, interestingly, almost the same problem as determining the algorithmic complexity of a sequence (see Sec. 5.4) by searching the smallest program which instructs a universal computer how to generate the given sequence. This problem has no general solution since the algorithmic complexity of a sequence is a proven uncomputable quantity.

Only collecting several different words can lead to identifying a code, with the further difficulty that the nested structure assumed according to the subsidiary hypothesis demands that the collected codewords belong to the same coding layer.

12.2.2 IDENTIFYING COMPONENT CODES OF THE NESTED SYSTEM

Identifying error-correction means would thus first imply acquiring an explicit knowledge of the several component codes of the nested system. Comparing genomes of closely related species could sometimes let know the constraints which characterize a code. For instance, the experimental comparison of different mutations affecting eukaryotic genes, some of which coding for snake venoms, can hopefully answer the question put in Sec. 10.5 and enable identifying this particular code for certain gene families.

Comparing genomes could first result in establishing a map of the component codes in the nested system. Some loci are much more variable than others, and some appear as almost invariant. A difficulty of this task is to determine the scale to be used. Should individual nucleotides be considered, or is a broader scale sufficient? Another point possibly worth studying would be to establish a correlation between the map of component codes and biological constraints. Establishing a map of the component codes in the nested system is clearly a very long process needing a great deal of experimental work. Even its completion would, however, be far from achieving the identification of genomic codes.

Studying species whose genomes notoriously resist damages due to radiation or drought, like *Deinococcus radiodurans* or Bdelloid rotifers, could possibly result in discovering especially efficient genomic error-correcting codes. The currently proposed explanations of their performance have recourse to templates, but no quantitative performance assessment is given for justifying them [81, 77, 41]. Indeed, the damages considered mainly consist of DNA 'backbone' breakings which cannot be repaired by merely using error-correcting codes, but it is unlikely that the events which break DNA do not also result in errors in the sequence of nucleotides, which need to be corrected.

12.2.3 IDENTIFYING REGENERATION MEANS

Even if the identification of certain codes is successful, the understanding of how they are used to regenerate parts of the genome is a most difficult, but mandatory, task. Indeed, an identified genomic error-correcting code has only a potential error-correcting ability. It is actually useful when, and only when, an algorithm implemented using the tools of molecular genetics is able to regenerate its words. The complexity of the decoding algorithms designed by engineers can easily be matched by processes of molecular genetics but the algorithms actually used for regenerating genomes are *a priori* unknown. Identifying how unknown algorithms are implemented is indeed a formidable task and the very method for doing so has to be invented. The ability of the ordinary tools of molecular genetics to perform elementary tasks of signal processing useful for genome regeneration should first be investigated. Many provisional assumptions followed by experiments intended to validate or disprove them will probably be necessary.

The genome regeneration process aims at producing a new genome subjected to all the constraints that a genome should obey. On the other hand, it should replicate the old genome which has possibly suffered errors. Since we likened the constraints of biological origin to soft codes, obeying constraints amounts to correct errors. Transposing in genetic terms the optimum regeneration rule we gave in Sec. 6.5.2, namely *choose the codeword the closest to the received word*, results in *choose the DNA sequence which obeys all the genomic constraints and is the closest to the genome to be regenerated.* We have seen in Sec. 6.6 that practical regeneration algorithms for convolutional codes actually operate symbol-by-symbol and moreover, in Sec. 6.7, that iterated soft input- soft output (SISO) decoding is especially efficient for regenerating turbocodes. The elementary mechanisms of template-replication already operate symbol-by-symbol. This suggests to look for more sophisticated, possibly iterated, regeneration mechanisms taking place during replication of prokaryotes or meiosis in sexed eukaryotes.

Among the constraints enumerated in Sec. 10.2.3, only some of them directly affect DNA strings. Other ones result from downstream constraints, for instance are induced by structural constraints which affect proteins. Still other constraints, the linguistic ones, have a more abstract character. It is easy to conceive that regeneration as defined takes account of the physical-chemical constraints directly affecting DNA. As regards the other ones, we must accept that they are in some way embodied in the molecular mechanism which performs regeneration. Identifying the regeneration means which specifically concern the DNA constraints then appears as a first, fairly easy step. Once this task is completed, that of identifying the other regeneration means will presumably be alleviated.

12.3 GENOME DISTINCTION AND CONSERVATION

The description of an error-correcting code as given in Ch. 6 shows that the possibility of distinguishing its elements, to be referred to for short as *distinction*, is its most important feature. A meaningful distance (Hamming's) has been introduced into the space of codewords, in terms of which the words which make up a code should be as separated from each other as possible. The set of distances between all pairs of words of a code is referred to as its *distance spectrum*. The channel errors which affect the transmitted word tend to blur the distinction between the received word and the codewords by diminishing their mutual distances. Decoding (or regeneration as we prefer to name it) consists of modifying the received word so as to restore the initial distance spectrum, hence the initial distinction (which may result in a regeneration error if too many channel errors occurred).

The analysis of sequences of nucleotides given by Forsdyke [37] may look entirely different, but almost all the reported genetic mechanisms result in *maintaining distinction*: for instance, insuring reproductive isolation between species or preserving gene integrity against recombination. Beyond the differences in vocabulary and points of view between separate disciplines, maintaining distinction may thus be thought of as a common basis of the theory of error-correcting codes and Forsdyke's evolutive bioinformatics. As an engineering discipline, the former is concerned with the problem of designing codes with an appropriate distance spectrum. The latter considers the discreteness of species and genes as prime biological facts and investigates the means which can maintain this discreteness. Both thus converge at the deepest level although their starting points are different. This level is that of *information theory*, which indeed may rightfully be referred to as the science of distinction.

Stated otherwise, the persistence of discreteness in the living world and the existence of genomic error-correcting means are two facets of one and the same thing. The reality of the former implies that of the latter. Discreteness is the price to be paid for conservation. Relevance of information theory as a basic background results, with the advantage of introducing quantitative methods and stating well defined limits, especially the channel capacity.

12.4 DIFFICULTIES OF CONCILIATING GENOMIC CODES AND SEXUAL REPRODUCTION

We were already led to conjecture that the peripheral layer of the nested system is left uncoded in order to make genomic encoding compatible with sexual reproduction. Then the genetic shuffling operated by crossing-over only affects the peripheral uncoded layer and the remainder of the genome is common to all members of a species.

Since regeneration errors result in genomes significantly different from the original ones, it is reasonable to think of these errors as a possible origin of species, hence that biological evolution is saltationist (as discussed in Sec. 11.4). On the one hand, it seems impossible to explain some biological facts unless assuming saltationism. To give a single example, Jean-Henri Fabre observed a female solitary wasp that stings a tarentula much bigger than her at a very precise point of its mouth so as to disable its venomous hooks [32]. She then paralyses the spider by stinging nervous ganglia and lays her egg inside it for feeding the larva after its hatching out. Were she not immediately successful in disabling the venomous hooks, she would be killed by the spider and her offspring would be lost. Can such an inherited behavior gradually appear? On the other hand, it is difficult to conciliate saltationism with sexual reproduction: how does Goldschmidt's 'hopeful monster' find a mate and produce viable offspring? This is an old objection to saltationism and the hypothesis that a genomic error-correcting code exists, which implies saltationism as an unavoidable consequence, does not make this objection stronger nor weaker. Too little is known as yet about the actual implementation of genomic error-correcting codes to address this problem.

Similar difficulties are met with 'horizontal genomic transfer' of any kind. This work has been exclusively devoted to the vertical transfer of genetic information, which is indeed essential in heredity. The compatibility of genomic error-correcting codes with horizontal genomic transfer is not obvious, although the onset of the nested system assumed to exist according to the subsidiary hypothesis necessarily relied on this kind of mechanisms.

CHAPTER 13

Conclusion and Perspectives

Reviewing in [20] the book by Ed Regis, *What is Life? Investigating the Nature of Life in the Age of Synthetic Biology* [68], Steven Benner wrote:

> "Because building something requires a deep understanding of its parts, synthesis also stops scientists from fooling themselves. Data are rarely collected neutrally during analyses by researchers, who may discard some, believing the data to be wrong if they do not meet their expectations. Synthesis helps manage this problem. Failures in understanding mean that the synthesis fails, forcing discovery and paradigm change in ways that analysis does not."

Synthesis being the engineers' job, this remark is an excellent plea for a close collaboration of biologists and engineers. It is intended to genetic engineering but actually applies to any instance where nature and engineers are faced with the same problems.

Communication engineering benefits from the conceptual framework of information theory. Literal communication of symbol sequences ('literal' meaning that semantics is ignored) is actually a mathematical problem, and information theory is just that branch of mathematics which deals with it. On the other hand, heredity consists of literally communicating sequences of nucleotides over time, hence information theory is fully relevant to it.

Information theory can bring to genetics its concepts and methods as well as its results. Maybe its most important concept is that of *channel capacity*, a key notion which states an impassable limit to any communication. Information theory proves that communication without errors (more precisely, with an arbitrarily small error rate) is possible over a channel despite the errors which affect it, provided the information rate is less than the channel capacity, a quantity which decreases as the channel error rate increases. However, the very means which enable 'errorless' communication hinder any communication at all beyond the channel capacity. Both the possibility of 'errorless' communication below the capacity and its impossibility above it, although counterintuitive, are fully confirmed by the engineers' experience, besides being theoretically proven.

Applying the concept of channel capacity to genetics shows that the template-replication paradigm is wrong since merely copying genomes does not ensure their conservation. Genomes must act as words of an error-correcting code which enables their *regeneration*. Current genetics and evolution theory are deeply rooted in the implicit assumption that DNA strings, and especially genes, are permanent objects (which nevertheless suffer mutations). Drawing the consequences of their intrinsic impermanence demands that the bases of these sciences be drastically reinterpreted, a truly formidable task. Once information theory and error-correcting codes are recognized as relevant to biology, they impose to the living world severe constraints and limits which can no longer be ignored.

Finding that genomes, far from being intrinsically permanent objects, need a dynamic process to be conserved is not so surprising although it refutes a major tenet of current biology. Indeed, all other biological entities owe their conservation to dynamic processes, and we just state that genomes are no exceptions in this respect. One may rather wonder why they have been believed intrinsically permanent. That the results of science are provisional is a too often forgotten lesson of history. Good research generally provides questions as well as answers, and answering the new questions often leads to questioning the former answers. Considering that the present state of knowledge enables explaining once and for all how nature solves a problem is therefore an illusion. How heredity works has been wrongly believed for more than half a century as understood. The engineering point of view expounded in this book shows that it is actually a very tough problem. Its solution necessarily involves sophisticated engineering tools that most biologists ignore, namely, error-correcting codes. Finding a gap that wide in the current knowledge warns to be modest and cautious!

Revealing the fallacy of the template-replication paradigm is only the negative contribution of this work. On the positive side it opens perspectives for future research, some of them at most cursorily dealt with above. Among their possible benefits, let us mention improving the methods of biology, renewing the understanding of basic biological facts, and prompting discoveries.

As regards methodology, information theory makes the problems of communication met in biology accessible to computation. For instance, many mechanisms assumed to perform 'proof-reading' or 'DNA repair' based on the availability of two or more copies of the genomic message are described in the literature without any assessment of their expected performance. Once identified as solutions to engineering problems, assessing their performance becomes possible. Then the inadequacy of some of them would become obvious. Computation would act as regards biological hypotheses the same role as natural selection as regards species. Information theory provides means of such a performance assessment, and checking that a device or process actually fulfils a required performance is a current engineering practice which would be most beneficial to biology. Information theory can more generally provide a conceptual framework for genetics and even for biology as a whole. Its abstractness sharply contrasts with the current methods of biology but hopefully complements them by providing broad and general perspectives. The potential benefits to biology are worth the efforts necessary to assimilate information theory. Maybe the fusion of biology and information theory will eventually result in the long term in an extended biology.

As regards the renewed understanding of biological facts, it has been shown above that information-theoretic arguments explain features as basic to life as the necessity of successive generations and the existence of discrete species organized according to a hierarchical taxonomy, which moreover coincides with the phylogeny. They suggest that evolution proceeds by jumps and is basically contingent. Even the mysterious trend of evolution towards an increased complexity, which conventional biology fails to explain, appears as a mere consequence of natural selection operating on genomic error-correcting codes. Probably many other biological features can be similarly interpreted.

However, there is as yet no direct experimental evidence of the existence of genomic error-correcting codes. Revealing its necessity prompts investigating how such codes are implemented, and especially how the genome regeneration they enable is performed. The novelty of this approach entails that even its methods have to be invented. The vision provided by information theory turns the old living world into a new world to be explored.

Bibliography

[1] B. Audit, C. Vaillant, A. Arneodo, Y. d'Aubenton-Carafa, and C. Thermes, "Long-range correlation between DNA bending sites: relation to the structure and dynamics of nucleosomes," *J. Mol. Biol.*, Vol. 316, 2002, pp. 903–918. DOI: 10.1006/jmbi.2001.5363

[2] O. Avery, M. McCarty, and C. MacLeod, "Studies of the chemical nature of the substance inducing the transformation of pneumococcal types. Induction of transformation by a desoxyribonucleic acid fraction isolated from Pneumococcus Type III," *J. Exp. Med.*, Vol. 79, 1944, pp. 137–158. DOI: 10.1084/jem.79.2.137

[3] L.R. Bahl, J. Cocke, F. Jelinek, and J. Raviv, "Optimal decoding of linear codes for minimizing symbol error rate," *IEEE Trans. Infor. Theor.*, Vol. IT-20, Mar. 1974, pp. 284–287. DOI: 10.1109/TIT.1974.1055186

[4] D. Baltimore, "Our genome unveiled," *Nature*, Vol. 409, No. 6822, Feb. 2001, pp. 814–816. DOI: 10.1038/35057267

[5] M. Barbieri, *The Organic Codes*, Cambridge University Press, Cambridge, 2003.

[6] G. Battail and M. Decouvelaere, "Décodage par répliques," *Annales Télécommunic.*, Vol. 31, No. 11-12, Nov.-Dec. 1976, pp. 387–404.

[7] G. Battail, M. Decouvelaere, and P. Godlewski, "Replication decoding," *IEEE Trans. Infor. Theor.*, Vol. IT-25, No. 3, May 1979, pp. 332–345. DOI: 10.1109/TIT.1979.1056035

[8] G. Battail, "Décodage pondéré optimal des codes linéaires en blocs I.- Emploi simplifié du diagramme du treillis," *Annales Télécommunic.*, Vol. 38, No. 11-12, Nov.-Dec. 1983, pp. 443–459.

[9] G. Battail, "Pondération des symboles décodés par l'algorithme de Viterbi," *Annales Télécommunic.*, Vol. 42, No. 1-2, Jan.-Feb. 1987, pp. 31–38.

[10] G. Battail, "Construction explicite de bons codes longs," *Annales Télécommunic.*, Vol. 44, No. 7-8, Jul.-Aug. 1989, pp. 392–404.

[11] G. Battail, "We can think of good codes, and even decode them," *Eurocode 92*, Udine, 26-30 Oct. 92. CISM courses and lectures No. 339, P. Camion, P. Charpin, and S. Harari, Eds., Springer, 1993, pp. 353–358.

[12] G. Battail, "On random-like codes," *Proc. 4-th Canadian Workshop on Information Theory*, Lac Delage, Québec, 28-31 May 1995. Lecture Notes in Computer Science No. 1133, Springer, 1996, pp. 76–94.

[13] G. Battail, "Does information theory explain biological evolution?", *Europhys. Lett.*, Vol. 40, No. 3, Nov. 1st, 1997, pp. 343–348. DOI: 10.1209/epl/i1997-00469-9

[14] G. Battail, "On Gallager's low-density parity-check codes," International Symposium on Information Theory, *Proc. ISIT 2000*, Sorrento, Italy, 25-30, juin 2000, p. 202, DOI: 10.1109/ISIT.2000.866500

[15] G. Battail, "Is biological evolution relevant to information theory and coding?", *Proc. ISCTA '01*, Ambleside, UK, Jul. 2001, pp. 343–351.

[16] G. Battail, "An engineer's view on genetic information and biological evolution," *Biosystems*, Vol. 76, 2004, pp. 279–290. DOI: 10.1016/j.biosystems.2004.05.029

[17] G. Battail, "Genetics as a communication process involving error-correcting codes," *J. Biosemiot.*, Vol. 1, No. 1, 2005, pp. 103–144.

[18] G. Battail, "Should genetics get an information-theoretic education?", *IEEE Eng. Med. Biol. Mag.*, Vol. 25, No. 1, Jan.-Feb. 2006, pp. 34–45. DOI: 10.1109/MEMB.2006.1578662

[19] G. Battail, "Information theory and error-correcting codes in genetics and biological evolution," in *Introduction to Biosemiotics*, Barbieri, M., Ed., Springer, 2006. DOI: 10.1007/1-4020-4814-9_13

[20] S. Benner, "Biology from the bottom up," *Nature*, Vol. 452, No. 7188, 10 avr. 2008, pp. 692–694, DOI: 10.1038/452692a

[21] C. Berrou, A. Glavieux, and P. Thitimajshima, "Near Shannon limit error-correcting coding and decoding : turbo-codes," *Proc. ICC'93*, Geneva, Switzerland, May 1993, pp. 1064–1070. DOI: 10.1109/ICC.1993.397441

[22] C. Berrou and A. Glavieux, "Near optimum error correcting coding and decoding: turbo codes," *IEEE Trans. Commun.*, Vol. 44, Oct. 1996, pp. 1261–1271, DOI: 10.1109/26.539767

[23] C. Berrou, "Métamorphoses et promesses du codage correcteur d'erreurs," Seminar organized for the election of Claude Berrou at the French Academy of Sciences, Jun. 2, 2008.

[24] C. Branden and J. Tooze, *Introduction to Protein Structure*, New York and London, Garland, 1991.

[25] G. Chaitin, *Metamaths!*, New York: Pantheon Books, 2005.

[26] E. Chargaff, "How genetics got a chemical education," *Ann. New York Acad. Sci.*, Vol. 325, 1979, pp. 345–360. DOI: 10.1111/j.1749-6632.1979.tb14144.x

[27] T.M. Cover and J.A. Thomas, *Elements of Information Theory*, New York: Wiley, 1991.

[28] P. David and S. Samadi, *La théorie de l'évolution*. Paris: Flammarion, 2000.

[29] R. Dawkins, *The Selfish Gene*, Oxford: Oxford University Press, 1976; new edition, 1989.

[30] R. Dawkins, *The Blind Watchmaker*, New York, Norton, Longman, 1986.

[31] P. Elias, "Coding for noisy channels," *IRE Conv. Rec., Part 4*, New York, 1955, pp. 37–46.

[32] J.-H. Fabre, *Souvenirs Entomologiques*, Paris: Delagrave, 1880.

[33] G.D. Forney, Jr, "The Viterbi algorithm," *Proc. IEEE*, Vol. 61, 1973, pp. 268–278. DOI: 10.1109/PROC.1973.9030

[34] D.R. Forsdyke, "Are introns in-series error-detecting sequences?", *J. Theor. Biol.*, Vol. 93, 1981, pp. 861–866. DOI: 10.1016/0022-5193(81)90344-1

[35] D.R. Forsdyke, `http://post.queensu.ca/ forsdyke/`

[36] D.R. Forsdyke, "Conservation of stem-loop potential in introns of snake venom phospholipase A_2 genes. An application of FORS-D analysis," *Mol. Biol. Evol.*, Vol. 12, 1995, pp. 1157–1165.

[37] D.R. Forsdyke, *Evolutionary Bioinformatics*, New York, Springer, 2006.

[38] R.E. Franklin and R.G. Gosling, "Molecular configuration in sodium thymonucleate," *Nature*, Vol. 171, No. 4356, 25 Apr. 1953. pp. 740–741. Reprinted in *Nature*, Vol. 421, No. 6921, Jan. 23, 2003, pp. 400–401.

[39] R.G. Gallager, "Low density parity-check codes," Cambridge, MA: MIT Press, 1963.

[40] R.G. Gallager, "A simple derivation of the coding theorem and some applications," *IEEE Trans. Infor. Theor.*, Vol. IT-13, No. 1, Jan. 1965, pp. 3–18. DOI: 10.1109/TIT.1965.1053730

[41] E. Gladyshev and M. Meselson, "Extraordinary resistance of Bdelloid rotifers to ionizing radiation," *Proc. Ntl. Acad. Sci. USA*, Vol. 105, 2008, pp. 5139–5144. DOI: 10.1073/pnas.0800966105

[42] M.J.E. Golay, "Notes on digital coding," *Proc. IRE*, Vol. 37, No. 6, Jun. 1949, p. 637.

[43] E. Guizzo, "Closing in on the perfect code," *IEEE Spectrum*, Vol. 41, No. 3 (INT), Mar. 2004, pp. 28–34. DOI: 10.1109/MSPEC.2004.1270546

[44] J. Hagenauer and P. Hoeher, "A Viterbi algorithm with soft-decision outputs and its applications," *Proc. GLOBECOM'89*, Nov. 1989, Dallas, pp. 1680–1686.

[45] R.W. Hamming, "Error detecting and error correcting codes," *Bell Syst. Tech. J.*, Vol. 29, No. 1, Jan. 1950, pp. 147–160.

[46] D.A. Huffman, "A method for the construction of minimum redundancy codes," *Proc. IRE,* Vol. 40, 1952, pp. 1098–1101. DOI: 10.1109/JRPROC.1952.273898

[47] International Human Genome Sequencing Consortium, "Initial sequencing and analysis of the human genome," *Nature*, No. 6822, Vol. 409, Feb. 15, 2001, pp. 860–921.

[48] E. Jablonka and M.J. Lamb, *Epigenetic Inheritance and Evolution. The Lamarckian Dimension*, Oxford: Oxford University Press, 1995.

[49] S.A. Kauffman, *The Origins of Order*, Oxford: Oxford University Press, 1993.

[50] A.I. Khinchin, *Mathematical Foundations of Information Theory*, New York, Dover 1957.

[51] A.N. Kolmogorov, "Logical basis for information theory and probability theory," *IEEE Trans. Infor. Theor.,* Vol. IT-14, No. 5, Sept. 1968, pp. 662–664. DOI: 10.1109/TIT.1968.1054210

[52] A.S. Kondrashov, "Direct estimate of human per nucleotide mutation rate at 20 loci causing Mendelian diseases," *Human Mutation*, Vol. 21, No. 1, Jan. 2003, pp. 12–27. DOI: 10.1002/humu.10147

[53] L.S. Liebovitch, Y. Tao, A.T. Todorov, and L. Levine, "Is there an Error Correcting Code in the Base Sequence in DNA?", *Biophys. J.*, Vol. 71, 1996, pp. 1539–1544.

[54] S.J. Lolle, J.L. Victor, J.M. Young, and R.E. Pruitt, "Genome-wide non-mendelian inheritance of extra-genomic information in *Arabidopsis*," *Nature*, Vol. 434, No. 7032, Mar. 24, 2005, pp. 505–509. DOI: 10.1038/nature03380

[55] F.J. MacWilliams and N.J.A. Sloane, *The Theory of Error-Correcting Codes*, Amsterdam: North Holland, 1977.

[56] R.N. Mantegna, S.V. Buldyrev, A.L. Goldberger, S. Havlin, C.-K. Peng, S. Simons, and H.E. Stanley, "Linguistic features of noncoding DNA sequences," *Phys. Rev. Lett.*, Vol. 73, 1994, pp. 3169–3172. DOI: 10.1103/PhysRevLett.73.3169

[57] J.L. Massey, *Threshold Decoding*, Cambridge, MA: MIT Press, 1963.

[58] J. Maynard Smith, "The theory of games and the evolution of animal conflicts," *J. Theorect. Biol.*, Vol. 47, 1974, pp. 209–221.

[59] J. Maynard Smith, "The idea of information in biology," *Quart. Rev. Biol.*, 74(4), 1999, 395–400.

[60] J. Maynard Smith and E. Szathmáry, *The Major Transitions in Evolution*, W.H. Freeman, 1995, New York, Oxford: Oxford University Press, 1997.

[61] O. Milenkovic, "The information processing mechanism of DNA and efficient DNA storage," DIMACS working group on theoretical advances in information recording, Mar. 22-24, 2004.

[62] B. Moher, "Spring theory," *Nature*, Vol. 448, No. 7157, Aug. 30, 2007, pp. 984–986. DOI: 10.1038/448984a

[63] M.E. Pembrey, L.O. Bygren, G. Kaasti, S. Edvinsson, K. Northstone, M. SJöström, and J. Golding. "Sex-specific, male-line transgenerational responses in humans," *Eur. J. Hum. Genet.*, Vol. 14, 2006, pp. 156–166. DOI: 10.1038/sj.ejhg.5201538

[64] M.W. Nachman, "Haldane and the first estimates of the human mutation rate," *J. Genet.*, Vol. 83, No. 3, Dec. 2004, pp. 231–233. DOI: 10.1007/BF02717891

[65] W.W. Peterson and E.J. Weldon, Jr., *Error-Correcting Codes*, 2nd ed., Cambridge, MA: MIT Press, 1972.

[66] J.R. Pierce, "The early days of information theory," *IEEE Trans. Infor. Theor.*, Jan. 1973, pp. 3–8. DOI: 10.1109/TIT.1973.1054955

[67] M. Radman, "Fidelity and infidelity," *Nature*, Vol. 413, No. 6852, Sept. 13, 2001, p. 115. DOI: 10.1038/35093178

[68] E. Regis, *What is Life? Investigating the Nature of Life in the Age of Synthetic Biology*, New York, Farrar, Straus and Giroux, 2008.

[69] M. Ridley, *Mendel's Demon: [Gene Justice and the Complexity of Life]*, London: Weidenfeld and Nicholson, 2000.

[70] M.E. Samuels, Private communication (Aug. 2007).

[71] D.B. Searls, "The language of genes," *Nature*, Vol. 420, No. 6912, Nov. 14, 2002, pp. 211–217. DOI: 10.1038/nature01255

[72] C.E. Shannon, "A mathematical theory of communication," *Bell Syst. Tech. J.*, Vol. 27, Jul. and Oct. 1948, pp. 379–457 and pp. 623–656.

[73] E. Szathmáry and J.M. Smith, "The major evolutionary transitions," in *Shaking the Tree: Readings from* Nature *in the History of Life*, H. Gee, Ed., 2000, pp. 32–47. The University of Chicago Press. DOI: 10.1038/374227a0

[74] A.J. Viterbi, "Convolutional codes and their performance in communication systems," *IEEE Trans. Commun. Tech.*, Vol. COM-19, Oct. 1971, pp. 751–772. DOI: 10.1109/TCOM.1971.1090700

[75] R.F. Voss, "Evolution of long-range fractal correlation and $1/f$ noise in DNA base sequences," *Phys. Rev. Lett.*, Vol. 68, Jun. 1992, pp. 3805–3808. DOI: 10.1103/PhysRevLett.68.3805

[76] J.D. Watson and F.H.C. Crick, "Molecular structure of nucleic acids," *Nature*, Vol. 171, No. 4356, Apr. 25, 1953, pp. 737–738. Reprinted in *Nature*, Vol. 421, No. 6921, Jan. 23, 2003, pp. 397–398.

[77] D.B. Mark Welch, J.L. Mark Welch, and M. Meselson, "Evidence for degenerate triploidy in Bdelloid retifers," *Proc. Ntl. Acad. Sci. USA*, Vol. 105, 2008, pp. 5145–5149.

[78] J.K. Wolf, "Efficient maximum likelihood decoding of linear block codes using a trellis," *IEEE Trans. Infor. Theor.*, Vol. IT-24, No. 1, Jan. 1978, pp. 76–80. DOI: 10.1109/TIT.1978.1055821

[79] H.P. Yockey, *Information Theory and Molecular Biology*, Cambridge: Cambridge University Press, 1992.

[80] H.P. Yockey, *Information Theory, Evolution, and the Origin of Life*, Cambridge: Cambridge University Press, 2005.

[81] K. Zahradka, D. Slade, A. Bailone, S. Sommer, D. Averbeck, M. Petranovic, A.B. Lindner, and M. Radman, "Reassembly of shattered chromosomes in D. radiodurans," *Nature*, Vol. 443, No. 7111, Oct. 5, 2006, pp. 569–573. DOI: 10.1038/nature05160

Biography

Gérard Battail was born in Paris, France, on June 5, 1932. He graduated at the Faculté des Sciences (1954) and the Ecole nationale supérieure des Télécommunications (ENST) in 1956, both in Paris. After his military duty, he joined the Centre national d'Etudes des Télécommunications (CNET) in 1959. He worked there on modulation systems and especially on frequency modulation, using fundamental concepts of information theory to understand its behaviour in the presence of noise, especially the threshold effect. In 1966, he joined the Compagnie française Thomson-Houston (later become Thomson-CSF) as a scientific advisor to technical teams designing radioelectric devices. There he interpreted channel coding as a diversity system for designing decoders, especially soft-input ones. He also worked on source coding, frequency synthesizers, mobile communication and other problems related to the design of industrial radiocommunication devices. In 1973, he joined ENST as a Professor. He taught there modulation, information theory and coding. He had also research activities in the same fields with special emphasis on adaptive algorithms as regards source coding and, for channel coding, on soft-in, soft-output decoding of product and concatenated codes. He proposed as a criterion for designing good codes the closeness of its distance distribution with respect to that of random coding instead of maximizing the minimum distance. These rather unorthodox views are now recognized as having paved the way to the invention of turbocodes by Berrou and Glavieux in the early 90s. After his retirement in 1997, he started working on applications of information theory to the sciences of nature. He especially investigated the role of information theory and error-correcting codes in genetics and biological evolution, showing that the conservation of genomes needs error-correcting means.

He applied for many patents, wrote many papers and participated in many symposia and workshops. He also authored a textbook on information theory published by Masson in 1997. He is a member of the Société de l'Electricité, de l'Electronique, des Technologies de l'Information et de la Communication (SEE) and of the Institute of Electrical and Electronics Engineers (IEEE). Before his retirement, he was a member of the editorial board of the *Annales des Télécommunications*. From 1990 to 1997, he was the French official member of Commission C of URSI (International Radio-Scientific Union). From June 2001 to May 2004, he served as Associate Editor at Large of the *IEEE Transactions on Information Theory*.

Index

[75] R.F. Voss, "Evolution of long-range fractal correlations and 1/f noise in DNA base sequences," *Phys. Rev. Lett.*, Vol. 68, No. 25, 1992, pp. 3805–3808.

[76] J.D. Watson and F.H.C. Crick, "Molecular structure of nucleic acids," *Nature*, Vol. 171, No. 4356, Apr. 25, 1953, pp. 737–738. Reprinted in *Nature* Vol. 421, No. 6921, Jan 23, 2003, pp. 397–398.

[77] D.B. Mark Welch, J.L. Mark Welch, and M. Meselson, "Evidence for degenerate tetraploidy in bdelloid rotifers," *Proc. Nat. Acad. Sci. USA*, Vol. 105, [illegible], pp. [illegible].

[78] J.K. Wolf, "Efficient maximum likelihood decoding of linear block codes using a trellis," *IEEE Trans. Inform. Theory*, Vol. IT-24, No. 1, Jan. 1978, pp. 76–80.

[79] H.P. Yockey, *Information Theory and Molecular Biology*, Cambridge: Cambridge University Press, 1992.

[80] H.P. Yockey, *Information Theory, Evolution, and the Origin of Life*, Cambridge: Cambridge University Press, 2005.

[81] K. Zahradka, D. Slade, A. Bailone, S. Sommer, D. Averbeck, M. Petranovic, A.B. Lindner, and M. Radman, "Reassembly of shattered chromosomes in *Deinococcus radiodurans*," *Nature*, Vol. 443, No. 7111, Oct. 5, 2006, pp. 569–573.